Nanotechnological Approaches in Food Microbiology

Nanotechnological Approaches in Food Microbiology

Edited By

Sanju Bala Dhull

Prince Chawla

Ravinder Kaushik

CRC Press is an imprint of the
Taylor & Francis Group, an **informa** business

First edition published 2020
by CRC Press
6000 Broken Sound Parkway NW, Suite 300, Boca Raton, FL 33487-2742

and by CRC Press
2 Park Square, Milton Park, Abingdon, Oxon, OX14 4RN

CRC Press is an imprint of Taylor & Francis Group, LLC

Library of Congress Cataloging-in-Publication Data

ISBN: 978-0-367-35944-7 (hbk)
ISBN: 978-0-429-34277-6 (ebk)

Typeset in Times
by SPi Global, India

Contents

Preface

Over the last decades, nanotechnology gained the centre of attraction in all aspects of modern science, and it has vital applications in the food chain, storage, quality monitoring, processing, preservation, and packaging. The global population is increasing rapidly; therefore, there is a requirement to produce food products in a more proficient, non-toxic, and sustainable way. Food scientists and microbiologists are interested in food safety and quality assurance to produce excellent-quality food free of food pathogens. Nanotechnology plays a significant role in fulfilling the desire of food microbiologists. In this context, scientists developed nanoemulsions, nanohydrogel, nano-starch-based biofilms, biogenic metal nanoparticles, and nano-antimicrobials for the eradication of harmful food pathogenic bacteria from a variety of food products. Furthermore, microorganisms develop a strategical response towards the existing antibiotic drugs by developing innate immunity which results in evasion of the process of examination. Although antibiotics have become the mainstay against the infection caused by microorganisms, however, the resistance of microorganisms to these antibiotics is alarming on the rise. The resistance of microorganisms against antibiotics results from the continuous use of antibiotics that allow the pathogenic microorganism to transform their genotype, resulting in the emergence of multidrug strains. The development of resistant strains emphasizes the need to develop effective therapeutic options. Recently, scientists have begun considering nanomaterials as an alternative to antibiotics due to their effective biocidal and immunopotentiating properties. The number of nano-components in conjugation with existing antibiotics, plant bioactive components, essential oils, and biopolymers for bacterial infections result in lowering of uptake dosage and, therefore, minimize the toxicity and reduce the probability of development of resistance. The synergistic approach of nanomaterials with natural components may serve as a complement to the existing therapies and may help control the serious bacterial infections. Therefore, the present book is explaining nanotechnological approaches towards the eradication of harmful effects of pathogenic microorganisms.

Editors

Sanju Bala Dhull, PhD, is presently working as senior assistant professor in Department of Food Science and Technology, Chaudhary Devi Lal University, Sirsa. She has more than 12 years of teaching and research experience. Her area of interests includes the synthesis and characterization of nanoparticles, nanoemulsions, antimicrobial activity of nano-formulations in food, and the characterization of biomolecules. She has published more than 30 research papers, 1 book, and 10 book chapters of national and international repute. Dr. Dhull has presented more than 20 research papers in various national and international conferences. She is an active member of the Association of Food Scientists and Technologists (India) and the Association of Microbiologists of India. She also serves as reviewer of several national and international journals

Prince Chawla, PhD, is currently working as assistant professor in Food Technology and Nutrition (School of Agriculture), Lovely Professional University, Phagwara, Punjab. He is an alumnus of Chaudhary Devi Lal University, Sirsa, and Shoolini University, Solan. He has a chief interest in mineral fortification, functional foods, protein modification, and the detection of adulterants from foods using nanotechnology. He has worked on Department of Biotechnology– and Department of Science and Technology–funded research projects and has 4 years of research experience. He has 3 patents, 2 books, 30 international research papers, and 7 international book chapters. Dr. Chawla is a recognized reviewer of more than 30 international journals and has reviewed numbers of the research and review articles.

Ravinder Kaushik, PhD, is presently working as senior assistant professor, University of Petroleum and Energy Studies. He also served at the School of Bioengineering and Food Technology, Shoolini University, Solan, for about 6 years. He is an alumnus of the National Dairy Research Institute, Karnal, and an eminent researcher whose chief interests lie in post-harvest technology, dairy processing, food processing, food chemistry, and new product development. He is the editor of two international journals, the *International Journal of Food and Fermentation* and the *International Journal of Food and Nutrition*, and a referee for more than 30 journals. He has 9 patents, 2 books, 3 compendiums, 36 international research papers, 11 national papers, 4 international book chapters, and 10 national book chapters. He has guided 27 master's students, 1 MPhil student, and 3 Ph.D. students. He is a member of the Science Advisory Board, USA, as well as appointed as an expert for assessing grant proposals by the National Center of Scientific and Technical Evaluation in Kazakhstan. He is also a member of the Mendeley Advisor Community.

Contributors

Akinbode A. Adedeji
Department of Biosystems and Agricultural Engineering
University of Kentucky
Lexington, Kentucky, USA

Arun Kumar Attkan
Department of Processing and Food Engineering
AICRP-PHET, Chaudhary Charan Singh Haryana Agricultural University
Hisar, Haryana, India

Shobhit Ambawat
Department of Food Technology
Guru Jambheshwar University of Science and Technology
Hisar, Haryana, India

Aarti Bains
Chandigarh Group of Colleges Landran
Mohali, Punjab

Kiran Bala
Department of Food Science and Technology
University College, Chaudhary Devi Lal University
Sirsa, Haryana, India

Anindita Behera
School of Pharmaceutical Science
Siksha 'O' Anusandhan (Deemed to be University)
Bhubaneswar, Odisha, India

Megha D. Bhatt
G. B. Pant University of Agriculture & Technology
Pantnagar, Uttarakhand, India

Deepesh Bhatt
Department of Biotechnology
Shree Ramkrishna Institute of Computer Education and Applied Sciences
Surat, India

Aradhita Barmanray
Department of Food Technology
Guru Jambheshwar University of Science and Technology
Hisar, Haryana, India

Rohit Biswas
Department of Food Technology and Nutrition
Lovely Professional University
Phagwara, Punjab, India

Naresh Butani
Department of Microbiology
Shree Ramkrishna Institute of Computer Education and Applied Sciences
Surat, Gujarat, India

Parveen Chauhan
Faculty of Applied Sciences and Biotechnology
Shoolini University
Solan, Himachal Pradesh, India

P.K. Chauhan
Faculty of Applied Sciences and Biotechnology
Shoolini University
Solan, Himachal Pradesh, India

Navnidhi Chhikara
Department of Food Technology
Guru Jambheshwar University of Science and Technology
Hisar, Haryana, India

Sook Chin Chew
Xiamen University Malaysia Campus
Sepang, Selangor, Malaysia

Anoop K. Dobriyal
Department of Zoology and Biotechnology
HNB Garhwal Central University, Pauri Campus
Uttarakhand, India

Kanika Dulta
Faculty of Applied Sciences and Biotechnology
Shoolini University
Solan, Himachal Pradesh, India

Mukesh Kumar Garg
Department of Processing and Food Engineering
AICRP-PHET, Chaudhary Charan Singh Haryana Agricultural University
Hisar, Haryana, India

Samriti Guleria
Department of Food Technology and Nutrition
Lovely Professional University
Phagwara, Punjab, India

Jaydip Jobanputra
Department of Biotechnology
Bhagwan Mahavir College of Science and Technology
Surat, Gujarat, India

Ambothi Kanagalakshimi
Department of Biochemistry and Biotechnology
Faculty of Science, Annamalai University
Annamalainagar, Tamil Nadu, India
and
Postgraduate and Research Department of Biochemistry
Government Arts College for Women
Krishnagiri, Tamil Nadu, India

Monika Kataria
Department of Bio & Nano Technology
Guru Jambheshwar University of Science and Technology
Hisar, Haryana, India

Ramandeep Kaur
Department of Food Science and Technology
Punjab Agriculture University
Ludhiana, India

Bababode Adesegun Kehinde
Department of Biosystems and Agricultural Engineering
University of Kentucky
Lexington, Kentucky, USA

Mohammed Azhar Khan
Faculty of Biotechnology
Shoolini University of Biotechnology and Management Sciences
Solan, Himachal Pradesh, India

Huma Khan
School of Biotechnology
Shoolini University
Solan Himachal Pradesh, India

Ashwani Kumar
Department of Food Technology and Nutrition
Lovely Professional University
Phagwara, Punjab, India

Mukul Kumar
Department of Food Technology and Nutrition
Lovely Professional University
Phagwara, Punjab, India

Nitin Kumar
Department of Processing and Food Engineering
AICRP-PHET, Chaudhary Charan Singh Haryana Agricultural University
Hisar, Haryana, India

Sunil Kumar
Department of Processing and Food Engineering
AICRP-PHET, Chaudhary Charan Singh Haryana Agricultural University
Hisar, Haryana, India

Pradyuman Kumar
Department of Food Engineering and Technology, SLIET
Sangrur, Punjab, India

Shailja Kumari
Faculty of Applied Sciences and Biotechnology
Shoolini University
Solan, Himachal Pradesh, India

Vaibhao Lule
Department of Dairy Microbiology
College of Dairy Technology
Warud (Pusad)

Atul Anand Mishra
Department of Food Process Engineering
Sam Higginbottom University of Agriculture, Technology, and Sciences
Naini, Allahabad, India

Bharti Mittu
National Institute of Pharmaceutical Education and Research
Mohali, Chandigarh, India

Raju Murali
Department of Biochemistry and Biotechnology
Faculty of Science, Annamalai University
Annamalainagar, Tamil Nadu, India
and
Postgraduate and Research Department of Biochemistry
Government Arts College for Women
Krishnagiri, Tamil Nadu, India

Dipsha Narang
Maharaja Lakshman Sen Memorial College
Sundernagar, Himachal Pradesh, India

Kar Lin Nyam
UCSI University
Kuala Lumpur, Malaysia

Santwana Padhi
KIIT Technology Business Incubator
KIIT Deemed to be University
Bhubaneswar, Odisha, India

Anil Panghal
Department of Processing and Food Engineering
AICRP-PHET, Chaudhary Charan Singh Haryana Agricultural University
Hisar, Haryana, India

Ruhi Pathania
Shoolini University
Solan, Himachal Pradesh, India

Priti Parmar
Navsari Agricultural University, India

Sneh Punia
Department of Food Science and Technology
Chaudhary Devi Lal University
Sirsa, Haryana, India

Vinayagam Ramachandran
Department of Biotechnology
Thiruvalluvar University, Serkadu
Vellore, Tamil Nadu, India

Rohit Sangwan
Institute of Biochemistry and Molecular Biology
University of Potsdam
Potsdam, Germany

Silvia Sequeira
Department of Dairy Microbiology
College of Dairy Technology
Warud, Pusad, India

Bhanu Sharma
Shoolini University
Solan, Himachal Pradesh, India

Ritu Sindhu
Centre of Food Science and Technology
Chaudhary Charan Singh Haryana Agricultural University
Hisar, Haryana, India

Ajay Singh
Department of Food Technology
Mata Gujri College
Fatehgarh Sahib, Punjab, India

Neha Singh
Sam Higginbottom University of Agriculture
Technology, and Sciences
Naini, Allahabad, India

Subramani Srinivasan
Department of Biochemistry and Biotechnology
Faculty of Science, Annamalai University
Annamalainagar, Tamil Nadu, India
and
Postgraduate and Research Department of Biochemistry
Government Arts College for Women
Krishnagiri, Tamil Nadu, India

Devarajan Raajasubramanian
Department of Botany
Faculty of Science, Annamalai University
Annamalainagar, Tamil Nadu, India
and
Department of Botany
Thiru. A. Govindasamy Government Arts College
Tindivanam, Tamil Nadu, India

Suk Kuan Teng
Xiamen University Malaysia Campus
Sepang, Selangor, Malaysia

Kiran Thakur
Faculty of Applied Sciences and Biotechnology
Shoolini University
Solan, Himachal Pradesh, India

Sudhir Kumar Tomar
Department of Dairy Microbiology
College of Dairy Technology
Warud, Pusad, India

Veerasamy Vinothkumar
Department of Biochemistry and Biotechnology
Faculty of Science, Annamalai University
Annamalainagar, Tamil Nadu, India

List of Abbreviations

%	Percent
+ve	Positive
°C	Degree centigrade
M	Micro
Ml	Microliter
A_{600}	Absorbance
AgNPs	Silver nanoparticles
Al_2O_3	Aluminium oxide
AMR	Antimicrobial resistance
ATP	Adenosine triphosphate
BSE	Backscatter electrons
CdS	Cadmium sulphate
Cm	Centimeter
CO_2	Carbon dioxide
CuO	Copper oxide
DAN	Diaminonapthotriazole
DLS	Dynamic Light Scattering
DNA	Deoxyribonucleic acid
DSC	Differential scanning calorimetry
E. coli	*Escherichia coli*
EDS	Energy-dispersive X-ray spectroscopy
EDX	Energy-dispersive X-ray spectroscopy
FAD	Flavin adenine dinucleotide
FAO	Food and Agriculture Organization
FDA	Food and Drug Administration
Fig	Figures
FMN	Flavin mononucleotide
FTIR	Fourier-transform infrared
G	Gram
GRAS	Generally recognized as safe
H	Hour
H_2O_2	Hydrogen peroxide
$HAuCl_4$	Chloroauric acid
HPMC	Hydroxypropyl methylcellulose
IR	Infrared
Kb	Kilo base
LDPE	Low-density polyethylene
LPS	Lipopolysaccharides
M	Molar
MgO	Magnesium oxide
MIC	Minimum inhibitory concentrations
min	Minute

mL	Milliliter
mm	millimeter
mM	Mill molar
MnO_2	Manganese oxide
NADH	Nicotinamide adenine dinucleotide
NADPH	Nicotinamide adenine dinucleotide phosphate
nm	Nanometre
NPs	Nanoparticles
O_2	Oxygen
OD	Optical density
PBS	Phosphate buffer saline
PBS	Polybutylene succinate
PCL	Polycaprolactone
PE	Primary electron beam
PET	Polyethylene terephthalate
PG	Peptidoglycan
PGA	Polyglycolide
pH	Potential of hydrogen
PLA	Polylactic or polygalactic acid
PNC	Poly-nanocomposite
PP	Polypropylene
PVOH	Polyvinyl alcohol
RNA	Ribonucleic acid
ROS	Reactive oxygen species
rpm	Revolution per minute
S. aureus	*Staphylococcus aureus*
SDS-PAGE	Sodium dodecyl sulphate-polyacrylamide gel electrophoresis
S.No.	Serial number
SEM	Scanning electron microscopy
Sp.	Species
SPR	Surface plasmon resonance
TGA	Thermo gravimetric analysis
TiO_2	Titanium oxide
UV	Ultraviolet
WHO	World Health Organization
XRD	X-ray diffraction
ZnO	Zinc oxide

1 Mathematical Models and Kinetic Studies for the Assessment of Antimicrobial Properties of Metal Nanoparticles

Rohit Biswas
Lovely Professional University, India
Neha Singh, and Atul Anand Mishra
Sam Higginbottom University of Agriculture, Technology, and Sciences, India
Prince Chawla
Lovely Professional University, India

CONTENTS

1.1 INTRODUCTION

Nanotechnology is a new concept of science and trending field also; it plays a dominant role in aspects of everyday life. The aim of nanotechnology is to the engineering of the functional system at the nano (10^{-9} m) scale. Nanotechnology is the branch of technology that deals with a matter which is converted to super small or billionths of metre or 10 angstroms or 1 nm, which essentially involves the manipulation of substance at the atom and molecule level (He et al., 2019). Nanotechnology is the combination of biological sciences, biotechnology, and chemistry to understand, manipulate, and fabricate devices based on particles that are nanometres in size across atoms and molecules. A variety of surprising and interesting aspects of utilization can be found at this scale of substance as the atom and molecules behave differently. Suppose, comparing a cricket ball with super-duper, tiny nanometre particles, that is like comparing a ball to the size of Earth. If we have a gold piece the size of 1 cm^2, it looks normal, shiny in a golden color. But when we reduce it to a smaller size, it looks red in water, and when it is reduced to a nanoscale, it changes to a green color. That is because changes in size lead to unexpected properties in which not only the color but also all kinds of physical and chemical properties are changed at the nanoscale.

Nanoparticles (NPs) are mainly based on their larger surface area, which can allow more atoms to interact with other materials. Due to the interaction, NPs provide stronger ability, more durability, and more effectiveness than their larger structures (called bulk). For example, take a brick of stone (which is used in the construction of buildings) and grind it to powder form; the surface area increases more than a brick. Therefore, they use less chemicals or any materials while working more effectively than their bulk form. Also, NPs must have a size of 1 to 100 nm. To understand very thoroughly about nano-size, let us take an example. Nano is similar to 1-nm-wide sugar molecule, 10 times larger than an atom,10 times smaller than a cell membrane (10 nm), 100 times smaller than a virus (1/10 μm), 1000 times smaller than bacteria (1 μm), 10,000 times smaller than red blood cells (10 μm),100,000 times smaller than a strand of hair(1/10 mm), 1 million times smaller than freckle (1 mm), 10 million times smaller than the width of one's pinky finger (1 cm), 100 million times smaller than the width of the palm (1 dm), and 1 billion times smaller than the tall child (1 m) (Enescu et al., 2019).

1.2 HISTORY

In 1960, the word *nano* was officially confirmed as the standard prefix, which was derived from Greek νᾶνος, meaning "dwarf". In 1959, American physicist Richard Feynman gave a speech on nanotechnology which was considered the earliest systematic discussion. It was titled "There's Plenty of Room at the Bottom". In his speech, he discussed the importance of controlling and manipulating the small scale, which eventually led to the understanding of strange phenomena occurring in complex scenarios. In 1974, Japanese scientist Norio Taniguchi, in his paper on production technology that creates objects and features on the minimized scale, used the term *nanotechnology*. Then in the 1980s, an IBM Zurich scientist developed the scanning tunneling microscope, followed by the atomic force microscope invention, which led to an exploration of an unprecedented atomic level. The availability of supercomputers at this time helped further stimulate NPs on a large scale which further provided an understanding about the material structure and their properties at the minimized nanoscale level (Drexler et al., 1986; as shown in Figure 1.1).

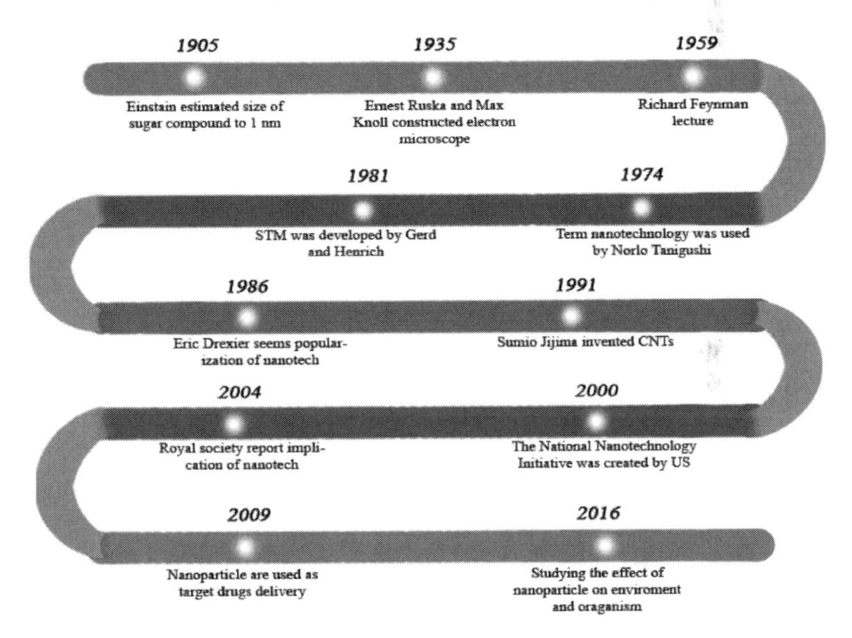

FIGURE 1.1 Short history of nanotechnology (Niska et al., 2018).

1.3 CLASSIFICATION OF NANOPARTICLES

Various aspects of classification can be taken into consideration in the case of NP as mainly three hierarchy of classification involve zero-, one-, two-, and three-dimensional classification (Hett, 2004; Abdullaeva, 2017).

1.3.1 Zero-Dimensional Nanoparticles

In the modern classification system of NPs, highly dispersive systems which include NPs and nano-power which are ultra-dispersive are mostly considered as zero-dimensional. They might be classified as having all dimensions, but their microscopic size and the negligible difference in all dimensions correspond to being considered as the zero-dimensional objects (Abdullaeva, 2017).

1.3.2 One-Dimensional Nanoparticles

One-dimensional NPs have had a in place in electronics, engineering, and chemistry for more than decades as in the form of thin-film or manufactured surfaces. In the field of solar cells or catalysis, there is a common practice of thin-film or monolayer generation of silicon which ranges from 1 to 100 nm in size. In this manner, the dimension is considered in one-way, itself neglecting another dimension and thus showing the one dimension in phases (Abdullaeva, 2017). Thin-film has a wide application in a variety of industries, such as storage systems, biological sensors, fiber optics, the magneto-optic, and optical device, to name a few (Pal et al., 2011).

1.3.3 Two-Dimensional Nanoparticles

Carbon nanotubes (CNTs) are NPs with two dimensions as the third dimension of this doesn't hinder the dispersity. CNTs are two-dimensional NPs which are cylindrical in shape having a diameter of 1 nm and 100 nm, which consist of a hexagonal network of carbon atoms in the form of graphite layer rolled up to provide the state shape. CNTs have astonishing mechanical, physical, and electrical properties characterizing them as unique material among others; they can be classified as single-walled or multi-walled (Köhler and Fritzsche, 2008). In their basic form, two dimensions are characterized as two dimensions which are perpendicular to each other mutually, without effecting the dispersity even with macro-length, as in case of fibres, threads, and capillaries (Abdullaeva, 2017). Depending on the carbon leaf arrangement, the conductivity of CNTs is decided as metallic or semiconductive. Due to the high density of current in these, they can reach significant current density, making CNTs superconductors. They are chemically quite stable and have a higher capacity for molecular absorption.

1.3.4 Three-Dimensional Nanoparticles

Fullerenes are spherical cages consisting of carbon atoms ranging from 28 to >100 atoms; these are generally formed by an sp-bonded dimensionless structure mainly with zero dimension (Abdullaeva, 2017). The hollow ball can represent in the form of pentagons and hexagons with interconnected carbons existing as the allotropic modification of carbons. These molecules have an intriguing property in terms of lubricants due to their non-combining molecules. They also show interesting electrical properties which can be useful in terms of solar cell manufacturing. Because it has an empty structure resembling various biologically active molecules, it can be used as a potential medical carrier for the treatment of various containments, as well as for targeted-based NP release (Abdullaeva, 2017; Tomalia, 2004).

1.4 NANOPARTICLES BASED ON ORIGIN

Apart from dimensions, NPs can be classified based on origin as natural or synthetic-based (Jeevanandam et al., 2018).

1.4.1 Natural Nanoparticles

NPs produced by biological sources or by anthropogenic activities are considered as natural NPs. The natural occurrence of NPs can be found on Earth in all three spheres, namely hydrosphere, atmosphere, and lithosphere (Hochella et al., 2015).

1.4.2 Synthetic Nanoparticles

Synthetic NPs are generally produced by a variety of man-made methods, such as mechanical grinding, exhaust, and smoke produced by engines, or synthesis by either physical, chemical, biological, or combination of all in the form of the hybrid method. Synthetic NPs can lead to a category of engineered NPs that exhibit distinctive properties, based on the source, methods, and condition utilized during synthesis, that significantly differ from that of natural NPs (Wagner et al., 2014; Sharma et al., 2015).

1.4.3 Nanoparticles Based on Material

1.4.3.1 Organic Nanoparticles

Dendrimers, liposomes, micelles, and ferritin are generally known as organic NPs or polymers; these are biodegradable and non-toxic, whereas some, such as micelles and liposomes, have a hollow cavity similar to fullerenes, which are also known as nanocapsules which are generally sensitive to thermal and electromagnetic changes. All these exclusive characteristics can be harnessed in term of drug delivery either in an absorbed form or an entrapped form (Wiener et al., 1994; Tiwari et al., 2008). Dendrimers are a well-researched subject as due to their being highly specialized, encapsulating a targeted drug and its delivery (Li et al., 2007).

1.4.3.2 Inorganic Nanoparticles

Inorganic NPs are those which are mainly void of carbon; generally, metal-and metal oxide–based NPs are categorized in this category.

Metal-based NPs are those which are directly synthesized from metal to NP sizes either by mechanical destruction or biological and chemical formation. These commonly used metal-based NPs are aluminium, cadmium, cobalt, copper, gold, iron, lead, silver, and zinc. These have distinctive properties as having a smaller size, ranging from 10 to 100 nm, and surface characteristics like high surface-to-volume ratio, charge, shape, color, and amorphous structure (Salavati-Niasari et al., 2008).

Metal oxide–based NPs are synthesized to modify the property from their respective metal-based NP; these are synthesized because of their higher reactivity and efficiency due to the oxide molecules' availability. In terms of metal oxide–based NPs, they have incomparable properties to their counterpart metal-based NPs. Some of the metal oxide–based NPs are aluminium oxide, iron oxide, silicon dioxide, titanium oxide, and zinc oxide (Tai et al., 2007).

Quantum dots (QDs) are similar to NPs but contain tiny droplets of electrons. Quantum drops are nanocrystals ranging from 2 to 10 nm in size and are capable as semiconductors. They can be synthesized from various semiconductor material via colloidal synthesis or electrochemistry. The most common being cadmium selenide (CdSe), cadmium telluride (CdTe), indium phosphide (InP) (Pal et al., 2011).

1.5 NANOPARTICLE USE IN ANTIMICROBIALS (BASICS)

Nano-reinforcement could be used to enhance the antimicrobial properties of the films, which ultimately extends the service life of the product. Moreover, antimicrobial agents in the nano form when reinforced into the polymer matrix hold improved activity owing to their high surface-to-volume ratio (Llorens et al., 2012). The selection of a suitable antimicrobial agent is based on its effect on the food product and the associated microorganism. Several studies have been conducted on nano-reinforcement of antimicrobials into the polymer matrix both in individual and combined forms (Jin and Gurtler, 2011; Zohri et al., 2013). To improve the antimicrobial properties, essential oils are also incorporated with other reinforcements. The addition of essential oils may affect the mechanical, physical, and barrier properties of the film. For instance, the addition of essential oil in hydroxypropyl methylcellulose decreased water activity by 20% and increased water vapour permeability and oxygen permeability by 38% and 40%, respectively (Klangmuang and Sothornvit, 2016). Apart from the antimicrobials discussed earlier, some inorganic nanofillers such as zinc oxide (Mallakpour and Behranvand, 2016), copper oxide (Oun and Rhim, 2017), silver (Franci et al., 2015), and copper (Chatterjee et al., 2014) also possess antimicrobial properties.

1.5.1 Titanium Dioxide–Based Nanocomposites

Titanium dioxide (TiO_2) possesses a property of a high refractive index and is resistant to discoloration. It tends to give opacity to papers, coatings, plastics, and films, among other items. In the cell membrane of microorganisms, polyunsaturated phospholipids are present. On the TiO_2 surface, the photocatalytic reaction of TiO_2 generates the hydroxyl radicals (_OH) and reactive oxygen species (ROS), ultimately resulting in the oxidation of polyunsaturated phospholipids, conclusively results into the inactivation of that microorganism. The exhibited photocatalytic property is stronger in nano form in comparison with that of the micro-sized particle. Many researchers (Babaei-Ghazvini et al., 2018; Oleyaei et al., 2016; Vejdan et al., 2016) have noted blending TiO_2NPs with suitable packaging material to increase its efficiency. Whey protein film was blended with TiO_2NPs to obtain a nanocomposite food package, and the improved mechanical strength of the film was observed. The variation of TiO_2 concentration affected the tensile strength and water barrier properties of the film (Zhou et al., 2009). Long et al. (2014) studied the disinfection of gram-negative and gram-positive bacteria, mainly *Salmonella typhimurium* and *Listeria monocytogenes*, respectively, as these are considered pathogenic bacteria found in meat and meat products, by TiO_2 photocatalytic effect by varying ultraviolet (UV) exposure time and concentration. Xing et al. (2012) observed antibacterial activity of the TiO_2-polyethylene film against *Escherichia coli* and *Staphylococcus aureus* and found that the nanocomposite film exhibited more effective antibacterial activity for *S. aureus*.

1.5.2 Silver-Based Nanocomposites

Silver NPs (AgNPs) possess distinctive properties such as thermal conductivity, catalytic activity, and chemical stability; they also exhibit antimicrobial and antifungal properties which lend to their application in various fields, such as textiles, medical applications, consumer products, and others (Akter et al., 2018; González et al., 2015). De Moura et al. (2012) prepared nanocomposite film by blending hydroxypropyl methylcellulose (HPMC) with AgNPs and found that the developed nanocomposite film had excellent mechanical and barrier properties; it also exhibited antimicrobial properties against an *E. coli* culture. It is also revealed that AgNPs act as antimicrobial agents and increase the mechanical properties of the film. This strengthens the point that AgNPs find their potential application in the field of food packaging. Zhou et al. (2011) developed the nanostructured low-density polyethylene (LDPE)/Ag_2O film bags, which decreased microbial spoilage in apple slices. After 6 days, the quality of the apple slices was found to be deteriorated when packaged in conventional LDPE bags while those slices stored at 5°C in an LDPE/Ag_2O bag were acceptable after 12 days. In a few cases, two or more antimicrobial reinforcements are also combined to produce the synergistic effect of microbial inactivation. It is also found that LDPE nanocomposite packaging materials containing silver and zinc oxide (ZnO) inactivated *Lactobacillus plantarum* in orange juice at 4°C (Emamifar et al., 2010). Besides, the AgNPs retain higher antimicrobial activity in *L. plantarum* when compared with ZnONPs, especially for longer storage times. Similarly, orange juice packed in a nanocomposite of LDPE filled with a powder containing $TiO_2$and nanosilver was found to have a longer shelf life without any deterioration (Emamifar et al., 2010). Moreover, in this context, Zhou et al. (2012) prepared nanosilver/gelatin/carboxymethyl chitosan hydrogel by a green and simple fabrication method, that is radiation-induced reduction and cross-linking. The hydrogels exhibited an antibacterial effect on *E. coli*.

1.5.3 Zinc Oxide–Based Nanocomposite

ZnO is extensively used in numerous applications such as food additives. ZnO has been regarded as generally recognized as safe (GRAS) material by the Food and

Drug Administration (FDA). Out of various known metals, ZnO has been proved to be extremely toxic towards diverse microorganisms (Espitia et al., 2012; Stoimenov et al., 2002). For the inactivation of foodborne pathogens, Li et al. (2009) performed on ZnO–polyvinyl chloride (PVC) and proved to be efficient towards *E. coli* and *S. aureus*. Akbar and Anal (2014) found highly against either of them when they used ZnONPs against two pathogens, *S. typhimurium* and *S. aureus*. Towards the food-borne pathogens present in ready-to-eat poultry meat, for an active packaging application, an ZnONP-loaded film of calcium alginate was formulated.

1.6 BARRIER PROPERTIES

Barrier property is one of the key parameters in selecting the packaging material for food packaging applications. Barrier properties denote permeability of gases such as oxygen, carbon dioxide, and dinitrogen; water vapor; and aroma compounds which determine the quality of the food. For instance, permeability to oxygen facilitates degradation through mechanisms such as corrosive phenomena, oxidations, and the modification of organoleptic properties (López et al., 2015). Knowledge of the diffusion/permeation behaviors of these molecules through packaging material is necessary for designing novel packaging material. The diffusion of gas/water vapour across packaging film is influenced by the following factors (Siracusa, 2012):

- Film structure
- Film permeability properties to a specific substance
- Physical properties of the film such as thickness and area
- Conditions of the packaging material and the environment (temperature)
- Difference in pressure and concentration gradient across the film

1.7 KINETIC MODEL FOR THE ANALYSIS OF THE ANTIMICROBIAL ACTIVITY OF NANOPARTICLES

Various studies have been carried out in terms of the antimicrobial property of NPs. Several metal NPs, such as silver, copper, zinc, carbon, and iron, either generated naturally or created artificially, have been used. Of the metals used, silver is the most extensively used metal in the form of NP for various functions on such being the antimicrobial property. Since ancient Greece, silver has been known for its effectiveness in cases of microbial activity, as it is effective in case of all microbes were basic antibiotics fail to action (Panáček et al., 2006; Beyth et al., 2015; Nallamuthu et al., 2013). NPs heavily depend on various physical factors such as the form of its availability and the dimension of the NP (Panáček et al., 2006). As stated, mainly particles with smaller dimensions in terms of diameter are comparatively more effective than those with larger diameters (Beyth et al., 2015). Based on various understandings in terms of growth of microbes, it can be either stated as the death or birth of microbes in any order to understand the basic effect of microbial load in any scenario (Pearl, 1927). Metal NPs can be synthesized using the various organic and chemical methods by reducing other compounds of metals. According to these generations of NP, the

NPs can be termed as organic and engineered NPs, respectively (Dinesh et al., 2012; Kheybari et al., 2010).

Almost all NPs are generally produced using a compound of metals and then reducing it to elemental NPs. In this scenario, bacteria play an important role in producing NPs; as stated by Klaus et al. (1999) and Joerger et al. (2000), bacteria named *Pseudomonas stutzeri* are capable of reducing the nitrate of silver in an aqueous solution of NPs in range of 2 to 200 nm. In contrast, engineered NPs are produced using the same techniques but with chemicals that act as reducing agents, such as sodium borohydride ($NaBH_4$), ascorbate, and citrate (Panáček et al., 2006). The engineered NP synthesized by man have a specific characteristic property in term of size, properties and behaviors.

1.8 EFFECT OF PARTICLE AND SIZE DETERMINATION

Various sizes have different effects on microbes, thus changing the attributes of action as antimicrobial agents. The most proliferated use of silver as an antimicrobial agent is documented by various researchers, who also mention various sizes and their consequential effect and mode of action. Whereas silver can range from 4 to 22 nm in particle size, which has various modes of action such as attaching to bacterial membrane and forming protein complexes, thus leading to lysis of the overall cell, or binding to the membrane, thus eventually rupturing the membrane to cause lysis of the cell (Sondi and Salopek-Sondi, 2004; Prema and Raju, 2009), in the case of gold-and-silver compounds, they can be as big as 209 nm, targeting the cell wall by creating a depression into it and causing the release of silver ions as NPs that attach to various elements, thus retarding the metabolic activity. In the case of gallium and zinc, they have a range of 28 to 305 nm and 12 to 2000 nm, respectively, in size in which zinc occurs in the form of an oxide, thus targeting cytokines and eliminating them from the system. Other NPs can be of manganese and titanium, both occurring in their oxide form, in a size range of 11 to 130 nm and mainly acting to oxidization of protein or preventing surface adhesion of the bacterial cell to the wall, thus preventing growth.

Various models can be utilized to determine the particle size of NPs whether synthesized organically or chemically. For calculation of particle size, the Scherrer equation can be used.

The particle size in case of an iron oxide NP (IONP) can be determined by the following equation:

$$\text{Particle size} = \frac{K\lambda}{\beta\cos\theta}, \tag{1.1}$$

where

K is the proportionality coefficient (in case of IONP it can be taken as 0.9),
λ is the wavelength of X-ray for the X-ray diffraction (XRD) equipment,
β is defined as the full width at half maximum (FWHM) in term of radians, and
θ is termed as Braggs angle.

The preceding equation can be used in terms of calculation of particle size using XRD data generated by using X' pert high score software with the option of search and match (Arakha et al., 2015). As per the research done by Ramani et al. (2012), which suggested that the particle size obtained using the Scherrer equation increased if there is an increase in the concentration of tri-n-propylamine during the synthesis of ZnONP using zinc acetate dihydrate to overcome the condition of increasing size, a modification was made in term of Scherrer equation by adding a lattice strain in consideration and then calculating the particle size by a Williamson–Hall plot followed by the application of the formula

$$\beta \mathrm{Cos}\theta = \frac{K\lambda}{\mathrm{L}} + \varepsilon \sin\theta, \tag{1.2}$$

where

β is the FWHM intensity of the diffraction line,
L is the crystallite size,
ε is the synthesized structure lattice strain,
K is the shape factor (in case of ZnO shape factor is 0.89), and
λ is the wavelength of the CuKα (1.54 Å).

1.9 ANTIMICROBIAL EFFICACY

The antimicrobial efficacy can be calculated using the basic formula as:

$$\mathrm{ABE}(\%) = \left(\left(\mathrm{V_c} - \mathrm{V_t}\right) / \mathrm{V_c}\right) \times 100. \tag{1.3}$$

V_c and V_t are the viable bacterial colonies count for blank control and test specimen, respectively. The following formula gives the CFU/ml reduction in case of microbial destruction (Ramani et al., 2012).

1.9.1 Kinetic Model for the Release Rate of Nanoparticles

The release of NPs is to be regulated to have a more controlled environment in terms of NP activity onto the microbial load. A monolithic matrix is defined as a polymeric system on a solid silicon matrix in which NPs are distributed homogenously. To further regulate and understand the release of NPs from the given matrix composite, three different kinetic models can be used.

The first is the basic first-order equation that gives a relation of the release of NPs depending on the concentration of the min the matrix in the form of an echo of the chemical first-order kinetic reaction (Cussler and Cussler, 2009):

$$\frac{\mathrm{M_t}}{\mathrm{M_\infty}} = \mathrm{M_s} \times \left(1 - \mathrm{e}^{-\mathrm{kt}}\right), \tag{1.4}$$

where

M_t is the amount of drug released in course of time t,
M_∞ is the total amount of drug in the matrix at the start,
M_s is the mass of molecule that will be released for an in infinite time,
K is the release constant, and
t is time.

The second model is derived from the Higuchi equation for the release of a drug from the insoluble matrix as a time-dependent process, further following Fick's law of diffusion as (Higuchi, 1963; Higuchi, 1961):

$$\frac{M_t}{M_\infty} = \frac{4}{l}\sqrt{\frac{D \times t}{\pi}} = k_H \times t^{0.5}, \tag{1.5}$$

where

l is the thickness of the matrix,
D is the diffusional release in cm^2/s,
M_t/M_∞ is the fraction of drug that is released, and
M_∞ is the absolute mass of NP within the matrix.

The third model for NP release is based on the Peppas, 1985 equation developed on Fick's second law. The release of NPs for Fickian and non-Fickian diffusional release for a short time solution from the thin film (less than 60% of total drug release) can be understood by

$$\frac{M_t}{M_\infty} = k_p \times t^n, \tag{1.6}$$

where k_p is the constant incorporating the drug and the characteristic of the neural system for macromolecules. The second and third models are both valid for perfect sink condition and less than 60% of total diffusion or till this limit. This all states the one-directional diffusion.

In case of NP release in a certain environment, where solubility is a constraint that has to be considered in terms of the release of NPs and the distribution of NPs within the surrounding system, it can be understood by the equation given by Higuchi (1960) as:

$$Q = (2A - C_s)\sqrt{\frac{Dt}{1 + \frac{2(A - C_s)}{C_s}}}, \tag{1.7}$$

where

Q is the amount of NPs absorbed per unit time t and per unit area,
A is the concentration of NPs expressed in units/cm^3,
C_s is the solubility of NPs in the release environment, and
D is the diffusion constant of NPs in the release environment.

1.9.2 Two-Step Release Profile

As the equation formulated by Hahn et al. (2011) shows the relation of time-dependent rate-limiting step to that of the fundamental release characteristics to provide a relation between both parameters; thus, the data for a two-step release can be plotted to a twofold exponential equation as:

$$\frac{M_t}{M_\infty} = b + M_{s1} \times \left(1 - e^{-\frac{t}{k_1}}\right) + M_{s2} \times \left(1 - e^{-\frac{t}{k_2}}\right). \tag{1.8}$$

It is also stated that the release of NPs in the first order is almost similar to the second order in the case of silver, due to which the relation does not seem to apply as intended as the silver ion release is very low in case of overall consideration which causes a deviation in measurement and converges to being undetectable. But in the case of copper, the twofold exponential fits, having a more notable relation with data compared to the first order of the reaction.

1.9.3 Model Based on the Arrhenius Equation

Zhang et al. (2011) derived a model regarding the release of silver NPs based on the hard-sphere collision theory. The sphere collision theory was majorly used to understand NP dissolution kinetics (Meulenkamp, 1998). So the proposed model for the silver oxidation stoichiometry of AgNPs is:

$$Ag_{(s)} + \frac{1}{2}O_2 + 2H^+_{(aq)} \leftrightarrow 2Ag^+_{(aq)} + H_2O_{(l)}. \quad (1.9)$$

Since the silver solubility is the same in the first-order and second-order reactions, as discussed earlier, the whole process can be summarized as in first-order kinetics as the size of the AgNP is relatively small which can act as a soluble reactant and further oxidation model generated onto Arrhenius equation as:

$$\gamma_{Ag^+} = kC_{AgNPs}\left[O_2\right]^{0.5}\left[H^+\right]^2, \quad (1.10)$$

where

$\gamma_{Ag}{}^+$ is AgNP release in mol/L.h,

k is the reaction rate constant in mol/h, and

AgNPs, O_2, and H^+ are the molar concentration silver ions, dissolved oxygen, and protons, respectively, in mol/L.

The preceding equation can be rewritten in term of Arrhenius equation incorporating frequency factor as:

$$\gamma_{Ag^+} = f\exp\left(\frac{-E_a}{k_B T}\right)\left[AgNPs\right]\left[O_2\right]^{0.5}\left[H^+\right]^2, \quad (1.11)$$

where

f is the frequency factor for the overall reaction,

E_a is the activation energy for the reaction to start,

k_B is the Boltzmann constant, and

T is the temperature.

The frequency factor is the frequency of reactant molecule collisions, as the number of times the molecule in a said system collide for the reaction to happen; furthermore, the frequency factor can be calculated using

$$f = \pi\sigma_{AB}^2\left(\frac{8\pi k_b T}{\mu_{AB}}\right)^{\frac{1}{2}} N_A, \quad (1.12)$$

where

σ_{AB} is the collision radius of the two reactants (mm),
μ_{AB} is the reduced mass of the system (g/mol), and
N_A is Avogadro number.

Since both the reactants are having a spherical dimension, the mass of both the reactant reacting onto each other can be categorized as a spherical mass only, which can be determined using the size of particle reacting as:

$$r_{AB} = r_A + r_B. \tag{1.13}$$

Since the spherical size can be ordered in terms of its radius, then σ_{AB} can be replaced by r. Furthermore, the μ_{AB} can be termed as the mass in term of m as the overall reduced mass for the reactant, whose average mass can be determined using

$$m_{AB} = \frac{m_A m_B}{m_A + m_B}, \tag{1.14}$$

where m_A, m_B, and m_{AB} are the molecular weights of both the reactants and final reduced weight, respectively. Further rearranging of both equations and incorporating the value m_{AB}, r_{AB}, and f in the model for absorption can be modelled as:

$$\gamma_{Ag^+} = \pi r^2 \left(\frac{8\pi k_B T}{m_B}\right)^{\frac{1}{2}} N_A \exp\left(\frac{-E_a}{k_B T}\right)\left[AgNPs\right]\left[O_2\right]^{0.5}\left[H^+\right]^2. \tag{1.15}$$

The molar concentration of the AgNP is similar to that of mass-based concentration, so it can be replaced by

$$\left[AgNPs\right] = \frac{\left[Ag\right]}{N_A \rho \left(\frac{4}{3}\right)\pi r^3}, \tag{1.16}$$

where ρ is the density of AgNPs in the matrix. Now incorporating the molar concentration of AgNP replacement within the model, it can be written as:

$$\gamma_{Ag^+} = \left(\frac{8\pi k_B T}{m_B}\right)^{\frac{1}{2}} \exp\left(\frac{-E_a}{k_B T}\right)\frac{\left[Ag\right]}{\rho\left(\frac{4}{3}\right)r}\left[O_2\right]^{0.5}\left[H^+\right]^2. \tag{1.17}$$

Furthermore, the preceding equation can be understood that the release of the NP is directly proportional to the silver, dissolved oxygen, and proton concentration and inversely proportional to the radius of NPs synthesized. Furthermore, it can also be understood from the equation that the increase in the temperature of the reaction could significantly increase the release of NPs.

1.9.4 Langmuir–Hinshelwood Mechanism of Adsorption

The product formation with the use of catalyst as end product follows a Langmuir–Hinshelwood mechanism, thus stating the conversion of substrate to the final product in which the NP acts as a catalyst, thus maintaining a fast reaction rate in terms of product formation and substrate consumption. There are two pathways of formation for the product: one being the direct formation as catalysing the substrate directly, creating the product without attaching to the product, or the substrate and other being the NP adsorption to the product till it leaves the substrate (Xu et al., 2009b). To understand the basic catalyst mechanism of NPs, the study of resazurin reduction is considered to develop kinetics for the same. The Langmuir adsorption isotherm states

$$n = n_T \theta_S, \tag{1.18}$$

where

n_T is the total number of sites available onto the surface for the catalytic reaction to happen and

θ_S is the fraction of catalytic surface occupied by the NPs.

The equation can also be stated as:

$$n = n_T \theta_S = \frac{n_T K_1 [S]}{1 + K_1 [S]}, \tag{1.19}$$

where K_I is the adsorption equilibrium constant.

Thus, following the single-molecule of kinetics analysis, we obtain the Langmuir–Hinshelwood equation for a single molecule as:

$$(\tau_{off})^{-1} = \frac{1}{\int_0^\infty \tau f_{off}(\tau) d\tau} = k n_T \theta_S = \frac{\gamma_{eff} K_1 [S]}{1 + K_1 [S]}, \tag{1.20}$$

where

$f_{off}(\tau)$ is the probability of density function and

$\gamma_{eff} = kn_T$ shows the activity of all the surface catalytic sites combined as one NP.

For the preceding equation, equating it with $P_{AumSn}(\tau)$ as an initial condition, the equation can be solved as:

$$f_{off}(\tau) = \frac{\gamma_{eff} K_1 [S]}{1 + K_1 [S]} \exp\left(-\frac{\gamma_{eff} K_1 [S]}{1 + K_1 [S]} \right). \tag{1.21}$$

In this equation, it can be understood that $f_{off}(\tau)$ is a single exponential decay function regardless of $\gamma_{eff} = (-kn_T)$ and K_1 with the substrate decay constant being $\frac{\gamma_{eff} K_1 [S]}{1 + K_1 [S]}$.
In the case of all the catalytic sites present on the surface being occupied for a reaction to occur, and also termed as saturating substrate, $\theta_s = 1$ and $f_{off}(\tau) = \gamma_{eff} \exp(-\gamma_{eff}\tau)$.

Since the product is formed in two ways as described earlier, it can be understood that the product is still attached during formation by either of the two ways in which it forms; in the earlier description, τ_{off} is the waiting time required for the stabilization of the NP before each product is formed (Xu et al., 2009a). And the time required for the product to be formed is denoted as τ_{on}. The inverse of the time function for the product formation shows the kinetics of the product dissociation rate, and it states the participation of the substrate in the product dissociation (Xu et al., 2008).

The product dissociation follows two pathways: one with the substrate and other without, thus following two pathways as described in the following.

The first step in Figure 1.2 shows the product formed onto the NP without any substrate attached, followed by the addition of a substrate to the NP at a rate of k_1 and the removal of substrate happens at the k_{-1} rate of dissociation; both the states happened to be on the state of reaction, thus given by τ_{on} and thus stated as the time required for the product formation and the attachment of the product onto the NP. Furthermore, the reaction happening after the attachment of the substrate is thus displacing the product from the NP surface, moving on to the off state of the reaction, in which the product is departed from the surface and only the NP and the substrate is left as one; this state of reaction is termed as τ_{off}.

Another state of reaction as shown in Figure 1.3, which states the reaction in which the product departs from the NP surface without any aid of the substrate or the substrate does not play any role in the removal of the product from the NP surface. This is governed by the rate of reaction as k_3. Finally, the state at the start of the reaction is termed as the on state, and the NP left after the product has departed from the surface is termed as the off state; in both cases, the NP is void of the substrate. Both states show basic maintenance of equilibrium to fulfill the void space formed due to product dissociation (Xu et al., 2009b).

Thus, the equation connecting the $(\tau_{on})^{-1}$ with the kinetics of product dissociation conventional approach, including the substrate role, can be understood by

$$(\tau_{on})^{-1} = \frac{1}{\int_0^{\infty} \tau f_{on}(\tau)\,d\tau} = \frac{k_2K_2[S]+k_3}{1+K_2[S]}, \tag{1.22}$$

where

k_2 is the rate constant for substrate assisted product dissociation,

k_3 is the rate constant for dissociation of the product without the substrate involvement, and

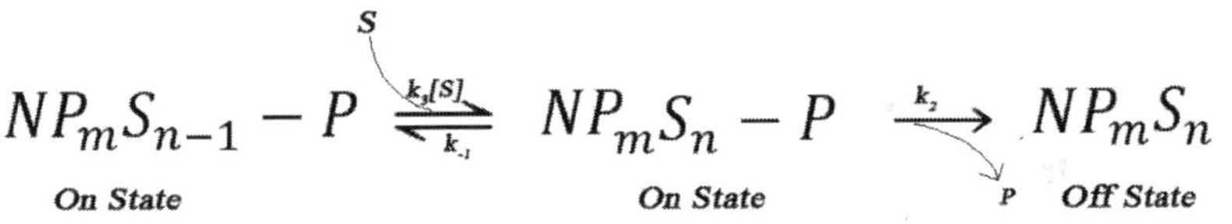

$$NP_mS_{n-1} - P \underset{k_{-1}}{\overset{k_1[S]}{\rightleftharpoons}} NP_mS_n - P \xrightarrow{k_2} NP_mS_n$$

On State — On State — P — Off State

FIGURE 1.2 Substrate action.

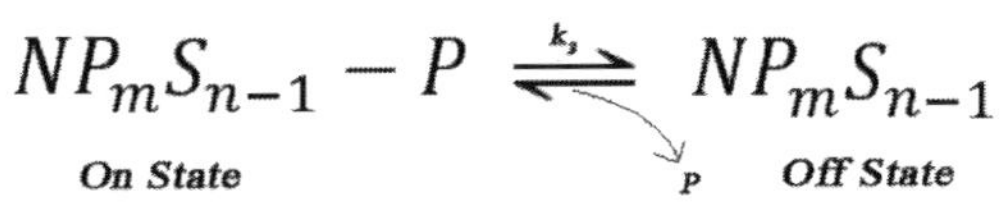

$$NP_mS_{n-1} - P \overset{k_3}{\rightleftharpoons} NP_mS_{n-1}$$

On State — P — Off State

FIGURE 1.3 Nanoparticle void of substrate action.

$K_2 = \frac{k_1}{k_1 + k_2}$ is the rate in the unit of per molar.

In this case, the substrate association in the dissociation of the product is zero, and there is only product dissociation from the NP without any assistant as [S] is 0; then the rate of product dissociation is directly dependent on the rate of product dissociation without substrate assistant, that is k_3 (Xu et al., 2008).

1.9.5 Gompertz and Logistics Modified Model

Microbial loss due to NPs can be calculated using Gompertz models and the logistics model by simple modification curated for the calculation of microbial death in the case of NPs (Chatterjee et al., 2015). The log-linear curve gives a basic relation to the inactivation of microbes with its kinetics. To fit the log-linear curve for microbial death, it is assumed that a negative and linear correlation exists between the NP's activity for microbial death and the number of cell deaths occurring within a given time (Bevilacqua et al., 2015). The basic log-linear curve is:

$$N = N_0 - kt, \tag{1.23}$$

where

N is the number of microbial cells after a given time,
N_0 is the initial number of microbial cells,
k is the inactivation rate, and
t can be any variable on which the inactivation of microbes depends (time, temperature, pressure).

This equation was further modified by Esty and Meyer (1922) and Esty and Williams (1924) to obtain an exponential equation for the microbial growth and death as:

$$N = N_0 e^{-kt}. \tag{1.24}$$

In the preceding equation, the growth or death can be termed as exponential change that might occur mainly in a log phase in terms of growth and the decline phase in terms of death; the death phase might be influenced depending on the concentration of the NPs which can be found by plotting the number of microbes with respect of time to get the curve fit for the increase or decrease. Furthermore, the log-linear relation of the equation can be simplified as:

$$\log N = \log N_0 - \frac{k_{max} t}{\ln 10}, \tag{1.25}$$

where

k_{max} is the maximum rate of growth or death for any given microbes.

Ball and Olson (1957) introduced the decimal log reduction by a basic relation of

$$k = \frac{\ln 10}{D}. \tag{1.26}$$

The value of k can be used in the following equation to be modified and termed as decimal log reduction of microbes, which is termed as D, indicating 90% destruction of microbes from the initial number:

$$\log N = \log N_0 - \frac{t}{D}. \tag{1.27}$$

1.9.6 Negative Sigmoid Model

Various authors have suggested that the inactivation kinetics of microbes can be described by the negative sigmoid or by incorporating the non-linear equation with three parameters included, that is a shoulder, an exponential death phase, and a tail population. The shoulder phase is described as the time when there is no cell decrease, the exponential death phase is when the rate of microbial death is maximum k_{max}, and the tail phase or population is the leftover microbes after the inactivation is done. Geeraerd et al. (2000) reported various reasons for the shoulder presence in death kinetics, such as the time till one cell is killed from the microbes clump, the time for the increase in the rate of destruction as compared to the rate of synthesis which counterbalances the death, and the protective covering formed outside the cell and its efficiency to protect it from damage. It is the time required for pre-inactivation injury to occur before the actual destruction happens. In the contrast, the tail population is composed of the leftover microbes after destruction, indicating the microbes having a higher resistance to various destruction parameters.

Geeraerd et al. (2000) and Peleg and Corradini (2011) defined a model including shoulder and tail as:

$$k = k_{max}\left(\frac{1}{1+C_c}\right)\left(1-\frac{N_{res}}{N}\right), \tag{1.28}$$

where

k_{max} is the inactivation rate,
N_{res} is the residual population, and
N and C_c are the two states.

Substituting an initial condition such as $C_c(0)$ (initial concentration of critical component) by $e^{k\,max\,SL}-1$ with SL (time unit) in this equation, then the parameter representing the shoulder length is:

$$N(t) = \left(N_0 - N_{res}\right)e^{-k_{max}t}\left(\frac{\left(e^{k_{max}SL}\right)}{1+\left(e^{k_{max}SL}-1\right)e^{-k_{max}t}}\right) + N_{res}. \tag{1.29}$$

This model has various benefits as it can be used for both the determination and mechanistic approaches, covering various situations as log-linear, shoulder/tail shape, and all with the combination of log decay, and it does not rely on the estimation of N(0).

1.9.7 The Logistic Model With a Shift in Lag Phase

Corradini and Peleg (2005) defined the logistic model with a shift in lag phase as either short or long duration depending on certain microbes and environmental conditions, which was further condensed to a linear or logistics growth ratio as:

$$Y(t) = a\left(\frac{1}{1+\exp(k(t_c - t))} - \frac{1}{1+\exp(kt_c)}\right), \tag{1.30}$$

where k and t_c are temperature- or condition-related coefficients.

This equation can be defined to understand the death of microbes in the case of NP action as a negative relation ratio of Y(t), which states the basic decrease in microbes, including the shoulder phase, where inactivation is in pre-stage.

In the preceding equation, the time for lag phase can be estimated by using Chatterjee et al. (2015) equation for the lag time, which utilizes the Gompertz model of growth incorporated with microbial death into the model, thus simplifying it as:

$$t_{lag} = -\frac{1}{r}\ln\left(c+\frac{1}{2b}+\sqrt{\frac{1}{4b^2}+\frac{c}{b}}\right) - \frac{2}{r\left(1+\sqrt{1+4bc}\right)}\left(1-e^{-b-\frac{d}{r}\ln\left(c+\frac{1}{2b}+\sqrt{\frac{1}{4b^2}+\frac{c}{b}}\right)+\frac{d}{r}+\frac{1}{2}+\sqrt{\frac{1}{4}+\frac{d}{r}}}\right), \tag{1.31}$$

where

d is death rate constant,

r is effective growth rate constant, and

$$c = \frac{d}{br}. \tag{1.32}$$

1.9.8 Weibull Model

Weibull described frequency distribution model which involved the time of failure in the mechanical system for the bacterial death time calculation as:

$$f(t) = \frac{\beta}{\alpha}\left(\frac{t}{\alpha}\right)^{\beta-1} * \exp\left(-\left(\frac{t}{\alpha}\right)^{\beta}\right). \tag{1.33}$$

In this equation,

β is the marker in Weibull distribution determining the failure rate. (According to the value of β, the distribution follows various laws: if $\beta = 2$, then normal law; if $\beta = 1$, then exponential law; and if $\beta < 1$, then asymptotic law.)

α is the time unit, as the value of α describes the shape of the distribution obtained as it is stretched to the right and the height decrease if the value increases (Couvert et al., 2005).

1.9.9 Biphasic Equation

In the case of NPs targeted for two species of microbes at the same moment of a span both having different phenotypes, in this case, Cerf (1977) proposed a model as:

$$y = y_o + \log\left[fe^{-k_{max1}t} + (1-f)e^{(-k_{max2}t}\right], \tag{1.34}$$

where

f is the population fraction of the first microbe,

k_{max1} is the death rate for the first microbes,

(1−f) is the faction of the second population, and

k_{max2} is the death rate for the second fraction of the population.

1.9.10 Alternative Death Models

If the momentary logarithmic inactivation rate, d $\log_{10}S(t)/dt$, under changing the concentration of agent used for inactivation, in our case, the NPs which correspond to the moment of death ratio with respect to time, we can write (Peleg, 2004) as:

$$\left|\frac{d\log_{10} S(t)}{dt}\right|_{C=const} = -b(C)nt^{n-1} \tag{1.35}$$

and

$$t^* = \left[-\frac{\log_{10} S(t)}{b(C)}\right]^{\frac{1}{n}}. \tag{1.36}$$

These two equations describe the inactivation rate with respect to the concentration and time with respect to the inactivation rate, respectively, to obtain the relation of both the time and concentration with respect to that of inactivation curve. Combining both equations will yield the required relation as (Peleg, 2004):

$$\frac{d\log_{10} S(t)}{dt} = -b\left[C(t)\right]n\left\{-\frac{\log_{10} S(t)}{b\left[C(t)\right]}\right\}^{\frac{n-1}{n}}. \tag{1.37}$$

In this equation, as understood by the relation of the inactivation curve and the concentration of NPs, it can be concluded that b(C) will still follow the log-logistic model, if the power n is also concentration-dependent, as it will be replaced by n[C(t)]. Although there is no exact set of experimental concentrations for the survival curve which will help us exactly determine b(C) and n or b(C) and n(C), even if n(C) is thought to be dependent on concentration, it will be considered for b(C) as:

$$b(C) = \ln\{1+\exp[k_c(C-C_c]\}, \tag{1.38}$$

where C_c is the mark of NP concentration at which the lethality accelerates.

1.10 MODELS FOR THE OPTIMIZATION OF THE CHEMICAL VAPOR SYNTHESIS OF NANOPARTICLES

NPs are the initiation of nanotechnology, and for the synthesis of NPs with similar characteristics every time, several parameters should be considered to optimize their final features, as discussed in the previous section. The smaller the size of an NP, the higher its mode of action for antimicrobial activity. To achieve certain specific parameter, the optimization of the process plays an important role in this. Various methods are involved in the synthesis of NPs, including several different approaches and techniques which can be majorly classified as solid or liquid/gaseous substrate synthesis, which further involves various other specific instrumental techniques for their synthesis.

1.10.1 Methods Using the Solid Precursor

This method typically implies condensation of gas vapourized from a solid surface in a specific gaseous condition to achieve supersaturation which eventually gives rise to homogenously size NPs. Some of the general methods include the following (Swihart, 2003):

- Inert gas condensation
- Pulsed laser ablation
- Spark discharge generation
- Ion sputtering

1.10.2 Methods Using Vapor or Liquid Precursor

These methods involve the substrate to be heated and mixed with the gaseous phase to achieve the supersaturation and the chemical reaction happens in the gaseous state itself thus initiating homogenous NP formation. The following are some of the techniques involving this method (Swihart, 2003):

- Chemical vapour synthesis
- Spray pyrolysis
- Laser pyrolysis
- Thermal plasma pyrolysis
- Flame synthesis
- Flame spray synthesis
- Low-temperature reactive synthesis

From this list of synthesis methods, the most commonly used is chemical vapour synthesis (CVS). CVS is a highly efficient and consistent type of synthesis in comparison to all other synthesis methods. CVS involves the deposition of NPs onto the surface by the chemical reaction of substrate heated in a gaseous environment, thus creating a layer of NPs on a homogenous substrate and of the best quality (Sun et al., 2010). To further maintain this condition and specify parameters so the same result is obtained every time and can be optimized for other material, some models can be taken into consideration (Lee et al., 2016; Kumar and Yadav, 2018).

Before going into the modelling, we need to understand some basic terms:

- Sintering—a technique of compacting solid material by heating or pressure without melting it to the initiation phase change (Koch and Friedlander, 1990)
- Collision radius—the radius of the agglomerate characterizes to its parameter as the radius of gyration (Megaridis and Dobbins, 1990)
- Sintering time—this describes the time of particle to agglomerate in one large chuck over the surface; it helps identify the morphology of the NP and the process (Kruis et al., 1993).
- Degree of agglomeration—it is the ratio of the actual versus the predicted agglomeration of NP (Danilenko et al., 2011).
- Residence time—it is the ratio calculated versus the actual flow through the reactor (Winterer, 2018).

The modelling of the process involves understanding how the particle is generated and its growth. Differential equations help in understanding the relationships in terms of heat transfer and heat production which occur during the chemical reaction was used to develop a model for the CVS process and are explained using finite the difference with the Runge–Kutta algorithm. First, since the NP is emitted from a substrate, the calculation of the decomposition of the substrate helps with understanding the NP's formation rate using single-order kinetics:

$$\frac{dN_{precursor}}{dt} = -k \times N_{precursor}, \tag{1.39}$$

where K is the reaction velocity constant for the NP formation or substrate decomposition:

$$k = k_0 \times \exp\left\{\frac{-E_A}{k_B T}\right\}, \tag{1.40}$$

where

k_0 is the pre-exponential factor, that is 3.96×10^5 s^{-1}, and

E_A/k_B is the activation enthalpy 8479 K (Okuyama et al., 1990).

Particle aggregation/coagulation is described with the assumption of all the particles dispersed as having the same dimension, as required, and all the particles are consistent in this regard. Also, particles formed at the start tend to have a round shape, as well as a smaller diameter, which increases the aggregation of the particles with each other. The rate of aggregation can be described as:

$$\frac{dN}{dt} = -\frac{1}{2}\beta N^2, \tag{1.41}$$

where

N is an aggregate concentration on the substrate as any given moment of time,

β is the collision frequency function for the coagulated particles, and coagulation kernel β describes the rate of aggregation with respect to the aggregate volume and aggregate radius. It implies two statements: (1) If the aggregate particle is one, then the radius is also reduced to the primary particle, and (2) if the particle colliding is more than one, then the radius of the aggregate is dependent on the radius of gyration/collision radius, which can be calculated as (Kruis et al., 1993):

$$\beta = 8\pi D r_c \left[\frac{r_c}{2r_c + \sqrt{2}\,g} + \frac{\sqrt{2}\,D}{c r_c} \right]^{-1}, \tag{1.42}$$

where

D is the particle diffusion constant,
c is the particle velocity,
g is the transition parameter, and
r_c is the radius of collision.

Furthermore, the radius of collision can be explained using Fuchs equation as described by Mountain et al. (1986), assuming the coagulation kernel in the free molecular and continuum regime to aggregate volume and area as:

$$r_c = \frac{3v}{a}\left(\frac{a^3}{36\pi v^2} \right)^{\frac{1}{D_f}}, \tag{1.43}$$

where

a is the area of the aggregate and
v is the volume of the aggregate.

Furthermore, the characteristic time for coagulation is calculated as:

$$\tau_{coll} = \frac{1}{\beta N}. \tag{1.44}$$

And the evolution of aerosol particle is estimated as:

$$\frac{dV}{dt} = -v_1 \frac{dN_{precursor}}{dt}. \tag{1.45}$$

The area of the aggregate evolving in form aerosol in the gas chamber during NP synthesis can be estimated as:

$$\frac{dA}{dt} = N\frac{da}{dt} + a\frac{dN}{dt}, \tag{1.46}$$

where

a is the single agglomerate of particle surface area and
N is the number density of the agglomerate on the substrate.

The sintering time is estimated using grain boundary diffusion:

$$\tau_{sintering} = \frac{k_B T \times d^4}{16\sigma w \Omega \gamma D_{gb}}, \tag{1.47}$$

where
σ is the numerical constant dependent on the sintering mechanism,
w is the grain boundary width,
Ω is the volume of the gas within the chamber,
γ is the interfacial enthalpy (0.35 J/m^2 for grain boundary energy), and
D_{gb} is the grain boundary diffusion coefficient, which is described as (Liao et al., 1997):

$$D_{gb} = D_0 \exp\left\{-\frac{Q}{k_B T}\right\}, \tag{1.48}$$

where
D_0 is the initial diffusion coefficient (1.5×10^{-7} m^2/s for O_2 diffusion).

The development of total surface area covered by the particle with coalescence is estimated as:

$$\frac{da}{dt} = -\frac{1}{\tau_{sintering}}\left(a - a_s\right), \tag{1.49}$$

where a_s is the area completely covered by the coalesced particle (Kruis et al., 1993).

The reaction enthalpy for the precursor decomposition and heat transfer, mainly convective, sums for the energy balance for the system:

$$\frac{dT}{dt} = \frac{\alpha}{c_p \rho L}\left(T_w - T\right) - \frac{\Delta_R H \times C_0 \times k \times \exp\{-k \times t\}}{c_p \rho}, \tag{1.50}$$

where
α is the heat transfer coefficient,
c_p is the specific heat capacity,
ρ is the density of gas,
$\Delta_R H$ is the reaction enthalpy, and
C_0 is the concentration of precursor at the start (Jakubith, 1991).

The residence time for NP within the gaseous phase is calculated by

$$\tau_{residence} = \frac{V}{v} \tag{1.51}$$

where
V is the volume of the reactor and
v is the volume of flow from the reactor.

The size of the particle formed at the start is computed with the help of surface area a and volume of the particle v:

$$d_{pri} = \frac{6v}{a}. \tag{1.52}$$

And the aggregate particle size is estimated by

$$d_{agg} = \left(\frac{6v}{a}\right)^{\frac{1}{3}}. \tag{1.53}$$

And the degree of agglomeration is estimated by

$$N_{agg} = \left(\frac{d_{agg}}{d_{pri}}\right)^{3}. \tag{1.54}$$

In the earlier discussion about the optimization of the model and finding specific characteristics of particle synthesized, using CVS is helpful in the case of two dimensions or more since for a one-dimension model, residence time and temperature cannot be taken into consideration, thus hindering the accuracy of the final prediction. But the target of the method is to optimize the path for improvement of particle characteristics, and the one-dimensional model can be fitted to predict corresponding trends. However, since the synthesis of NPs is a highly randomizing type of process, the experimental trend may not be as clear as the predicted one but will help generate a deeper insight for the characteristics of NPs that will be synthesized.

1.11 CONCLUSION

Models are tools that help correlate different work and understand various aspects of a certain process quantitatively. In terms of NPs, where the effect, especially the antimicrobial effect, is highly unpredictable, model fitting in this term to study the effect of various NP and their effect plays an important role. As discussed within the chapter, various origins of NPs have different effects and efficiency in terms of natural and synthetic NPs, both being of the same element although still exhibiting significantly different antimicrobial effects depending on the synthesis condition and scenario. In this sense, engineered NPs and the control over their function, efficiency, and the path they follow can be manipulated and controlled accordingly. This helps predict and fit them into a model so that more could be employed for specified purposes. As discussed, there are various models studied, some derived whereas some are empirically based on research work conducted. Both have their specific applications and purposes, the most common being the Arrhenius equation, which offers simplified results and understanding. Since current research has advanced, several other models have shown phenomenal results in terms of NP antimicrobial efficiency and should be considered. All these have in common work basically on the growth rate of microbes which then, taken as the negative slope, can be considered as the death rate, and the death of certain microbes can be estimated using the same principle. The Weibull model provides a promising solution to mechanical failure in the case of either activation or incorporation of NPs within the system to act, thus including the time of failure, to further enhance and help in obtaining an approach closer to a practical scenario.

REFERENCES

Abdullaeva, Z. (2017). *Nano-and biomaterials: compounds, properties, characterization, and applications*. John Wiley & Sons. Weinheim, Germany.

Akter, M., Sikder, M. T., Rahman, M. M., Ullah, A. A., Hossain, K. F. B., Banik, S., ... and Kurasaki, M. (2018). A systematic review on silver nanoparticles-induced cytotoxicity: physicochemical properties and perspectives. *Journal of Advanced Research*, 9, 1–16.

Arakha, M., Pal, S., Samantarrai, D., Panigrahi, T. K., Mallick, B. C., Pramanik, K., ... and Jha, S. (2015). Antimicrobial activity of iron oxide nanoparticle upon modulation of nanoparticle-bacteria interface. *Scientific Reports*, 5, 14813.

Babaei-Ghazvini, A., Shahabi-Ghahfarrokhi, I., and Goudarzi, V. (2018). Preparation of UV-protective starch/kefiran/ZnO nanocomposite as a packaging film: characterization. *Food Packaging and Shelf Life*, 16, 103–111.

Ball, C. O., and Olson, F. C. W. (1957). *Sterilization in food technology. Theory, practice, and calculations*. McGraw-Hill Book Co. New York.

Bevilacqua, A., Speranza, B., Sinigaglia, M., and Corbo, M. R. (2015). A focus on the death kinetics in predictive microbiology: benefits and limits of the most important models and some tools dealing with their application in foods. *Foods*, 4(4), 565–580.

Beyth, N., Houri-Haddad, Y., Domb, A., Khan, W., and Hazan, R. (2015). Alternative antimicrobial approach: nano-antimicrobial materials. *Evidence-Based Complementary and Alternative Medicine*, 2015, 246012.

Cerf, O. (1977). A review tailing of survival curves of bacterial spores. *Journal of Applied Bacteriology*, 42(1), 1–19.

Chatterjee, A. K., Chakraborty, R., and Basu, T. (2014). Mechanism of antibacterial activity of copper nanoparticles. *Nanotechnology*, 25(13), 135101.

Chatterjee, T., Chatterjee, B. K., Majumdar, D., and Chakrabarti, P. (2015). Antibacterial effect of silver nanoparticles and the modeling of bacterial growth kinetics using a modified Gompertz model. *Biochimica et Biophysica Acta (BBA)-General Subjects*, 1850(2), 299–306.

Corradini, M. G., and Peleg, M. (2005). Estimating non-isothermal bacterial growth in foods from isothermal experimental data. *Journal of Applied Microbiology*, 99(1), 187–200.

Couvert, O., Gaillard, S., Savy, N., Mafart, P., and Leguérinel, I. (2005). Survival curves of heated bacterial spores: effect of environmental factors on Weibull parameters. *International Journal of Food Microbiology*, 101(1), 73–81.

Cussler, E. L., and Cussler, E. L. (2009). *Diffusion: mass transfer in fluid systems*. Cambridge University Press. New York.

Danilenko, I., Konstantinova, T., Pilipenko, N., Volkova, G., and Glasunova, V. (2011). Estimation of agglomeration degree and nanoparticles shape of zirconia nanopowders. *Particle & Particle Systems Characterization*, 28(1–2), 13–18.

De Moura, M. R., Mattoso, L. H., and Zucolotto, V. (2012). Development of cellulose-based bactericidal nanocomposites containing silver nanoparticles and their use as active food packaging. *Journal of Food Engineering*, 109(3), 520–524.

Dinesh, R., Anandaraj, M., Srinivasan, V., and Hamza, S. (2012). Engineered nanoparticles in the soil and their potential implications to microbial activity. *Geoderma*, 173, 19–27.

Emamifar, A., Kadivar, M., Shahedi, M., and Soleimanian-Zad, S. (2010). Evaluation of nano-composite packaging containing Ag and ZnO on shelf life of fresh orange juice. *Innovative Food Science & Emerging Technologies*, 11(4), 742–748.

Enescu, D., Cerqueira, M. A., Fucinos, P., and Pastrana, L. M. (2019). Recent advances and challenges on applications of nanotechnology in food packaging. A literature review. *Food and Chemical Toxicology*, 134, 110814.

Espitia, P. J. P., Soares, N. D. F. F., dosReis Coimbra, J. S., de Andrade, N. J., Cruz, R. S., and Medeiros, E. A. A. (2012). Zinc oxide nanoparticles: synthesis, antimicrobial activity and food packaging applications. *Food and Bioprocess Technology*, 5(5), 1447–1464.

Esty, J., and Meyer, K. (1922). The heat resistance of the spores of *B. botulinus* and allied anaerobes. XI. *The Journal of Infectious Diseases*, 31(6), 650–664. Retrieved April 7, 2020, from www.jstor.org/stable/30082503

Esty, J., and Williams, C. (1924). Heat resistance studies: I. a new method for the determination of heat resistance of bacterial spores. *The Journal of Infectious Diseases*, 34(5), 516–528. Retrieved April 7, 2020, from www.jstor.org/stable/30081352

Franci, G., Falanga, A., Galdiero, S., Palomba, L., Rai, M., Morelli, G., and Galdiero, M. (2015). Silver nanoparticles as potential antibacterial agents. *Molecules*, 20(5), 8856–8874.

Geeraerd, A. H., Herremans, C. H., and Van Impe, J. F. (2000). Structural model requirements to describe microbial inactivation during a mild heat treatment. *International Journal of Food Microbiology*, 59(3), 185–209.

Hahn, A., Brandes, G., Wagener, P., and Barcikowski, S. (2011). Metal ion release kinetics from nanoparticle silicone composites. *Journal of Controlled Release*, 154(2), 164–170.

He, X., Deng, H., and Hwang, H. M. (2019). The current application of nanotechnology in food and agriculture. *Journal of Food and Drug Analysis*, 27(1), 1–21.

Hett, A. (2004). *Nanotechnology: small matter. Many unknowns*. Swiss Reinsurance Company, Zurich.

Higuchi, T. (1960). Physical chemical analysis of percutaneous absorption process from creams and ointments. *Journal of the Society of Cosmetic. Chemist*, 11, 85–97.

Higuchi, T. (1961). Rate of release of medicaments from ointment bases containing drugs in suspension. *Journal of Pharmaceutical Sciences*, 50(10), 874–875.

Higuchi, T. (1963). Mechanism of sustained-action medication. Theoretical analysis of rate of release of solid drugs dispersed in solid matrices. *Journal of Pharmaceutical Sciences*, 52(12), 1145–1149.

Hochella, M. F., Spencer, M. G., and Jones, K. L. (2015). Nanotechnology: nature's gift or scientists' brainchild?. *Environmental Science: Nano*, 2(2), 114–119.

Jakubith, M. (1991). *Chemische Verfahrenstechnik: Einführung in Reaktionstechnik und Grundoperationen*. Wiley-VCH. Verlag GmbH & Co. KGaA. Weinheim.

Jeevanandam, J., Barhoum, A., Chan, Y. S., Dufresne, A., and Danquah, M. K. (2018). Review on nanoparticles and nanostructured materials: history, sources, toxicity and regulations. *Beilstein Journal of Nanotechnology*, 9(1), 1050–1074.

Jin, T., and Gurtler, J. B. (2011). Inactivation of *Salmonella* in liquid egg albumen by antimicrobial bottle coatings infused with allyl isothiocyanate, nisin and zinc oxide nanoparticles. *Journal of Applied Microbiology*, 110(3), 704–712.

Joerger, R., Klaus, T., and Granqvist, C. G. (2000). Biologically produced silver–carbon composite materials for optically functional thin-film coatings. *Advanced Materials*, 12(6), 407–409.

Klangmuang, P., and Sothornvit, R. (2016). Barrier properties, mechanical properties and antimicrobial activity of hydroxypropyl methylcellulose-based nanocomposite films incorporated with Thai essential oils. *Food Hydrocolloids*, 61, 609–616.

Klaus, T., Joerger, R., Olsson, E., and Granqvist, C. G. (1999). Silver-based crystalline nanoparticles, microbially fabricated. *Proceedings of the National Academy of Sciences*, 96(24), 13611–13614.

Koch, W., and Friedlander, S. K. (1990). The effect of particle coalescence on the surface area of a coagulating aerosol. *Journal of Colloid and Interface Science*, 140(2), 419–427.

Köhler, M., and Fritzsche, W. (2008). *Nanotechnology: an introduction to nanostructuring techniques*. John Wiley & Sons-VCH. Weinheim.

Kruis, F. E., Kusters, K. A., Pratsinis, S. E., and Scarlett, B. (1993). A simple model for the evolution of the characteristics of aggregate particles undergoing coagulation and sintering. *Aerosol Science and Technology*, 19(4), 514–526.

Kumar, K., and Yadav, B. C. (2018). An overview on the importance of chemical vapour deposition technique for graphene synthesis. *Advanced Science, Engineering and Medicine*, 10(7–8), 760–763.

Lee, H. C., Liu, W. W., Chai, S. P., Mohamed, A. R., Lai, C. W., Khe, C. S., … and Hidayah, N. M. S. (2016). Synthesis of single-layer graphene: a review of recent development. *Procedia Chemistry*, 19, 916–921.

Li, Y., Cheng, Y., and Xu, T. (2007). Design, synthesis and potent pharmaceutical applications of glycodendrimers: a mini review. *Current Drug Discovery Technologies*, 4(4), 246–254.

Li, X., Xing, Y., Jiang, Y., Ding, Y., and Li, W. (2009). Antimicrobial activities of ZnO powder-coated PVC film to inactivate food pathogens. *International Journal of Food Science & Technology*, 44(11), 2161–2168.

Liao, S. C., Mayo, W. E., and Pae, K. D. (1997). Theory of high pressure/low temperature sintering of bulk nanocrystalline TiO_2. *Acta Materialia*, 45(10), 4027–4040.

Llorens, A., Lloret, E., Picouet, P. A., Trbojevich, R., and Fernandez, A. (2012). Metallic-based micro and nanocomposites in food contact materials and active food packaging. *Trends in Food Science & Technology*, 24(1), 19–29.

Long, M., Wang, J., Zhuang, H., Zhang, Y., Wu, H., and Zhang, J. (2014). Performance and mechanism of standard nano-TiO_2 (P-25) in photocatalytic disinfection of foodborne microorganisms–*Salmonella typhimurium* and *Listeria monocytogenes*. *Food Control*, 39, 68–74.

Mallakpour, S., and Behranvand, V. (2016). Nanocomposites based on biosafe nano ZnO and different polymeric matrixes for antibacterial, optical, thermal and mechanical applications. *European Polymer Journal*, 84, 377–403.

Megaridis, C. M., and Dobbins, R. A. (1990). Morphological description of flame-generated materials. *Combustion Science and Technology*, 71(1–3), 95–109.

Meulenkamp, E. A. (1998). Size dependence of the dissolution of ZnO nanoparticles. *The Journal of Physical Chemistry B*, 102(40), 7764–7769.

Mountain, R. D., Mulholland, G. W., and Baum, H. (1986). Simulation of aerosol agglomeration in the free molecular and continuum flow regimes. *Journal of Colloid and Interface Science*, 114(1), 67–81.

Nallamuthu, I., Parthasarathi, A., and Khanum, F. (2013). Thymoquinone-loaded PLGA nanoparticles: antioxidant and anti-microbial properties. *International Current Pharmaceutical Journal*, 2(12), 202–207.

Niska, K., Zielinska, E., Radomski, M. W., and Inkielewicz-Stepniak, I. (2018). Metal nanoparticles in dermatology and cosmetology: interactions with human skin cells. *Chemico-biological Interactions*, 295, 38–51.

Okuyama, K., Ushio, R., Kousaka, Y., Flagan, R. C., and Seinfeld, J. H. (1990). Particle generation in a chemical vapor deposition process with seed particles. *AIChE Journal*, 36(3), 409–419.

Oleyaei, S. A., Almasi, H., Ghanbarzadeh, B., and Moayedi, A. A. (2016). Synergistic reinforcing effect of TiO_2 and montmorillonite on potato starch nanocomposite films: thermal, mechanical and barrier properties. *Carbohydrate Polymers*, 152, 253–262.

Oun, A. A., and Rhim, J. W. (2017). Carrageenan-based hydrogels and films: Effect of ZnO and CuO nanoparticles on the physical, mechanical, and antimicrobial properties. *Food Hydrocolloids*, 67, 45–53.

Pal, S. L., Jana, U., Manna, P. K., Mohanta, G. P., and Manavalan, R. (2011). Nanoparticle: an overview of preparation and characterization. *Journal of Applied Pharmaceutical Science*, 1(6), 228–234.

Panáček, A., Kvitek, L., Prucek, R., Kolář, M., Večeřová, R., Pizúrová, N., ... and Zbořil, R. (2006). Silver colloid nanoparticles: synthesis, characterization, and their antibacterial activity. *The Journal of Physical Chemistry B*, 110(33), 16248–16253.

Pearl, R. (1927). The growth of populations. *The Quarterly Review of Biology*, 2(4), 532–548. Retrieved April 7, 2020, from www.jstor.org/stable/2808218

Peleg, M. (2004). Analyzing the effectiveness of microbial inactivation in thermal processing. In *Improving the thermal processing of foods*, Philip Richardson (pp. 411–426). Woodhead Publishing. Sawston, UK.

Peleg, M., and Corradini, M. G. (2011). Microbial growth curves: what the models tell us and what they cannot. *Critical Reviews in Food Science and Nutrition*, 51(10), 917–945.

Peppas, N. (1985). Analysis of Fickian and non-Fickian drug release from polymers. *Pharmaceutica Acta Helvetiae*, 60(4), 110–111.

Prema, P., and Raju, R. (2009). Fabrication and characterization of silver nanoparticle and its potential antibacterial activity. *Biotechnology and Bioprocess Engineering*, 14(6), 842–847.

Ramani, M., Ponnusamy, S., and Muthamizhchelvan, C. (2012). From zinc oxide nanoparticles to microflowers: a study of growth kinetics and biocidal activity. *Materials Science and Engineering: C*, 32(8), 2381–2389.

Salavati-Niasari, M., Davar, F., and Mir, N. (2008). Synthesis and characterization of metallic copper nanoparticles via thermal decomposition. *Polyhedron*, 27(17), 3514–3518.

Sharma, V. K., Filip, J., Zboril, R., and Varma, R. S. (2015). Natural inorganic nanoparticles–formation, fate, and toxicity in the environment. *Chemical Society Reviews*, 44(23), 8410–8423.

Siracusa, V. (2012). Food packaging permeability behaviour: a report. *International Journal of Polymer Science*, 2012. 1–11.

Sondi, I., and Salopek-Sondi, B. (2004). Silver nanoparticles as antimicrobial agent: a case study on *E. coli* as a model for Gram-negative bacteria. *Journal of Colloid and Interface Science*, 275(1), 177–182.

Stoimenov, P. K., Klinger, R. L., Marchin, G. L., and Klabunde, K. J. (2002). Metal oxide nanoparticles as bactericidal agents. *Langmuir*, 18(17), 6679–6686.

Sun, Z., Yan, Z., Yao, J., Beitler, E., Zhu, Y., and Tour, J. M. (2010). Growth of graphene from solid carbon sources. *Nature*, 468(7323), 549.

Swihart, M. T. (2003). Vapor-phase synthesis of nanoparticles. *Current Opinion in Colloid & Interface Science*, 8(1), 127–133.

Tai, C. Y., Tai, C. T., Chang, M. H., and Liu, H. S. (2007). Synthesis of magnesium hydroxide and oxide nanoparticles using a spinning disk reactor. *Industrial & Engineering Chemistry Research*, 46(17), 5536–5541.

Tiwari, D. K., Behari, J., and Sen, P. (2008). Application of nanoparticles in waste water treatment, *World Applied Sciences Journal*, 3, 417–433.

Tomalia, D. A. (2004). Birth of a new macromolecular architecture: dendrimers as quantized building blocks for nanoscale synthetic organic chemistry. *Aldrichimica Acta*, 37(2), 39–57.

Vejdan, A., Ojagh, S. M., Adeli, A., and Abdollahi, M. (2016). Effect of TiO_2 nanoparticles on the physico-mechanical and ultraviolet light barrier properties of fish gelatin/agar bilayer film. *LWT—Food Science and Technology*, 71, 88–95.

Wagner, S., Gondikas, A., Neubauer, E., Hofmann, T., and von derKammer, F. (2014). Spot the difference: engineered and natural nanoparticles in the environment—release, behavior, and fate. *Angewandte Chemie International Edition*, 53(46), 12398–12419.

Xing, Y., Li, X., Zhang, L., Xu, Q., Che, Z., Li, W., ... and Li, K. (2012). Effect of TiO_2 nanoparticles on the antibacterial and physical properties of polyethylene-based film. *Progress in Organic Coatings*, 73(2–3), 219–224.

Xu, W., Kong, J. S., and Chen, P. (2009a). Probing the catalytic activity and heterogeneity of Au-nanoparticles at the single-molecule level. *Physical Chemistry Chemical Physics*, 11(15), 2767–2778.

Xu, W., Kong, J. S., Yeh, Y. T. E., and Chen, P. (2008). Single-molecule nanocatalysis reveals heterogeneous reaction pathways and catalytic dynamics. *Nature materials*, 7(12), 992–996.

Xu, W., Shen, H., Liu, G., and Chen, P. (2009b). Single-molecule kinetics of nanoparticle catalysis. *Nano Research*, 2(12), 911–922.

Zhang, W., Yao, Y., Sullivan, N., and Chen, Y. (2011). Modeling the primary size effects of citrate-coated silver nanoparticles on their ion release kinetics. *Environmental Science & Technology*, 45(10), 4422–4428.

Zhou, J. J., Wang, S. Y., and Gunasekaran, S. (2009). Preparation and characterization of whey protein film incorporated with TiO_2 nanoparticles. *Journal of Food Science*, 74(7), N50–N56.

Zhou, L., Lv, S., He, G., He, Q., and Shi, B. I. (2011). Effect of Pe/Ag_2O nano-packaging on the quality of apple slices. *Journal of Food Quality*, 34(3), 171–176.

Zhou, Y., Zhao, Y., Wang, L., Xu, L., Zhai, M., and Wei, S. (2012). Radiation synthesis and characterization of nanosilver/gelatin/carboxymethyl chitosan hydrogel. *Radiation Physics and Chemistry*, 81(5), 553–560.

Zohri, M., Shafiee Alavidjeh, M., Mirdamadi, S. S., Behmadi, H., Hossaini Nasr, S. M., Eshghi Gonbaki, S., and Jabbari Arabzadeh, A. (2013). Nisin-loaded chitosan/alginate nanoparticles: a hopeful hybrid biopreservative. *Journal of Food Safety*, 33(1), 40–49.

2 Mushroom Extract–Reduced Metal Nanoparticles

An Effective Approach Against Food Pathogenic Bacteria

Aarti Bains
Chandigarh Group of Colleges Landran, India

Dipsha Narang
Maharaja Lakshman Sen Memorial College, India

Prince Chawla
Lovely Professional University, India

Sanju Bala Dhull
Chaudhary Devi Lal University, India

CONTENTS

2.1 INTRODUCTION

Nanoparticles (NPs) are ultra-fine particles with distinct chemistry and morphologic dimensions (Chawla et al., 2020). NPs constitute of physiochemical and optoelectronic features, therefore, can keep the electrons within the particles that are smaller than delocalized electrons present in large amount (Sudha et al., 2018; Chawla et al., 2018). The synthesis of NPs results in the formation of functional NPs; these functional NPs are extensively used for different purposes like diagnosis, therapeutic, catalysis, electronics, and photonics (Lu et al., 2007). Formerly, NPs were synthesized by mainly two techniques physical and wet chemistry; however, alternate to these two processes, green synthesis techniques that include the use of eco-friendly and nontoxic reagents are in great demand (Chung et al., 2016). In this context, extract prepared from various plants is majorly used for the synthesis of metal oxide NPs, in addition to plant-extract microorganisms such as bacteria, fungi, and algae, which are extensively used to carry out the process of green synthesis (Chaudhari et al., 2016; Chawla et al., 2019). Living organisms used for the synthesis of metal oxide NPs have an advantage over chemically synthesized metal oxide NPs as they constitute special traits that provide resistance mechanisms against heavy metals due to their exposure to xenobiotic compounds (Banerjee and Rai, 2018). The resistance mechanism could be due to the presence of primary and secondary metabolites, as well as extracellular macromolecules. All these factors work together intracellularly or extracellularly in an accompanying manner to achieve the detoxification process, and that leads to the process of biosynthesis of NPs (Siddiqi and Husen, 2016; Agarwal et al., 2020). The extracellular compounds responsible for resistance mechanisms in living cells are diverse and, hence, can be utilized for the synthesis of NPs that have different morphology, size dimensions, and functional properties (Sirelkhatim et al., 2015). The macromolecules, including vitamins, enzymes, and polysaccharides produced extracellularly or intracellularly, have a high reduction potential and thus synthesize NPs by reducing metal ions or semiconductor ions present in them in an effective manner (Ravindran et al., 2013; Pathania et al., 2018). The microorganism includes both prokaryotes and eukaryotes that have complexity in their tissues and organs which are being utilized for the benignant biosynthesis process. Due to complexity, these organisms are responsible for obtaining different types of NPs. Prokaryotes, including both bacteria and actinomycetes, have been extensively used for the synthesis of NPs. Likewise, both unicellular and multicellular eukaryotic organisms devoid of hierarchy in their organ structure also shown excellent results in the synthesis of NPs (Kalia and Kaur, 2018). Among eukaryotes, both macrofungi and microfungi consist of peculiar characteristics of secretion of enzymes extracellularly and therefore show promising results in the synthesis of NPs. The synthesis of NPs by extracellular enzymes is mediated by proteins. These proteins act as reducing and capping agents and provide a reduction in downstream isolation and purification process during the synthesis of NPs (Sinha et al., 2015). Fungi generate a huge mycelial biomass that offers a large surface area for the generation of dissolved ions. These dissolved ions when generated provide an interface for the synthesis of NPs (Kalia and Kaur, 2018). Synthesis of NPs via mushrooms gained intense interest among researchers in the recent few years as they constitute a wide range of bioactive compounds (Roy et al., 2019). These bioactive compounds provide a huge potential for the synthesis of NPs. In addition to bioactive compounds, both edible and non-edible mushrooms consist of numerous proteins and polysaccharides which synthesize both organic and inorganic NPs through the intracellular and extracellular pathway (Owaid, 2019). The NPs which are synthesized via fungi are highly stable, are water-soluble, and have good dispersion properties which have proved to be beneficial (Iravani et al., 2014; Dhull et al., 2019). NPs synthesized using mushrooms, therefore, is a promising technology for synthesis of NPs is stable, eco-friendly, and non-toxic nature (Sharma et al., 2019).

2.2 GREEN SYNTHESIS OF NPs

The synthesis of NPs has gained importance in the field of nanotechnology in the recent era. The synthesis of NPs is done to obtain NPs of the desired shape, size, and functionalities and to carry out this process top-down and top-up methods are used (Wang et al., 2015). These two methods are the fundamental principles usually used in the existing literature. The top-down method includes the preparation of NPs through different approaches like lithographic, ball milling, etching, and sputtering. The bottom-up approach includes two methods for the synthesis of NPs; the first one includes chemical deposition, the sol-gel process, spray and laser pyrolysis, molecular condensation, and aerosol process; the latter includes the process of green synthesis. Among these two methods of the bottom-up approach, green synthesis of nanometal, ions have become a common and effective approach in the recent era. This method utilizes the principle that includes the removal and use of hazardous chemical compounds in the manufacture of NPs (Singh et al., 2018). The NPs thus synthesized are highly stable, are small in size, and have a high surface area–to–volume ratio, surface modifiability, excellent magnetic properties, and biocompatibility (Sen et al., 2013). In green chemistry, biological materials, including green plants, are microorganisms that are incorporated to carry out the synthesis of NPs. The microorganisms used include bacteria, fungi, and algae (Malik et al., 2014; Gurunathan et al., 2014). The use of microorganisms for the synthesis of nanometal ions is preferred over plant extract as the microorganisms can be cultured and preserved for constant use (Gurunathan et al., 2015) The microorganisms have diverse nature and among this diversity mushrooms, the macrofungi are found abundant in nature. Mushrooms have been reported as one of the rich sources of bioactive compounds bearing distinct biological activities (Mohanta et al., 2018). Mushrooms secret extracellular compounds, proteins, and polysaccharides in addition to the bioactive compounds. These compounds can reduce metal ions under optimized

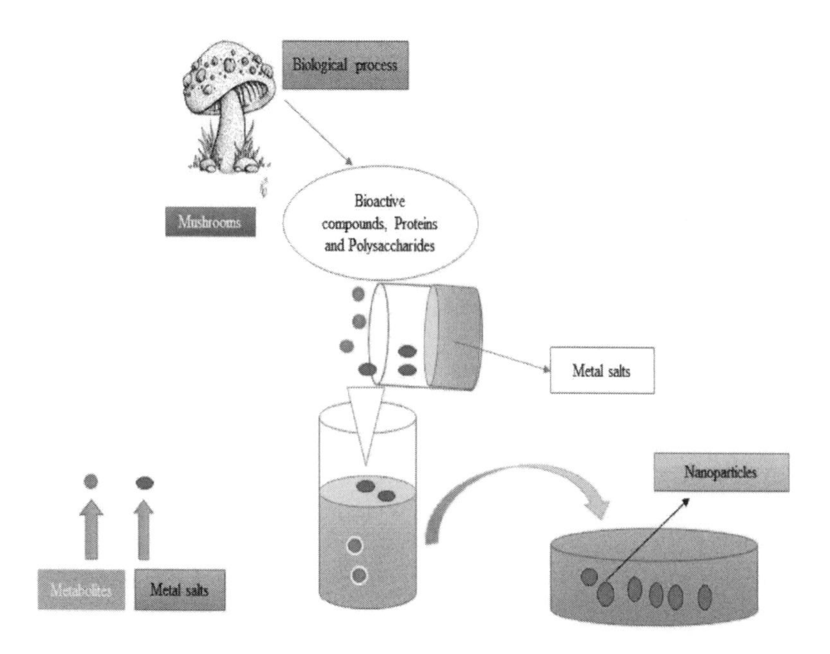

FIGURE 2.1 Synthesis of nanoparticles from mushroom (Mycosynthesis).

conditions and therefore are highly stable small in size and non-toxic in nature. The NPs thus obtained by mycosynthesis are referred to as biogenic NPs (Figure 2.1).

2.3 MECHANISM OF MYCOSYNTHESIS OF NANOPARTICLES

Riboflavin is a water-soluble vitamin and is sensitive to light. It is known to be the precursor of two coenzymes FMN (flavin mononucleotide) and FAD (flavin adenine dinucleotide) which catalyse oxidation-reduction reaction inside the cells of living beings (Bhat et al., 2013). Mushrooms are known to be a good source of riboflavin; it is therefore hypothesized by various scientists that riboflavin protein may be responsible for the reduction of metals ions into NPs (Kalia and Kaur, 2018). When an extract of mushroom dissolved with the metal salt solution is exposed to light, the riboflavin precursor coenzymes get excited and act as electron donors or oxidizers. This reaction provides a significant insight mechanism for the conversion of metal salts dissolved in the extract to nanometals (Rogers, 2010). The amino acids present in mushrooms also have the potential for the reduction of metal salts into NPs. Cysteine amino acid having an S-H group played a significant role in the reduction of metal salt especially Ag metal salt into NPs. The reduction of hydrogen radical and one electron from the β-carbon of cysteine amino acid occurs in the presence of enzyme nicotinamide adenine dinucleotide phosphate (NADPH). Another enzyme that plays an essential role in the reduction of metal salt into NPs is α-NADPH-dependent nitrate enzyme (Skimming et al., 2013). Besides this, the mycelial culture of mushrooms has the potential to secrete metabolites when they are exposed to

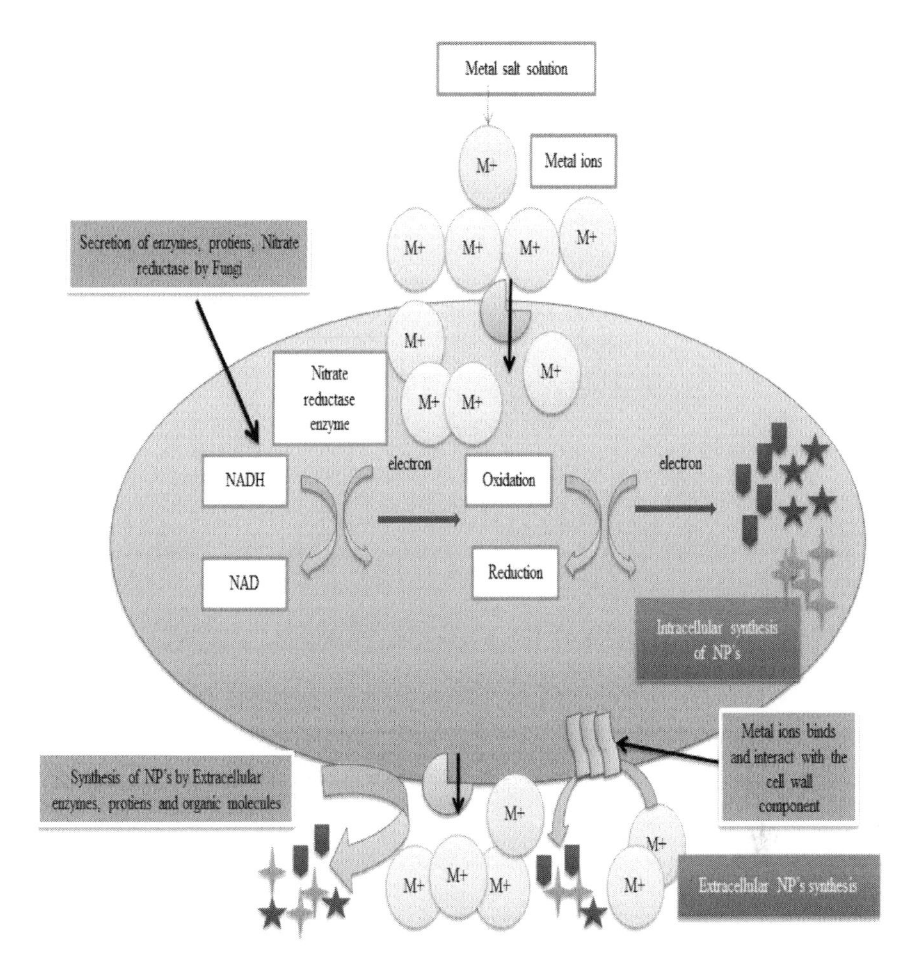

FIGURE 2.2 Mechanism of synthesis of nanoparticles from mushroom extract.

metal salts (Figure 2.2). These metabolites result in the reduction of metal salts into non-toxic solid metal NPs (Castro et al., 2014). Although the mechanism of synthesis of NPs is hypothesized, still there is a need to explore the mechanism to explain the synthesis of NPs.

2.3.1 Effect of Different Factors Upon the Synthesis of Nanoparticles

The different factors that affect the process of mycosynthesis include temperature, the pH of the reaction medium, the reduction agent, capping molecules, the concentration of substrate, and incubation time. Mushrooms can synthesize extracellular proteins. These extracellular proteins play an important role in the reduction of metal and capping of NPs. These proteins, when stabilized, increase the shelf life of NPs and their spectrum of application (Kulkarni and Muddapur, 2014). The protonation of amino acids is affected by the pH of the reaction medium; these amino acids take

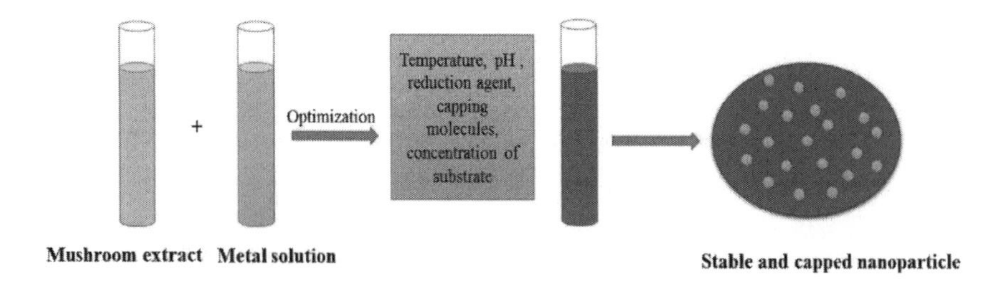

FIGURE 2.3 Effect of different factors on the synthesis of high yield nanoparticles.

part in the stabilization and formation of NPs of different sizes and shape (Figure 2.3). The temperature of the reaction medium also plays an important role in the development of the size, shape, and polydispersity of NPs (Sanghi and Verma, 2010).

2.4 CHARACTERIZATION OF SYNTHESIZED NANOPARTICLES

There is variation in the size and shape of NPs synthesized by mycosynthesis, so there is a need to characterize the physical, structural, and optical properties of synthesized NPs (Mercy et al., 2013). The different techniques are used to characterized NPs and these are discussed in the following subsections.

2.4.1 Chemical Characterization

The first step after the synthesis of NPs is to study the chemical composition of the compounds. Therefore, after the synthesis of NPs, they are studied to evaluate based on their chemical characterization. Chemical characterization involves the characterization of NPs based on their amount, quality crystalline properties, and for analysis of elements present in the compound. Therefore, based on these properties, chemical characters include optical spectroscopy and electron spectroscopy for characterizing the NPs.

2.4.2 Optical Spectroscopy

This method of characterization is used to study the interaction of metal ions with electromagnetic radiations. The electronic structure is revealed by absorption or emission experiments with visible and ultraviolet (UV) light as shown in Figure 2.3. Raman spectroscopy and infrared (IR) spectroscopy are used to study the vibrational properties of photons. Optical spectroscopy, being non-destructive and having high-resolution properties, is used for the characterization of material (Nemeth et al., 2013).

2.4.2.1 Optical Absorption Spectroscopy

UV-visible Spectroscopy

This technique is used to determine the presence and to quantify the number of metabolites present in the reaction mixture (Giusti and Wrolstad, 2001). The organic compounds present in the reaction mixture absorb energy in the form of light when exposed to UV or visible radiations. This light energy is used to transfer electrons from either bonding or non-bonding orbitals into one of the empty anti-bonding orbitals. As a result, electrons from both bonded or non-bonded orbitals get excited and get transferred to an empty orbital. The electron gain energy from the light obtains from UV and visible radiation to jump from one orbital to another. The larger the distance travelled by electrons to reach the orbital, the more energy required than for a small distance (Ratner and Castner, 2009). The light energy of a particular wavelength associated with UV and Visible radiation when absorbed by the molecules results in transferring/jumping off the electrons from one orbital to another (Gomes Silva et al., 2011). This relationship between the frequency of light absorbed and its energy is given as:

$$E = h\,\nu, \tag{2.1}$$

where E is energy of each quantum of light,
h is Planck's constant, and
ν is the frequency of light.

A UV-visible spectrophotometer is used for the detection of molecules and inorganic ions in solution. The detection is usually done based on the intensity of light and wavelength of the molecules (Wallin et al., 2009). Therefore, the intensity of light passing through the reference material is measured by the wavelength of the light emitted by the material through the spectrophotometer. Hence, according to Beer–Lambert's law,

$$A = \log_{10} I_0/I. \tag{2.2}$$

Here,
A is absorbance,
I is intensity of light passes through the sample, and
I_0 is intensity of light absorbed by the sample.

Therefore, the preceding equation states that the ability of the material to absorb radiation is expressed as a common logarithm of the reciprocal of transmittance of the substance (Lindkvist et al., 2013). NPs are prepared by the synthesis of metals such as gold, silver, copper, zinc iron, and many more. These metals have properties to interact with the specific wavelength of light. Hence, UV-visible spectroscopy plays an important role in the quantification of compounds present in metal NPs after synthesis (Vidhu and Philip, 2014). Besides these, plasmonic NPs, depending on their size and shape, absorb radiations of visible to the near-IR region. Due to the absorption at this region, NPs result in collective oscillations of electrons which are present on their electrons' surface. This is the process of oscillation of electrons is known as surface plasmon resonance (SPR) (Zhu et al., 2009b). SPR results in the dispersion of NPs. This dispersion gives one or more peaks which are further used to determine the shape, size, and size distribution of the NPs. Therefore, the characterization of plasmonic NPs loaded onto polymer microgels is done by using UV-visible spectroscopy (Begum et al., 2018). The growth of plasmonic NPs inside microgels and the growth of the polymeric network around plasmonic NPs can be studied by UV-visible spectroscopy. During the growth of plasmonic NPs in microgels, the surface plasmon band is red-shifted, and in the case of growth of polymeric network around plasmonic NPs, there is an increase in the thickness of the polymeric shell, which increases the refractive index of the medium around the NPs; hence, the plasmonic band shifts to a higher wavelength (Farooqi et al., 2015). The UV-visible spectroscopy is also used to confirm the stability of NPs in microgels. The measurement of SPR wavelength as a function of time of storage is the base to investigate the stability of NPs (Kumar et al., 2011).

2.4.2.2 Photoluminescence Spectroscopy

The photoluminescence technique is used to observe the impurities and crystalline quality of the compounds and to quantify the amount of disorder present in a system (Salome et al., 2014). In this technique, the substance absorbs electromagnetic radiation in the visible region by exciting one of its electrons and emits the light when the electron returns to a lower energy state (Xia et al., 2014). Electromagnetic radiations when absorbed raise electrons from their valance band to conduction band across the forbidden energy gap. This happens when electromagnetic radiations have a greater energy than the bandgap energy (Kaur and Verma, 2010). In the process of photo excitation, electrons lose their excess energy and return to the valence band before entering into the lowest energy level in the conduction band. The energy which is lost by electrons is converted into luminescent photons emitted from the material. Therefore, the energy emitted by electromagnetic radiations is a direct measure of bandgap energy (Kumar et al., 2016). The process of excitation of electromagnetic radiation, followed by the emission of electromagnetic radiation, is known as photoluminescence. The photoluminescence of metal NPs is considered an important application for characterization in recent times. Photoluminescence of NPs is considered a better option for optical labelling applications than are fluorescent molecules as it is free of photobleaching and photo blinking (Syamchand and Sony, 2015). For the number of plasmonic NPs that include nanosphere, nanorods, bipyramids, nanocubes, and nanoshells, single-photon and multiphoton excitation of photoluminescence has been reported (Simon et al., 2017). These reports throw light on the different properties and mechanisms of metal NPs and benefit their future applications. However, there is an inadequate investigation of the photoluminescence properties of complex-structure NPs than those of simple shapes.

2.4.2.3 Energy-Dispersive X-Ray Spectroscopy

This technique is used for the analysis of element and their chemical characterization. The energy-dispersive X-ray spectroscopy (EDS) system consists of an X-ray detector that detects the energy spectra and these energy spectra are collected by software and displayed on the monitor (Russ, 2013). These detectors consist of crystals that ionize and absorb energy, yielding free electrons. These electrons become inductive and produce an electrical charge bias. Hence, the energy of individual X-rays gets converted into an electrical voltage of proportional size due to absorption. This electrical voltage corresponds to the characteristics of X-rays of the element. In the EDS system, the electron strikes the sample through the source where there occurs excitation in the sample which is due to the elements present in it. There is a present Coulombic field in atoms due to which electrons get decelerated when an approach to them. The electrons lose energy as a result of deceleration, which appears as a photon and is known as bremsstrahlung, or "braking radiation" (Klockenkämper and Von Bohlen, 2014). The continuous spectrum of bremsstrahlung occurs which gradually becomes more intense, and therefore, the peak of intensity shifts towards a higher frequency. The shifting of the peak towards a higher frequency is due to the increase in the change of energy of the decelerated particles (Alabastri et al., 2013). The intensity of peaks is related to the concentration of the emitting particles present in the sample. The NPs used in chemotherapeutic agents to improve their therapeutic performance is detected by EDS.

2.5 STRUCTURAL CHARACTERIZATION

2.5.1 X-Ray Diffraction Technique

This technique of characterization is used to analyse the crystalline material structure. In this technique, different crystalline phases present in the material are identified which reveal the chemical composition of the material. The reference database is used to identify the phases achieved by acquired data (Herrera and Videla, 2009). The X-ray beam is directed at a sample; when it passes through it, its intensity is measured as a function of outgoing direction. Once the beam gets separated, it forms a diffraction pattern which indicates the sample's crystalline structure (Lin et al., 2017). The X-ray diffraction (XRD) technique is most extensively used for the characterization of metal NPs. The colloidal solution is dried to a powder form before characterization and results are represented as volume-average values (Hassellöv et al., 2008). The NPs composition is determined by comparing both the intensity and position of the peaks with a reference pattern. For the particles whose size is less than 3 nm, the XRD peaks are too broad. The broadening of peaks is due to the lattice strain and crystallite size (Mittemeijer and Welzel, 2013) The smaller domain in particles, whether it is single or multiple, is present where all moments are aligned in the same direction; therefore, the size derived from this technique is usually bigger than magnetic size (Akbarzadeh et al., 2012).

2.5.2 Scanning Electron Microscopy

Scanning electron microscopy (SEM) is a high-resolution technique and is used to characterize the size and shape of the particles. In this technique, the preparation of sample and images acquisition is quick and simple therefore it proves excellent for the characterization of NPs (Heinrich, 2010). The primary electrons (PE) beam causes the bombardment of the sample to create secondary electrons (SEs) and backscatter electrons (BSEs). The secondary electrons and the BSEs are responsible for the creation of the SEM image. The SEM bears a maximum accelerating voltage of about 30 kV. The secondary electrons result from the bombardment have lower energy below 50 eV; therefore, they have a small escape depth, in a range of few nanometres, and produce high-resolution micrographs. The BSEs, due to higher kinetic energy, escape at about half the depth of the interaction volume in the sample and, therefore, have poorer spatial resolution with superior compositional contrast

(Vladár and Hodoroaba, 2020). SEs are further characterized into three categories: SE_1, SE_2, and SE_3. The point where the PE has an impact on the sample SE_1 is generated, high electrons present within the sample generate SE_2, and SE_3 are generated when the BSEs hit the inner surface of the microscope specimen chamber (Suga et al., 2014). Inside the electron column, through lens secondary electrons are collected with excellent efficiency at an optimal distance and provide higher spatial resolution (Zhu et al., 2009b). The SEM image of the NP sample surface morphology is at the high spatial resolution to a sub-nanometre scale. This technique can assess the purity, homogeneity, degree of dispersion, and extent of aggregation of NPs due to the collection of transmitted electrons by detectors present in the device (Laborda et al., 2016). This technique prevents the destruction of the sample in E-SEM mode and hence has proved beneficial for polymeric NPs (Tamayo et al., 2016).

2.5.3 Transmission Electron Microscopy

This technique is used to measure size, grain size, size distribution, and morphology of NPs. Transmission electron microscopy (TEM) provides a much higher resolution and uses an electron beam to image the particles (Nie and Kumacheva, 2008). In this technique, a filament that has an extremely high potential gradient or a heated filament is used to emit electrons. The optical axis of microscope uses a condenser lens and focuses to small thin, coherent beam along which the monochromatic beam move. The beam knocks out high-angle electrons and is restricted by condenser aperture. The beam then delivers strikes to the specimen; as a result, a part of the beam gets transmitted, which is focused by the objective lens into an image (Hattar et al., 2014). The objective aperture blocks high-angle diffracted electrons to enhance the contrast. The selected-area aperture, on the other hand, orderly arranges the atom in the sample to examine the periodic diffraction of electrons. The intermediate and projector lenses then enable the image to pass down the column. The light is generated when the image strikes the phosphor image screen; this allows the user to see the image (Egerton, 2011). Where fewer electrons were transmitted is represented by the darker area of the image, and where more electrons have passed is represented by the bright area of the sample. TEM has two image modes: bright field and dark field. Bright field mode has an aperture placed in the back focal plane of the objective lens which allows only the direct beam to pass through it (Tang and Yang, 2017). This direct beam becomes weak when interacting with the specimen and results in the formation of the image. Hence, an image results due to the contribution of mass thickness and diffraction contrast. Thick areas are enriched by heavy atoms, and there is a dark contrast in the crystalline area. The occurrence of the phenomenon of contrast forming results in interruption of image. In the dark field mode, aperture blocks the direct beam and allows only one or more diffracted beams to pass through it. When these diffracted beams strongly interact with the specimen, the planar defects, stacking faults, or particle size can be obtained easily (Cowley, 1986).

The size distribution of synthesized NPs is studied by images obtained from TEM. The size distribution of NPs depends upon the reduction method and elemental composition (Srivastava et al., 2011).

2.5.4 Atomic Force Microscopy

Atomic force microscopy (AFM) is used to study different properties that include morphology, size, texture, and surface roughness of the particles (Kumar and Rao, 2012). It is used for three-dimensional characterization of NPs and also provides different geometrics structures and magnetic behaviour of the NPs (Lee et al., 2012). In AFM, the force between a sharp probe tip (<10) and the sample surface within the distance between 0.2- and 10-nm probe sample separation is measured (De Oliveira et al., 2012). The probe is attached to a cantilever, where deflection occurs upon interaction, which is measured by the laser beam reflection. This whole process is termed as a "beam bounce" method. The deflection of cantilever measures the topography of the surface. The topography of the surface is represented in the form of different peaks represented by different colour gradients or by greyscale. These peaks are helpful in the identification and measurements of parameters that are under investigation (Yang et al., 2011).

2.5.5 Differential Scanning Calorimetry

Differential scanning calorimetry (DSC) is used to measure thermodynamic data, including heat capacity, enthalpy, and entropy, as well as kinetic data that include reaction rate and activation energies of nanoparticle samples (Gill et al., 2010; Kaushik et al., 2015). In DSC, the differential flow of heat between the sample (T_S) and reference (T_R) materials are monitored and recorded as a function of temperature. The heating of reference cells and samples are performed independently, that is $\Delta T = T_S - T_R = 0$, and the difference in heat flow is monitored as a function of temperature. The maintenance of the temperature of the sample as material goes through a transition phase depends on the nature of the transition, whether it is endothermic or exothermic. The amount of excess heat released or absorbed due to the sample transition represents the difference in energy needed to match the temperature of the sample to reference. The information obtained by following this characterization method can be described by the following equation (Zheng, 2018):

$$\frac{dH}{dt} = mCp\frac{dT}{dt}. \tag{2.3}$$

Here, dH/dt is the heat flow signal, m is the mass, Cp is the heat capacity of sample, and dT/dt is the rate of heating.

This technique can also be used to observe physical changes that include melting, crystallization, and transition through transitions.

2.6 APPLICATION OF MYCONANOPARTICLES AGAINST FOOD PATHOGENIC BACTERIA

2.6.1 Food Pathogenic Bacteria and Their Effect on Health

Foodborne bacteria are the microorganisms that are present in food hence cause foodborne diseases (Sharma et al., 2016). The Food and Drug Administration has categorized foodborne bacteria according to the specific food consumed (Fao and Isric, 2010).

TABLE 2.1
Source of Different Foodborne Bacteria

Microorganism	Carrier Food
Campylobacter sp	Raw or undercooked poultry
Salmonella typhimurium	Meat, poultry, eggs, sprouts, vegetables
Shigella and *Escherichia coli*	Meat and unpasteurized milk, contaminated water
Clostridium botulinum	Home-canned food
Clostridium perfringens, Yersinia, Vibrio cholera, V. vulnificus, V. parahaemolyticus, Staphylococcus aureus, Bacillus spp., Listeria monocytogenes	In uncooked meats, vegetables, unpasteurized milk, and soft cheese, refrigerated and ready to eat foods

TABLE 2.2
Different Foodborne Bacteria and Disease Caused by Them

Foodborne Bacteria	Infection Caused by Bacteria
Campylobacter sp	Cause diarrhea
Salmonella typhimurium	Cause salmonellosis
Shigella and *E. coli*	Shigellosis
Clostridium botulinum	Produces a toxin which causes botulism
Clostridium perfringens	Release toxins that cause diarrhea
Yersinia enterocolitica	Cause yersiniosis
Vibrio cholera	Cause cholera
V vulnificus	Food poisoning
V parahaemolyticus	Cause gastrointestinal illness in human
Staphylococcus aureus	Produce toxin that causes food poisoning
Bacillus spp	Cause food poisoning
Listeria monocytogenes	Cause listeriosis

Most of the foodborne pathogenic bacteria are prevalent in foods that include raw, uncooked food, eggs, vegetables, unpasteurized milk, contaminated water, home-canned food, soft cheese, and refrigerated and ready-to-eat foods, as shown in Table 2.1. These bacteria can produce harmful toxins which result in food poisoning, followed by vomiting, diarrhoea, fever, cholera, and gastrointestinal illness, as shown in Table 2.2.

2.6.2 Resistance of Foodborne Bacteria Against Antibiotics

The resistance of disease-causing bacteria has become an alarming threat (Sharma et al., 2016. It has been observed by the World Health Organization (WHO) that the level of antibiotic resistance reached so high that it results in 700,000 deaths each year globally (World Health Organization, 2014). The resistance of foodborne pathogenic bacteria to antibiotics results due to the extensive use of antibiotics in agriculture to promote growth, for the prevention of disease in farm animals, to increase the production of food, and for human health (Done et al., 2015). There are two resistance mechanisms which help spread the bacterial population. These are vertical gene transfer and horizontal gene transfer, shown in Figures 2.4 (a), (b), and (c). In vertical gene transfer, there is a genetic error that accumulates in the genetic material, that is chromosomes or plasmid of the bacterial cells and the presence of efflux pump. The efflux pumps are ingenious weapons which are proteins located in the membranes of bacteria. These pumps recognize and expel drugs that break the cell membrane of the bacteria to gain entry inside it (Wilson et al., 2019). The vertical transfer is also known as intrinsic resistance. In horizontal gene transfer or acquired resistance, bacteria gain new genes on their mobile genetic elements that include plasmids, insertion sequences, integrons, phage-related elements, and transposons,

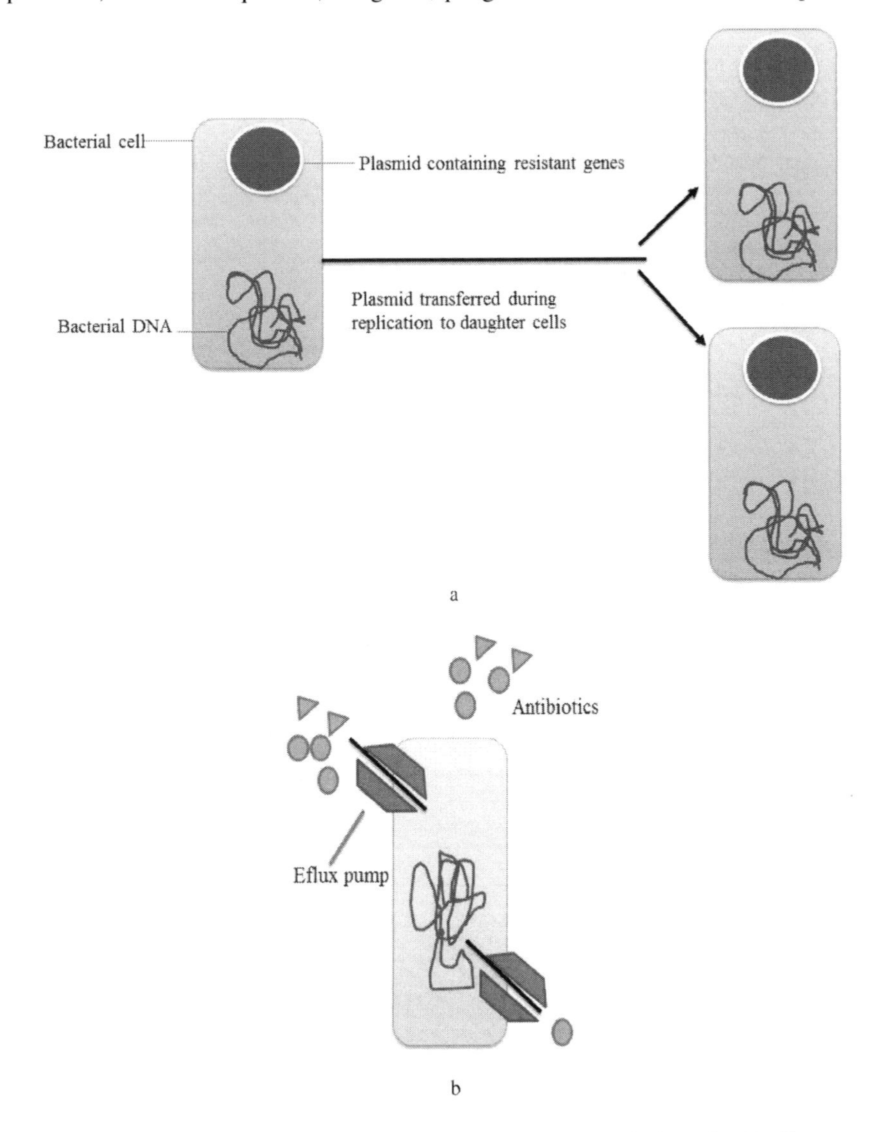

FIGURE 2.4 Mechanism of resistance of bacteria to antibiotics: (a) vertical transfer, (b) efflux pump.

(Continued)

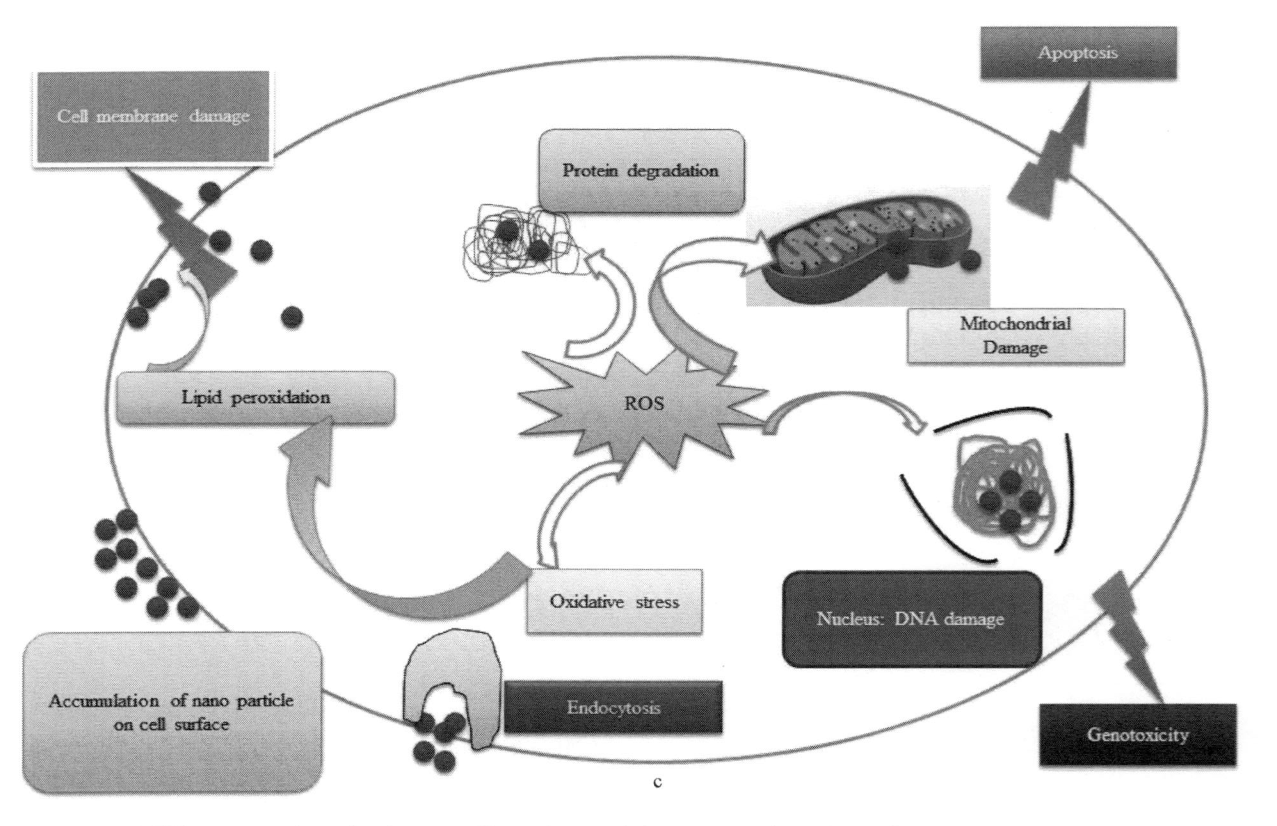

FIGURE 2.4 (CONTINUED) Mechanism of resistance of bacteria to antibiotics: (c) horizontal transfer.

therefore, occurs within or bacterial species (Brown-Jaque et al., 2015). To overcome this problem metal nanoparticle could be a better approach as microorganisms do not show any resistivity towards the NPs. In the recent era, green synthesis of NPs, which includes the use of biological compounds for the production, is highly appreciable (Ahmad et al., 2013). Among these biological compounds, the synthesis of NPs from different metals, with mushrooms extensively studied by several researchers as shown in Table 2.3. The process of synthesis of NPs from mushrooms is known as mycosynthesis and NPs thus formed are known as myconanoparticles.

TABLE 2.3
Mushrooms Strains Used for Mycosynthesis of Nanoparticles

Mushrooms	Metal Nanoparticles	References
Trametes versicolor	Cds and Ag metal nanoparticles	Sanghi and Verma (2009)
Ganoderma enigmaticum	Ag metal nanoparticles	Gudikandula et al. (2017)
Schizophyllum commune	Ag metal nanoparticles	Rafie et al. (2012)
Flammulina velutipes	Ag metal nanoparticles	Narayanan et al. (2015)
Volvariella volvacea	Au, Au-Ag	Philip (2009)
Agaricus bisporus	Ag metal nanoparticles	Narasimha et al. (2011)
Pleurotus florida	Ag metal nanoparticles	Bhat et al. (2011)
Pleurotuss ajor-caju	Ag metal nanoparticles	Musa et al. (2017)
Ganoderma neo-japonicum,	Ag metal nanoparticles	Gurunathan et al. (2014)
Ganoderma lucidum	Ag metal nanoparticles	Karwa et al. (2011)
Inonotus obliquus	Ag metal nanoparticles	Nagajyothi et al. (2014)
Tricholoma crassum	Ag metal nanoparticles	Ray et al. (2011)
Tricholoma matsutake	Ag metal nanoparticles	Anthony et al. (2014)
Calocyble indica	Ag metal nanoparticles	Gurunathan et al. (2015)
Agaricus bisporus, Pleurotus sapidus, Inonotus obliquus, Ganoderma spp. and *Flammulina velutipes*	Au metal nanoparticles	Khan et al. (2014); Lee et al. (2012); Sarkar et al. (2013)
Pleurotus florida	Au metal nanoparticles	Sen et al. (2013)
Pleurotuso streatus, Ganoderma lucidum, and *Grifola frondosa*	Selenium nanoparticles	Vetchinkina et al. (2016)
Cordyceps sinensis	Selenium nanoparticles	Xiao et al. (2017)
Pleurotus ostreatus	CdS	Borovaya et al. (2015)
Ganoderma lucidum	CdS	Raziya et al. (2006)
Pleurotus ostreatus	ZnS	Senapati and Sarkar (2014)
Pleurotus sp	Fe	Mazumdar and Haloi (2017)
Pleurotus ostreatus, Lentinula edodes, Ganoderma lucidum, and Grifola frondosa	Au, Ag, Se, and Si nanoparticles	Vetchinkina et al. (2018)
Pleurotus giganteus	Ag nanoparticles	Debnath et al. (2019)

2.6.3 Mechanism of Action of Myconanoparticles Against Food Pathogenic Bacteria

NPs show excellent activity against pathogenic microorganisms. The mode of action of their mechanism is proposed by several researchers in different theories as shown in Figure 2.5. The NPs anchor to the bacterial cell wall and penetrate inside the cell. Here, inside the cell, the NPs result in a change in the structure permeability of cell membrane and lead to cell death (Farkhani et al., 2014). NPs, having extremely large surface areas, provide better contact with the microorganisms and, hence, show good antimicrobial activity (Ahmad et al., 2013). Some NPs can dissipate the proton motive force at the bacterial cell membrane. This dissipation of proton motive force results in the blocking of oxidative phosphorylation (Desbois and Smith, 2010). NPs result in the production of free radicals by the membrane, which becomes porous and leads to cell death. The bacterial cell consists of DNA bases rich in sulfur and phosphorus, and these biomaterials get to interact with metal NPs. The NPs, upon interaction, act on these bases and damage the DNA, which leads to the death of the

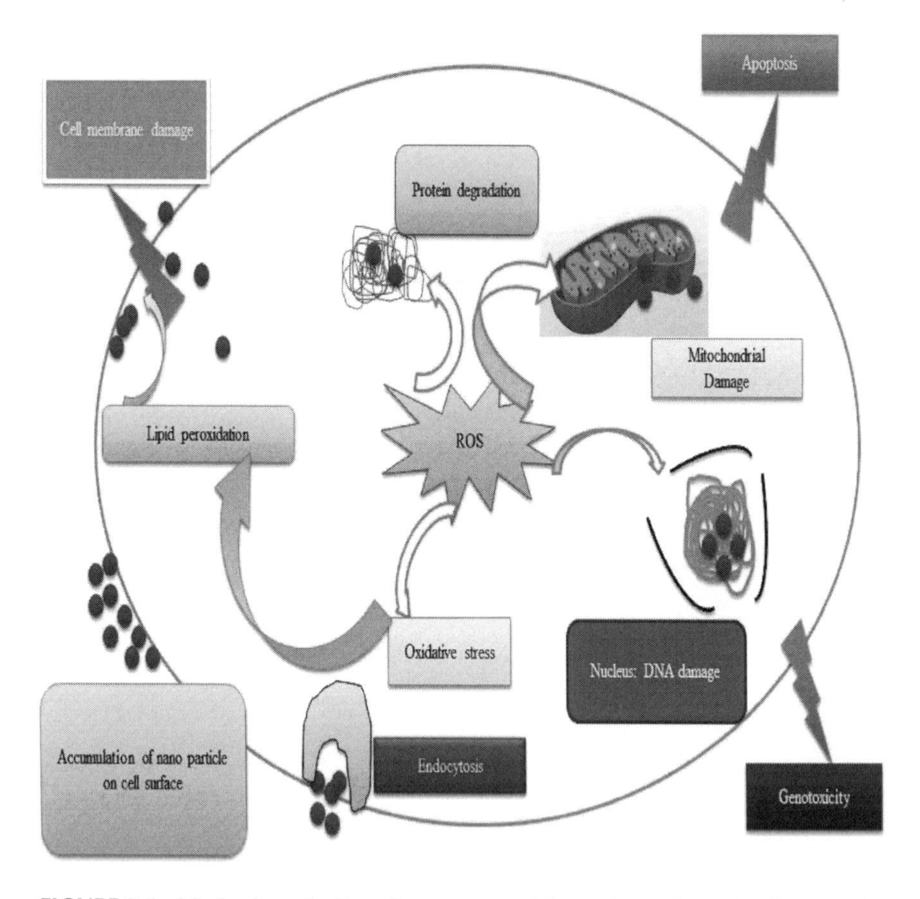

FIGURE 2.5 Mechanism of action of myconanoparticles against pathogenic microorganism.

TABLE 2.4
Myconanoparticles and Their Antibacterial Properties

Nanoparticles Synthesized from Mushroom	Antibacterial Properties	References
Ag nanoparticles synthesized from *Agaricus bisporus*	*E. coli*, *S. aureus*, *Pseudomonas spp.*, *Bacillus spp.*	Narasimha et al. (2011)
Ag nanoparticles synthesized from *Agaricus bisporus* and *Ganoderma lucidum*	*E. coli*, and *S. aureus*	Sriramulu and Sumathi (2017)
Ag nanoparticles synthesized from *Ganoderma sessiliforme*	*E. coli*, *Bacillus subtilis*, *Streptococcus faecalis*, *Listeria innocua*, and *Micrococcus luteus*	Mohanta et al. (2018)
Au nanoparticles synthesized from *Lignosus rhinocerotis*	*Pseudomonas aeruginosa*, *Escherichia coli*, *Staphylococcus aureus*, and *Bacillus sp.*	Katas et al. (2019)
Ag nanoparticles synthesized from *Pleurotus giganteus*	*E. coli*, *P. aeruginosa*, *Bacillus subtilis*, and *S. aureus*	Debnath et al. (2019)

cell (Durán et al., 2016). NPs are known as modulators of bacterial signal transduction and, therefore, cause signal transduction inhibition by dephosphorylating the peptides substrates on tyrosine residue (Dakal et al., 2016). NPs collapse the membrane potential of the bacterial cell membrane and decrease the adenosine triphosphate (ATP) level by inhibiting the activity of enzyme ATPase. NPs also inhibit the binding of tRNA, along with the subunit of ribosomes (Cui et al., 2012).

Mycosynthesis focuses to produce NPs having the variable size, shape, and chemical composition, as well their potential to use for the benefit of humans. NPs synthesized from the mushrooms due to their stability in size, shape, and free from toxicity act as excellent antimicrobial agents. Myconanoparticles, therefore, are studied for their antimicrobial properties. Silver and gold NPs synthesized from *Agaricus bisporus*, *Ganoderma lucidum*, *Ganoderma sessiliforme*, *Pleurotus giganteus*, and *Lignosus rhinocerotis* showed antimicrobial activity against both gram-positive including *Escherichia coli*, *Staphylococcus aureus*, *Pseudomonas aeruginosa*, *Bacillus spp. Listeria innocua*, and *Micrococcus luteus* (Narasimha et al., 2011; Anthony et al., 2014; Borovaya et al., 2015; Sriramulu and Sumathi, 2017; Mohanta et al., 2018; Katas et al., 2019; Gudikandula et al., 2017; Debnath et al., 2019), as shown in Table 2.4.

2.7 CONCLUSION

The mycosynthesis of NPs of metal include the use of non-toxic and eco-friendly reagents and therefore, these NPs have a great advantage over chemically synthesized NPs. Mushrooms consist of primary and secondary metabolites, as well as extracellular enzymes, that provide a resistance mechanism against heavy metal ions. The metabolites and extracellular enzymes work together and result in detoxification of metal ions and lead to the process of synthesis of NPs with different morphology, size dimensions, and functional properties which are stable and non-toxic. In conclusion, the synthesis of NPs from mushrooms, being stable and non-toxic, can be utilized as a therapeutic agent for the treatment of different diseases.

REFERENCES

Agarwal, A., Pathera, A. K., Kaushik, R., Kumar, N., Dhull, S. B., Arora, S., and Chawla, P. (2020). Succinylation of milk proteins: influence on micronutrient binding and functional indices. *Trends in Food Science and Technology*, 97, 254–264.

Ahmad, T., Wani, I. A., Manzoor, N., Ahmed, J., and Asiri, A. M. (2013). Biosynthesis, structural characterization and antimicrobial activity of gold and silver nanoparticles. *Colloids and Surfaces B: Biointerfaces*, 107, 227–234.

Akbarzadeh, A., Samiei, M., and Davaran, S. (2012). Magnetic nanoparticles: preparation, physical properties, and applications in biomedicine. *Nanoscale Research Letters*, 7(1), 144.

Anthony, K. J. P., Murugan, M., Jeyaraj, M., Rathinam, N. K., and Sangiliyandi, G, (2014). Synthesis of silver nanoparticles using pine mushroom extract: a potential antimicrobial agent against *E. coli* and *B. subtilis*. *Journal of Industrial and Engineering Chemistry*, 20(4), 2325–2331

Banerjee, K. and Rai, V. R. (2018). A review on mycosynthesis, mechanism, and characterization of silver and gold nanoparticles. *BioNanoScience*, 8(1), 17–31.

Begum, R., Farooqi, Z. H., Naseem, K., Ali, F., Batool, M., Xiao, J., and Irfan, A. (2018). Applications of UV/Vis spectroscopy in characterization and catalytic activity of noble metal nanoparticles fabricated in responsive polymer microgels: a review. *Critical Reviews in Analytical Chemistry*, 48(6), 503–516.

Bhat, R., Deshpande, R., Ganachari, S. V., Huh, D. S., and Venkataraman, A. (2011). Photo-irradiated biosynthesis of silver nanoparticles using edible mushroom *Pleurotus florida* and their antibacterial activity studies. *Bioinorganic Chemistry and Application*, 2011, 1–7.

Bhat, R., Sharanabasava, V. G., Deshpande, R., Shetti, U., Sanjeev, G., and Venkataraman, A. (2013). Photo-bio-synthesis of irregular shaped functionalized gold nanoparticles using edible mushroom *Pleurotus florida* and its anticancer evaluation. *Journal of Photochemistry and Photobiology B: Biology*, 125, 63–69.

Borovaya, M., Pirko, Y., Krupodorova, T., Naumenko, A., Blume, Y., and Yemets, A, (2015). Biosynthesis of cadmium sulphide quantum dots by using *Pleurotus ostreatus* (Jacq.) P. Kumm. *Biotechnol Biotechnol Equip*, 29(6), 1156–1163

Brown-Jaque, M., Calero-Cáceres, W., and Muniesa, M. (2015). Transfer of antibiotic-resistance genes via phage-related mobile elements. *Plasmid*, 79, 1–7.

Castro, L., Blázquez, M. L., González, F. G., and Ballester, A. (2014). Mechanism and applications of metal nanoparticles prepared by bio-mediated process. *Reviews in Advanced Sciences and Engineering*, 3(3), 199–216.

Chaudhari, P. S., Damahe, A., and Kumbhar, P. (2016). Silver nanoparticles-A review with focus on green synthesis. *International Journal of Pharma Research & Review*, 5, 14–28.

Chawla, P., Kaushik, R., Shiva Swaraj, V. J., and Kumar N. (2018). Organophosphorus pesticides residues in food and their colorimetric detection. *Environmental Nanotechnology, Monitoring and Management*, 10, 292–307.

Chawla, P., Kumar, N., Kaushik, R., and Dhull, S.B. (2019). Synthesis, characterization and cellular mineral absorption of Gum Arabic stabilized nanoemulsion of *Rhododendron arboreum* flower extract. *Journal of Food Science and Technology*, 56(12), 5194–5203,

Chawla, P., Kumar, N., Bains, A., Dhull, S. B. Kumar, M., Kaushik, R., and Punia, S. (2020). Gum Arabic capped copper nanoparticles: characterization and their utilization. *International Journal of Biological Macromolecules*, 146, 232–242.

Chung, I. M., Park, I., Seung-Hyun, K., Thiruvengadam, M., and Rajakumar, G. (2016). Plant-mediated synthesis of silver nanoparticles: their characteristic properties and therapeutic applications. *Nanoscale Research Letters*, 11(1), 40.

Cowley, J. M. (1986). Electron microscopy of surface structure. *Progress in Surface Science*, 21(3), 209–250.

Cui, Y., Zhao, Y., Tian, Y., Zhang, W., Lü, X., and Jiang, X. (2012). The molecular mechanism of action of bactericidal gold nanoparticles on *Escherichia coli*. *Biomaterials*, 33(7), 2327–2333.

Dakal, T. C., Kumar, A., Majumdar, R. S., and Yadav, V. (2016). Mechanistic basis of antimicrobial actions of silver nanoparticles. *Frontiers in Microbiology*, 7, 1831.

De Oliveira, R. R. L., Albuquerque, D. A. C., Cruz, T. G. S., Yamaji, F. M., and Leite, F. L. (2012). Measurement of the nanoscale roughness by atomic force microscopy: basic principles and applications. *Atomic force microscopy-imaging, measuring and manipulating surfaces at the atomic scale*, 147–175, Intech, Brazil.

Debnath, G., Das, P., and Saha, A. K. (2019) Green synthesis of silver nanoparticles using mushroom extract of *Pleurotusgiganteus*: characterization, antimicrobial, and α-amylase inhibitory activity. *BioNanoScience*, 9:611–619.

Desbois, A. P., and Smith, V. J. (2010). Antibacterial free fatty acids: activities, mechanisms of action and biotechnological potential. *Applied Microbiology and Biotechnology*, 85(6), 1629–1642.

Dhull, S. B., Anju, M., Punia, S., Kaushik, R., and Chawla, P. (2019). Application of Gum Arabic in Nanoemulsion for Safe Conveyance of Bioactive Components. In: Prasad R., Kumar V., Kumar M., and Choudhary D. (eds) *Nanobiotechnology in Bioformulations. Nanotechnology in the Life Science*. Springer, Cham. Chapter 3, pp. 85–98. DOI. org/10.1007/978-3-030-17061-5-3.

Done, H. Y., Venkatesan, A. K., and Halden, R. U. (2015). Does the recent growth of aquaculture create antibiotic resistance threats different from those associated with land animal production in agriculture? *The AAPS Journal*, 17(3), 513–524.

Durán, N., Durán, M., De Jesus, M. B., Seabra, A. B., Fávaro, W. J., and Nakazato, G. (2016). Silver nanoparticles: a new view on mechanistic aspects on antimicrobial activity. *Nanomedicine: Nanotechnology, Biology and Medicine*, 12(3), 789–799.

Egerton, R. F. (2011). *Electron energy-loss spectroscopy in the electron microscope*. Springer Science & Business Media, Canada.

Fao, I. and Isric, I. (2010). JRC. 2009. *Harmonized world soil database (version 1.1). Food and Agriculture Organization, Rome, Italy and International Institute for Applied Systems Analysis, Laxenburg, Austria.*

Farkhani, S. M., Valizadeh, A., Karami, H., Mohammadi, S., Sohrabi, N., and Badrzadeh, F. (2014). Cell penetrating peptides: efficient vectors for delivery of nanoparticles, nanocarriers, therapeutic and diagnostic molecules. *Peptides*, 57, 78–94.

Farooqi, Z. H., Khan, S. R., Begum, R., Kanwal, F., Sharif, A., Ahmed, E., … and Ijaz, A. (2015). Effect of acrylic acid feed contents of microgels on catalytic activity of silver nanoparticles fabricated hybrid microgels. *Turkish Journal of Chemistry*, 39(1), 96–107.

Gill, P., Moghadam, T. T., and Ranjbar, B. (2010). Differential scanning calorimetry techniques: applications in biology and nanoscience. *Journal of Biomolecular Techniques: JBT*, 21(4), 167.

Giusti, M. M. and Wrolstad, R. E. (2001). Characterization and measurement of anthocyanins by UV-visible spectroscopy. *Current Protocols in Food Analytical Chemistry*, (1), F1–F2.

Gomes Silva, C., Juárez, R., Marino, T., Molinari, R., and García, H. (2011). Influence of excitation wavelength (UV or visible light) on the photocatalytic activity of titania containing gold nanoparticles for the generation of hydrogen or oxygen from water. *Journal of the American Chemical Society*, 133(3), 595–602.

Gudikandula, K., Vadapally, P., and Charya, M. S. (2017). Biogenic synthesis of silver nanoparticles from white rot fungi: their characterization and antibacterial studies. *OpenNano*, 2, 64–78.

Gurunathan, S., Han, J., Park, J. H., and Kim, J. H. (2014) A green chemistry approach for synthesizing biocompatible gold nanoparticles. *Nanoscale Research Letters*, 9(1), 248.

Gurunathan, S., Park, J. H., Han, J. W., and Kim, J. H. (2015) Comparative assessment of the apoptotic potential of silver nanoparticles synthesized by *Bacillus tequilensis* and *Calocybe indica* in MDA-MB-231 human breast cancer cells: targeting p53 for anticancer therapy. *International Journal of Nanomedicine* 10, 4203.

Hassellöv, M., Readman, J. W., Ranville, J. F., and Tiede, K. (2008). Nanoparticle analysis and characterization methodologies in environmental risk assessment of engineered nanoparticles. *Ecotoxicology*, 17(5), 344–361.

Hattar, K., Bufford, D. C., and Buller, D. L. (2014). Concurrent in situ ion irradiation transmission electron microscope. *Nuclear Instruments and Methods in Physics Research Section B: Beam Interactions with Materials and Atoms*, 338, 56–65.

Heinrich, H. (2010). *Electron Microscopy for Functional Nanostructures: Processing, Characterization, and Applications*, 414, Springer, Florida.

Herrera, L. K. and Videla, H. A. (2009). Surface analysis and materials characterization for the study of biodeterioration and weathering effects on cultural property. *International Biodeterioration & Biodegradation*, 63(7), 813–822.

Iravani, S., Korbekandi, H., Mirmohammadi, S. V., and Zolfaghari, B. (2014). Synthesis of silver nanoparticles: chemical, physical and biological methods. *Research in Pharmaceutical Sciences*, 9(6), 385.

Kalia, A. and Kaur, G. (2018). Biosynthesis of Nanoparticles Using Mushrooms. In *Biology of Macrofungi* (pp. 351–360). Bhim Pratap Singh Lallawmsanga Ajit Kumar Passari Springer, Cham.

Katas, H., Lim, C. S., Azlan, A. Y. H. N., Buang, F., and Busra, M. F. M. (2019). Antibacterial activity of biosynthesized gold nanoparticles using biomolecules from *Lignosusrhinocerotis* and chitosan. *Saudi Pharmaceutical Journal*, 27(2), 283–292.

Kaushik, R., Sachdeva, B. and Arora, S. (2015). Heat stability and thermal properties of calcium fortified milk. *Cyta-Journal of Food*, 13(2), 305–311.

Kaushik, R., Swami, N., Sihag, M., and Ray, A. (2015). Isolation, characterization of wheat gluten and its regeneration properties. *Journal of Food Science and Technology*, 52(9), 5930–5937.

Klockenkämper, R. and Von Bohlen, A. (2014). *Total-reflection X-ray fluorescence analysis and related methods*. John Wiley and Sons, Germany.

Kulkarni, N., and Muddapur, U. (2014). Biosynthesis of metal nanoparticles: a review. *Journal of Nanotechnology*, 2014, 1-9.

Kumar, A., Dhoble, S. J., Peshwe, D. R., Bhatt, J., Terblans, J. J., and Swart, H. C. (2016). Crystal structure, energy transfer mechanism and tunable luminescence in Ce^{3+}/Dy^{3+} co-activated $Ca_{20}Mg_3Al_{26}Si_3O_{68}$ nanophosphors. *Ceramics International*, 42(9), 10854–10865.

Kumar, V. G., Gokavarapu, S. D., Rajeswari, A., Dhas, T. S., Karthick, V., Kapadia, Z., … and Sinha, S. (2011). Facile green synthesis of gold nanoparticles using leaf extract of antidiabetic potent *Cassia auriculata. Colloids and Surfaces B: Biointerfaces*, 87(1), 159–163.

Kumar, B. R. and Rao, T. S. (2012). AFM studies on surface morphology, topography and texture of nanostructured zinc aluminum oxide thin films. *Digest Journal of Nanomaterials and Biostructures*, 7(4), 1881–1889.

Laborda, F., Bolea, E., Cepriá, G., Gómez, M. T., Jiménez, M. S., Pérez-Arantegui, J., and Castillo, J. R. (2016). Detection, characterization and quantification of inorganic engineered nanomaterials: a review of techniques and methodological approaches for the analysis of complex samples. *Analytica Chimica Acta*, 904, 10–32.

Lee, G., Lee, H., Nam, K., Han, J. H., Yang, J., Lee, S. W., … and Kwon, T. (2012). Nanomechanical characterization of chemical interaction between gold nanoparticles and chemical functional groups. *Nanoscale Research Letters*, 7(1), 608.

Lin, F., Liu, Y., Yu, X., Cheng, L., Singer, A., Shpyrko, O. G., … and Yang, X. Q. (2017). Synchrotron X-ray analytical techniques for studying materials electrochemistry in rechargeable batteries. *Chemical Reviews*, 117(21), 13123–13186.

Lindkvist, M., Granåsen, G., and Grönlund, C. (2013). Coherent derivation of equations for differential spectroscopy and spatially resolved spectroscopy: an undergraduate tutorial. *Spectroscopy Letters*, 46(4), 243–249.

Lu, A. H., Salabas, E. E., and Schüth, F. (2007). Magnetic nanoparticles: synthesis, protection, functionalization, and application. *Angewandte Chemie International Edition*, 46(8), 1222–1244.

Malik, P., Shankar, R., Malik, V., Sharma, N., and Mukherjee, T. K. (2014). Green chemistry based benign routes for nanoparticle synthesis. *Journal of Nanoparticles*, 2014, 1-14.

Mercy, A., Selvaraj, R. S., Boaz, B. M., Anandhi, A., and Kanagadurai, R. (2013). Synthesis, structural and optical characterisation of cadmium sulphide nanoparticles. *Journal of Pure and Applied Physics*, 52, 448–452.

Mittemeijer, E. J. and Welzel, U. (Eds.). (2013). *Modern diffraction methods*. John Wiley & Sons, Germany.

Mohanta, Y. K., Nayak, D., Biswas, K., Singdevsachan, S. K., Abd Allah, E. F., Hashem, A., and Mohanta, T. K. (2018). Silver nanoparticles synthesized using wild mushroom show potential antimicrobial activities against food borne pathogens. *Molecules*, 23(3), 655.

Narasimha G., Praveen B., Mallikarjuna K., and Deva Prasad Raju B., (2011) Mushrooms *(Agaricus bisporus)* mediated biosynthesis of sliver nanoparticles, characterization and their antimicrobial activity. *International Journal of Nano Dimension*, 2:29–36

Nemeth, A., Hannesschläger, G., Leiss-Holzinger, E., Wiesauer, K., and Leitner, M. (2013). Optical coherence tomography–applications in non-destructive testing and evaluation. *Optical Coherence Tomography*, 6 163–186.

Nie, Z. and Kumacheva, E. (2008). Patterning surfaces with functional polymers. *Nature materials*, 7(4), 277–290.

Owaid, M. N. (2019). Green synthesis of silver nanoparticles by Pleurotus (oyster mushroom) and their bioactivity. *Environmental Nanotechnology, Monitoring & Management*,12, 100256.

Pathania, R., Khan, H., Kaushik, R., and Khan, M. A. (2018). Essential oil nanoemulsions and their antimicrobial and food applications. *Current Research Nutrition and Food Science*, 6(3), 1–16.

Ratner, B. D. and Castner, D. G. (2009). Electron spectroscopy for chemical analysis. *Surface Analysis: The Principal Techniques*, 2, 374–381.

Ravindran, A., Chandran, P., and Khan, S. S. (2013). Biofunctionalized silver nanoparticles: advances and prospects. *Colloids and Surfaces B: Biointerfaces*, 105, 342–352.

Rogers, K. (Ed.). (2010). *The Chemical Reactions of Life: From Metabolism to Photosynthesis*. Britannica Educational Publishing, New York.

Roy, A., Bulut, O., Some, S., Mandal, A. K., and Yilmaz, M. D. (2019). Green synthesis of silver nanoparticles: biomolecule-nanoparticle organizations targeting antimicrobial activity. *RSC Advances*, 9(5), 2673–2702.

Russ, J. C. (2013). *Fundamentals of Energy Dispersive X-Ray Analysis: Butterworths Monographs in Materials*. Butterworth-Heinemann, USA.

Sanghi, R., and Verma, P. (2009a). A facile green extracellular biosynthesis of CdS nanoparticles by immobilized fungus. *Chemical Engineering Journal* 155(3), 886–891

Sanghi, R., and Verma, P. (2009b). A facile green extracellular biosynthesis of CdS nanoparticles by immobilized fungus. *Chemical Engineering Journal*, 155(3), 886–891.

Sanghi, R., and Verma, P. (2010). pH dependent fungal proteins in the 'green' synthesis of gold nanoparticles. *Advanced Materials Letters*, 1(3), 193–199.

Sarkar, J., Roy. SK., Laskar, A., Chattopadhyay, D., and Acharya. K. (2013). Bioreduction of chloroaurate ions to gold nanoparticles by culture filtrate of *Pleurotus sapidus* Quel. *Materials Letters*, 92:313–316.

Sen, I. K. Maity, K., and Islam, S. S., (2013). Green synthesis of gold nanoparticles using a glucan of an edible mushroom and study of catalytic activity. *Carbohydrpolym*, 91(2): 518–528.

Senapati, U.S., and Sarkar, D., (2014). Characterization of biosynthesized zinc sulphide nanoparticles using edible mushroom *Pleurotus ostreatus*. *Indian Journal of Physics*, 88(6), 557–562.

Sharma, D., Kanchi, S., and Bisetty, K. (2019). Biogenic synthesis of nanoparticles: a review. *Arabian Journal of Chemistry*, 12(8), 3576–3600.

Sharma, S., Kaushik, R., Sharma, S., Chouhan, P., and Kumar, N. (2016). Effect of herb extracts on growth of probiotic cultures. *Indian Journal of Dairy Science*, 69(3), 1–8.

Sharma, S., Kaushik, R., Sharma, P., Sharma, R., Thapa A., and Indumathi, KP (2016). Antimicrobial activity of herbs against *Yersinia enterocolitica*. *The Annals of the University Dunarea de Jos of Galati—Food Technology*, 40 (2), 119–134.

Siddiqi, K. S. and Husen, A. (2016). Green synthesis, characterization and uses of palladium/platinum nanoparticles. *Nanoscale Research Letters*, 11(1), 482.

Singh, J., Dutta, T., Kim, K. H., Rawat, M., Samddar, P., and Kumar, P. (2018). 'Green' synthesis of metals and their oxide nanoparticles: applications for environmental remediation. *Journal of Nanobiotechnology*, 16(1), 84.

Singh, R., Thakur, P., Thakur, A., Kumar, H., Chawla, P., Jigneshkumar V. R., Kaushik, R., and Kumar, N. (2020). Colorimetric sensing approaches of surface modified gold and silver nanoparticles for detection of residual pesticides: a Review. *International Journal of Environmental Analytical Chemistry*. doi:10.1080/03067319. 2020.1715382

Sinha, S. N., Paul, D., Halder, N., Sengupta, D., and Patra, S. K. (2015). Green synthesis of silver nanoparticles using fresh water green alga *Pithophora oedogonia* (Mont.) Wittrock and evaluation of their antibacterial activity. *Applied Nanoscience*, 5(6), 703–709.

Sirelkhatim, A., Mahmud, S., Seeni, A., Kaus, N. H. M., Ann, L. C., Bakhori, S. K. M., and Mohamad, D. (2015). Review on zinc oxide nanoparticles: antibacterial activity and toxicity mechanism. *Nano-Micro Letters*, 7(3), 219–242.

Sriramulu, M. and Sumathi, S. (2017). Photocatalytic, antioxidant, antibacterial and anti-inflammatory activity of silver nanoparticles synthesised using forest and edible mushroom. *Advances in Natural Sciences: Nanoscience and Nanotechnology*, 8(4), 045012.

Srivastava, S., Awasthi, R., Gajbhiye, N. S., Agarwal, V., Singh, A., Yadav, A., and Gupta, R. K. (2011). Innovative synthesis of citrate-coated superparamagnetic Fe_3O_4 nanoparticles and its preliminary applications. *Journal of Colloid and Interface Science*, 359(1), 104–111.

Sudha, P. N., Sangeetha, K., Vijayalakshmi, K., and Barhoum, A. (2018). Nanomaterials history, classification, unique properties, production and market. In *Emerging Applications of Nanoparticles and Architecture Nanostructures* (pp. 341–384). Elsevier, USA.

Suga, M., Asahina, S., Sakuda, Y., Kazumori, H., Nishiyama, H., Nokuo, T., … and Cho, M. (2014). Recent progress in scanning electron microscopy for the characterization of fine structural details of nano materials. *Progress in solid State Chemistry*, 42(1–2), 1–21.

Syamchand, S. S. and Sony, G. (2015). Multifunctional hydroxyapatite nanoparticles for drug delivery and multimodal molecular imaging. *Microchimica Acta*, 182(9–10), 1567–1589.

Tamayo, L., Azócar, M., Kogan, M., Riveros, A., and Páez, M. (2016). Copper-polymer nanocomposites: An excellent and cost-effective biocide for use on antibacterial surfaces. *Materials Science and Engineering: C*, 69, 1391–1409.

Tang, C. Y. and Yang, Z. (2017). Transmission electron microscopy (TEM). In *Membrane Characterization* (pp. 145–159). Elsevier, USA.

Vetchinkina, E., Loshchinina, E., Kursky, V., and Nikitina, V. (2013). Reduction of organic and inorganic selenium compounds by the edible medicinal basidiomycete *Lentinula edodes* and the accumulation of elemental selenium nanoparticles in its mycelium. *Journal of Microbiology*, 51(6):829–835

Vidhu, V. K. and Philip, D. (2014). Catalytic degradation of organic dyes using biosynthesized silver nanoparticles. *Micron*, 56, 54–62.

Vladár, A. E. and Hodoroaba, V. D. (2020). Characterization of nanoparticles by scanning electron microscopy. In *Characterization of Nanoparticles* (pp. 7–27). Elsevier.

Wallin, S., Pettersson, A., Östmark, H., and Hobro, A. (2009). Laser-based standoff detection of explosives: a critical review. *Analytical and Bioanalytical Chemistry*, 395(2), 259–274.

Wang, S., Chen, Z. G., Cole, I., and Li, Q. (2015). Structural evolution of graphene quantum dots during thermal decomposition of citric acid and the corresponding photoluminescence. *Carbon*, 82, 304–313.

Wilson, B. A., Winkler, M., and Ho, B. T. (2019). *Bacterial Pathogenesis: A Molecular Approach*. John Wiley & Sons, London, United Kingdom.

World Health Organization. (2014). *Global Status Report on Noncommunicable Diseases 2014 (No. WHO/NMH/NVI/15.1)*. World Health Organization, Geneva.

Yang, Y., Wang, L., Zambaldi, C., Eisenlohr, P., Barabash, R., Liu, W., and Bieler, T. R. (2011). Characterization and modeling of heterogeneous deformation in commercial purity titanium. *JOM*, 63(9), 66.

Zheng, W. (2018). *High-throughput Magnetic Patterning with Laser Direct Writing* (Doctoral dissertation, Northeastern University Boston).

Zhu, H., Chen, X., Zheng, Z., Ke, X., Jaatinen, E., Zhao, J., and Wang, D. (2009). Mechanism of supported gold nanoparticles as photocatalysts under ultraviolet and visible light irradiation. *Chemical communications*, (48), 7524–7526.

Zhu, Y., Inada, H., Nakamura, K., and Wall, J. (2009). Imaging single atoms using secondary electrons with an aberration-corrected electron microscope. *Nature Materials*, 8(10), 808–812.

3 Antimicrobial Efficacy of Neem Extract–Stabilized Metal Nanoparticles

Huma Khan
Shoolini University of Biotechnology and Management Sciences, India
Monika Kataria
Guru Jambheshwar University of Science and Technology, India
Mohammed Azhar Khan
Shoolini University of Biotechnology and Management Sciences, India

CONTENTS

3.1 INTRODUCTION

In modern science, nanotechnology is the emerging field attributing several applications in various aspects such as biology, medicine, drug delivery, health care, and material science, engineering, cosmetics, paints, agriculture, food coatings (Albrecht et al., 2006; Veerannaet al., 2013; Doria et al., 2017; Seil and Webster, 2012). It is the fields of science, engineering, and technology that deal with small structures and small-sized materials whose dimensions are about 1 to 100 nm (Mandal et al., 2010; Dixon et al., 2011). The unit nanometre is a descendent of its prefix *nano–*, from a Greek word meaning "extremely small"; therefore, the term *nanotechnology* can be well defined as the manipulation of materials by certain chemical, physical or biological processes at the nanoscale with specific properties (Zargar et al., 2011; Baker and Satish, 2012). Several kinds of nanoparticles (NPs) exist in nature, and they have remarkable applications in their respective fields. In chemistry, this series of sizes has been linked with micelles, aggregates of several molecules, polymer molecules, and colloids (Dhull et al., 2019). In electrical engineering and physics, NPs have been linked to the behaviour of electrons of nanoscale structures and quantum behaviour. Biology and biochemistry have been intensely associated with nanoscience as components of the cells, such as DNA (deoxyribonucleic acid), RNA (ribonucleic acid), and subcellular organelles, are considered as nanostructures which are most remarkable structures of biology (Whitman, 2007; Lowe, 2000; Tsapis et al., 2002). Nanoscience deals with the probability to study the magnitudes of spatial confinement on electron behaviour and offers an opportunity to explore the complications related with the surface or interfacial nature of the molecule (Zhang et al., 1994). In the synthesis of nanoparticles, medicinal plants play a crucial role in many applications (Chawla et al., 2019). Therefore, the use of various medicinal plants has been explored. *Azadirachta indica* is a plant that is abundantly found in Asian countries and grows very fast, reaching a height of 15 to 20 m normally and up to 35 to 40 m under highly favourable conditions. It is an evergreen tree, but during extreme conditions, such as prolonged drought, it may shed most of its leaves. The trunk is moderately straight and short and may reach a width of 1.5 to 3.5 m. The bark is hard, scaly or fissured, and reddish-brown to whitish-grey. The developed root system consists of a taproot as well as lateral roots (Schmutterer and Singh, 1995). The composition of fresh neem leaves with respect to various chemical constituents is well known (Van Der Veen et al., 1997; Zakir et al., 2008). Schmutterer and Singh (1995) has given the average chemical constituents of the *A. indica* leaves. The properties of the various segments of the *A. indica* tree are due to a multitude of ingredients (Biswas et al., 2002a). The main chemical ingredients consist of three or four related components. These belong to a general class of natural products called "triterpenes", or, more specifically, "limonoids"; the main ones present in Neem are azadirachtin, meliantriol, salannin, nimbidin, and nimbin (Sadeghian and Mortazaienezhad, 2007; Mak-Mensah and Firempong, 2011). This chapter is aimed at the potential role of various parts of the *A. indica* plant, viz. leaves, root, and bark, to biogenically synthesize different kinds of NPs with distinctive properties and antimicrobial activities.

3.1.1 Different Types of Nanoparticles

In general, NPs are broadly categorized into the categories as follow:

3.1.1.1 Metal Nanoparticles

The metal NPs are used to define nanosized metals with dimensions such as thickness, width, or length within the size ranging from 1 to 100 nm. The metallic NPs have a large surface area–to–volume ratio as compared to other bulk materials (Thakkar et al., 2010).

They have large surface energy and several low coordination sites such as corners and edges. There are several types of metal NP are synthesized by different chemical, physical, and biological synthesis methods such as copper (Cu), silver (Ag), gold (Au), nickel (Ni), zinc (Zn), platinum (Pt), palladium (Pd), magnesium (Mg), iron (Fe), and silicon (Si).

3.1.1.2 Metal Oxide Nanoparticles

Metal oxide NPs play a significant role in many areas such as material sciences, physics, chemistry, and biology. Metal oxide NPs have a tremendously small size and large surface area–to–volume ratio, as compared to the bulk counterparts, and have-several applications due to these unique properties (Singh et al., 2020). Metal oxide NPs possess different properties such as magnetic, catalytic, anti-inflammatory, and antimicrobial properties. These have been also used to improve the stability of polymeric products as pigments and to improve the appearance (Niederberger and Pinna, 2009). The most important metal oxides are zinc oxide (ZnO), titanium oxide (TiO_2), copper oxide (CuO), magnesium oxide (MgO), aluminium oxide (Al_2O_3), iron oxide (Fe_3O_4 and Fe_2O_3) and manganese oxide (MnO_2). Among these properties, the microbicidal property holds a significant place having a lot of potentials to be utilized in human and animal medicine (Stoimenov et al., 2002). Metal oxide NPs display distinctive chemical and physical properties as a consequence of their reduced size and increased surface area. To stay structurally and mechanically stable, NPs must possess low surface free energy, and materials in the nanostructure form may exhibit greater stability as compared to their bulk equivalents (Azam et al., 2012).

3.1.1.3 Carbon-Based Nanostructure

The carbon-based nanostructures are important types of nanoparticles and have different types such as graphene, nanotubes, nanofibers, and diamond (Durgun et al., 2008). They possess a larger surface area for the deposition of conducting polymer and/or metal oxide NPs which aids the efficient ion diffusion phenomenon and contributes towards higher specific capacitance of carbon-based composite material with excellent cyclic stability. They have smaller dimensions, higher chemical stability, and excellent thermal conductivity with low resistivity (Chen and Huang, 2006).

3.1.1.4 Polymeric Nanoparticles

The polymeric NPs are defined as particulate dispersion and solid particles with a size in the range of 10 to 1000 nm. The polymeric NPs have been extensively studied as particulate carriers in various fields such as pharmaceutical and medical fields (Kumari et al., 2010). They are also used for the drug delivery system. Several polymeric NPs are synthesized from biodegradable and biocompatible polymers. Several types of drugs are dissolved in polymeric NPs. They increase the stability of any volatile pharmaceuticals agents. They are also used in cancer therapy and in the delivery of vaccines (Nagavarma et al., 2012).

3.2 GREEN SYNTHESIS OF NANOPARTICLES

The fabrication of NPs by chemical and physical methods possesses certain disadvantages, such as it is cost-effective and leads to the contamination of the ecosystem as it dispenses toxic by-products into the environment and thus contaminates the surroundings. To circumvent these limitations, a novel approach, that is green synthesis chemistry, is being explored (Dahl et al., 2007; Herlekar et al., 2014; Virkutyte and Varma, 2011; Siddiqi and Husen, 2016). This approach has directed the interests of researchers around the world to develop the non-toxic process for nanofabrication. Therefore, numerous methodologies have been devised for the biogenic production of NPs from their corresponding metal salts (de Lima et al., 2012; Wood et al., 2007; Thakkar et al., 2010). The biogenic synthesis of NPs is used not only because of the reduced hazardous effect on the environment but also because it produces large numbers of NPs which are non-toxic and have well-defined dimensions and morphologies (Dahl et al., 2007; Anastas and Zimmerman 2007). Hence, the biogenic method for fabricating NPs by leaves, fruits, plant tissue, plant extracts, whole plant, algae, fungi, and bacteria has emerged. The biogenic synthesis of NPs using the whole plant, plant extracts, algae, fungi, and bacteria are environment-friendly and cost-effective method as well as synthesized NPs possess well-defined dimensions and morphologies (Luangpipat et al., 2011; Singaravelu et al., 2007; Hernandez et al., 2008).

Bio-fabrication of various metal/metallic NPs by employing the extracts of plant components is cheaper than other methods based on the microbial processes and the whole plant (Kaler, 2011; Bankar et al., 2010; Beattie and Haverkamp, 2011).

The conventional NP synthesis processes use hazardous chemicals and various solvents in the fabrication process for stabilizing the NPs by adding the synthetic capping or preservatives agents. Therefore, the applications of such conventional methods are not considered in clinical and biomedical fields (De et al., 2010).

In the green synthesis method, the phytochemicals, such as alkaloid, flavonoid, saponins, and terpenoid, present in plants act as natural synthesizing and/or capping and stabilizing agents (Chawla et al., 2019, 2020; Jain et al., 2010).

3.3 TECHNIQUES FOR THE CHARACTERIZATION OF NANOPARTICLES

NPs are generally characterized by their size, morphology, stability, and dispersity using various advanced techniques, such as ultraviolet (UV)-visible spectrophotometry, dynamic light scattering (DLS), zeta potential, Fourier-transformed infrared (FTIR) spectrophotometry, scanning electron microscopy (SEM), and transmission electron microscopy (TEM) (Agarwal et al., 2020). UV-visible spectroscopy is one of the most commonly used techniques for characterizing the metallic NP (Chawla et al., 2020). The visible region of the electromagnetic spectrum (300–800 nm) is used for characterizing NPs (Feldheim and Foss, 2002; Chawla et al., 2018). The UV-visual spectrum of silver NPs (AgNPs) depicts the absorption maxima in the range of 400 to 450 nm, while for gold NPs (AuNPs), it falls in the range of 500 to 550 nm (Huang and Yang, 2004). Similarly, the absorption spectrum of copper NPs shows maximum absorption in the range of 550 to 600 nm (Ramyadevi et al., 2011).

DLS also constitutes an important tool for roughly estimating the size and dispersity of the NPs in solution. A DLS system usually employs a laser technique. The intensity of the laser scattered after passing through the dilute sample is measured and used for calculating the average size and polydispersity index of the NPs solution using the Stokes–Einstein equation (Jose et al., 2019). Zeta potential is a measure of the stability of the NPs, and the value is calculated by measuring electrophoretic mobility (Patel and Agrawal, 2011).

Electron microscopy constitutes a very powerful tool for studying the accurate size and morphology of the NPs. SEM deciphers the surface morphology and shape of the NPs while TEM further reveals the quantitative aspects size, size distribution, and morphology (Lin et al., 2014). FTIR spectroscopy, on the other hand, provides information about the biomolecules or functional groups attached to the surface of the NP during synthesis. It also gives information about whether the biomolecules are involved in biological reduction during the synthesis process (Zhang et al., 2016; Rana et al., 2018).

3.4 APPLICATIONS OF NANOPARTICLES

NPs are fairly a fascinating material having already found many applications because of their large surface area–to–volume ratio and extraordinary magnetic properties. NPs exhibit sorbent properties, which were successfully tested on the removal of organic dyes and toxic inorganic metal pollutants from the wastewater of industries. Magnetite NPs have been used in magnetic recording media, such as tapes and hard disk drives (HDD). High susceptibility and low coactivity of superparamagnetic iron oxide nanoparticles (SPIONs) and their biocompatibility and biodegradability show great potential for biomedical applications.

Nanoparticles have been used in various areas of biomedical sciences, biotechnology, and environmental sciences. The biomedical applications are primarily based on utilizing certain properties, viz. hyperthermia, drug delivery, magnetic targeting, magnetic separation, magnetic resonance imaging (MRI) contrast, and antimicrobials (Gupta and Gupta, 2005).

3.4.1 Magnetic Separation

It is a simple, rapid, and efficient tool for the quickly isolating the target molecules, cells, and cell organelles from crude samples such as blood, bone marrow, tissue homogenates, urine, stools, and other biological materials and complex biological mixtures and (Olsvik et al., 1994).

3.4.2 Hyperthermia Treatment

It is a gifted therapy for cancer in which the temperature of the tumour tissue is raised slightly above physiological temperature, that is about 42 to 46, artificially (Britt and Kalow, 1970). This is based on the fact that when magnetic materials are exposed to an alternating-current (AC) magnetic field, it generates heat. The heating happens due to magnetic losses in the form of hysteresis loss. Magnetic materials can be used as bulk as well as in the form of small particles. But bulk magnetic materials have been replaced by small magnetic particles, preferably nonmagnetic materials because of their favourable magnetic properties to hyperthermia treatment and ease of administration in the target tissues. Iron oxide particles are regularly used in hospitals as a contrast agent for MRI. Tracking and monitoring of stem cells after transplantation in vivo can provide important information for determining the efficacy of stem cell therapy (Babes et al., 1999).

3.4.3 Magnetic Nanoparticles

Magnetic NPs attached with drugs have been found as a prospective field of targeted drug delivery, whereby the distribution of drug attached with magnetic NPs within the body is manipulated by the external magnetic field (Jain et al., 2005). However, if magnetic particles attached to drugs can be concentrated into the desired area by an external magnetic field and thereafter exposed to the AC magnetic field, a more effective therapeutic outcome is expected. Liposome has been extensively used for the delivery of drugs. Liposomes have also been used for the efficient delivery of magnetic particles known as magneto liposomes (Hodenius et al., 2002). Targeting therapeutic agents is necessary for maximum utilization, thereby reducing the dose and avoid unwanted adverse effects due to higher doses.

3.4.4 Antibacterial Agents

Antimicrobial agents are very significant in water disinfection, the textile industry, food packaging, and medicine. Organic compounds used for disinfection purposes possess certain limitations being toxic to the human body as one of them (Pathania et al., 2018). Therefore, metallic NPs have gathered plenty of interest to be utilized as an inorganic, less toxic disinfectant. The inorganic nanostructured materials with effective surface modification show excellent microbicidal activity by locally destroying bacteria without causing any damage to the surrounding tissues. Iron oxide NPs (IONPs) have been widely preferred as antibacterial agents because of biodegradable nature, low cytotoxicity, and reactive surfaces that can be modified with biocompatible coatings. The mechanism of antimicrobial property of NP lies in the fact that they possess large surface areas relative to the bulk, which efficiently cover the microorganisms and reduce the oxygen supply for their respiration (Casillas-Martinez et al., 2005).

3.4.5 Heavy Metals and Dye Removal From Water

Water resources become critically important to living things, but there is a major environmental concern due to increasing pollution from industrial wastewater (Wang et al., 2013). Numerous industries, for example pulp, paper, textiles, and plastics, consume dyes and chemicals to process their products and require a large amount of water. Consequently, organic compounds, heavy metals, and other hazardous materials contaminate water. The contaminants make deleterious impacts on terrestrial and aquatic ecosystems. The removal of contaminants in wastewater by adsorption, Magnetic IONPs coated with suitable surfactants show promising help (Bailey et al., 1999). Currently, surface-modified IONPs afford an alternative in bioremediation processes for the removal of Cu (II), Cr (VI), Ni from water by adsorption (Barakat, 2011). More than 900 different chemicals have been used in the textile industry;

mostly, they are dyes and transfer agents. Natural and artificial dyes are used in the textile industry for economical and efficiency reasons. It creates problems for the environment because many artificial dyes are highly stable molecules, recalcitrant to be degraded by light, biological, and chemical treatment. Some commercial synthetic dyes usually have unreported large complex structures. Therefore, the release of dye wastewater can be one of the major problems because it contains numerous noxious waste such as acid or caustic, dissolved solid toxic compounds like heavy metals and colour.

3.5 *AZADIRACHTA INDICA* MEDIATED SYNTHESIS OF METALLIC NANOPARTICLES

Plants are the source of many phytoconstituent and fewer side effects. Neem (*A. indica*) is a quite commonly available plant and abundant in nature. *A. indica*, generally known as neem, is an evergreen tree belongs to family *Meliaceae.* It is extensively distributed in Asia, Africa, and other semi-tropical and tropical areas of the world (Ghimeray et al., 2009b). Its leaf extract has been found to have divergent applications in several medical fields, such as drugs and medicine. It acts as anti-inflammatory, antimalarial, antifungal, anti-diabetic, antibacterial, and antiviral and is particularly recommended for skin diseases (Sharma et al., 2010; Mgbemena et al., 2010; Maragathavalli et al., 2012). In India, the leaves of *A. indica* are desiccated and used as an insect repellent in the tropical regions to keep away the mosquitoes. It has been identified for its insecticide activity against more than 400 insect pests.

A. indica leaves have also been used to treat several skin-related diseases such as psoriasis, eczema, and others (Ghimeray et al., 2009a; Sharma et al., 2010). Neem is composed more than 250 natural components, such as salanin, azadirachtin, valassin, meliacin, gedunin, nimbin, and several other by-products (Girish and Bhat, 2008). Quercetin and β-sitosterol were the first-ever-purified polyphenolic flavonoids from fresh neem leaves and possessed excellent antifungal and antibacterial properties (Alzohairy, 2016). Table 3.1 summarizes the bioactive compounds present in various parts of the neem plant.

The aqueous extract of neem is capable of synthesizing a variety of NPs, for example zinc oxide, gold, silver, copper, iron/flavanones, and terpenoids present in neem, which play an imperative role in synthesizing, as well as stabilizing, NPs by capping (Banerjee et al., 2014).

There are numerous reports on the synthesis of silver (Chand et al., 2019; Mohanaparameswari et al., 2019; Ramar and Ahamed, 2018), copper/copper oxide (Abhiman et al., 2018; Ansilin et al., 2016), zinc oxide (Sharma and Oudhia, 2016; Bhuyan et al., 2015), iron/iron oxide (Pattanayak and Nayak, 2013; Taib et al., 2018; Zambri et al., 2019), gold (Thirumurugan et al., 2010; Bindhani and Panigrahi, 2014; Shankar et al., 2004), nickel oxide (Helan et al., 2016), platinum (Thirumurugan et al., 2016), and titanium dioxide (Thakur et al., 2019) using neem leaf extract. Shankar et al. (2004), reported the biological synthesis of Au-Ag bimetallic NPs using neem leaf broth. In another study, Amrutham et al. (2020) devised a cheap, high-yielding, single-step, and novel microwave irradiation method for synthesizing palladium NPs using neemgum.

TABLE 3.1
Bioactive Compound Found in Various Parts of *Azadirachta indica*

S. No	Plant Parts	Bioactive Compounds	Biological Properties	Chemical Structure	References
1	Seeds	Azadirachtin	Insecticide, Antibacterial	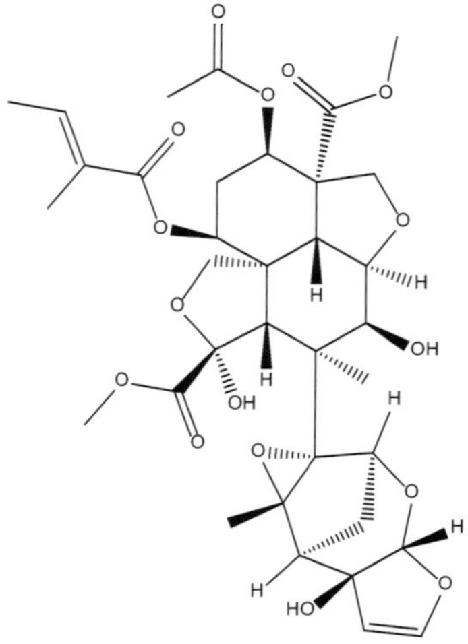	Schaaf et al., 2000

S. No	Plant Parts	Bioactive Compounds	Biological Properties	Chemical Structure	References
2		Nimbidin	Anti-inflammatory, hypoglycaemic, antibacterial, antifungal	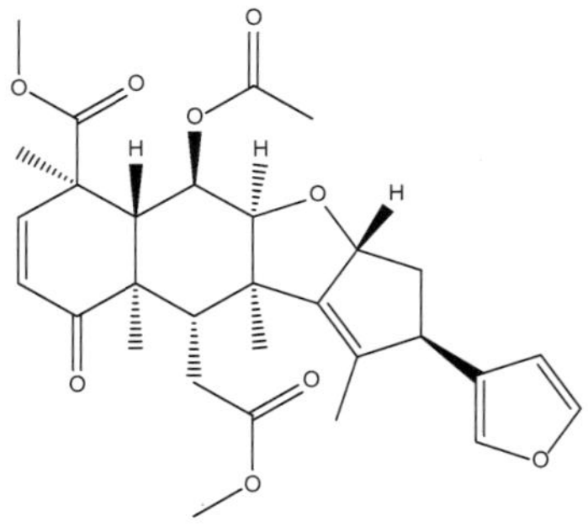	Sarsaiya et al., 2019
3		Nimbolide	Anticancer, antibacterial, and antifungal	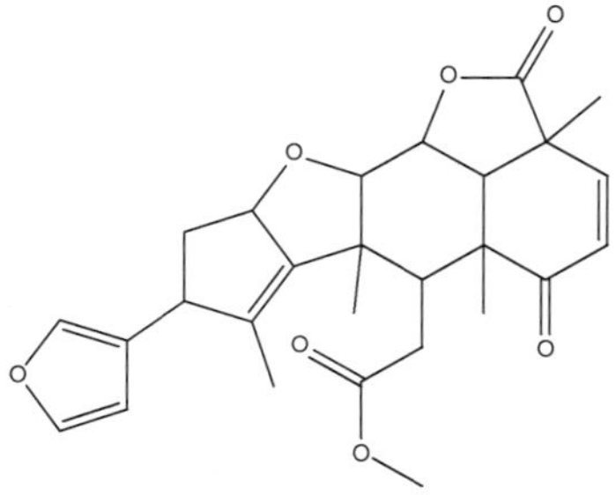	Sarah et al., 2019

(*Continued*)

TABLE 3.1 (Continued)
Bioactive Compound found in Various Parts of *Azadirachta indica*

S. No	Plant Parts	Bioactive Compounds	Biological Properties	Chemical Structure	References
4		Gedunin	Antifungal and antimalarial	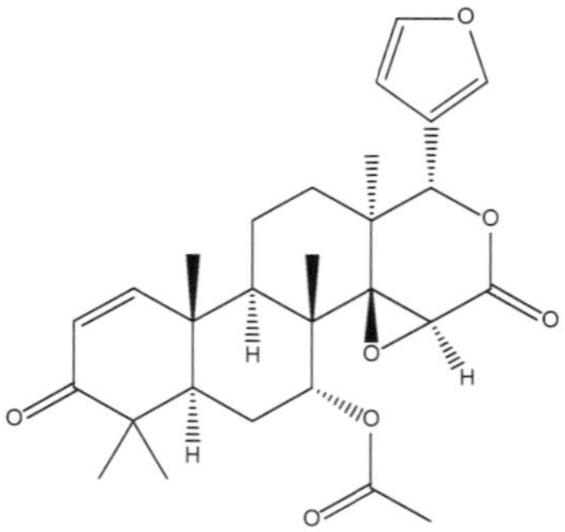	Biswas et al., 2002a
5		Mahmoodin	Antibacterial	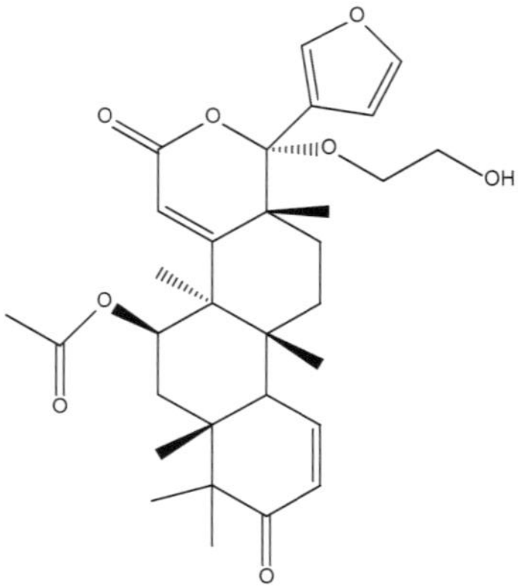	Jones et al., 1994

S. No	Plant Parts	Bioactive Compounds	Biological Properties	Chemical Structure	References
6		Azadirachtin	Antibacterial and anticancer	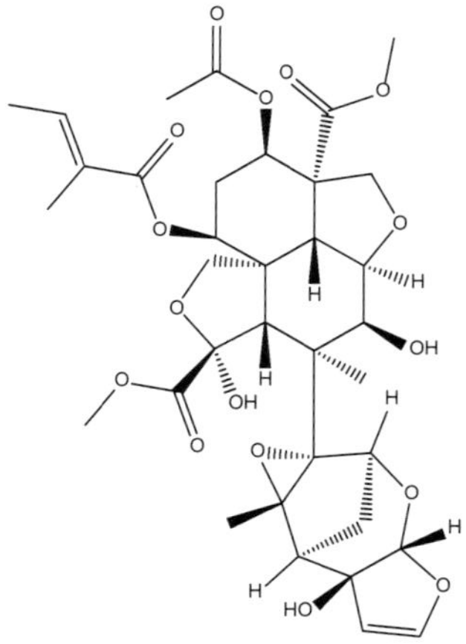	Khalid et al., 1989
7	Bark	Gallic acid	Anti-inflammatory and immunomodulatory	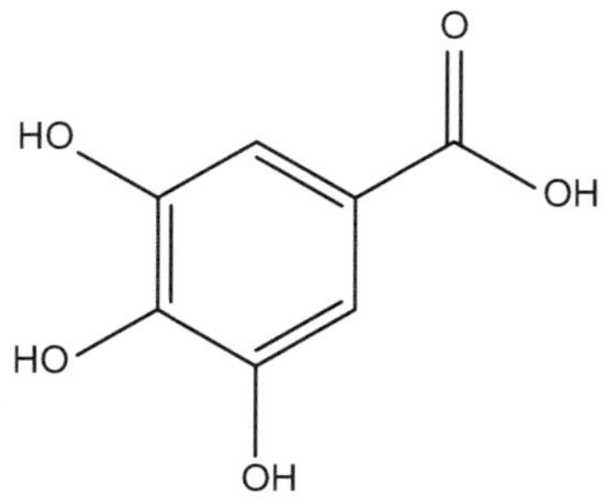	Biswas et al., 2002a

(Continued)

TABLE 3.1 (Continued)
Bioactive Compound found in Various Parts of *Azadirachta indica*

S. No	Plant Parts	Bioactive Compounds	Biological Properties	Chemical Structure	References
8		Margolone	Antibacterial	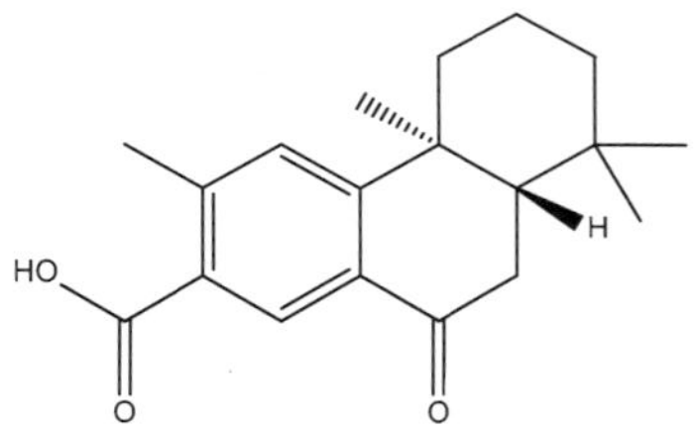	Biswas et al., 2002b
9		Quercetin	Antioxidant activity	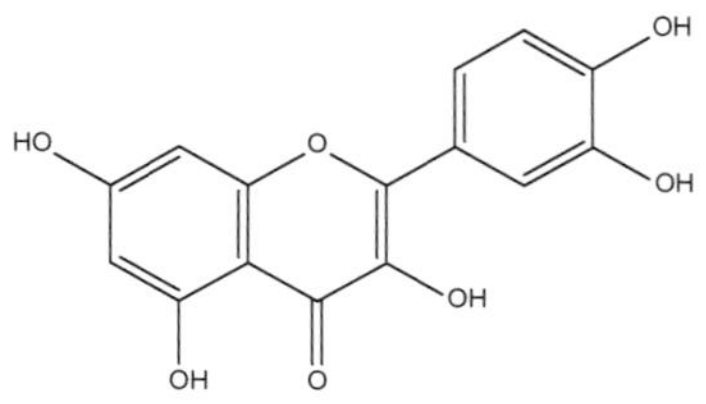	
10	Root	Quercetin	Antioxidant	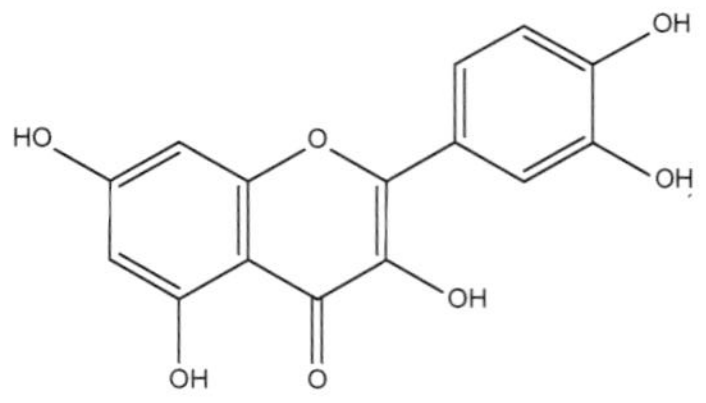	Rao et al., 2018

S. No	Plant Parts	Bioactive Compounds	Biological Properties	Chemical Structure	References
11		Rutin	Antioxidant	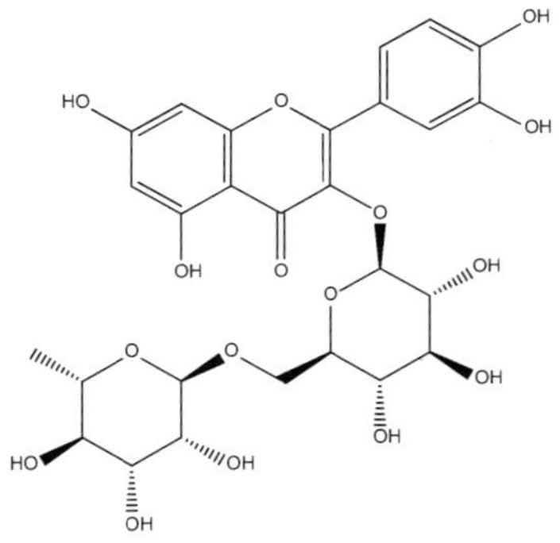	Rao et al., 2018
12		Melicitrin	Antioxidant	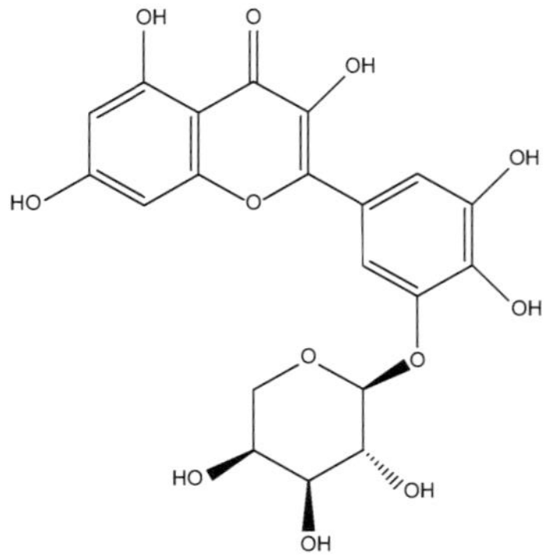	Rao et al., 2018

(Continued)

TABLE 3.1 (Continued)
Bioactive Compound found in Various Parts of *Azadirachta indica*

S. No	Plant Parts	Bioactive Compounds	Biological Properties	Chemical Structure	References
13	Leaf	Salannol	Pesticides and cytotoxic	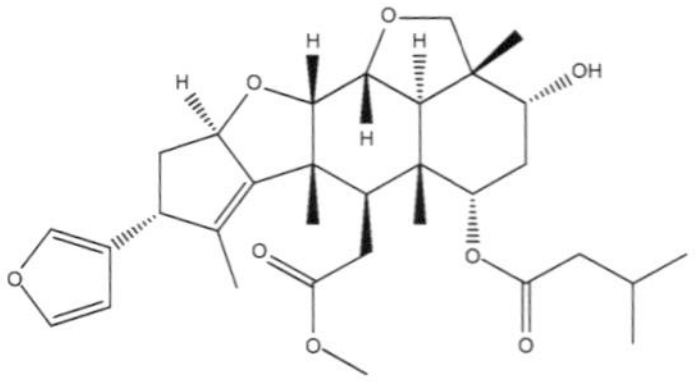	Koul et al., 2004
14		Nimbinene	Skin disease	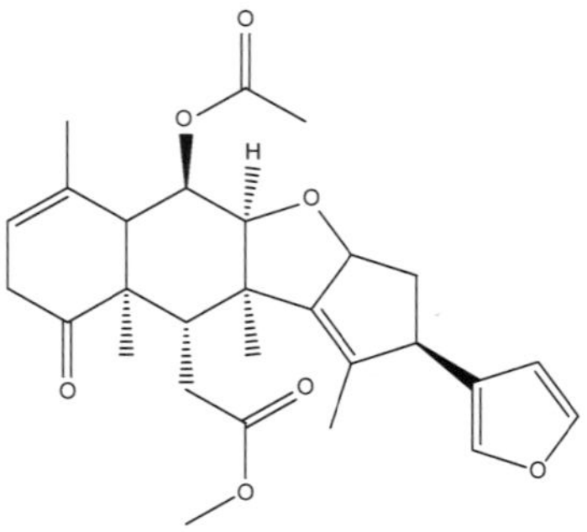	Rao et al., 2018
15		β-Sitosterol	Antifungal	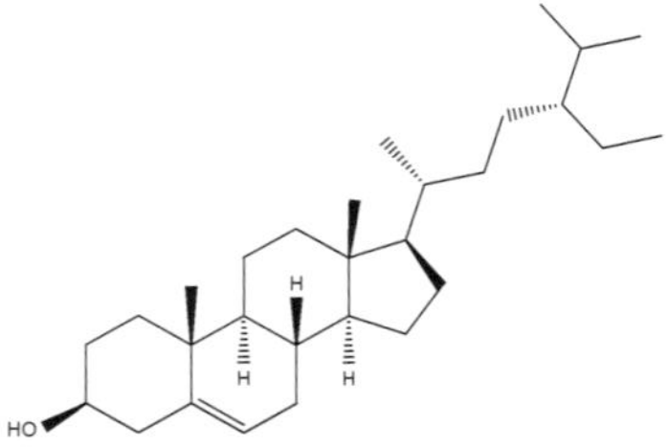	

S. No	Plant Parts	Bioactive Compounds	Biological Properties	Chemical Structure	References
16		Azadirachtol	Anti-inflammatory and antibacterial	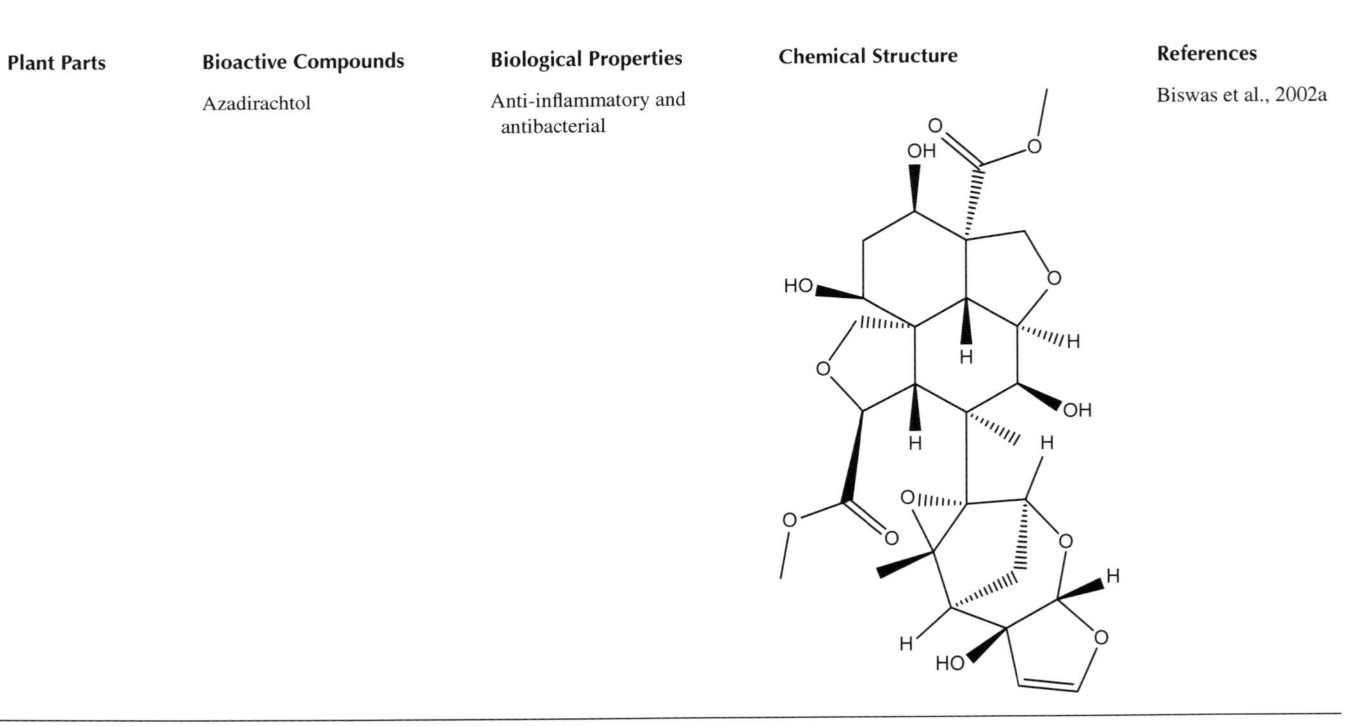	Biswas et al., 2002a

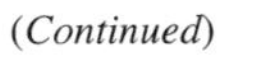

(*Continued*)

TABLE 3.1 (Continued)
Bioactive Compound found in Various Parts of *Azadirachta indica*

S. No	Plant Parts	Bioactive Compounds	Biological Properties	Chemical Structure	References
17		Nimbandiol	Antihyperglycemic, antifungal, and antimutagenic		Sarah et al., 2019
18		β-nimolactone	Antigastric and antibacterial		Sarah et al., 2019
19		α-nimolactone	Anticancerous	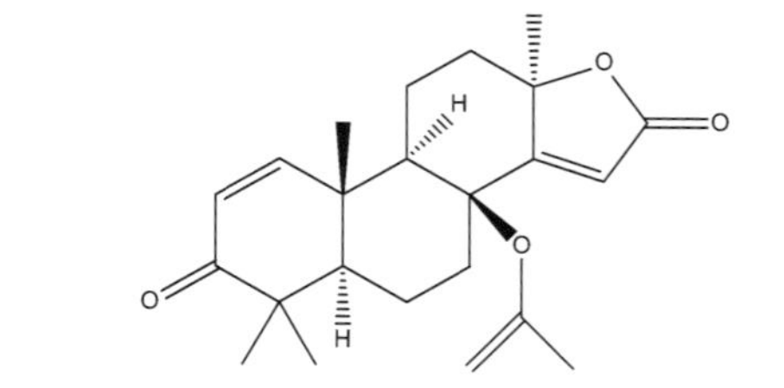	Sarah et al., 2019

S. No	Plant Parts	Bioactive Compounds	Biological Properties	Chemical Structure	References
20		Epoxyazadiradione	Anti-inflammatory	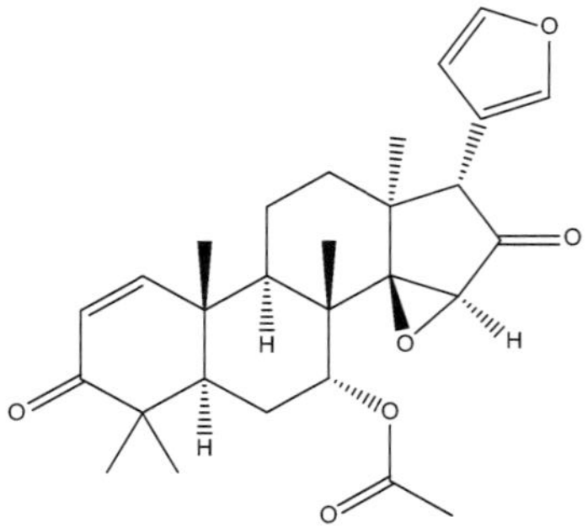	Alam et al., 2012
21		Salanin	Anti-insecticidal, anti-helminthic, and antibacterial	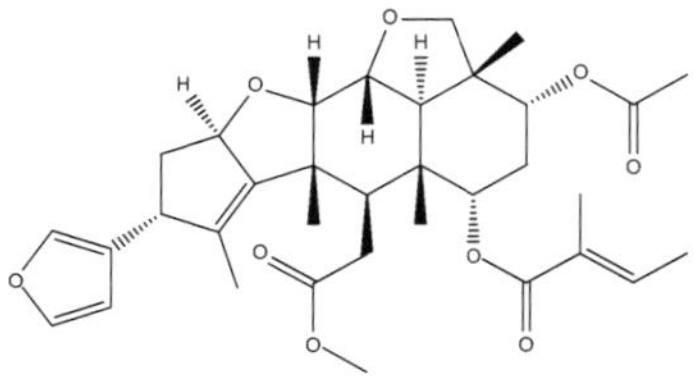	Sarsaiya et al., 2019

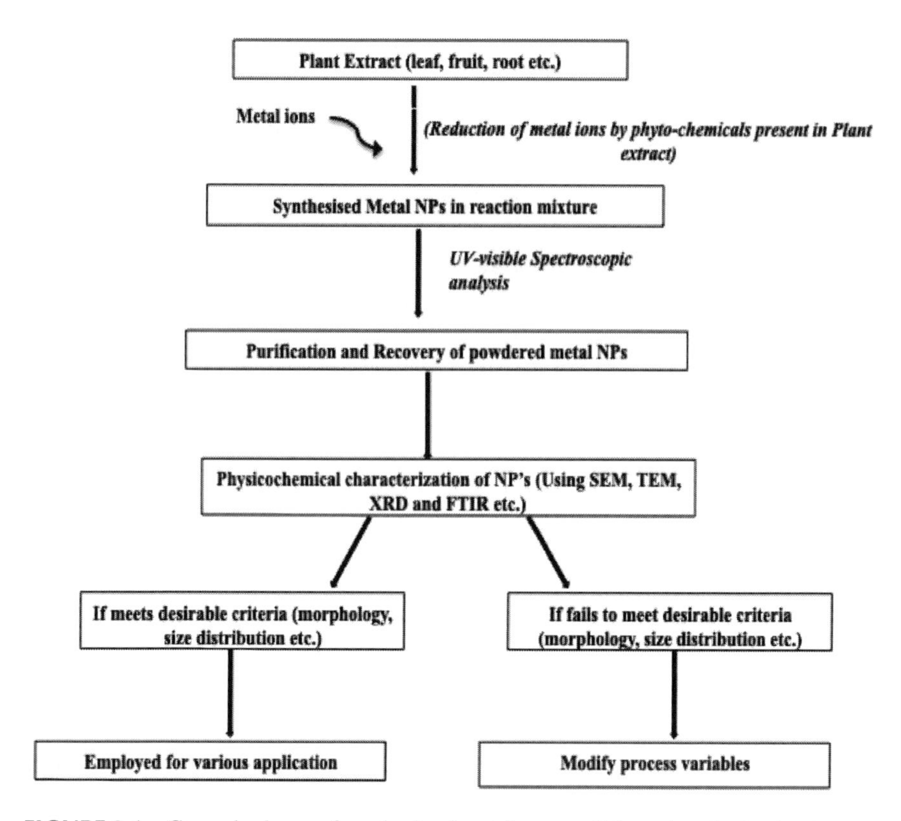

FIGURE 3.1 General scheme of synthesis of metal nanoparticles using plant extracts.

The general method of synthesizing metal NPs involves the preparation of the aqueous plant extract (whole-plant extract or plant-part extract; e.g. leaf, fruit, stem, root) followed by the addition of metal salt to the same. Figure 3.1 represents a generalized scheme for the synthesis of metal NPs using plant extracts.

Whole-leaf extracts are loaded with polyphenols such as flavonoids, which are powerful reducing agents for reducing inorganic salts (Park et al., 2011); therefore, leaf extracts are preferred for the synthesis. The majority of the research on metal/metal oxide NPs biosynthesis using *A. indica* reports the use of leaf extract as a reducing agent.

The most widely utilized method of green synthesis using plants generally involves the treatment of metal salts with aqueous extract and investigating various parameters such as pH, time, temperature, the concentration of respective metal salt, and plant extract (Dwivedi and Gopal, 2010).

The bio-reduction of metal salts is considered to be brought about by the flavonoids, terpenoids, or other biomolecules present in the plant extract. The size and size distribution of the metal NPs are directly related to the reducing capacity of the biocompounds present in the plant extract. A strong reducing biomolecule tends to rapidly reduce the metal ions, forming smaller NPs (Roy et al., 2019). Bioactive compounds (flavonoids, terpenoids) serve the dual purpose by bringing about the bio-reduction of metal salts and the stabilization of as-synthesized NPs by capping or surface modification. The report of the synthesis of silver NPs by Asimuddin et al. (2018) using neem leaf extract also supports the dual action of the extract. The FTIR spectra of the neem leaf extract and synthesized AgNPs depict the presence of hydroxyl, aldehyde, phenolic, and carboxylic groups, indicating the presence of strong reducing phyto-molecules, such as terpenoids, flavonoids, and polyphenols. It was further concluded that the hydroxyl groups might have possibly played a part in reducing silver ions.

In another study, Singh et al. (2020), investigated the synthesis of ZnO particles using the extract of neem leaves. The FTIR spectra of the plant extract and bio-synthesized ZnO NPs, suggested the involvement of functional groups such as amines, alcohols ketones, flavones, polyols, and terpenoids in synthesis and stabilization. However, the exact mechanism involved in flavonoid- or polyol-mediated bio-reduction still needs to be deciphered.

3.6 ANTIMICROBIAL POTENTIAL OF BIOSYNTHESIZED NANOPARTICLES

As discussed earlier, neem (*A. indica*) has been employed for the bio-fabrication of several metals, metaloxide, and hybrid metallic NPs. The metallic NPs have attracted the attention of researchers worldwide due to their amazing antimicrobial application. The antimicrobial potential of various metal (Ag, Cu, Au, Pt) and metal oxide NPs (CuO, ZnO, TiO_2, Fe_2O_3) has been reported in numerous studies. Therefore, the same has also been implemented in various biomedical applications such as wound dressings, bone cement, and dental materials (Wang et al., 2017). The NPs synthesized using neem extract in various studies have shown excellent results against various pathogenic microorganisms, including the multidrug-resistant bacteria as well. Algebaly et al. (2020) reported the biogenic synthesis of silver NPs using neem leaf extract and the antimicrobial activity of the NPs were tested against *Pseudomonas aeruginosa*, *Escherichia coli*, and *Staphylococcus aureus*. The synthesized NPs were potentially able to inhibit the growth of the bacteria completely and the SEM analysis of the treated bacteria showed significant membrane damage to the bacteria. In another study to explore the microbicidal potential of AgNP, Saranya et al. (2016) phytogenically synthesized AgNPs using *A. indica* leaf extract. The 40- to 50-nm-sized NPs were then used for testing antimicrobial activity against pathogenic bacteria, viz. *Salmonella enterica*, *S. aureus* and *Streptococcus agalactiae*, and fungi, viz. *Malassezia pachydermatis* and *M. globosa*. The synthesized NPs were able to completely inhibit the growth of tested bacteria as well as fungi. Roy et al. (2017) described in their study that silver NPs synthesized using neem leaf extract were more effective against *E. coli* as compared with gram-positive bacteria. Silver has been known to inhibit microbial growth since ancient times, and in the form of NPs, the antimicrobial potential of the same is enhanced manifolds. Not only silver but also copper, copper oxide, iron, iron oxide, zinc oxide, and titanium oxide NPs have also been synthesized using neem extract and evaluated for their antimicrobial potential.

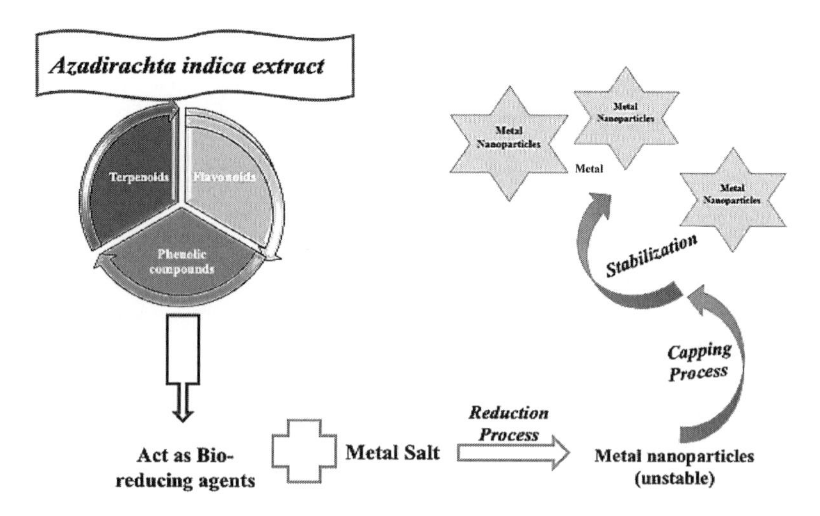

FIGURE 3.2 Schematic diagram showing green synthesis and the stabilization of metal nanoparticles using neem extract.

The copper NPs (CuNPs) synthesized by Abhiman et al.(2018) using *A. indica* leaf extract were found to inhibit the growth of pathogenic bacteria, viz. with *Bacillus cereus*, *S. aureus*, *E. coli*, and *Klebsiella pneumonia* used as test organisms. Similarly, copper oxide NPs (CuONPs) synthesized using neem leaf extract completely inhibited the growth of *E. coli* (Sharma and Oudhia, 2016). Bhuyan et al. (2015), explored the synthesis of ZnONPs and their potential to inhibit the growth of pathogenic bacteria. It was concluded in their study that biosynthesized ZnO significantly inhibited the growth of *E. coli, Streptococcus pyogenes*, and *S. aureus*, and the efficacy increased with increasing the NP concentration used. In addition to this, the gram-positive bacteria appeared to be more susceptible to ZnONPs than were gram-negative bacteria. Poopathi et al. (2015) revealed that the high viability of the toxic action of silver NPs synthesized from *A. indica* against mosquito vectors (*Aedes aegypti* and *Culex. quinquefasciatus*). Table 3.2 lists the NPs synthesized using neem extract and their antimicrobial activity against various microorganisms.

3.6.1 Mechanism of Antimicrobial Activity

The growing use of metallic NPs in biomedical applications has led the scientists to explore the science behind their bactericidal action. The ability of NPs to alter the metabolism of the bacteria in several ways (Slavin et al., 2017) implies that they can be used in eliminating disease-causing bacteria. The potential of NPs to inhibit *S. aureus* biofilm by altering gene expression was also shown in a study published by Zhao and Ashraf in 2016. This implies how useful metal NPs can prove to be in biomedical applications. Recently, NPs have also emerged as a potential candidate to combat multidrug-resistant microorganisms caused due to improper/excessive use of antibiotics (Singh et al., 2014).

TABLE 3.2
Antimicrobial Activity of *Azadirachta indica* Mediated Biosynthesized Nanoparticles (NPs)

S.No	Metallic NP synthesized	Plant Part Used	Organism Inhibited by NPs	References
1	Ag	Leaf extract	*Pseudomonas aeruginosa*, *Escherichia coli*, *Staphylococcus aureus*	Algebaly et al. (2020)
2	Ag	Leaf extract	*S. aureus* and *E. coli*	Mohanaparameswari et al. (2019)
3	Ag	Leaf extract	*S. aureus*	Chand et al. (2019)
4	Ag	Fruit juice	*S. aureus* and *E. coli*	Ramar and Ahamed (2018)
5	Ag	Leaf extract	*Salmonella typhi* and *E. coli*	Sheikh et al. (2017)
6	Ag	Leaf extract	*Salmonella enterica*, *S. aureus*, *Streptococcus agalactiae*, *Malassezia pachydermatis*, and *M. globosa*	Saranya et al. (2016)
7	Ag	Bark	*E. coli* and *Bacillus subtilis*	Nayak et al. (2016)
8	Ag	Neem gum	*Staphylococcus enteritidis* and *Bacillus cereus*	Velusamy et al. (2015)
9	Cu	Leaf extract	*B. cereus*, *S. aureus*, *E. coli*, and *Klebsiella pneumonia*	Abhiman et al. (2018)
10	CuO	Leaf extract	*E. coli*	Sharma et al. (2018)
11	ZnO	Leaf extract	*E. coli*	Sharma and Oudhia (2016)
12	ZnO	Leaf extract	*E. coli*, *Streptococcus pyogenes*, and *S. aureus*	Bhuyan et al. (2015)
13	Fe	Leaf extract	*E. coli*, *Pseudomonas aeruginosa*, and *S. aureus*	Devatha et al. (2018)
14	NiO	Leaf extract	*S. aureus* and *E. coli*	Helan et al. (2016)
15	TiO_2	Leaf extract	*E. coli*, *B. subtilis*, *S. typhi*, and *K. pneumonia*	Thakur et al. (2019)

For effective bactericidal action, the contact of NPs with the bacterial cell is a prerequisite, which may be achieved by any of the forms of interaction, including van der Waals interactions, receptor-ligand binding, and hydrophobic interactions (Wang et al., 2017). Once the NPs succeed in crossing the cell membrane, they start altering the bacterial basic metabolism by interacting with the DNA, enzymes, ribosomes, and lysosomes, leading to several outcomes, including oxidative stress and distorted cell membrane permeability, to name a few. The exact mechanism by which the NPs act against bacteria remains to be unravelled. The following are the most hypothesized and accepted mechanisms for the antimicrobial action of the metallic NPs. Bacterial cell walls and membranes act as shields against the external environment. The bacterial cell wall particularly keeps the bacteria intact. NPs interact with bacterial cell walls according to the nature of its components. It has been seen that the NPs exert a greater inhibitory action on gram-negative bacteria as compared to the gram-positive ones. This has been linked to the difference in their cell walls. Gram-positive bacteria possess thicker peptidoglycan (PG) layer studded with teichoic acid while gram-negative bacteria possess thinner peptidoglycan layer surrounded by an additional layer of lipopolysaccharides (LPS). This arrangement favours the facilitation of NPs across the PG layer. Also, due to the presence of the LPS layer, the negative charge is higher on the gram-negative cell wall, which can easily attract positively charged ions released by NPs and lead to disruption of the cell wall. In an investigation to explore the bactericidal action of AgNPs on *E. coli*, Li and his colleagues (2010) performed permeability studies that implied an efflux of essential sugars and proteins and the distortion of respiratory enzymes. The TEMs and SEMs depicted the destruction of bacterial cell walls, causing the death of the bacteria. Oxidative stress induced by the reactive oxygen species (ROS) generated by nanoparticles is one of the most important mechanisms rendering the bacteria susceptible to NPs. It is one of the major factors damaging the integrity of the bacterial cell membrane and altering permeability behaviour (Cheloni et al., 2016). Superoxide radical (O^{-2}), hydrogen peroxide (H_2O_2), the hydroxyl radical (·OH), and singlet oxygen (O_2) are the ROSs that cause damage to the bacterial cell in varying amount. ZnONPs can possess the ability to produce OH and H_2O_2 free radicals while CuO NPs can generate all of them (Malka et al., 2013). The generation and scavenging of the ROS are in equilibrium under normal conditions, but surplus production of ROS induces oxidative stress, culminating into the damaging number of cell components (Li et al., 2012; Peng et al., 2013). Metal oxides such as CuO, TiO_2, and ZnO produce excellent antibacterial activity as they are capable of generating ROS (Singh et al., 2014). Das et al. (2017) studied the antimicrobial action of green-synthesized silver NPs on *E. coli* and *S. aureus*. It was concluded that ROS generation was associated with the treatment of bacteria with AgNPs and that the consequent oxidative stress caused the death of the bacteria. Another crucial mechanism involved in the lethal action of metal NPs on bacteria is based on the interference of the synthesis of DNA and bacterial proteins. It has been shown that silver nanoparticles disintegrate the bacterial DNA as they interact with phosphorus and sulfur-containing compounds. Also, nanoparticles can disrupt signal transduction pathways by dephosphorylating the peptide substrate on tyrosine residue in gram-positive bacteria, thus obstructing growth. The binding ability of AgNPs to the mercapto (-SH) group leads to the denaturation of bacterial proteins (Kim et al.,

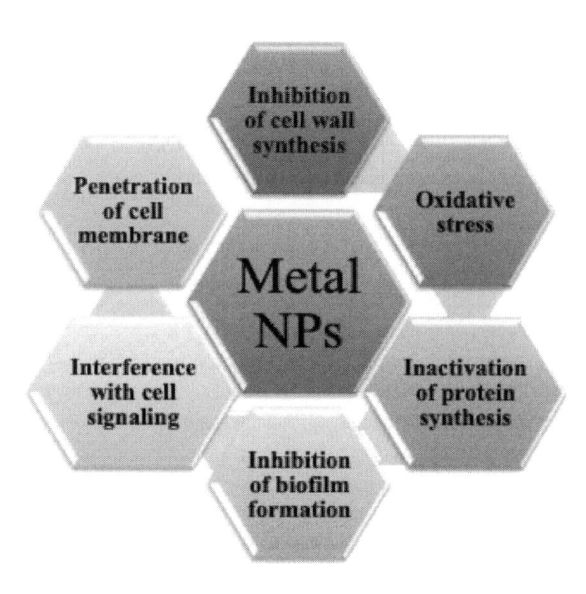

FIGURE 3.3 Mechanism of antimicrobial action of metal nanoparticles.

2009). CuNPs also act similarly by affecting DNA replication and transcription and obstructing growth by interacting with the mercapto (-SH) group (Nisar et al., 2019).

The antimicrobial activity of NPs is greatly dependent on their size as well. There is a direct correlation between the size and the antimicrobial potential of the NPs. Smaller NPs show enhanced antimicrobial activity. The larger surface area offers greater contact while the smaller size facilitates the penetration into the bacterial cell wall (Gurunathan et al., 2014). NPs having size of less than 50 nm possess efficient antimicrobial activity while those having a size if 10 to 15 nm possess superior microbicidal action (Roy et al., 2019). A study by Panáček et al. (2006) reported the synthesis of silver NPs using four saccharides, viz. glucose, galactose, maltose, and lactose. The antimicrobial activity of the NPs synthesized using glucose and galactose was found to be higher than those synthesized using maltose and lactose. This might have resulted due to the smaller size of NPs synthesized using glucose and galactose than those synthesized using maltose and lactose.

3.7 CONCLUSION

The present chapter is focused on *A. indica* (neem)–mediated synthesis of various metallic NPs such as silver, copper, copper oxide, gold, iron, iron oxide, zinc oxide, platinum, palladium, bimetallic, and others. The plant extract is loaded with numerous phytochemicals viz. alkaloids, saponins, flavonoids, and terpenoids, among others, which enable the same to biologically reduce inorganic metal ions leading to the synthesis of various stable, bio-capped NPs. Most of the metallic nanoparticles including silver, copper, copper oxide, and zinc oxide, among others, exhibit significant microbicidal potential against various pathogenic gram-positive and negative bacteria and fungi. The availability of the *Azadirachta* plant in abundance, the ability of derived extracts to non-toxically reduces numerous inorganic ions, and the cost-effectiveness of the synthesis process render the plant as a potential candidate for fabrication variety of stable, antimicrobial metallic NPs for application in diverse fields.

REFERENCES

Abhiman, G. R., Devi, P. S., and Vijayalakshmi, K. A. (2018). Green synthesis with antibacterial investigation of Copper nanoparticles using Azadirachta indica (neem) leaf extract. *Themed Section: Science and Technology*, 4(8), 697–701.

Agarwal, A., Pathera, A. K., Kaushik, R., Kumar, N., Dhull, S. B., Arora, S., and Chawla, P. (2020). Succinylation of milk proteins: Influence on micronutrient binding and functional indices. *Trends in Food Science and Technology*, 97, 254–264.

Alam, A., Haldar, S., Thulasiram, H. V., Kumar, R., Goyal, M., Iqbal, M. S., ... and Pal, U. (2012). Novel anti-inflammatory activity of epoxyazadiradione against macrophage migration inhibitory factor inhibition of tautomerase and proinflammatory activities of macrophage migration inhibitory factor. *Journal of Biological Chemistry*, 287(29), 24844–24861.

Albrecht, M. A, Evans, C. W, and Raston, C. L (2006). Green chemistry and the health implications of nanoparticles. *Green Chemistry*, 8(5), 417–432.

Algebaly, A. S., Mohammed, A. E., Abutaha, N., and Elobeid, M. M. (2020). Biogenic synthesis of silver nanoparticles: antibacterial and cytotoxic potential. *Saudi Journal of Biological Sciences*, *27,* 1340–1351.

Alzohairy, M. A. (2016). Therapeutics Role of *Azadirachta indica* (Neem) and their active constituents in diseases prevention and treatment. *Evidence-based Complementary and Alternative Medicine: eCAM*, 2016, 7382506.

Amrutham, S., Maragoni, V., and Guttena, V. (2020). One-step green synthesis of palladium nanoparticles using neem gum (*Azadirachta indica*): Characterization, reduction of Rhodamine 6G dye and free radical scavenging activity. *Applied Nanoscience*, 12, 1–7.

Anastas, P. T., and Zimmerman, J. B. (2007). Green nanotechnology. In *Why We Need a Green Nano Award and How to Make it Happen*. Washington, DC: Woodrow Wilson International Center for Scholars.

Ansari, M. A., Khan, H. M., Khan, A. A., Cameotra, S. S., Saquib, Q., and Musarrat, J. (2014). Interaction of Al_2O_3 nanoparticles with *Escherichia coli* and their cell envelope biomolecules. *Journal of Applied Microbiology*, 116(4):772–783.

Ansilin, S., Nair, J. K., Aswathy, C., Rama, V., Peter, J., and Persis, J. J. (2016). Green synthesis and characterisation of copper oxide nanoparticles using *Azadirachta indica* (Neem) leaf aqueous extract. *Journal of Nanoscience and Technology*, 221–223.

Asimuddin, M., Shaik, M. R., Adil, S. F., Siddiqui, M. R. H., Alwarthan, A., Jamil, K., and Khan, M. (2018). *Azadirachta indica* based biosynthesis of silver nanoparticles and evaluation of their antibacterial and cytotoxic effects. *Journal of King Saud University-Science*, 32(1), 648–656.

Azam, A., Ahmed, A. S., Oves, M., Khan, M. S., Habib, S. S., and Memic, A. (2012). Antimicrobial activity of metal oxide nanoparticles against Gram-positive and Gram-negative bacteria: A comparative study. *International Journal of Nanomedicine*, 7, 6003.

Babes, L., Denizot, B., Tanguy, G., Le Jeune, J. J., and Jallet, P. (1999). Synthesis of iron oxide nanoparticles used as MRI contrast agents: A parametric study. *Journal of Colloid and Interface Science*, 212(2), 474–482.

Bailey, S. E., Olin, T. J., Bricka, R. M., and Adrian, D. D. (1999). A review of potentially low-cost sorbents for heavy metals. *Water Research*, 33(11), 2469–2479.

Baker, S., and Satish, S. (2012). Endophytes toward a vision in synthesis of nanoparticle for future therapeutic agents. *International Journal of Bio-Inorganic Hybrid Nanomaterials*, 1(2), 67–77.

Banerjee, P., Satapathy, M., Mukhopahayay, A., and Das, P. (2014). Leaf extract mediated green synthesis of silver nanoparticles from widely available Indian plants: Synthesis, characterization, antimicrobial property and toxicity analysis. *Bioresources and Bioprocessing*, 1(1), 3.

Bankar, A., Joshi, B., Kumar, A. R., and Zinjarde, S. (2010). Banana peel extract mediated novel route for the synthesis of silver nanoparticles. *Colloids and Surfaces A: Physicochemical and Engineering Aspects*, 368(1–3), 58–63.

Barakat, M. A. (2011). New trends in removing heavy metals from industrial wastewater. *Arabian Journal of Chemistry*, 4(4), 361–377.

Beattie, I. R., and Haverkamp, R. G. (2011). Silver and gold nanoparticles in plants: Sites for the reduction to metal. *Metallomics*, 3(6), 628–632.

Bhuyan, T., Mishra, K., Khanuja, M., Prasad, R., and Varma, A. (2015). Biosynthesis of zinc oxide nanoparticles from *Azadirachta indica* for antibacterial and photocatalytic applications. *Materials Science in Semiconductor Processing*, 32, 55–61.

Bindhani, B. K., and Panigrahi, A. K. (2014). Green synthesis of gold nanoparticles using neem (*Azadirachta indica* L.) leaf extract and its biomedical applications. *International Journal of Advanced Biotechnology and Research*, 5, 457–464.

Biswas, K., Chattopadhyay, I., Banerjee, R. K., and Bandyopadhyay, U. (2002). Biological activities and medicinal properties of neem (*Azadirachta indica*). *Current Science-Bangalore*, 82(11), 1336–1345.

Britt, B. A., and Kalow, W. (1970). Malignant hyperthermia: A statistical review. *Canadian Anaesthetists Society Journal*, 17(4), 293–315.

Brunet, L., Lyon, D. Y., Hotze, E. M., Alvarez, P. J., and Wiesner, M. R. (2009). Comparative photoactivity and antibacterial properties of C_{60} fullerenes and titanium dioxide nanoparticles. *Environmental Science & Technology*, 43(12), 4355–4360.

Casillas-Martinez, L., Gonzalez, M. L., Fuentes-Figueroa, Z., Castro, C. M., Nieves-Mendez, D., Hernandez, C., Ramirez, W., Sytsma, R. E., Perez-Jimenez, J., and Visscher, P. T. (2005). Community structure, geochemical characteristics and mineralogy of a hypersaline microbial mat, Cabo Rojo, *PR. Geomicrobiology Journal*, 22(6), 269–281.

Chaloupka, K., Malam, Y., and Seifalian, A. M. (2010). Nanosilver as a new generation of nanoproduct in biomedical applications. *Trends in Biotechnology*, 28(11):580–588.

Chand, K., Abro, M. I., Aftab, U., Shah, A. H., Lakhan, M. N., Cao, D., … and Mohamed, A. M. A. (2019). Green synthesis characterization and antimicrobial activity against *Staphylococcus aureus* of silver nanoparticles using extracts of neem, onion and tomato. *RSC Advances*, 9(30), 17002–17015.

Chawla, P., Kaushik, R., Shiva Swaraj, V. J., and Kumar, N. (2018). Organophosphorus pesticides residues in food and their colorimetric detection. *Environmental Nanotechnology, Monitoring and Management*, 10, 292–307.

Chawla, P., Kumar, N., Kaushik, R., and Dhull, S.B. (2019). Synthesis, characterization and cellular mineral absorption of gum arabic stabilized nanoemulsion of *Rhododendron arboreum* flower extract. *Journal of Food Science and Technology*, 56(12), 5194–5203.

Chawla, P., Kumar, N., Bains, A., Dhull, S. B., Kumar, M., Kaushik, R., and Punia, S. (2020). Gum Arabic capped copper nanoparticles: Characterization and their utilization. *International Journal of Biological Macromolecules*, 146, 232–242.

Cheloni, G., Marti, E., and Slaveykova, V. I. (2016). Interactive effects of copper oxide nanoparticles and light to green alga *Chlamydomonas reinhardtii*. *Aquatic Toxicology*, 170, 120–128.

Chen, Q., and Huang, Z. (2006). Carbon based nanostructures. In *Micro manufacturing and Nanotechnology* Springer, Berlin, Heidelberg, 247–274.

Dahl, J. A., Maddux, B. L., and Hutchison, J. E. (2007). Toward greener nanosynthesis. *Chemical Reviews*, 107(6), 2228–2269.

Das, B., Dash, S. K., Mandal, D., Ghosh, T., Chattopadhyay, S., Tripathy, S., ... and Roy, S. (2017). Green synthesized silver nanoparticles destroy multidrug resistant bacteria via reactive oxygen species mediated membrane damage. *Arabian Journal of Chemistry*, 10(6), 862–876.

de Lima, R., Seabra, A. B., and Durán, N. (2012). Silver nanoparticles: A brief review of cytotoxicity and genotoxicity of chemically and biogenically synthesized nanoparticles. *Journal of Applied Toxicology*, 32(11), 867–879.

De, D., Mandal, S. M., Gauri, S. S., Bhattacharya, R., Ram, S., and Roy, S. K. (2010). Antibacterial effect of lanthanum calcium manganate (La_0. $67Ca_0$. $33MnO_3$) nanoparticles against *Pseudomonas aeruginosa* ATCC 27853. *Journal of Biomedical Nanotechnology*, 6(2), 138–144.

Devatha, C. P., Jagadeesh, K., and Patil, M. (2018). Effect of Green synthesized iron nanoparticles by *Azardirachta Indica* in different proportions on antibacterial activity. *Environmental Nanotechnology, Monitoring and Management*, 9, 85–94.

Dhull, S. B., Anju, M., Punia, S., Kaushik, R., and Chawla, P. (2019). Application of gum Arabic in nanoemulsion for safe conveyance of bioactive components. In: Prasad R., Kumar V., Kumar M., Choudhary D. (eds) *Nanobiotechnology in Bioformulations. Nanotechnology in the Life Science*. Springer, Cham. Chapter 3, pp. 85–98. DOI. org/10.1007/978-3-030-17061-5-3.

Dixon, M. B., Falconet, C., Ho, L., Chow, C. W., O'Neill, B. K., and Newcombe, G. (2011). Removal of cyanobacterial metabolites by nanofiltration from two treated waters. *Journal of Hazardous Materials*, 188(1–3), 288–295.

Doria, G., Conde, J., Veigas, B., Giestas, L., Almeida, C., Assunção, M., Rosa, J., and Baptista, P. V. (2017). Noble metal nanoparticles for biosensing applications. *Sensors*, 12(2), 1657–1687.

Durgun, E., Ciraci, S., and Yildirim, T. (2008). Functionalization of carbon-based nanostructures with light transition-metal atoms for hydrogen storage. *Physical Review B*, 77(8), 085405.

Emami-Karvani, Z., and Chehrazi, P. (2011). Antibacterial activity of ZnO nanoparticle on gram-positive and gram-negative bacteria. *African Journal of Microbiology Research*, 5(12), 1368–1373.

Feldheim, D. L., and Foss, C. A. (2002). *Metal Nanoparticles: Synthesis, Characterization, and Applications*. Boca Raton, FL: CRC Press.

Ghimeray, A. K., Jin, C. W., Ghimire, B. K., and Cho, D. H. (2009). Antioxidant activity and quantitative estimation of azadirachtin and nimbin in *Azadirachta indica* A. Juss grown in foothills of Nepal. *African Journal of Biotechnology*, 8(13), 281–289.

Gupta, A. K., and Gupta, M. (2005). Synthesis and surface engineering of iron oxide nanoparticles for biomedical applications. *Biomaterials*, 26(18), 3995–4021.

Gurunathan, S., Han, J. W., Kwon, D. N., and Kim, J. H. (2014). Enhanced antibacterial and anti-biofilm activities of silver nanoparticles against gram-negative and gram-positive bacteria. *Nanoscale Research Letters*, 9(1), 373.

Helan, V., Prince, J. J., Al-Dhabi, N. A., Arasu, M. V., Ayeshamariam, A., Madhumitha, G., ... and Jayachandran, M. (2016). Neem leaves mediated preparation of NiO nanoparticles and its magnetization, coercivity and antibacterial analysis. *Results in Physics*, 6, 712–718.

Herlekar, M., Barve, S., and Kumar, R. (2014). Plant-mediated green synthesis of iron nanoparticles. *Journal of Nanoparticles*, 54, 1–9.

Hernandez, Y., Nicolosi, V., Lotya, M., Blighe, F. M., Sun, Z., De, S., McGovern, I. T., Holland, B., Byrne, M., Gun'Ko, Y. K., and Boland, J. J. (2008). High-yield production of graphene by liquid-phase exfoliation of graphite. *Nature Nanotechnology*, 3(9), 563.

Hodenius, M., De Cuyper, M., Desender, L., Müller-Schulte, D., Steigel, A., and Lueken, H. (2002). Biotinylated stealth® magnetoliposomes. *Chemistry and Physics of Lipids*, 120(1–2), 75–85.

Huang, H. and Yang, X. (2004). Synthesis of polysaccharide-stabilized gold and silver nanoparticles: A green method. *Carbohydrate Research*, 339(15), 2627–2631.

Jain, K. K. (2005). Nanotechnology in clinical laboratory diagnostics. *Clinica Chimica Acta*, 358(1–2), 37–54.

Jain, T. K., Morales, M. A., Sahoo, S. K., Leslie-Pelecky, D. L., and Labhasetwar, V. (2005). Iron oxide nanoparticles for sustained delivery of anticancer agents. *Molecular Pharmaceutics*, 2(3), 194–205.

Jin, T., and He, Y. (2011). Antibacterial activities of magnesium oxide (MgO) nanoparticles against foodborne pathogens. *Journal of Nanoparticle Research*, 13(12), 6877–6885.

Jones, I. W., Denholm, A. A., Ley, S. V., Lovell, H., Wood, A., and Sinden, R. E. (1994). Sexual development of malaria parasites is inhibited in vitro by the neem extract azadirachtin, and its semi-synthetic analogues. *FEMS Microbiology Letters*, 120(3), 267–273.

Jose, N., Deshmukh, G. P., and Ravindra, M. R.. (2019). Dynamic Light Scattering: Advantages and Applications. *Acta Scientific Nutritional Health*, 3(3), 50–52.

Kaler, S. G. (2011). ATP7A-related copper transport diseases – emerging concepts and future trends. *Nature Reviews Neurology*, 7(1), 15.

Khalid, S. A., Duddeck, H., and Gonzalez-Sierra, M. (1989). Isolation and characterization of an antimalarial agent of the neem tree *Azadirachta indica*. *Journal of Natural Products*, 52(5), 922–927.

Kim, J., Lee, J., Kwon, S., and Jeong, S. (2009). Preparation of biodegradable polymer/silver nanoparticles composite and its antibacterial efficacy. *Journal of Nanoscience and Nanotechnology*, 9(2), 1098–1102.

Koul, O., Singh, G., Singh, R., Singh, J., Daniewski, W. M., and Berlozecki, S. (2004). Bioefficacy and mode-of-action of some limonoids of salannin group from *Azadirachta indica* A. Juss and their role in a multicomponent system against lepidopteran larvae. *Journal of Biosciences*, 29(4), 409–416.

Kumari, A., Yadav, S. K., and Yadav, S. C. (2010). Biodegradable polymeric nanoparticles based drug delivery systems. *Colloids and Surfaces B: Biointerfaces*, 75(1), 1–8.

Kunkalekar, R. K., Naik, M. M., Dubey, S. K., and Salker, A. V. (2013). Antibacterial activity of silver-doped manganese dioxide nanoparticles on multidrug-resistant bacteria. *Journal of Chemical Technology & Biotechnology*, 88(5), 873–877.

Kurtycz, P., Karwowska, E., Ciach, T., Olszyna, A., and Kunicki, A. (2013). Biodegradable polylactide (PLA) fiber mats containing Al_2O_3-Ag nanopowder prepared by electrospinning technique-antibacterial properties. *Fibers and Polymers*, 14(8), 1248–1253.

Li, W. R., Xie, X. B., Shi, Q. S., Zeng, H. Y., You-Sheng, O. Y., and Chen, Y. B. (2010). Antibacterial activity and mechanism of silver nanoparticles on *Escherichia coli*. *Applied Microbiology and Biotechnology*, 85(4), 1115–1122.

Li, Y., Zhang, W., Niu, J., and Chen, Y. (2012). Mechanism of photogenerated reactive oxygen species and correlation with the antibacterial properties of engineered metal-oxide nanoparticles. *ACS Nano*, 6(6), 5164–5173.

Lin, P. C., Lin, S., Wang, P. C., and Sridhar, R. (2014). Techniques for physicochemical characterization of nanomaterials. *Biotechnology Advances*, 32(4), 711–726.

Lowe, C. R. (2000). Nanobiotechnology: The fabrication and applications of chemical and biological nanostructures. *Current Opinion in Structural Biology*, 10(4), 428–434.

Luangpipat, T., Beattie, I. R., Chisti, Y., and Haverkamp, R. G. (2011). Gold nanoparticles produced in a microalga. *Journal of Nanoparticle Research*, 13(12), 6439–6445.

Mak-Mensah, E. E., and Firempong, C. K.(2011). Chemical characteristics of toilet soap prepared from neem (*Azadirachta indica* A. Juss) seed oil. *Asian Journal of Plant Science and Research*, 1(4), 1–7.

Malka, E., Perelshtein, I., Lipovsky, A., et al. (2013). Eradication of multi-drug resistant bacteria by a novel Zn-doped CuO nanocomposite. *Small*, 9(23), 4069–4076.

Maragathavalli, S., Brindha, S., Kaviyarasi, N. S., Annadurai, B., and Gangwar, S. K. (2012). Antimicrobial activity in leaf extract of neem (*Azadirachta indica* Linn.). *International Journal of Science and Nature*, 3(1), 110–113.

Mgbemena, I. C., Opara, F. N., Ukaoma, A., Ofodu, C., and Ogbuagu, D. H. (2010). Prophylactic potential of lemon grass and neem as antimalarial agents. *Journal of American Science*, 6(8), 20–26.

Mittal, A. K., Chisti, Y., and Banerjee, U. C. (2013). Synthesis of metallic nanoparticles using plant extracts. *Biotechnology Advances*, 31(2), 346–356.

Mohanaparameswari, S., Balachandramohan, M., and Murugeshwari, P. (2019). Bio synthesis and characterization of silver nanoparticles by leaf extracts of Moringa oleifere leaf, *Azardica indica* (Neem) leaf, bamboo leaf, and their antibacterial activity. *Materials Today: Proceedings*, 18, 1783–1791.

Nagavarma, B. V., Yadav, H. K., Ayaz, A., Vasudha, L. S., and Shivakumar, H. G. (2012). Different techniques for preparation of polymeric nanoparticles-a review. *Asian Journal of Pharm ceutical and Clinical Research*, 5(3), 16–23.

Nayak, D., Ashe, S., Rauta, P. R., Kumari, M., and Nayak, B. (2016). Bark extract mediated green synthesis of silver nanoparticles: Evaluation of antimicrobial activity and antiproliferative response against osteosarcoma. *Materials Science and Engineering: C*, 58, 44–52.

Nisar, P., Ali, N., Rahman, L., Ali, M., and Shinwari, Z. K. (2019). Antimicrobial activities of biologically synthesized metal nanoparticles: An insight into the mechanism of action. *JBIC Journal of Biological Inorganic Chemistry*, 24(7), 929–941.

Olsvik, O., Popovic, T., Skjerve, E., Cudjoe, K. S., Hornes, E., Ugelstad, J., and Uhlen, M. (1994). Magnetic separation techniques in diagnostic microbiology. *Clinical Microbiology Reviews*, 7(1), 43–54.

Padil, V. V., and Cerník, M. (2013). Green synthesis of copper oxide nanoparticles using gum karaya as a biotemplate and their antibacterial application. *International Journal of Nanomedicine*, 8, 889.

Panáček, A., Kvitek, L., Prucek, R., Kolář, M., Večeřová, R., Pizúrová, N., ... and Zbořil, R. (2006). Silver colloid nanoparticles: Synthesis, characterization, and their antibacterial activity. *The Journal of Physical Chemistry B*, 110(33), 16248–16253.

Park, Y., Hong, Y. N., Weyers, A., Kim, Y. S., and Linhardt, R. J. (2011). Polysaccharides and phytochemicals: A natural reservoir for the green synthesis of gold and silver nanoparticles. *IET Nanobiotechnology*, 5(3), 69–78.

Patel, V.R., and Agrawal, Y. K. (2011). Nanosuspension: An approach to enhance solubility of drugs. *Journal of Advanced Pharmaceutical Technology & Research*, 2(2), 81.

Pathania, R., Khan, H., Kaushik, R., and Khan, M. A. (2018). Essential oil nanoemulsions and their antimicrobial and food applications. *Current Research Nutrition and Food Science*, 6(3), 1–16.

Pattanayak, M., and Nayak, P. L. (2013). Green synthesis and characterization of zero valent iron nanoparticles from the leaf extract of *Azadirachta indica* (Neem). *World Journal of Nano Science & Technology*, 2(1), 6–9.

Peng, Z., Ni, J., Zheng, K. et al. (2013). Dual effects and mechanism of TiO_2 nanotube arrays in reducing bacterial colonization and enhancing $C_3H_{10}T_{1/2}$ cell adhesion. *International Journal of Nanomedicine*, 8, 3093–3105.

Poopathi, S., De Britto, L. J., Praba, V. L., Mani, C., and Praveen, M. (2015). Synthesis of silver nanoparticles from *Azadirachta indica*—a most effective method for mosquito control. *Environmental Science and Pollution Research*, 22(4), 2956–2963.

Pratt, A. S., and Smith, P. R. (1989). Inventors; Johnson Matthey PLC, Assignee. Antimicrobial compositions consisting of metallic silver combined with Titanium oxide or Tantalum oxide. United States patent US 4,849,223.

Ramar, K., and Ahamed, A. J. (2018). Hydrothermal green synthesis of silver nanoparticles using *Azadirachta indica* A. juss fruit juice for potential antibacterial activity. *Journal of Nanoscience and Technology*, 519–523.

Ramyadevi, J., Jeyasubramanian, K., Marikani, A., Rajakumar, G., Rahuman, A. A., Santhoshkumar, T., … and Marimuthu, S. (2011). Copper nanoparticles synthesized by polyol process used to control hematophagous parasites. *Parasitology Research*, 109(5), 1403–1415.

Rana, B., Kaushik, R., Kaushal, K., Kaushal, A., Gupta S., Upadhay, N., Rani, P., and Sharma P. (2018). Application of biosensors for determination of physicochemical properties of zinc fortified milk. *Food Bioscience*, 21, 117–124.

Rao, P. S., Subramanayam, G., and Sridhar, P. R. (2018). Flavonol glycosides from *Azadirachta indica* L. *Drug Invent Today*, 10, 1421–1426.

Ren, G., Hu, D., Cheng, E. W., Vargas-Reus, M. A., Reip, P., and Allaker, R. P. (2009). Characterisation of copper oxide nanoparticles for antimicrobial applications. *International Journal of Antimicrobial Agents*, 33(6), 587–590.

Roy, A., Bulut, O., Some, S., Mandal, A. K., and Yilmaz, M. D. (2019). Green synthesis of silver nanoparticles: Biomolecule-nanoparticle organizations targeting antimicrobial activity. *RSC Advances,* 9(5), 2673–2702.

Roy, P., Das, B., Mohanty, A., and Mohapatra, S. (2017). Green synthesis of silver nanoparticles using *Azadirachta indica* leaf extract and its antimicrobial study. *Applied Nanoscience*, (8), 843–850.

Sadeghian, M. M., and Mortazaienezhad, F. (2007). Investigation of compounds from *Azadirachta indica* (neem). *Asian Journal of Plant Sciences*, 6(2), 444–445.

Sarah, R., Tabassum, B., Idrees, N., and Hussain, M. (2019). Bioactive compounds isolated from neem tree and their applications. *Natural Bioactive Compounds*. Doi: 10.1007/978-981-13-7154-7_17.

Saranya, S., Vijayarani, K., Ramya, K., Revathi, K., and Kumanan, K. (2016). Synthesis and characterization of silver nanoparticles using *Azadirachta indica* leaf extract and their anti-fungal activity against malassezia species. In *Journal of Nano Research* (Vol. 43, pp. 1–10). Trans Tech Publications Ltd.

Sarsaiya, S., Shi, J., and Chen, J. (2019). A comprehensive review on fungal endophytes and its dynamics on Orchidaceae plants: Current research, challenges, and future possibilities. *Bioengineered*, 10(1), 316–334.

Sawai, J., Kojima, H., Igarashi, H., Hashimoto, A., Shoji, S., Sawaki, T., Hakoda, A., Kawada, E., Kokugan, T., and Shimizu, M. (2000). Antibacterial characteristics of magnesium oxide powder. *World Journal of Microbiology and Biotechnology*, 16(2), 187–194.

Schaaf, O., Jarvis, A. P., Van Der Esch, S. A., Giagnacovo, G., and Oldham, N. J. (2000). Rapid and sensitive analysis of azadirachtin and related triterpenoids from neem (*Azadirachta indica*) by high-performance liquid chromatography–atmospheric pressure chemical ionization mass spectrometry. *Journal of Chromatography A*, 886(1–2), 89–97.

Schmutterer, H., and Singh, R. P. (1995). List of insect pests susceptible to neem products. *The Neem Tree: Azadirachta indica A. Juss and Other Meliaceae Plants*. VCH, New York, 326–325.

Seabra, A. B., Pelegrino, M. T., and Haddad, P. S. (2017). Antimicrobial applications of superparamagnetic iron oxide nanoparticles: Perspectives and challenges. In *Nanostructures for Antimicrobial Therapy,* 12, 531–550.

Seil, J. T., and Webster, T. J. (2012). Antimicrobial applications of nanotechnology: Methods and literature. *International Journal of Nanomedicine*, 7, 2767.

Shaikh, T. N., Chaudhari, S., Patel, B., and Poonia, M. P. (n.d.). Characterization of green synthesized silver nanoparticles using *Azadirachta Indica* (Neem) leaf extract, 3(5), 209–210.

Shankar, S. S., Rai, A., Ahmad, A., and Sastry, M. (2004). Rapid synthesis of Au, Ag, and bimetallic Au core-Ag shell nanoparticles using Neem (*Azadirachta indica*) leaf broth *Journal of Colloid and Interface Science*, 275(2), 496–502.

Sharma, S., and Oudhia, A. (2016). Green synthesis of ZnO NPs from various parts of *Azhadirachta indica* (neem) plant as biotemplates for anti-bacterial applications. In *AIP Conference Proceedings* (Vol. 1728, No. 1, p. 020032). Chattisgarh, India, AIP Publishing LLC.

Sharma, M. C., Sharma, S., and Kohli, D. V. (2010). In vitro studies of the use of some medicinal herbals leaves against antidepressant, analgesic activity, and anti-inflammatory activity. *Digest Journal of Nanomaterials and Biostructures*, 5, 131–134.

Shrivastava, S., Bera, T., Roy, A., Singh, G., Ramachandrarao, P., and Dash, D. (2007). Characterization of enhanced antibacterial effects of novel silver nanoparticles. *Nanotechnology*, 18(22), 225103.

Siddiqi, K. S., and Husen, A. (2016). Fabrication of metal nanoparticles from fungi and metal salts: Scope and application. *Nanoscale Research Letters*, 11(1), 98.

Singaravelu, G., Arockiamary, J. S., Kumar, V. G., and Govindaraju, K. (2007). A novel extracellular synthesis of monodisperse gold nanoparticles using marine alga, *Sargassum wightii* Greville. *Colloids and Surfaces B: Biointerfaces*, 57(1), 97–101.

Singh, V., and Chauhan, D. (2014). Phytochemical evaluation of aqueous and ethanolic extract of neem leaves (*Azadirachta indica*). *Indo American Journal of Pharmaceutical Research*, 4(12), 5943–5948.

Singh, R., Smitha, M. S., and Singh, S. P. (2014). The role of nanotechnology in combating multi-drug resistant bacteria. *Journal of Nanoscience and Nanotechnology*, 14(7), 4745–4756.

Singh, R., Thakur, P., Thakur, A., Kumar, H., Chawla, P., Jigneshkumar, V. R., Kaushik, R., and Kumar, N. (2020). Colorimetric sensing approaches of surface modified gold and silver nanoparticles for detection of residual pesticides: A review. *International Journal of Environmental Analytical Chemistry*, 118, 1–7.

Sirelkhatim, A., Mahmud, S., Seeni, A., Kaus, N. H., Ann, L. C., Bakhori, S. K., Hasan, H., and Mohamad, D. (2015). Review on zinc oxide nanoparticles: Antibacterial activity and toxicity mechanism. *Nano-Micro Letters*, 7(3), 219–242.

Slavin, Y. N., Asnis, J., Häfeli, U. O., and Bach, H. (2017). Metal nanoparticles: Understanding the mechanisms behind antibacterial activity. *Journal of Nanobiotechnology*, 15(1), 65.

Stoimenov, P. K., Klinger, R. L., and Marchin, G. L., and Klabunde, K. J. (2002). Metal oxide nanoparticles as bactericidal agents. *Langmuir*, 18(17), 6679–6686.

Taib, N. I., Latif, F. A., Mohamed, Z., and Zambri, N. D. S. (2018). Green synthesis of iron oxide nanoparticles (Fe 3 O 4-NPs) using *Azadirachta indica* aqueous leaf extract. *International Journal of Engineering and Technology*, 7, 9–13.

Thakkar, K. N., Mhatre, S. S., and Parikh, R. Y. (2010). Biological synthesis of metallic nanoparticles. *Nanomedicine: Nanotechnology, Biology and Medicine*, 6(2), 257–262.

Thakur, B. K., Kumar, A., and Kumar, D. (2019). Green synthesis of titanium dioxide nanoparticles using Azadirachta indica leaf extract and evaluation of their antibacterial activity. *South African Journal of Botany*, 124, 223–227.

Thirumurugan, A., Aswitha, P., Kiruthika, C., Nagarajan, S., and Christy, A. N. (2016). Green synthesis of platinum nanoparticles using *Azadirachta indica*–An eco-friendly approach. *Materials Letters*, 170, 175–178.

Thirumurugan, A., Jiflin, G. J., Rajagomathi, G., Tomy, N. A., Ramachandran, S., and Jaiganesh, R. (2010). "Biotechnological synthesis of gold nanoparticles of *Azadirachta indica* leaf extract." *International Journal of Biological and Technology*, 1, 75–77.

Tsapis, N., Bennett, D., Jackson, B., Weitz, D. A., and Edwards, D. A. (2002). Trojan particles: Large porous carriers of nanoparticles for drug delivery. *Proceedings of the National Academy of Sciences*, 99(19), 12001–12005.

Van Der Veen, A. J., Talwar, S, and Paulraj, A. (1997). A subspace approach to blind space-time signal processing for wireless communication systems. *IEEE Transactions on Signal Processing*, 45(1), 173–190.

Veeranna, S., Burhanuddin, A., Khanum, S., Narayan, S. L., and Pratima, K. (2013). Biosynthesis and antibacterial activity of silver nanoparticles. *Research Journal of Biotechnology*, 8(1), 11–17.

Velusamy, P., Das, J., Pachaiappan, R., Vaseeharan, B., and Pandian, K. (2015). Greener approach for synthesis of antibacterial silver nanoparticles using aqueous solution of neem gum (*Azadirachta indica* L.). *Industrial Crops and Products*, 66, 103–109.

Virkutyte, J., and Varma, R. S. (2011). Green synthesis of metal nanoparticles: Biodegradable polymers and enzymes in stabilization and surface functionalization. *Chemical Science*, 2(5), 837–846.

Wang, L., Hu, C., and Shao, L. (2017). The antimicrobial activity of nanoparticles: Present situation and prospects for the future. *International Journal of Nanomedicine*, 12, 1227.

Wang, S. L., Xu, X. R., Sun, Y. X., Liu, J. L., and Li, H. B. (2013). Heavy metal pollution in coastal areas of South China: A review. *Marine Pollution Bulletin*, 76(1–2), 7–15.

Whitman, J. (2007). The governance of nanotechnology. *Science and Public Policy*, 34(4), 273–283.

Zakir, H. A., Subbarao, G. V., Pearse, S. J., Gopalakrishnan, S., Ito, O., Ishikawa, T., Kawano, N., Nakahara, K., Yoshihashi, T., Ono, H., and Yoshida, M. (2008). Detection, isolation and characterization of a root-exuded compound, methyl 3-(4-hydroxyphenyl) propionate, responsible for biological nitrification inhibition by sorghum (Sorghum bicolor). *New Phytologist*, 180(2), 442–451.

Zambri, N. D. S., Taib, N. I., Abdul Latif, F., and Mohamed, Z. (2019). Utilization of Neem Leaf Extract on Biosynthesis of Iron Oxide Nanoparticles. *Molecules*, *24*(20), 3803.

Zargar, M., Hamid, A. A., Bakar, F. A., Shamsudin, M. N., Shameli, K., Jahanshiri, F., and Farahani, F. (2011). Green synthesis and antibacterial effect of silver nanoparticles using *Vitex negundo* L. *Molecules*, 16(8), 6667–6676.

Zhang, X. F., Liu, Z. G., Shen, W., and Gurunathan, S. (2016). Silver nanoparticles: Synthesis, characterization, properties, applications, and therapeutic approaches. *International Journal of Molecular Sciences*, 17(9), 1534.

Zhang, J. Z., O'Neil, R. H., and Roberti, T. W. (1994). Femtosecond studies of photoinduced electron dynamics at the liquid-solid interface of aqueous CdS colloids. *The Journal of Physical Chemistry*, 98(14), 3859–3864.

4 Metal Nanoparticles of Microbial Origin and Their Antimicrobial Applications in Food Industries

Vaibhao Lule
College of Dairy Technology, India
Sudhir Kumar Tomar and Silvia Sequeira
ICAR-National Dairy Research Institute, India

CONTENTS

4.1 INTRODUCTION

Nanotechnology ("nanotech") is the art and science of manipulating matter at the atomic or molecular scale and holds the promise of providing significant improvements in technologies for protecting the environment. The birth of nanotechnology is attributed to physicist Richard Feynman, who suggested the possibility of manipulation of individual atoms as a more powerful form of synthetic chemistry than those used at the time In 1974, Norio Taniguchi first used the word *nanotechnology* in the context of an ion sputter machine to refer production technology to get the extra-high accuracy and ultra-fine dimensions, that is the preciseness and fineness in the order of 1 nm (Chawla et al., 2018). However, the term *nano* is adapted from the Greek word meaning "dwarf". When used as a prefix, it implies 10^{-9}. A nanometre is one-billionth of a metre, or roughly the length of three atoms side by side. A DNA molecule is 2.5 nm wide; a protein, approximately 50 nm; and a flu virus. about 100 nm. A human hair is approximately 10,000 nm thick. The National Nanotechnology Initiative (NNI) defines nanotechnology as the manipulation of matter with at least one dimension sized from 1 to 100 nanometers (Chawla et al., 2019). In the 21st century, nanotechnology has become one of the key technologies influencing science on a global scale. The engineering with nanomaterials is an enabling technology that has opened up new avenues of research and development in several fields, including medicine, cosmetics, agriculture, and food, and is being used as a means for comprehending how physicochemical characteristics of nano-sized substances can change the structure, texture, and quality of foodstuffs. Modern science is focused on exploiting some of the special properties expressed by nanomaterials to generate useful technologies for the benefit of humanity (Agarwal et al., 2020). Current applications of nanotechnology in the food sector, pharmaceutical, cosmetics, nutraceuticals, and water purification, among others, have made it prevalent as a commercial commodity in many aspects of modern life (Pathania et al., 2018). Accordingly, there is an essential requirement to develop high-yield, low-cost, non-toxic, simple, and environmentally benign procedures for the green synthesis of nanoparticles (NPs; Singh et al., 2020). Consequently, the biological approach for the synthesis of NPs is important, especially in employing microorganisms as novel green "nanofactories" (Dhull et al., 2019). The adoption of biological methods in the synthesis of NPs is expected to yield novel structural entities of the desired morphology.

4.2 ROUTES FOR THE SYNTHESIS OF NANOSTRUCTURES

There are different routes for synthesizing nanostructures which are mainly categorized into chemical routes and biological routes.

4.2.1 Chemical Synthesis

Chemical method of synthesis is valuable as it takes a tiny time for synthesis of a large number of NPs. Nevertheless, in this method, capping agents are necessary for the size stabilization of the NPs (Chawla et al., 2020). Following are some of the wet chemical techniques that are used for the synthesis of metal NPs by chemical means (Tieshi et al., 2009):

- Microemulsions
- Solvent-extraction reduction
- Chemical oxidation/reduction
- Sol-gel
- Coprecipitation
- Hydrothermal/solvothermal
- Ultraviolet (UV) irradiation
- Template-assisted

4.2.2 Biological Synthesis

The biosynthesis approach typically employs whole living organisms for the synthesis of bio-inorganic materials. Although solution-based chemical methods enjoy a long history dating back to the pioneering work of Faraday on the synthesis of aqueous gold colloids, biosynthesis is still largely in the "discovery phase" wherein different nanomaterials are synthesized using microorganisms like fungi, bacteria, algae, and plants (Bansal et al., 2012). Microorganisms have been employed for an eon of time towards remediation of toxic metals due to their inherent capability to withstand high concentrations of heavy metal ions through specific resistance mechanisms (Valls and De Lorenzo, 2002), but the possibility of exploring these organisms for nanomaterials synthesis is a relatively recent phenomenon. A few early reports in this area encompassed organisms known to create specific functional materials in natural habitats, for example silica in diatoms (Parkinson and Gordon, 1999), gold in algal and bacterial cells (Hosea et al., 1986), cadmium sulfide (CdS) in bacteria and yeast (Dameron et al., 1989), zinc sulfide (ZnS) in sulfate-reducing bacteria (Temple and Le Roux, 1964) and magnetite in magnetotactic bacteria (Blakemore et al., 1979). Although using diatoms or magnetotactic bacteria to synthesize nanomaterials in our laboratories might sound interesting, this is not necessarily highly appealing from a fundamental perspective since these organisms are already known to create these specific inorganic materials during their natural growth.

An important and even more interesting aspect of the biological synthesis is the use of living organisms for the synthesis of those inorganic materials, which these organisms are not known to encounter during their natural growth environments (e.g. gold, silver, oxide nanomaterials, etc.). These observations were the source of inspiration for using microorganisms for deliberate synthesis of a range of nanomaterials (intracellularly or extracellularly; Thakkar et al., 2010), including bacteria for the synthesis of gold (Kashefi et al., 2001), silver (Ramanathan et al., 2011), palladium (Windt et al., 2005), gold–silver (Au-Ag) alloy (Nair and Pradeep, 2002), CdS (Kang et al.,

2008), ZnS (Bai et al., 2006), iron sulfide (Mann et al., 1990) and magnetite (Mann et al., 1984); algae for the synthesis of gold and silver (Lengke et al., 2007); and fungi for the synthesis of gold (Shankar et al., 2003), silver (Balaji et al., 2009), silica, titania, zirconia, barium titanate (Bansal et al., 2006), CdS (Ahmad et al., 2002), and Au-Ag alloy NPs (Senapati et al., 2005). One important aspect outlined in chemical synthesis routes is the ability to control shape and size, which confer unique properties to these particles. To compete with chemical methods, monodispersed gold NPs were synthesized using an extremophilic actinomycete, *Thermonospora* sp. (Ahmad et al., 2003a, b). The synthesis of metal NPs using biological systems has been in vogue in the recent past, with substantial evidence proving its superiority over the chemical route. The formation of NPs by this method is extremely rapid, requires no toxic chemicals and the NPs are stable for several months (Tejo Prakash et al., 2009).

4.3 BIOSYNTHESIS OF METAL NANOPARTICLES USING BACTERIA

Among the milieu of natural resources, prokaryotic bacteria have been known to interact with metals and are explored for biotechnological applications such as ore leaching and metal recovery. One of the reasons for the "bacterial preference" for NP synthesis is their relative ease of manipulation. Bacteria isolated from different habitats and nutritional modes have been employed for the synthesis of metal NPs either intracellularly or extracellularly. Among the different biological entities employed for the biosynthesis of NPs, bacteria have received the most attention and are preferred due to their ability to withstand high concentrations of heavy metal ions, ease of culturing, manipulation of genetic make-up, and downstream processing (Parikh et al., 2011).

Most of the research in the field of bacteria mediated biosynthesis has been concentrated on the synthesis of noble metal NPs that dates back to the 1970s, when *edomicrobium*-like budding bacteria was shown to accumulate lacelike gold-decorated structures, and it was postulated that in the near future, the "Midas-gene" could be isolated, cloned, and over-expressed for the fast synthesis of gold NPs. Today, after about two decades of research in the area, a repertoire of bacterial species isolated from different families in the bacterial classification system have been reported for the synthesis of gold NPs. This ability of bacteria to synthesize NPs was attributed to the presence of potential anionic sites on the cell wall which includes teichoic phosphodiester groups, free carboxylic groups of the peptidoglycan layer, and the sugar hydroxyl groups from wall polymers and amide groups of the peptide chains that bind and reduce gold ions to their Au^0 forms (Rayman and MacLeod, 1975; Mandal et al., 2006). Most of the bacterial species isolated for the synthesis of gold NPs are on isotropic spherical or quasi-spherical NPs. But bacterial isolates like *Escherichia coli* (Park et al., 2010), *Lactobacilli* (Nair and Pradeep, 2002), *Rhodopseudomonas capsulate* (He et al., 2008), *Bacillus licheniformis* (Kalishwaralal et al., 2009), and *Shewanella algae* (Konishi et al., 2007) have been reported for the synthesis of anisotropic particles ranging from triangular or hexagonal plates to nanowires and nanocubes.

In addition to gold NPs, the ability of bacterial systems to synthesize NPs from the platinum group of metals (PGM) including palladium and platinum NPs (PtNPs) has also been reported. Although the reports for this group are scarce, this shows the potential of bacterial species to synthesize a range of metal NPs with a wide size range (Hennebel et al., 2011). Besides PGM, bacteria have also been shown to survive in high concentrations of tellurium and selenium ions and have evolved cellular mechanisms to convert the highly toxic ionic states of these metalloids to their zero-valent forms.

Interestingly, the cellular mechanism for the bioremediation of these metalloids has been well studied with reports of reductase enzymes present in metalloid resistant bacteria mediating the reduction of selenate/selenite and tellurate/tellurite (Stolz et al., 2006). Even more interesting is the fact that some of the organisms reducing these ions have shown the ability to control NP shapes, including, for example, selenium nanorods by *Pseudomonas alacliphila* (Zang et al., 2011) and nano-rosettes of tellurium by *B. beveridgei* (Baesman et al., 2009). Recently, the even more interesting ability of *Enterobacter sp.* was shown towards the synthesis of mercury NPs. This ability of bacteria to stabilize mercury NPs is important because elemental mercury is known to be volatile (Sinha et al., 2011).

In addition to the biosynthesis of a range of metals and metalloids like gold, platinum, selenium, and tellurium NPs, the biosynthesis of silver NPs (AgNPs) encompasses a large population of bacterial species with the first reports only dating back to 1992 where silver-resistant *Pseudomonas stutzeri* was shown to accumulate AgNPs (Slawson et al., 1992). Unlike gold, in which the resistance mechanism was only recently reported, ionic silver is known to be toxic to bacterial cells and genes conferring silver resistance have been studied and reported for bacterial survival in high silver concentration environments (Parikh et al., 2008). Although silver biosynthesis is well studied, most studies report isotropic NPs except *Pseudomonas stutzeri* AG259, where a few triangular plates were found to accumulate in the periplasmic space of the bacterium (Klaus et al., 1999). Among the first reports of intracellular semiconductor NP synthesis, *E. coli*, when incubated with cadmium chloride ($CdCl_2$) and sodium sulfide (Na_2S) spontaneously, formed CdS semiconductor nanocrystals (NC) which showed that the formation of NCs was markedly affected by physiologic parameters. The entry into the stationary phase increased the yield by 20-fold (Sweeney et al., 2004). In line with these observations, it was found that *Clostridium thermoaceticum* precipitates CdS at the cell surface as well as in the medium when exposed to $CdCl_2$ in the presence of cysteine hydrochloride as a source of sulfide in the growth medium (Cunningham and Lundie, 1993).

Sulfate-reducing bacteria synthesized magnetic iron sulfide (FeS) NPs about 20 nm in size on their surfaces which were separated from the solution by a high gradient field of 1 Tesla. Bacterially produced FeS is an adsorbent for a wide range of heavy metals and some anions. Furthermore, magnetite is a common product of bacterial iron reduction and could be a potential physical indicator of biological activity in geological settings (Watson et al., 1999). *Acinetobacter*, a non-magnetotactic bacterium, was employed for magnetite NP synthesis. In prior studies, biosynthesis of magnetite was found to be extremely slow (often requiring 1 week) under strictly anaerobic conditions. *Acinetobacter* spp. were capable of magnetite synthesis by reaction with suitable aqueous iron precursors under fully aerobic conditions. Importantly, the extracellular magnetite NPs showed excellent magnetic properties

TABLE 4.1
Metallic Nanoparticles Synthesized by Different Bacteria

Microorganism	Type of Nanoparticle	Location	Size Range (nm)	Reference
Pseudomonas stutzeri	Ag	Intracellular	~200	Klaus et al. (1999)
Morganella sp.	Ag	Extracellular	20–30	Parikh et al. (2008)
Plectonema boryanum (Cyanobacteria)	Ag	Intracellular	1–10 1–100	Lengke et al. (2007)
Escherichia coli	CdS	Intracellular	2–5	Sweeney et al. (2004)
Clostridium thermoaceticum	CdS	Intracellular and extracellular	—	Cunningham and Lundie (1993)
Actinobacter spp.	Magnetite	Extracellular	10–40	Bharde et al. (2005)
Shewanella algae	Au	Intracellular, pH 7 Extracellular, pH 1	10–20 50–500	Konishi et al. (2004)
Rhodopseudomonas capsulata	Au	Extracellular, pH 7 Extracellular, pH 4	10–20 50–400	Shiying et al. (2007)
Escherichia coli DH5α	Au	Intracellular	25–33	Liangwei et al. (2007)
Thermomonospora sp.	Au	Extracellular	8	Ahmad et al. (2003)
Rhodococcus sp.	Au	Intracellular	5–15	Ahmad et al. (2003a)
Klebsiella pneumoniae	Ag	Extracellular	5–32	Shahverdi et al. (2007)
Pseudomonas aeruginosa	Au	Extracellular	15–30	Husseiney et al. (2007)
Shewanella oneidensis	Uranium (IV)	Extracellular	—	Marshall et al. (2007)
Lactobacillus spp. from yoghurt	Ag and TiO_2	Extracellular	10–25 10–70	Jha and Prasad (2010)
Lactobacillus sporogens	ZnO	Extracellular	5–15	Prasad and Jha (2000)
S. thermophilus	Se(0)	Intracellular	50–500	Eszenyi et al. (2011)
Lactobacillus spp.	TiO_2	Extracellular	8–35	Jha et al. (2011)
Lactobacillus spp.	Ag and Au	Intracellular	—	Nair and Pradeep (2002)
Lactobacillus fermentum	Ag	Extracellular	11.2	Sintubin et al. (2009)

(Bharde et al., 2005). Bacteria synthesized in a variety of metallic NPs with different size is tabulated in Table 4.1.

Although the field of biosynthesis has been much explored, achieving shape control is still one of the biggest challenges with very few reports entailing shape control (Ramanathan et al., 2011). Even in the case where anisotropic shapes have been achieved, it only reports the outcomes of exposure to heavy metal ions to bacteria, without making any deliberate efforts to control the bacterial growth kinetics to achieve shape control. Another interesting aspect is that, although a wide range of genera has been reported for the biosynthesis of metal NPs, in all cases, typically only a few species of those particular genera have shown the ability to biosynthesize NPs.

Thus, to survive in environments containing high levels of metals, organisms have adapted by evolving mechanisms to cope with them. This mechanism mainly involves altering the chemical nature of the toxic metal so that it no longer causes toxicity, resulting in the formation of NPs of the metal concerned. Thus NPs formation is the "by-product" of a resistance mechanism against a specific metal, and this can be used as an alternative way of producing them.

4.3.1 Metal Nanoparticle Synthesis by Lactic Acid Bacteria

Lactic acid bacteria (LAB) are one of the most important groups of microorganisms (a predominant form of life on the earth) which maintain a balanced ecosystem in the gastrointestinal tract of humans and proven to be generally recognized as safe (GRAS; Kaushik et al., 2017; Bhandari et al., 2018). LAB are prokaryotes in terms of cellular organization, and they are gram-positive (a thick peptidoglycan cell wall) bacteria showing facultative anaerobic properties, which probably makes them suitable candidate microorganisms for the biosynthesis of metal as well as oxide NP. Like most of the bacteria, they have a negative electrokinetic potential, which readily attracts the cations, and this step probably acts as a trigger of the procedure of biosynthesis (Jha et al., 2009a, Jha and Prasad, 2010). They have adapted to the food ecological niche by developing important technological characteristics like rapid acidification of the food matrix, specialized proteolytic and lipolytic capabilities to release nutrients, and so on (Pfeiler and Klaenhammer, 2007; Indumathi et al., 2015). Beyond their technological roles, LAB are also being successfully explored and employed as cell factories for the food-grade production of biomolecules like bacteriocins, vitamins, exopolysaccharides, and polyols, among others. LAB are also useful as probiotics for humans and animals, with a wide range of physiological effects. Some genera of LAB like *Lactobacilli*, *Bifidobacteria*, and others, per se, have been recognized to be endowed with the ability to bind, uptake, and biotransform metal ions from the medium (Mrvcic et al., 2012). This property provides important physiological, technological, and nutritional implications for both LAB and humans. The enrichment with selected heavy metals alters their physicochemical properties and can be used for potentiating their probiotic and health-promoting attributes (Bomba et al., 2002).

4.3.2 Yeast in Nanoparticle Synthesis

Among the eukaryotes, yeasts have been explored mostly for the fabrication of semiconductor NPs. Intracellular synthesis of CdSNCs from *Candida glabrata* was reported by Dameron et al. (1989). The reduction of Cd+ ion to CdSNCs was mediated via the degradation of the Cd–PC complex (PC refers to phytochelatins having a repeat sequence of (γ-Glu-Cys)n Gly where $n = 2$–6; Reese and Winge, 1988; Dameron et al., 1989). Similarly, when *Schizosaccharomyces pombe* was challenged by Cd+ ions, the result was the intracellular synthesis of biogenic CdSNCs in the size range of 1 to 1.5 nm exhibiting an absorbance maximum at 305 nm (Kowshik et al., 2002). The findings from this study also suggest that the biogenesis of CdSNCs was growth phase-dependent and that the yeast cells in the mid-log phase of growth resulted in maximum NCs synthesis. The mechanism involves an enzyme phytochelatin synthase which synthesizes phytochelatins (PC), helping to chelate the cytoplasmic Cd+ ions to phytochelatins, forming a low-molecular-weight Cd–PC complex. With the help of HMT1 (an adenosine triphosphate [ATP]–binding cassette-type vacuolar membrane protein), these complexes are transported across the vacuolar membrane. With the addition of sulfide, high molecular-weight $PCCdS_2$– forms, allowing them to sequester into the vacuole (Reese and Winge, 1988; Ortiz et al., 1995). The involvement of phytochelatins in bio-reduction was also confirmed through size-exclusion chromatography. The cadmium was attached to a protein fraction of molecular weight 25 to 67 kDa which corresponds to the theoretical molecular weight of CdSNPs of 35 kDa coated with phytochelatins (Krumov et al., 2007). The crystallites exhibited size-dependent tenability of the fluorescence spectrum, enabling easy recognition and ideal diode characteristics (Kowshik et al., 2002). Further extracellular synthesis of AgNPs by silver-tolerant yeast strain MKY3 has been reported. The NPs obtained were of size 2 to 5 nm and formed when the yeast cells were challenged with Ag+ ions in the exponential growth phase, separated through differential thawing of the samples (Kowshik et al., 2003). Yeast cells have also demonstrated the size- and shape-controlled synthesis of gold NPs (AuNPs) by optimizing different parameters for growth and cellular activities of the cell (Gericke and Pinches, 2006). The bio-reduction potential of non-conventional tropical marine yeast *Yarrowia lipolytica* was also demonstrated (Agnihotri et al., 2009). AuNPs were synthesized at a diverse range of pH values (2, 7, and 9), whereby acidic pH favoured NCs while basic pH resulted in NPs of 15 nm in size, associated with the cell wall. In a recent report, the biosynthesis of AgNPs and AuNPs were achieved using an extremophilic yeast strain isolated from acid mine drainage in Portugal. The authors investigated the growth potential of the isolate in the presence of metal ions, the ability of biomass, and the culture supernatant for NP synthesis. The findings suggest that isolate responded well in the presence of Ag+ ions rather than high levels of Au+ ions. The reaction was carried out at 22°C, resulting in AgNPs of less than 20 nm with characteristic absorbance at 420 nm, whereas synthesized AuNPs were of a size between 30 and 100 nm, with an absorbance of 550 nm. The proteins present in the supernatant were found to be responsible for the formation and stabilization of the AgNPs, while the involvement of the cell wall was essential for AuNPs synthesis (Mourato et al., 2011). Similarly, biogenic AuNPs were synthesized using yeast *Hansenula anomala* in the presence of gold salt as the precursor and amine-terminated polyamidoamine dendrimer (G4 and G5) as a stabilizer, with the average size of the synthesized particles 14 nm and 40 nm, respectively. The potential of AuNPs to function as an antimicrobial agent and as biological ink to be used in fingerprint analysis has also been investigated (Kumar et al., 2011). *Candida guilliermondii* were reported to synthesize gold and AgNPs extracellularly (Mishra et al., 2011). The NPs exhibited distinct surface plasmon peaks at 530 nm for AuNPs and 425 nm for AgNPs. The particles were spherical and well dispersed with face-centered cubic structures in the size range of 50 to 70 nm and 10 to 20 nm, respectively. The authors also investigated the antimicrobial potential of the NPs against five pathogenic bacteria; they found that biogenic NPs exhibited potent antimicrobial activity, especially against *Staphylococcus aureus*, while the synthetic NPs showed no inhibitory effect against any of the pathogenic strains (Singh et al., 2015).

4.3.3 Viruses in Nanoparticle Synthesis

Intriguingly, a few reports on the virus-mediated assembly of majorly semiconducting nanocrystals have also been published recently (Shenton et al., 1999; Lee et al., 2002; Mao et al., 2003, 2004; Banerjee et al., 2005; Slocik et al., 2005). In this context, the hollow protein tube of the tobacco mosaic virus (TMV) was used as a template for the synthesis of a range of nanotubes through different processes, namely iron oxides by oxidative hydrolysis, CdS and lead(II) sulfide (PbS) by co-crystallization, and silicon oxide (SiO_2) by sol-gel condensation. Glutamate and aspartate present on the external surface of the virus assisted the assembly of particles over the protein template (Shenton et al., 1999). Genetically engineered M13 bacteriophage-based liquid crystal system was used for the assembly of zinc sulfide (ZnS) NCs of 10 to 20 nm in diameter (Lee et al., 2002). The peptides were selected through the pIII phage display library for their ability to nucleate ZnS and later expressed in M13 bacteriophage to form the basis of the self-ordering system. When challenged with ZnS solution precursors, the A7 phage resulted in A7-ZnS biofilm formation of 15 mm thickness, which aligns to form the liquid crystal system. Additionally, the fabrication of viral film was a reversible process, and the biofilm was found to be stable at room temperature for about 7 months without losing their ability to infect a bacterial host with minimal titer loss. The authors also suggested that the potential of these genetically engineered viruses with specific recognition, as well as a liquid crystalline self-ordering system, can be harnessed to create newer pathways to organize electronic, optical, and magnetic materials and store high-density engineered DNA (Lee et al., 2002). Similarly, genetically controlled synthesis of quantum dot nanowires (including heterostructures and superlattices) was reported using self-assembled viral capsids of genetically engineered viruses as biological templates following peptide-templated growth mechanism (Singh et al., 2015). The peptides A7 (Cys-Asn-Asn-Pro-Met-His-Gln-Asn-Cys) and J140 (Ser-Leu-Thr-Pro-Leu-Thr-Thr-Ser-His-Leu-Arg-Ser) were selected by using a pIII phage display library for their ability to nucleate ZnS and CdSNCs, respectively, and were expressed as pVIII fusion proteins into the crystalline capsid of the M13 bacteriophages. In the presence of semiconductor precursor solutions, these organized template peptides (A7/J140-pVIIIM13) synthesized ZnS NCs on the viral capsid with a hexagonal wurtzite structure of a size between 3 and 5 nm or wurtzite CdS assembled as nanowires of 20 nm. Furthermore, heterogeneous nanowires (ZnS–CdS) were also obtained with a dual peptide virus engineered to express A7 and J140 on the same viral capsid (Mao et al., 2003). Apart from ZnS and CdS NCs, viral assembly of ferromagnetic alloys (CoPt and FePt) has also been reported using genetically engineered M13 bacteriophages. The specific peptides for each NP nucleation were selected through an evolutionary screening process and expressed on the highly ordered filamentous capsid of the M13 bacteriophage. The peptides were identified as FP12 (HNKHLPSTQPLA) for FePt and CP7 (CNAGDHANC) for CoPt systems. The obtained nanowires were crystalline and had a one-dimensional (1D) structure of 10 nm ± 5% in diameter (Mao et al., 2004).

Banerjee et al. (2005) demonstrated the phase-controlled synthesis of ZnS nanocrystals and nanowires of average size 4 nm at room temperature using the zinc finger-like peptides as a template consisting of VAL-CYS-ALA-THR-CYS-GLUGLN-ILE-ALA-ASP-SER-GLN-HIS -ARG-SER-HIS-ARG-GLN-MET-VAL, M1 peptide sequences, synthesized based on the peptide motif of the influenza virus matrix protein M1. The change in pH was essential to obtain the desired phase and size control and for the number of nucleation sites in M1 peptides to grow ZnS nanocrystals, thereby tuning the bandgap of the resulting nanotube. In a comprehensive mechanistic study, Slocik et al. (2005) demonstrated AuNP synthesis using wild-type and engineered viral template of Cowpea chlorotic mottle virus as unmodified SubE (yeast) and engineered HRE peptide epitopes (AHHAHHAAD) as (HRE)-SubE and wild type. The viral capsid enabled the bioreduction of tetrachloroaurate ($AuCl_4$), with the help of surface tyrosine residues, to AuNPs with an average size of 9.2 to 23.8 nm.

4.3.4 Fungi in Nanoparticle Synthesis

Due to their strong tolerance and biosorption ability for metals, fungi are taking the center stage in bio-NP research. Various authors have observed that fungi are easy to handle during the synthesis of NPs in terms of flow pressure as well as agitation in the bioreactor. The concept of formation of these NPs is mainly based on the enzyme capable of reduction. Furthermore, the secretions of these extracellular reductive and capping proteins are quite common in fungi, making downstream processing quite easy (Narayanan and Sakthivel, 2010). Various studies have been reported for the synthesis of NPs through fungi, in both intracellular and extracellular environments (Singh et al., 2015).

Intracellular synthesis of AuNPs using the biomass of *Verticillium sp.*, upon exposure to aqueous $HAuCl_4$ solution, was reported (Mukherjee et al., 2001a). The gold-loaded biomass exhibited a surface plasmon resonance (SPR) at about 550 nm, and transmission electron microscopy (TEM) analysis revealed the average size of AuNPs to be 20 ± 8 nm. The NPs were located on the cell wall with spherical, triangular, and hexagonal morphology and in the cytoplasmic membrane with a quasi-hexagonal morphology of the fungal mycelia. Similarly, the authors also demonstrated the intracellular synthesis of AgNPs using the same species and suggested that the gold and silver ions initially bind to the cell surface via electrostatic interaction. Upon successful adsorption, the enzymes present on the cell wall reduce the metal ion to zero valency metal nuclei and subsequently to NPs (Mukherjee et al., 2001b). Shankar et al. (2003) reported AuNPs synthesis from an endophytic fungus *Colletotrichum sp.* isolated from the leaves of the geranium plant (*Pelargonium graveolens*). The NPs were polydispersed, spherical, and within the size range of 8 to 40 nm. Fourier-transform infrared (FTIR) spectroscopy analysis revealed strong bands at 1658, 1543, and 1240 cm^{-1} which correspond to amide I, II, and III bands of proteins. The stability of AuNPs was speculated to be conferred by glutathiones binding either through free amine group or cysteine residues (Gole et al., 2001). Ahmad et al. (2005) reported the reaction-condition-related synthesis of AuNPs using the alkalotolerant fungus *Trichothecium sp.* biomass. It was observed that under stationary conditions the biomass resulted in extracellular AuNPs of 5 to 200 nm in size, while under shaking condition intracellular gold NPs of 10 to 25 nm in size were synthesized on the cell wall as well as on the cytoplasmic membrane of the fungus.

The NPs were of face-centered cubic (FCC) structure and demonstrated a varied morphology of polydispersed, spherical, rod-like, and triangular. The process was enzymatically catalysed by the enzymes released by the fungus into the medium. Extracellular biogenesis of gold NPs was demonstrated by Durán et al. (2005) using *Fusarium oxysporum*. The NPs were of 20 to 50 nm in size at 28°C, and the reduction of Ag+ ions was influenced by a nitrate-dependent reductase and a quinone electron shuttle process. When silver ions were incubated with the culture filtrate of *Fusarium semitectum*, AgNPs of 10 to 60 nm in size of spherical shape and polydispersity were synthesized (Basavaraja et al., 2008). *Trichoderma asperellum*, a non-pathogenic fungus and a known biocontrol agent, has also shown the potential to reduce silver ions to AgNPs in the size range of 13 to 18 nm with well-defined morphology and stability, with a pseudo-zero-order kinetic mechanism (Mukherjee et al., 2008). Further extracellular synthesis of spherical AgNPs of a size range of 60 to 80 nm using silver nitrate as a precursor by the fungal filtrate of *Phoma glomerata* has been reported (Birla et al., 2009). The authors also investigated the antibacterial activity of the AgNPs against *E. coli*, *P. aeruginosa*, and *S. aureus*. Shaligram et al. (2009) reported the extracellular synthesis of stable AgNPs using the fungus *Penicillium brevicompactum* WA 2315.

The green biosynthesis of AgNPs using cell-free filtrate of *Aspergillus flavus* NJP 08 has been attempted recently (Jain et al., 2011). The authors stressed that the two major proteins were responsible for the synthesis and stability of AgNPs formed in the extracellular environment. Through ammonium sulfate precipitation and dialysis, two major proteins of molecular weight 32 and 35 kDa were purified from the extracellular filtrate. SDS-PAGE results suggested that the enzyme reductase (probably a 32-kDa protein) was responsible for the synthesis of AgNPs from the aqueous silver ions. The role of fungal secreted proteins as capping ligands for imparting stability to AgNPs after their synthesis has also been advocated. In another study, biosynthesis of mono- and bimetallic AgNPs and gold AuNPs was observed with non-pathogenic filamentous fungus *Neurospora crassa* using different ratios of silver and gold ions by Castro-Longoria et al. (2011). It was observed that AgNPs were formed both extracellularly and intracellularly while AuNPs were only formed intracellularly. The metallic NPs synthesized were observed throughout the cell area and were mainly spherical/ellipsoidal, with an average size of 11 nm for AgNPs and 32 nm for AuNPs. Recently Das et al. (2012) reported biosynthesis of AuNPs using the cell-free protein extract of *Rhizopus oryzae* that served as both a reducing and stabilizing agent. The NPs exhibited a characteristic absorption band at 538 nm, as observed by UV-visual spectroscopy. The bio-NPs were well dispersed without agglomeration, of a size of 20 nm, and were stable for 3 months at room temperature. The findings of the study also suggest that the carboxyl and amino groups of phosphoproteins were responsible for the reduction and subsequent stabilization of AuNPs. In addition to gold and silver, the biogenic potential of fungus has also been explored for other metals and oxide NPs. Attempts have also been made to synthesize semiconductor NPs from fungal species. When exposed to the $CdSO_4$ solution, the biomass free filtrate of *Fusarium oxysporum* leads to the production of stable CdSNPs extracellularly, which was confirmed by the bright yellow colour of the solution after the reaction. UV-visible spectroscopic analysis showed a characteristic absorption peak for CdSNPs at 450 nm. TEM analysis reported the well-dispersed CdS NPs of size 5 to 20 nm. However, CdSNPs were not formed when the fungal biomass was exposed to aqueous $CdNO_3$ solution, even for an extended time, suggesting the possible induction of sulfate reductase enzyme by the presence of $CdSO_4$ into the solution. Furthermore, four protein bands were found in the aqueous extract of the fungal biomass through polyacrylamide gel electrophoresis, suggesting the possible involvement of the proteins in the biogenesis of CdSNCs and that the stability was due to the attachment of proteins on the surface of NPs, thereby preventing their coalescence (Ahmad et al., 2002).

Extracellular biosynthesis of spherical PtNPs in the size range 5 to 30 nm by *F. oxysporum* has been attempted recently, using hexachloroplatinic acid (H_2PtCl_6) as precursor at 25 to 27°C. The synthesis of NPs was due to the proteins present in the medium and PtNPs were characterized by the appearance of an absorption band at 270 to 280 nm. The findings suggest that the synthesis and stability of NPs were governed by the enzyme-mediated process (Syed and Ahmad, 2012). Similarly, different fungal species isolated from rhizospheric soil of plants thriving around a zinc mine in India were screened for their metal tolerance and ability to synthesize zinc oxide NPs (ZnONPs) under ambient conditions (Jain et al., 2013). Results indicated that the isolate *Aspergillus aeneus* NJP 12 exhibited the maximum tolerance to zinc ions and resulted in extracellular biogenesis of ZnO NPs.

Extracellular synthesis of magnetite NPs from *F. oxysporum* and *Verticillium sp.* using ferricyanide/ferrocyanide as a precursor at room temperature has been reported (Bharde et al., 2006). The magnetite obtained from *F. oxysporum* was of quasi-spherical morphology, of 20 to 50 nm in size, and attached with fungal proteins; magnetite obtained from *Verticillum* sp. demonstrated cubo-octahedral morphology and a size range of 10 to 40 nm. The hydrolysis of anionic precursor and capping of the magnetite NPs was mediated by the secretion of cationic proteins of molecular weights 55 kDa and 13 kDa. The biogenic potential of *F. oxysporum* was further extended towards synthesizing zirconia NPs, which have important technological applications. The fungus at room temperature was challenged by the aqueous solution of potassium hexafluorozirconate (K_2ZrF_6), resulting in the extracellular reduction of zirconium hexafluoride (ZrF_6) anions to crystalline zirconia NPs with the help of cationic protein of molecular weight of 24 to 28 kDa (Bansal et al., 2004). This fungus was also reported to synthesize strontianite ($SrCO_3$) nanocrystals of needle-like, quasi-linear morphology using aqueous Sr^{3+} ions and carbonate ions supplied by the fungus itself (Rautaray et al., 2004). This procedure was therefore referred to as "total biological synthesis". Similarly, this fungus also synthesized silica and crystalline titania NPs using SiF_6^{2-} and TiF_6^{2-} anionic complexes as precursors at room temperature (Bansal et al., 2005).

Recently, cerium oxide NPs were synthesized by the mycelia of thermophilic fungus *Humicola* sp. using aqueous solution of cerium (III) nitrate hydrate ($CeN_3O_9{\cdot}6H_2O$) as a precursor. The NPs exhibited strong absorption bands at 300 and 400 nm, were spherical, had a size range of 12 to 20 nm, and were naturally stabilized by the proteins secreted by the fungus (Khan and Ahmad, 2013).

4.4 NANOPARTICLE SYNTHESIS PROCESS

The exact mechanism for the synthesis of nanoparticles using biological agents has not yet been determined, as different biological agents react differently with metal ions, and there are also different biomolecules responsible for the synthesis of nanoparticles. Also, the mechanism for intra- and extracellular synthesis of nanoparticles is different in various biological agents. Due to their complex structure organization, microorganisms have evolved various methods to counter the heavy metal stress through metabolism-dependent and metabolism-independent processes. Metabolism-dependent uptake of metal ions is often the result of the active defence system of the microbe, which becomes triggered in the presence of the toxic metals. The formation of nanoparticles is mediated by viable cells and is always intracellular. In metabolism-independent processes, the uptake of metal ions is however mediated by the physicochemical interaction between the metal and the functional groups present on the microbial cell surface. This interaction is based on ion exchange, oxidation reduction, physical adsorption, and chemical sorption, which are not dependent on cell metabolism. The process is rapid and can be reversible (Kuyucak and Volesky, 1988). The microbial cell wall is mainly composed of polysaccharides, glycoproteins, and glycolipids that interact with the metals (Geesey and Jang, 1990). Apart from these, the cell wall of microbes also harbours various metal-binding components, such as carboxyl, sulfate, phosphate, and amino groups (McLean and Beveridge, 1990). Various metal-binding proteins and peptides are known that have a strong affinity towards metal binding and become induced by their presence nearby.

Among these, the most extensively studied are metallothioneins and metal γ-glutamyl peptides (phytochelatins). These are short peptides that are involved in heavy metal detoxification in algae, plants, and some fungi and yeasts (Mehra and Winge, 1991; Gadd, 1993). Essential aspects of bio-NP formation are the macromolecules DNA, protein, or peptide template that serves as the locus of control for morphology and the congregation procedure of NPs. It was suggested by Niemeyer (2001) that it is the electrostatic and topographic properties of biological macromolecules, and their derived supramolecular complexes, that influence the synthesis and assembly of organic and inorganic components. A wide range of biological entities such as DNA (Mirkin et al., 1996; Braun et al., 1998), protein cages (Wong et al., 1998), viroid capsules (Douglas and Young, 1998), bacterial rhapidosomes (Pazirandeh et al., 1992), biolipid tubules (Archibald and Mann, 1993), bacterial S-layers (Shenton et al., 1997), and multicellular superstructures (Davis et al., 1997) have already been exploited for the template-mediated synthesis of inorganic NPs and supramolecular structures. The use of protein cages as the mini-bioreactor for NPs synthesis serves as an ideal template for confining particle growth and assembly with homogenous distribution; it also acts as a stabilizer to avert particle aggregation. Due to the extensive use of proteins as specific biomolecular recognition elements in the synthesis of bio-NPs, these will serve as the "factory of the future" for the nanofabrication of spatially defined aggregates via the bottom-up assembly of NPs (Mann et al., 2000). The enzymatic route to biosynthesis has also been well established, where oxidoreductase, nicotinamide adenine dinucleotide (NADH)/nicotinamide adenine dinucleotide phosphate (NADPH)–dependent reductase, nitrate/nitrite reductase, sulfate, and sulfite reductase, hydrolase, cysteine desulfhydrase, and hydrogenases are the major classes of the enzyme involved in mediating the reduction of metal ions to NPs (Ahmad et al., 2002; Durán et al., 2005; Bharde et al., 2006; He et al., 2007; Kalishwaralal et al., 2008; Bai et al., 2009; Jha et al., 2009b; Nangia et al., 2009; Riddin et al., 2009).

4.5 CHARACTERIZATION OF NANOPARTICLES

The obtained NPs need to be characterized to validate their physicochemical characteristics using a range of diverse techniques, including UV-visual spectroscopy, dynamic light scattering (DLS), zeta potential measurement, X-ray diffractometry (XRD), FTIR spectroscopy; x-ray photoelectron spectroscopy (XPS), atomic force microscopy (AFM), and scanning and transmission electron microscopy (SEM, TEM). The advent of these advanced techniques has helped to resolve different parameters such as particle shape, size, surface area, size dispersion, crystallinity, composition, and scattering properties. The initial confirmation of the formed NPs is ascertained by visual observation for the colour change, and the extinction spectra of metallic NPs is recorded by UV-visual spectroscopy, allowing the concentration and aggregation level to be estimated. Second, through DLS techniques, the particle size distribution of the NPs can be determined. To identify the functional groups present on the NP's surface and to ascertain the chemical composition, the FTIR technique is used. Finally, the crystallinity of NPs is determined by X-ray diffraction and knowledge of the particle size, shape, morphology, height, volume, and the three-dimensional image is determined by TEM, SEM, and AFM, respectively.

4.6 NEED FOR GREEN SYNTHESIS

The synthesis of metallic NPs is an active area of academic and, more important, "application research" in nanotechnology. Biosynthesis of NPs is a kind of bottom-up approach where the main reaction occurring is reduction/oxidation. A variety of chemical and physical procedures could be used for the synthesis of metallic NPs. The need for the biosynthesis of NPs rose as the physical and chemical processes were costly. Chemical methods are fraught with many problems, including the use of toxic solvents, generation of hazardous by-products, and high energy consumption. Often, the chemical synthesis method leads to the presence of some of the toxic chemicals absorbed on the surface that may harm the medical applications. So there was an essential need to develop environmentally benign procedures for the synthesis of metallic NPs. A promising approach to achieve this objective was to exploit the array of biological resources in nature. This was not an issue when it comes to biosynthesized NPs via the green synthesis route.

Among the key advantages that the biological approach has over traditional chemical and physical NP synthesis methods is the biological capacity to catalyse reactions in aqueous media at standard temperature and pressure. Production in aqueous media under standard conditions leads to many cost advantages, in terms of both capital equipment and operating expenses, especially in the purchase and disposal of solvents and other consumable reagents. Biosynthesis can be implemented in nearly any setting and at any scale. Also, extensive investment in biotechnology know-how for optimized production of food, pharmaceuticals, and fuels informs NP biosynthesis techniques. One important drawback of NP biosynthesis methods is the requirement in some applications to purify the sample or to separate the NPs from the biological material used in their synthesis. Thus, green synthesis provides an advancement over chemical and physical methods as it is cost-effective and environmentally-friendly and is easily scaled up for large-scale synthesis, and in this method, there is no need to use high pressure, energy, temperature, and toxic chemicals.

4.7 FACTORS ASSOCIATED WITH MICROBIAL SYNTHESIS OF METAL NANOPARTICLES

Major drawbacks associated with the biosynthesis of NPs using bacteria are tedious purification steps and poor understanding of the mechanisms. The important challenges frequently encountered in the biosynthesis of NPs are to control the shape and size of the particles and to achieve the monodispersity in the solution phase. It seems that several important technical challenges must be overcome before this green bio-based method will be a successful and competitive alternative for the industrial synthesis of NPs. An important challenge is scaling up for production-level processing. Furthermore, little is known about the mechanistic aspects, and information in this regard is necessary for the economic and rational development of NP biosynthesis. The important aspects which might be considered in the process of producing well-characterized NPs are as follows.

4.7.1 Selection of the Best Microorganism

To choose the best candidates, researchers have focused on some important intrinsic properties of the bacteria including growth rate, enzyme activities, and biochemical pathways. Choosing a good candidate for NP production depends on the application we expect from the resulting NPs. For instance, one may need to synthesis NPs with smaller sizes or specific shapes, or it might be important to synthesize NPs within less time (Iravani, 2011; Korbekandi et al., 2009).

4.7.2 Selection of the Biocatalyst State

It seems that the bacterial enzymes (the biocatalysts) are the major agents in NP synthesis. The biocatalysts can be used as either of whole cells, crude enzymes, and purified enzymes. It seems that using culture supernatant or cell extract of the cell could increase the rate of reaction. However, these NPs did not show long-term stability. Moreover, the release of NPs from the cells was an important aspect that might be considered in the case of intracellularly produced NPs. Most of the reactions responsible for NP synthesis seem to be bioreductions. In bioreduction processes, we need the coenzymes (e.g., NADH, NADPH, FAD (Flavin Adenine Dinucleotide), etc.) to be supplied in stoichiometric amounts. As they are expensive, the use of whole cells is preferred because the coenzymes will be recycled during the pathways in live whole cells (Korbekandi et al., 2009).

4.7.3 Optimal Conditions for Cell Growth and Enzyme Activity

We need to produce greater amounts of the enzymes which can be accomplished by the synthesis of more biomass. Thus, the optimization of growth conditions is very important. The nutrients, inoculum size, pH, light, temperature, buffer strength, and mixing speed should be optimized. The induction of the responsible enzymes seems to be crucial as well. The presence of the substrates or related compounds in sub-toxic levels from the beginning of the growth would increase the activity. Harvesting time is important in the case of using whole cells and crude enzymes. Therefore, it might be necessary to monitor the enzyme activity during the time course of growth (Korbekandi et al., 2009).

4.7.4 Optimal Reaction Conditions

It is better to harvest the cells (the biocatalysts) to remove unwanted residual nutrients and metabolites to avoid adverse reactions and provide a cleaner medium for better and easier analysis. To use bacteria for synthesis of NP in the industrial scale, the yield and the production rate are important issues to be considered. Therefore, we need to optimize the bio-reduction conditions in the reaction mixture. The substrate concentration (to be in sub-toxic level for the biocatalyst), the biocatalyst concentration, the electron donor (and its concentration), exposure time, pH, temperature, buffer strength, mixing speed, and light need to be optimized. The researchers have used some complementary factors such as visible light or microwave irradiation and boiling which could affect the morphology, size, and rate of reaction. It seems that by optimization of these critical parameters, highly stable NPs with desired sizes and morphologies can be achieved. Also, the purification, isolation, and stabilization of the produced NPs are very important, and challenges in this regard must be overcome. Researchers have focused their attention on finding optimal reaction conditions and cellular mechanisms involved in the bioreduction of metal ions and synthesis of NPs (Iravani, 2011; Iravani et al., 2014; Iravani and Zolfaghari, 2013; Korbekandi et al., 2013; Korbekandi and Iravani, 2013; Korbekandi et al., 2009; Korbekandi et al., 2012; Iravani et al., 2014a).

4.7.5 Extraction and Purification Processes

The extraction and purification of the produced metal NPs from bacteria (intercellular or extracellular synthesis) for further applications are not well investigated, but studies are moving towards solving these problems and finding the best ways. To release the intracellularly produced NPs, additional processing steps such as ultrasound treatment or reaction with suitable detergents are required. This can be exploited in the recovery of precious metals from mine wastes and metal leachates. Biomatrix metal NPs could also be used as catalysts in various chemical reactions. This will help to retain the NPs for continuous usage in bioreactors. Physicochemical methods including freeze-thawing, heating processes, and osmotic shock can be used to extract the produced NPs from the cells. But it seems that these methods may interfere with the structure of NPs, and aggregation, precipitation, and sedimentation could happen. These may change the shape and size of NPs and interfere with the suitable properties of them. Moreover, enzymatic lysis of the microbial cells containing intracellular NPs can be used, but this method is expensive, and it cannot be used in up-scalable and industrial production of NPs. It seems that surfactants and organic solvents can be used for both extraction and stabilization of NPs, but these chemical materials are toxic, expensive, and hazardous. It should be noted that in the case of extracellular production of NPs, centrifuge could be used for extraction and purification of NPs, but aggregation might happen.

4.7.6 Stabilization of the Produced Nanoparticles

Researchers have illustrated that the NPs produced by these eco-friendly bio-based approaches, showed interesting stability without any aggregations even for many weeks at room temperature (Wen et al., 2009; Shankar et al., 2004). The stability of these NPs might be due to the proteins and enzymes secreted by the microorganisms. Thus, it seems that these green approaches can be used for the synthesis of highly stable NPs.

4.7.7 Scaling Up the Laboratory Process to the Industrial Scale

Optimization of the reaction conditions may lead to the enhanced biosynthesis of NPs. Biological protocols could be used for the synthesis of highly stable and well-characterized NPs when critical aspects, such as types of organisms, inheritable and genetic properties of organisms, optimal conditions for cell growth and enzyme activity, optimal reaction conditions, and selection of the biocatalyst state, have been considered. The size and morphology of the NPs can be controlled by altering the aforementioned reaction conditions (optimal reaction conditions section). The industrial-scale synthesis of metal NPs using biomass needs some processes, including seed culture, inoculation of the seed into the biomass, harvesting the cells, synthesis of NPs by adding metal ions to the cells, separation of cells by filtration, homogenization of the cells to isolate the produced NPs, stabilization of the NPs, product formulation, and quality control (Iravani, 2011; Iravani et al., 2014; Iravani and Zolfaghari, 2013; Korbekandi et al., 2013; Korbekandi and Iravani, 2013; Korbekandi et al., 2009; Korbekandi et al., 2012; Iravani et al., 2014a).

4.8 APPLICATION OF NANOPARTICLES IN FOOD AND PHARMACEUTICAL INDUSTRY

There are widespread applications of NPs such as pharmaceuticals, cosmetics, food and beverages, agriculture, surface coating, and polymers, among others (see Figure 4.1). A few of them are discussed here.

4.8.1 Nanoparticles as a Potent Antimicrobial Agent

The AgNPs synthesized using an endophytic fungus, *Pestalotia* sp., isolated from leaves of *Syzygium cumini* has antibacterial activity against human pathogens, that is *S. aureus* and *S. typhi* (Raheman et al., 2011). AgNPs showed powerful bactericidal

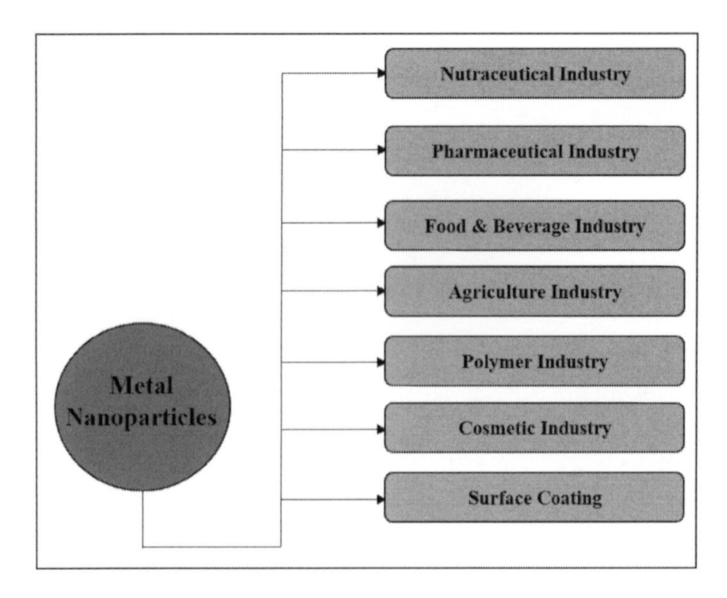

FIGURE 4.1 Application of metal nanoparticles.

potential against both gram-positive and gram-negative bacteria. Numbers of AgNPs are used against pathogenic bacteria. The bactericidal prospective of AgNPs against the multidrug-resistant bacteria are also investigated (Rai et al., 2012; Morones et al., 2005). NPs in electrochemical sensors and biosensors A set of forms of NPs such as oxide, metal, and semiconductor NPs have been utilized for constructing electrochemical sensors and biosensors, and these NPs play diverse roles in different sensing systems. The significant functions provided by NPs comprise the immobilization of biomolecules, the catalysis of electrochemical reactions, the improvement of electron transfer among electrode surfaces and proteins, labelling of biomolecules, and still acting as the reactant. The exclusive chemical and physical properties of NPs make them enormously appropriate for designing new and enhanced sensing devices, particularly electrochemical sensors and biosensors. The gold AuNPs are most frequently used for the immobilization of proteins (Liu et al., 2003). Xiao et al. (1999) initially attached AuNPs to gold electrodes modified with cysteamine monolayer and then effectively immobilized horseradish peroxidase on these NPs. An additional type of biomolecules, DNA, can also be immobilized with NPs and used for the creation of electrochemical DNA sensors. In command to immobilize DNA onto the surfaces of NPs, the DNA strands are frequently modified with meticulous functional groups that can work together powerfully with convinced NPs (Cai et al., 2001).

4.8.2 Nanoparticles for Detection and Destruction of Pesticides

Pesticides are hazardous to both human beings and the environment, contaminating drinking and surface water. The unique properties of NPs allow their use in the detection and destruction of pesticides (Rai et al., 2015). The large surface area–to–volume ratio property of NPs plays a crucial role in the catalytic reactions used to degrade pesticides (Aragay et al., 2012). The optical properties of NPs are related to their size and surface-induced changes in electronic structure, which helps in the detection of pesticides. For the destruction of pesticides, a photocatalytic oxidation method employing titanium NPs is used (Aragay et al., 2012).

4.8.3 Nanoparticles in Drug Delivery

Metal NPs with magnetic properties work as an effective molecular carrier for gene separation and show the promising application in drug delivery (Bava et al., 2013). For drug delivery, magnetic NPs are injected into the drug molecule which is to be delivered; these particles are then guided towards the chosen site under a localized magnetic field. These magnetic carriers can carry large doses of drugs (Lu et al., 2007; Perez-Martinez et al., 2012). Similarly, silica-coated NPs are also used in drug delivery due to their high stability, surface properties, and compatibility. Silica NPs are also used in biological applications such as artificial implants (Dikpati et al., 2012; Perez-Martinez et al., 2012; Rai et al., 2015).

4.8.4 Nanoparticles in Medicine and Healthcare

NPs have been utilized newly to develop the present imaging techniques for in vivo diagnosis of biomedical disorders. Presently, iron oxide NPs are being used in patients for both diagnosis and therapy, leading to more effective medication with less unfavourable effects (Gao et al., 2008). An exclusive, susceptible, and greatly explicit immunoassay system based on the aggregation of gold AuNPs that are coated with protein antigens, in the attendance of their corresponding antibodies, was also developed (Thanh and Rosenzweig, 2002). NPs, as drug delivery systems, are capable to uplift the several crucial properties of free drugs, such as solubility, in vivo stability, pharmacokinetics, biodistribution, and enhancing their efficiency (Allen and Cullis, 2004). In this facet, NPs could be used as potential drug delivery systems, owing to their advantageous characteristics. As an illustration of cellular delivery, mixed monolayer protected gold clusters were oppressed for in vitro delivery of a hydrophobic fluorophore (Hong et al., 2006). Pandey and Khuller (2007) designed NP for the growth of oral drug delivery system and recommended that nano-encapsulation may be useful for developing an appropriate oral dosage form for streptomycin and other antibiotics that are, if not, injectable (Pandey and Khuller, 2007). Elechiguerra et al. (2005) demonstrated the interaction of metal NPs with viruses and explained that AgNPs experience a size-dependent interaction with HIV-1; the NPs of 1 to 10 nm are close to the virus. The usual spatial understanding of the attached NPs, the centre-to-centre space among NPs, and the bare sulfur-bearing residues of the glycoprotein knobs suggested that, through favoured binding, the silver AgNPs prohibited the HIV-1 virus from binding to host cells. Currently, the majority of imaging studies using AuNPs are carried out in cell culture (Elechiguerra et al., 2005). The functional cellular imaging about single molecules has been reported by Peleg et al. (1999), the captivating benefit of the enhanced second harmonic signal by antibody-conjugated gold nanospheres. The exploitation of NPs in cosmetics and medicine coating is widely increased day by day. The metal oxides in NPs, such as zinc oxide and titanium dioxide, now emerge on the component records of household products, as general and assorted as cosmetics, sunscreens, toothpaste, and medicine (Yu and Li, 2011).

4.8.5 Nanoparticles in Cancer Treatment

AuNPs have shown potential in the treatment of cancer (Bhattacharya and Mukherjee, 2008; Chauhan et al., 2011). Vascular endothelial growth factor (VEGF) acts as a potent angiogenic factor and blood vessel permeabilizing agent after ligand binding to VEGF receptors (VEGFRs) on endothelial cells. Blocking the interaction of the VEGF with its receptors could be a possible way to inhibit angiogenesis. Quantum dots are luminescent crystals that allow specific drugs such as proteins, oligonucleotides, and siRNA (small interfering RNA) to penetrate targeted cancer cells in the central nervous system; they are therefore utilized for imaging in biological crystals. However, the toxicity issues of quantum dots is a major obstacle in its medical application to humans (Dikpati et al., 2012; Rai et al., 2015).

4.8.6 Nanoparticles in Agriculture

Nanotech delivery systems for pests, nutrients, and plant hormones: In the proficient use of agricultural natural assets such as water, nutrients, and chemicals during precision farming, nanosensors and nano-based smart delivery systems are user-friendly. It makes the use of nanomaterials and global positioning systems with satellite imaging of fields, farm supervisors might distantly detect crop pests or facts of stress such as drought. Nanosensors disseminated in the field can sense the existence of plant viruses and the level of soil nutrients. To put aside fertilizer consumption and to minimize environmental pollution, nano-encapsulation slow-release fertilizers have also become a style (DeRosa et al., 2010). To check the quality of agricultural manufacture, nano–bar codes and nano processing could be used. Li et al. (2005) used the idea of grocery barcodes for economical, proficient, rapid, and effortless decoding and recognition of diseases. They created microscopic probes or nano–bar codes that may perhaps tag multiple pathogens in a farm, which may simply be detected using any fluorescent-based tools (Li et al., 2005). Through nanotechnology, scientists are capable to study plant's regulation of hormones such as auxin, which is accountable for root growth and seedling organization. Nanosensors have been developed that react with auxin. This is a step forward in auxin research, as it helps scientists know how plant roots acclimatize to their environment, particularly to marginal soils (McLamore et al., 2010).

4.8.7 Nanotechnology for Crop Biotechnology

Nanocapsules can facilitate successful incursion of herbicides through cuticles and tissues, allowing slow and regular discharge of the active substances. This can act as "magic bullets", containing herbicides, chemicals, or genes that target exacting plant parts to liberate their substance (Perez-de-Luque and Rubiales 2009). Torney et al. (2007) have exploited a 3-nm mesoporous silica NP in delivering DNA and chemicals into isolated plant cells. Mesoporous silica NP is chemically coated and act as containers for the genes delivered into the plants and triggers the plant to take the particles through the cell walls, where the genes are put in and activated in a clear-cut and controlled way, without any toxic side effects. This technique, first, has been applied to establish DNA fruitfully to tobacco and corn plants (Torney et al., 2007).

4.9 CONCLUSION

In conclusion, the present chapter provides a brief overview of the use of microbes in the biosynthesis of metal NPs. Bio-based approaches are still in the development stages, but this approach can serve as a possible alternative to the more popular physical and chemical methods that are currently prevalent. Furthermore, biologically synthesized NPs in comparison with chemically synthesized ones are more polydisperse. However, in the case of biosynthesized NPs, most experienced problems are stability and aggregation of NPs and control of crystal growth, shape, size, and size distribution. Mechanistic aspects related to bio-nanotechnology have not been studied yet. Thus, more interventions are needed to know the exact mechanisms of reaction and to identify the bacterial enzymes and proteins which involve NP biosynthesis. It will be interesting to produce NPs at a large scale with the help of bacteria as it does not involve any hazardous, toxic, and expensive chemical materials for synthesis and stabilization processes.

REFERENCES

Agarwal, A., Pathera, A. K., Kaushik, R., Kumar, N., Arora, S., and Chawla, P. (2020). Succinylation of milk proteins: Influence on micronutrient binding and functional indices. *Trends in Food Science and Technology*, 97, 254–264.

Agnihotri, M., Joshi, S., Kumar, A. R., Zinjarde, S., and Kulkarni, S. (2009). Biosynthesis of gold nanoparticles by the tropical marine yeast *Yarrowia lipolytica* NCIM 3589. *Materials Letters*, 63(15), 1231–1234.

Ahmad, A., Mukherjee, P., Mandal, D., Senapati, S., Khan, M. I., Kumar, R., and Sastry, M. (2002). Enzyme mediated extracellular synthesis of CdS nanoparticles by the fungus, *Fusarium oxysporum*. *Journal of the American Chemical Society*, 124(41), 12108–12109.

Ahmad, A., Senapati, S., Khan, M. I., Kumar, R. and Sastry, M. (2003b). Extracellular biosynthesis of monodisperse gold nanoparticles by a novel extremophilic actinomycete, *Thermomonospora* sp. *Langmuir*, 19, 3550–3553.

Ahmad, A., Senapati, S., Khan, M. I., Kumar, R., and Sastry, M. (2005). Extra-/intracellular biosynthesis of gold nanoparticles by an alkalotolerant fungus, *Trichothecium* sp. *Journal of Biomedical Nanotechnology*, 1(1), 47–53.

Ahmad, A., Senapati, S., Khan, M. I., Ramani, R., Srinivas, V., and Sastry, M. (2003a). Intracellular synthesis of gold nanoparticles by a novel alkalotolerant actinomycete, *Rhodococcus* species. *Nanotechnology*, 14, 824–828.

Allen, T. M., and Cullis, P. R. (2004). Drug delivery systems: Entering the mainstream. *Science*, 303, 1818–1822.

Aragay, G., Pino, F., and Merkoci, A. (2012). Nanomaterials for sensing and destroying pesticides. *Chemical Reviews*, 112, 5317–5338.

Archibald, D. D., and Mann, S. (1993). Template mineralization of self-assembled anisotropic lipid microstructures. *Nature*, 364(6436), 430–433.

Baesman, S., Stolz, J., Kulp, T., and Oremland, R. (2009). Enrichment and isolation of *Bacillus beveridgei* sp. nov., a facultative anaerobic alkaliphile from Monolake, California, that respires oxyanions of tellurium, selenium and arsenic. *Extremophiles*, 13, 695–705.

Bai, H.J., Zhang, Z.M., and Gong, J. (2006). Biological synthesis of semiconductor zinc sulfide nanoparticles by immobilized *Rhodobacter sphaeroides*. *Biotechnology Letters*, 28(14), 1135–1139.

Bai, H. J., Zhang, Z. M., Guo, Y., and Yang, G. E. (2009). Biosynthesis of cadmium sulfide nanoparticles by photosynthetic bacteria *Rhodopseudomonas palustris*. *Colloids and Surfaces B: Biointerfaces*, 70(1), 142–146.

Balaji, D.S., Basavaraja, S., Deshpande, R., Mahesh, D.B., Prabhakar, B.K., and Venkataraman, A. (2009). Extracellular biosynthesis of functionalized silver nanoparticles by strains of *Cladosporium cladosporioides* fungus. *Colloids and Surfaces B: Biointerfaces*, 68(1), 88–92.

Banerjee, I. A., Yu, L., and Matsui, H. (2005). Room-temperature wurtzite ZnS nanocrystal growth on Zn finger-like peptide nanotubes by controlling their unfolding peptide structures. *Journal of the American Chemical Society*, 127(46), 16002–16003.

Bansal, V., Bharde, A., Ramanathan, R., and Bhargava, S.K. (2012). Inorganic materials using 'unusual' microorganisms. *Advances in Colloid and Interface Science*, 1, 179–182.

Bansal, V., Poddar, P., Ahmad, A., and Sastry, M.J. (2006). Room-temperature biosynthesis of ferroelectric barium titanate nanoparticles. *Journal of the American Chemical Society*, 128(36), 11958–11963.

Bansal, V., Rautaray, D., Ahmad, A., and Sastry, M. (2004). Biosynthesis of zirconia nanoparticles using the fungus *Fusarium oxysporum*. *Journal of Materials Chemistry*, 14(22), 3303–3305.

Bansal, V., Rautaray, D., Bharde, A., Ahire, K., Sanyal, A., Ahmad, A., and Sastry, M. (2005). Fungus-mediated biosynthesis of silica and titania particles. *Journal of Materials Chemistry*, 15(26), 2583–2589.

Basavaraja, S., Balaji, S., Lagashetty, A., Rajasab, A., and Venkataraman, A. (2008). Extracellular biosynthesis of silver nanoparticles using the fungus *Fusarium semitectum*. *Materials Research Bulletin*, 43(5), 1164–1170.

Bava, A., Cappellini, Ffv., Pedretti, E., Rossi, F., Caruso, E., Vismara, E., Chiriva-Internati, M., Bernardini, G., and Gornati, R. (2013). Heparin and carboxymethyl-chitosan metal nanoparticles: An evaluation on their cytotoxicity. *BioMed Research International*, 2013: Article ID 314091, doi: http://dx.doi.org/10.1155/2013/314091.

Bhandari, L., Chawla, P., Dhull, S.B., Sadh, P.K., and Kaushik, R. (2018). Health benefits of yoghurt-cheese. *Plant Biotechnology: Recent Advancements and Developments*, Springer, Chapter 11, pp 259–268.

Bharde, A., Rautaray, D., Bansal, V., Ahmad, A., Sarkar, I., Yusuf, S. M., Sanyal, M., and Sastry, M. (2006). Extracellular biosynthesis of magnetite using fungi. *Small*, 2(1), 135–141.

Bharde, A., Wani, A., Shouche, Y., Pattayil, A., Bhagavatula, L., and Sastry, M. (2005). Bacterial aerobic synthesis of nanocrystalline magnetite. *Journal of the American Chemical Society*, 127, 9326–9327.

Bhattacharya, R., and Mukherjee, P. (2008). Biological properties of naked nanoparticles. *Advanced Drug Delivery Reviews*, 60, 1289–1306.

Birla, S., Tiwari, V., Gade, A., Ingle, A., Yadav, A., and Rai, M. (2009). Fabrication of silver nanoparticles by *Phoma glomerata* and its combined effect against *Escherichia coli*, *Pseudomonas aeruginosa* and *Staphylococcus aureus*. *Journal of & Letters in Applied Microbiology*, 48(2), 173–179.

Blakemore, R.P., Maratea, D., and Wolfe, R.S. (1979). Isolation and pure culture of a freshwater magnetic spirillum in chemically defined medium. *Journal of Bacteriology*, 140(2), 720–729.

Bomba, A., Nemcova, R., Mudronova, D., and Guba, P. (2002). The possibilities of potentiating the efficacy of probiotics. *Trends in Food Science and Technology*, 13(4), 121–126
Braun, E., Eichen, Y., Sivan, U., and Ben-Yoseph, G. (1998). DNA-templated assembly and electrode attachment of a conducting silver wire. *Nature*, 391(6669), 775–778.
Cai, H., Xu, C., He, P., and Fang, Y. (2001). Colloid Au-enhanced DNA immobilization for the electrochemical detection of sequence-specific DNA. *Journal of Electroanalytical Chemistry*, 510, 78–85.
Castro-Longoria, E., Vilchis-Nestor, A. R., and Avalos-Borja, M. (2011). Biosynthesis of silver, gold and bimetallic nanoparticles using the filamentous fungus *Neurospora crassa*. *Colloids and Surfaces B: Biointerfaces*, 83(1), 42–48.
Chauhan, A., Zubair, S., Tufail, S., Sherwani, A., Sajid, M., Raman, S. C., Azam, A., and Owais, M. (2011). Fungus-mediated biological synthesis of gold nanoparticles: Potential in detection of liver cancer. *International Journal of Nanomedicine*, 6, 2305–2319.
Chawla, P., Kaushik, R., Shiva Swaraj, V.J., and Kumar N. (2018). Organophosphorus pesticides residues in food and their colorimetric detection. *Environmental Nanotechnology, Monitoring and Management*, 10, 292–307.
Chawla, P., Kumar, N., Kaushik, R., and Dhull, S. B. (2019). Synthesis, characterization and cellular mineral absorption of gum arabic stabilized nanoemulsion of *Rhododendron arboreum* flower extract. *Journal of Food Science and Technology*, 56(12), 5194–5203.
Chawla, P., Kumar, N., Bains, A., Dhull, S. B., Kumar, M., Kaushik, R., and Punia, S. (2020). Gum Arabic capped copper nanoparticles: Characterization and their utilization. *International Journal of Biological Macromolecules*, 146, 232–242.
Cunningham, D.P., and Lundie, L.L. (1993). Precipitation of cadmium by *Clostridium thermoaceticum*. *Applied and Environmental Microbiology*, 9, 7–14.
Dameron, C.T., Reese, R.N., Mehra, R.K., Kortan, A.R., Carroll, P.J., Steigerwald, M.L., Brus, L.E., and Winge, D.R. (1989). Biosynthesis of cadmium sulphide quantum semiconductor crystallites. *Nature*, 338, 596–597.
Das, S. K., Dickinson, C., Lafir, F., Brougham, D. F., and Marsili, E. (2012). Synthesis, characterization and catalytic activity of gold nanoparticles biosynthesized with *Rhizopus oryzae* protein extract. *Green Chemistry*, 14(5), 1322–1334.
Davis, S. A., Burkett, S. L., Mendelson, N. H., Mann, S. (1997). Bacterial templating of ordered macrostructures in silica and silica-surfactant mesophases. *Nature*, 385, 420–423.
DeRosa, M. C., Monreal, C., Schnitzer, M., Walsh, R., and Sultan, Y. (2010). Nanotechnology in fertilizers. *Nature Nanotechnology*, 5, 91.
Dhull, S.B., Anju, M., Punia, S., Kaushik, R., and Chawla, P. (2019). Application of gum Arabic in nanoemulsion for safe conveyance of bioactive components. In: Prasad R., Kumar V., Kumar M., and Choudhary D. (eds) *Nanobiotechnology in Bioformulations. Nanotechnology in the Life Science*. Springer, Cham. Chapter 3, pp. 85–98. DOI.org/10.1007/978-3-030-17061-5-3.
Dikpati, A., Madgulkar, A. R., Kshirsagar, S. J., Bhalekar, M. R., and Chahal, A. S. (2012). Targeted Drug delivery to CNS using nanoparticles. *Journal of Advanced Pharmaceutical Science and Technology*, 2(1), 179–191.
Douglas, T., and Young, M. (1998). Host-guest encapsulation of materials by assembled virus protein cages. *Nature*, 393(6681), 152–155.
Durán, N., Marcato, P. D., Alves, O. L., De Souza, G. I., and Esposito, E. (2005). Mechanistic aspects of biosynthesis of silver nanoparticles by several *Fusarium oxysporum* strains. *Journal of Nanobiotechnology*, 3(8), 1–7.
Elechiguerra, J.L., Burt, J.L., Morones, J.R., Camacho-Bragado, A., Gao, X., et al. (2005). Interaction of silver nanoparticles with HIV-1. *Journal of Nanobiotechnology*, 3, 6.
Eszenyi, P., Sztrik, A., Babka, B., and Prokisch, J. (2011). Elemental, nano-sized (100–500 nm) selenium production by probiotic lactic acid bacteria. *International Journal of Bioscience, Biochemistry and Bioinformatics*, 1(2), 148.

Gadd, G.M. (1993). Interactions of fungi with toxic metals. *New Phytologist*, 124(1), 25–60.

Gao, L., Zhang, D., and Chen, M. (2008). Drug nanocrystals for the formulation of poorly soluble drugs and its application as a potential drug delivery system. *Journal of Nanoparticle Research*, 10, 845–862.

Geesey, G. and Jang, L. (1990). Extracellular polymers for metal binding. In *Microbial Mineral Recovery* (eds H.L. Ehrlich and C.L. Brierley), pp. 223–247. New York: McGraw-Hill.

Gericke, M. and Pinches, A. (2006). Biological synthesis of metal nanoparticles. *Hydrometallurgy*, 83(1), 132–140.

Gole, A., Dash, C., Soman, C., Sainkar, S., Rao, M., and Sastry, M. (2001). On the preparation, characterization, and enzymatic activity of fungal protease-gold colloid bioconjugates. *Bioconjugate Chemistry*, 12(5), 684–690.

He, S., Guo, Z., Zhang, Y., Zhang, S., Wang, J., and Gu, N. (2007). Biosynthesis of gold nanoparticles using the bacteria *Rhodopseudomonas capsulata*. *Materials Letters*, 61(18), 3984–3987.

He, S., Zhang, Y., Guo, Z., and Gu, N. (2008). Biological synthesis of gold nanowires using extract of *Rhodopseudomonas capsulata*. *Biotechnology Progress*, 24(2), 476–480.

Hennebel, T., Van Nevel, S., Verschuere, S., De Corte, S., De Gusseme, B., Cuvelier, C., Fitts, J., van der Lelie, D., Boon, N., and Verstraete, W. (2011). Palladium nanoparticles produced by fermentatively cultivated bacteria as catalyst for diatrizoate removal with biogenic hydrogen. *Applied Microbiology and Biotechnology*, 91(5), 1435–1445.

Hong, R., Han, G., Fernández, J. M., Kim, B. J., and Forbes, N. S., et al. (2006). Glutathione mediated delivery and release using monolayer protected nanoparticle carriers. *Journal of the American Chemical Society*, 128, 1078–1079.

Hosea, M., Greene, B., McPerson, R., Henzel, M., Alexander, M., and Darnall, W. (1986). Accumulation of elemental gold on the alga *Chlorella vulgaris*. *Inorganica Chimica Acta*, 123(3), 161–165.

Husseiney, M.I., Abd El-Aziz, M., Badr, Y., and Mahmoud, M.A. (2007). Biosynthesis of gold nanoparticles using *Pseudomonas aeruginosa*. *Spectrochimica Acta Part A*, 67, 1003–1006.

Indumathi, K.P., Kaushik, R., Arora, S., and Wadhwa, B.K. (2015). Evaluation of iron fortified Gouda cheese for sensory and physicochemical attributes. *Journal of Food Science and Technology*, 52(1), 493–499.

Iravani, S. (2011). Green synthesis of metal nanoparticles using plants. *Green Chemistry*, 13(10), 2638–2650.

Iravani, S., Korbekandi, H., Mirmohammadi, S.V., and Zolfaghari, B. (2014a). Synthesis of silver nanoparticles: Chemical, physical, and biological methods. *Research in Pharmaceutical Sciences*, 9, 385–406.

Iravani, S., Korbekandi, H., Mirmohammadi, S.V., and Mekanik, H. (2014). Plants in nanoparticle synthesis. *Reviews in Advanced Sciences and Engineering*, 3(3), 261–274.

Iravani, S. and Zolfaghari, B. (2013). Green synthesis of silver nanoparticles using *Pinus eldarica* bark extract. *BioMed Research International*, 5.

Jain, N., Bhargava, A., Majumdar, S., Tarafdar, J., and Panwar, J. (2011). Extracellular biosynthesis and characterization of silver nanoparticles using *Aspergillus flavus* NJP08: A mechanism perspective. *Nanoscale*, 3(2), 635–641.

Jain, N., Bhargava, A., Tarafdar, J. C., Singh, S. K., and Panwar, J. (2013). A biomimetic approach towards synthesis of zinc oxide nanoparticles. *Applied Microbiology and Biotechnology*, 97(2), 859–869.

Jha, A. K. and Prasad, K. (2010). Green synthesis of silver nanoparticles using Cycas leaf. *ternational Journal of Green Nanotechnology: Physics and Chemistry*, 1(2), 110–117.

Jha, A. K., Prasad, K., and Kulkarni, A. R. (2009a). Synthesis of TiO_2 nanoparticles using microorg*anisms*. *Colloids and Surfaces B: Biointerfaces*, 71(2), 226–229.

Jha, A. K., Prasad, K., and Prasad, K. (2009b). Biosynthesis of Sb_2O_3 nanoparticles: A low-cost green approach. *Biotechnology Journal*, 4(11), 1582–1585.

Kalishwaralal, K., Deepak, V., Ram Kumar Pandian, S., and Gurunathan, S. (2009). Biological synthesis of gold nanocubes from *Bacillus licheniformis*. *Bioresource Technology*, 100(21), 5356–5358.

Kalishwaralal, K., Deepak, V., Ramkumarpandian, S., Nellaiah, H., and Sangiliyandi, G. (2008). Extracellular biosynthesis of silver nanoparticles by the culture supernatant of *Bacillus licheniformis*. *Materials Letters*, 62(29), 4411–4413.

Kang, S. H., Bozhilov, K. N., Myung, N. V., Mulchandani, A., and Chen, W. (2008). Microbial synthesis of CdS nanocrystals in genetically engineered *E. coli*. *Angewandte Chemie International Edition Engl*, 47(28), 5186–5189.

Kashefi, K., Tor, J.M., Nevin, K.P., and Lovley, D.R. (2001). Reductive precipitation of gold by dissimilatory Fe(III)-reducing bacteria and archaea. *Applied and Environmental Microbiology*, 67(7), 3275–3279.

Kaushik, R., Sachdeva, B. and Arora, S. (2017). Effect of calcium and vitamin D_2 fortification on quality characteristics of dahi. *International Journal of Dairy Technology*, 70(2), 269–276.

Khan, S.A. and Ahmad, A. (2013). Fungus mediated synthesis of biomedically important cerium oxide nanoparticles. *Materials Research Bulletin*, 48, 4134–4138.

Klaus, T., Joerger, R., Olsson, E., and Granqvist, C.G. (1999). Silver-based crystalline nanoparticles, microbially fabricated. *Proceedings of the National Academy of Sciences of the United States of America*, 96, 13611–13614.

Konishi, Y., Ohno, K., Saitoh, N., Nomura, T., and Nagamine, S. (2004). Microbial synthesis of gold nanoparticles by metal reducing bacterium. *Transactions of the Materials Research Society of Japan*, 29, 2341–2343.

Konishi, Y., Tsukiyama, T., Tachimi, T., Saitoh, N., Nomura, T., and Nagamine, S. (2007). Bioreductive deposition of platinum nanoparticles on the bacterium *Shewanella algae*. *Electrochimica Acta*, 53, 186–192.

Korbekandi, H., Ashari, Z., Iravani, S., and Abbasi, S. (2013). Optimization of biological synthesis of silver nanoparticles using *Fusarium oxysporum*. *Iranian Journal of Pharmaceutical Research*, 12(3), 289–298.

Korbekandi, H., and Iravani, S. (2013). Biological synthesis of nanoparticles using algae. In: *Green Biosynthesis of Nanoparticles: Mechanisms and Applications*, (eds M. Rai, and C. Posten), pp. 53–60. Wallingford, UK: CABI.

Korbekandi, H., Iravani, S., and Abbasi, S. (2009). Production of nanoparticles using organisms. *Critical Reviews in Biotechnology*, 29(4), 279–306.

Korbekandi, H., Iravani, S., and Abbasi, S. (2012). Optimization of biological synthesis of silver nanoparticles using *Lactobacillus casei* subsp. Casei. *Journal of Chemical Technology & Biotechnology*, 87(7), 932–937.

Kowshik, M., Ashtaputre, S., Kharrazi, S., Vogel, W., Urban, J., Kulkarni, S., and Paknikar, K. (2003). Extracellular synthesis of silver nanoparticles by a silver-tolerant yeast strain MKY3. *Nanotechnology*, 14(1), 95.

Kowshik, M., Deshmukh, N., Vogel, W., Urban, J., Kulkarni, S., and Paknikar, K. (2002). Microbial synthesis of semiconductor CdS nanoparticles, their characterization, and their use in the fabrication of an ideal diode. *Biotechnology and Bioengineering*, 78(5), 583–588.

Krumov, N., Oder, S., Perner-Nochta, I., Angelov, A., and Posten, C. (2007). Accumulation of CdS nanoparticles by yeasts in a fed-batch bioprocess. *Journal of Biotechnology*, 132(4), 481–486.

Kumar, S., Amutha, R., Arumugam, P., and Berchmans, S. (2011). Synthesis of gold nanoparticles: An ecofriendly approach using *Hansenula anomala*. *ACS Applied Materials & Interfaces*, 3(5), 1418–1425.

Kuyucak, N. and Volesky, B. (1988). Biosorbents for recovery of metals from industrial solutions. *Biotechnology Letters*, 10(2), 137–142.

Lee, S.-W., Mao, C., Flynn, C. E., and Belcher, A. M. (2002). Ordering of quantum dots using genetically engineered viruses. *Science*, 296(5569), 892–895.

Lengke, M., Fleet, M., and Southam, G. (2007). Biosynthesis of silver nanoparticles by filamentous cyanobacteria from a silver(I) nitrate complex. *Langmuir*, 10, 1021–1030.

Li, Y., Cu, Y. T., and Luo, D. (2005). Multiplexed detection of pathogen DNA with DNA based fluorescence nanobarcodes. *Nature Biotechnology*, 23, 885–889.

Liangwei, D., Hong, J., Xiaohua, L., and Erkang, W. (2007). Biosynthesis of gold nanoparticles assisted by *Escherichia coli* DH5α and its application on direct electrochemistry of hemoglobin. *Electrochemistry Communication*, 9, 1165–1170.

Liu, S., Leech, D., and Ju, H. (2003). Application of colloidal gold in protein immobilization, electron transfer, and biosensing. *Analytical Letters*, 36, 1–19.

Lu, A. H., Salabas, E. L., and Schuth, F. (2007). Magnetic nanoparticles: Synthesis, protection, functionalization and application. *Angewandte Chemie International Edition*, 46, 1222–1244.

Mandal, D., Bolander, M. E., Mukhopadhyay, D., Sarkar, G., and Mukherjee, P. (2006). The use of microorganisms for the formation of metal nanoparticles and their application. *Applied Microbiology and Biotechnology*, 69(5), 485–492.

Mann, S., Frankel, R.B., and Blakemore, R.P. (1984). Structure, morphology and crystal growth of bacterial magnetite. *Nature*, 310, 405–407.

Mann, S., Shenton, W., Li, M., Connolly, S., and Fitzmaurice, D. (2000). Biologically programmed nanoparticle assembly. *Advanced Materials*, 12(2), 147–150.

Mann, S., Sparks, N.H.C., Frankel, R.B., Bazylinski, D.A., and Jannasch, H.W. (1990). Biomineralization of ferromagnetic greigite and iron pyrite. *Nature*, 343, 258–261.

Mao, C., Flynn, C.E., Hayhurst, A., Sweeney, R., Qi, J., Georgiou, G., Iverson, B., and Belcher, A.M. (2003). Viral assembly of oriented quantum dot nanowires. *Proceedings of the National Academy of Sciences of the United States of America*, 100(12), 6946–6951.

Mao, C., Solis, D. J., Reiss, B. D., Kottmann, S. T., Sweeney, R. Y., Hayhurst, A., Georgiou, G., Iverson, B., and Belcher, A.M. (2004). Virus-based toolkit for the directed synthesis of magnetic and semiconducting nanowires. *Science*, 303(5655), 213–217.

Marshall, M., Beliaev, A., Dohnalkova, A., David, W., Shi, L., Wang, Z., et al. (2007). c-Type cytochrome-dependent formation of U(IV) nanoparticles by *Shewanella oneidensis*. *PLoS Biology*, 4(8), 1324–1333.

McLamore, E. S., Diggs, A., Calvo Marzal, P., Shi, J., Blakeslee, J. J. et al. (2010). Noninvasive quantification of endogenous root auxin transport using an integrated flux microsensor technique. *The Plant Journal*, 63, 1004–1016.

McLean, R. J. C., and Beveridge, T. J. (1990). Metal-binding capacity of bacterial surfaces and their ability to form mineralized aggregates. In *Microbial Mineral Recovery* (eds H.L. Ehrlich and C.L. Brierley), pp. 185–222. New York: McGraw-Hill.

Mehra, R. K. and Winge, D. R. (1991). Metal ion resistance in fungi: Molecular mechanisms and their regulated expression. *Journal of Cellular Biochemistry*, 45(1), 30–40.

Mirkin, C. A., Letsinger, R. L., Mucic, R. C., and Storhoff, J. J. (1996). A DNA-based method for rationally assembling nanoparticles into macroscopic materials. *Nature*, 382, 607–609.

Mishra, A., Tripathy, S. K., and Yun, S. I. (2011). Bio-synthesis of gold and silver nanoparticles from *Candida guilliermondii* and their antimicrobial effect against pathogenic bacteria. *Journal of Nanoscience and Nanotechnology*, 11(1), 243–248.

Morones, J. R., Elechiguerra, J. L., Camacho, A., Holt, K., Kouri, J. B. et al. (2005). The bactericidal effect of silver nanoparticles. *Nanotechnology*, 16, 2346–2353.

Mourato, A., Gadanho, M., Lino, A. R., and Tenreiro, R. (2011). Biosynthesis of crystalline silver and gold nanoparticles by extremophilic yeasts. *Bioinorg Chem Appl*, (2011), 1–8.

Mrvcic, J., Stanzer, D., Solic, E., and Stehlik-Tomas, V. (2012). Interaction of lactic acid bacteria with metal ions: Opportunities for improving food safety and quality. *World Journal of Microbiology and Biotechnology*, 28(9), 2771–2782.

Mukherjee, P., Ahmad, A., Mandal, D., Senapati, S., Sainkar, S. R., Khan, M. I., Ramani, R., Parischa, R., Ajayakumar, P. V., Alam, M., Sastry, M., and Kumar, R. (2001a). Bioreduction of $AuCl_4$- ions by the fungus, *Verticillium* sp. and surface trapping of the gold nanoparticles formed. *Angewandte Chemie International Edition*, 40(19), 3585–3588.

Mukherjee, P., Ahmad, A., Mandal, D., Senapati, S., Sainkar, S. R., Khan, M. I., Ramani, R., Parischa, R., Ajaykumar, P. V., Alam, M., Kumar, R. and Sastry, M. (2001b). Fungus-mediated synthesis of silver nanoparticles and their immobilization in the mycelial matrix: A novel biological approach to nanoparticle synthesis. *Nano Letters*, 1(10), 515–519.

Mukherjee, P., Roy, M., Mandal, B., Dey, G., Mukherjee, P., Ghatak, J., Tyagi, A., and Kale, S. (2008). Green synthesis of highly stabilized nanocrystalline silver particles by a non-pathogenic and agriculturally important fungus *T. asperellum*. *Nanotechnology*, 19(7), 075103.

Nair, B. and Pradeep, T. (2002). Coalescence of nanoclusters and formation of submicron crystallites assisted by *Lactobacillus* strains. *Crystal Growth & Design*, 2, 293–298.

Nangia, Y., Wangoo, N., Sharma, S., Wu, J.-S., Dravid, V., Shekhawat, G., Raman Suri, C. (2009). Facile biosynthesis of phosphate capped gold nanoparticles by a bacterial isolate *Stenotrophomonas maltophilia*. *Applied Physics Letters*, 94(23), 233901–233903.

Narayanan, K. B., and Sakthivel, N. (2010). Biological synthesis of metal nanoparticles by microbes. *Advances in Colloid and Interface Science*, 156(1), 1–13.

Niemeyer, C. M. (2001). Nanoparticles, proteins, and nucleic acids: Biotechnology meets materials science. *Angewandte Chemie International Edition*, 40(22), 4128–4158.

Ortiz, D. F., Ruscitti, T., McCue, K. F., and Ow, D. W. (1995). Transport of metal-binding peptides by HMT1, a fission yeast ABC-type vacuolar membrane protein. *Journal of Biological Chemistry*, 270(9): 4721–4728.

Pandey, R., and Khuller, G. K. (2007). Nanoparticle-based oral drug delivery system for an injectable antibiotic—streptomycin. Evaluation in a murine tuberculosis model. *Chemotherapy*, 53, 437–441.

Parikh, R. Y., Ramanathan, R., Coloe, P. J., Bhargava, S. K., Patole, M. S., Shouche, Y. S., and Bansal, V. (2011). Genus-wide physicochemical evidence of extracellular crystalline silver nanoparticles biosynthesis by *Morganella* spp. *PLoS One*, 6, 21401.

Parikh, R. P., Singh, S., Prasad, B. L. V., Patole, M. S., Sastry, M., and Shouche, Y. S. (2008). Extracellular synthesis of crystalline silver nanoparticles and molecular evidence of silver resistance from *Morganella* sp.: Towards understanding biochemical synthesis mechanism. *Chembiochem*, 9(9), 1415–1422.

Park, T. J., Lee, S. Y., Heo, N. S., and Seo, T. S. (2010). In vivo synthesis of diverse metal nanoparticles by recombinant *Escherichia coli*. *Angewandte Chemie International Edition*, 49(39), 7019–7024.

Parkinson, J. and Gordon, R. (1999). Beyond micromachining: The potential of diatoms. *Trends in Microbiology*, 17(5), 190–196.

Pathania, R., Khan, H., Kaushik, R., and Khan, M. A. (2018). Essential oil nanoemulsions and their antimicrobial and food applications. *Current Research in Nutrition and Food Science*, 6(3), 1–16.

Pazirandeh, M., Baral, S., and Campbell, J. (1992). Metallized nanotubules derived from bacteria. *Biomimetics*, 1(1), 41–50.

Peleg, G., Lewis, A., Linial, M., and Loew, L.M. (1999). Nonlinear optical measurement of membrane potential around single molecules at selected cellular sites. *Proceedings of the National Academy of Sciences of the United States of America*, 96, 6700–6704.

Pérez-de-Luque, A. and Rubiales, D. (2009). Nanotechnology for parasitic plant control. *Pest Management Science*, 65, 540–545.

Perez-Martinez, F. C., Carrion, B., and Cena, V. (2012). The use of nanoparticles for gene therapy in the nervous system. *Journal of Alzheimer's Disease*, 31: 697–710.

Pfeiler, E. A. and Klaenhammer, T. R. (2007). The genomics of lactic acid bacteria. *Trends in Microbiology*, 15, 546–553.

Prasad, K. and Jha, A. K. (2009). ZnO nanoparticles: Synthesis and adsorption study. *Natural Science*, 1(02), 129.

Raheman, F., Deshmukh, S., Ingle, A., Gade, A., and Rai, M. (2011). Silver nanoparticles: Novel antimicrobial agent synthesized from a endophytic fungus *Pestalotia* sp. isolated from leaves of *Syzygium cumini* (L.). *Nano Biomedicine and Engineering*, 3, 174–178.

Rai, M., Maliszewska, I., Ingle, A., Gupta, I., and Yadav, A. (2015). Diversity of microbes in synthesis of metal nanoparticles. In *Bio-Nanoparticles: Biosynthesis and Sustainable Biotechnological Implications* (ed O. V. Singh). John Wiley & Sons, Inc., Hoboken, NJ. doi: 10.1002/9781118677629.ch1.

Rai, M. K., Deshmukh, S. D., Ingle, A. P., and Gade, A. K. (2012). Silver nanoparticles: The powerful nanoweapon against multidrug-resistant bacteria. *Journal of Applied Microbiology*, 112, 841–852.

Ramanathan, R., O'Mullane, A.P., Parikh, R.Y., Smooker, P.M., Bhargava, S.K., and Bansal, V. (2011). Bacterial kinetics-controlled shape-directed biosynthesis of silver nanoplates using *Morganella psychrotolerans*. *Langmuir*, 27(2), 714–719.

Rautaray, D., Sanyal, A., Adyanthaya, S. D., Ahmad, A., and Sastry, M. (2004). Biological synthesis of strontium carbonate crystals using the fungus *Fusarium oxysporum*. *Langmuir*, 20(16), 6827–6833.

Rayman, M. K., and MacLeod, R. A. (1975). Interaction of Mg-2+ with peptidoglycan and its relation to the prevention of lysis of a marine pseudomonad. *Journal of Bacteriology*, 122(2), 650–659.

Reese, R., and Winge, D. (1988). Sulfide stabilization of the cadmium-gamma-glutamyl peptide complex of *Schizosaccharomyces pombe*. *Journal of Biological Chemistry*, 263(26), 12832–12835.

Riddin, T., Govender, Y., Gericke, M., and Whiteley, C. (2009). Two different hydrogenase enzymes from sulphate-reducing bacteria are responsible for the bioreductive mechanism of platinum into nanoparticles. *Enzyme and Microbial Technology*, 45(4), 267–273.

Senapati, S., Ahmad, A., Khan, M.I., Sastry, M., and Kumar, R. (2005). Extracellular biosynthesis of bimetallic Au-Ag alloy nanoparticles. *Small*, 1(5), 517–520.

Shahverdi, A. R., Fakhimi, A., Shahverdi, H. R., and Minaian, S. (2007). Synthesis and effect of silver nanoparticles on the antibacterial activity of different antibiotics against *Staphylococcus aureus* and *Escherichia coli*. *Nanomedicine*, 3, 168–171.

Shaligram, N. S., Bule, M., Bhambure, R., Singhal, R. S., Singh, S. K., Szakacs, G., and Pandey, A. (2009). Biosynthesis of silver nanoparticles using aqueous extract from the compactin producing fungal strain. *Process Biochemistry*, 44(8), 939–943.

Shankar, S. S., Ahmad, A., Pasricha, R., and Sastry, M. (2003a). Geranium leaf associated biosynthesis of silver nanoparticles. *Journal of Materials Chemistry*, 19(6), 1627–1631.

Shankar, S. S., Rai, A., Ahmad, A., and Sastry, M. (2004). Rapid synthesis of Au, Ag, and bimetallic Au core-Ag shell nanoparticles using Neem (*Azadirachta indica*) leaf broth. *Journal of Colloid and Interface Science*, 275(2), 496–502.

Shankar, S. S., Ahmad, A., Pasricha, R., and Sastry, M. (2003b). Bioreduction of chloroaurate ions by geranium leaves and its endophytic fungus yields gold nanoparticles of different shapes. *Journal of Materials Chemistry*, 13(7), 1822–1826.

Shenton, W., Douglas, T., Young, M., Stubbs, G., and Mann, S. (1999). Inorganic-organic nanotube composites from template mineralization of tobacco mosaic virus. *Advanced Materials*, 11(3), 253–256.

Shenton, W., Pum, D., Sleytr, U.B., and Mann, S. (1997). Synthesis of cadmium sulphide superlattices using self-assembled bacterial S-layers. *Nature*, 389(6651), 585–587.

Shiying, H., Zhirui, G., Zhanga, Y., Zhanga, S., Wanga, J., and Ning, G. (2007). Biosynthesis of gold nanoparticles using the bacteria *Rhodopseudomonas capsulata*. *Materials Letters*, 61(18), 3984–3987.

Singh, R., Thakur, P., Thakur, A., Kumar, H., Chawla, P., Jigneshkumar V. R., Kaushik, R., and Kumar, N. (2020). Colorimetric sensing approaches of surface modified gold and silver nanoparticles for detection of residual pesticides: A review. *International Journal of Environmental Analytical Chemistry*, DOI:10.1080/03067319.2020.1715382

Singh, S., Vidyarthi, A. S., and Dev, A. (2015). Microbial synthesis of nanoparticles, In *Bio-Nanoparticles: Biosynthesis and Sustainable Biotechnological Implications* (ed O. V. Singh), John Wiley & Sons, Inc., Hoboken, NJ. doi: 10.1002/9781118677629.ch8

Sinha, R., Karana, R., Sinha, A., Khare, S.K. (2011). Interaction and nanotoxic effect of ZnO and Ag nanoparticles on mesophilic and halophilic bacterial cells. *Bioresource Technology*, 102, 1516–1520.

Sintubin, L., De Windt, W., Dick, J., Mast, J., Van Der Ha, D., Verstraete, W., and Boon, N. (2009). Lactic acid bacteria as reducing and capping agent for the fast and efficient production of silver nanoparticles. *Applied Microbiology and Biotechnology*, 84(4), 741–749.

Slawson, R. M., Trevors, J. T., and Lee, H. (1992). Silver accumulation and resistance in *Pseudomonas stutzeri*. *Archives of Microbiology*, 158(6), 398–404.

Slocik, J. M., Naik, R. R., Stone, M. O., and Wright, D. W. (2005). Viral templates for gold nanoparticle synthesis. *Journal of Materials Chemistry*, 15(7), 749–753.

Stolz, J. F., Basu, P., Santini, J. M., and Oremland, R. S. (2006). Arsenic and selenium in microbial metabolism. *Annual Review of Microbiology*, 60, 107–130.

Sweeney, R.Y., Mao, C., Gao, X., Burt, J.L., Belcher, A.M., Georgiou, G., and Iverson, B.L. (2004). Bacterial biosynthesis of cadmium sulfide nanocrystals. *Chemistry & Biology*, 11(11), 1553–1559.

Syed, A., and Ahmad, A. (2012). Extracellular biosynthesis of platinum nanoparticles using the fungus *Fusarium oxysporum*. *Colloids and Surfaces B: Biointerfaces*, 97, 27–31.

Tejo Prakash, N., Sharma, N., Prakash, R., Raina, K. K., and Fellowes, J. (2009). Aerobic microbial manufacture of nanoscale selenium: Exploiting nature's bionanomineralization potential. *Biotechnology Letters*, 31, 1857–1862.

Temple, K. L., and Le Roux, N.W. (1964). Syngenesis of sulphide ores: Desorption of adsorbed metal ions and their precipitation as sulphides. *Economic Geology*, 59, 647–655.

Thakkar, K. N., Mhatre, S.S., and Parikh, R.Y. (2010). Biological synthesis of metallic nanoparticles. *Nanomedicine: Nanotechnology, Biology and Medicine*, 6(2), 257–262.

Thanh, N. T., and Rosenzweig, Z. (2002). Development of an aggregation-based immunoassay for anti-protein A using gold nanoparticles. *Analytical Chemistry*, 74, 1624–1628.

Tieshi, C., Yongming, L., Yuping, Z., Jun, Y., Yuanming, Z., and Yu, T. (2009). Sphere-like nano-needle congeries of selenium prepared in reverse micelle solution of Tx-100-SDS. *CJI Journal*, 11(3), 13.

Torney, F., Trewyn, B. G., Lin, V. S., and Wang, K. (2007). Mesoporous silica nanoparticles deliver DNA and chemicals into plants. *Nature Nanotechnology*, 2, 295–300.

Valls, M. and De Lorenzo, V. (2002). Exploiting the genetic and biochemical capacities of bacteria for the remediation of heavy metal pollution. *FEMS Microbiology Reviews*, 26(4), 327–338.

Watson, J. H., Ellwood, D. C., Soper, A. K., and Charnock, J. (1999). Nanosized strongly magnetic bacterially-produced iron sulfide materials. *Journal of Magnetism and Magnetic Materials*, 203, 69–72.

Wen, L., Lin, Z., Gu, P. et al. (2009). Extracellular biosynthesis of monodispersed gold nanoparticles by a SAM capping route. *Journal of Nanoparticle Research*, 11(2), 279–288.

Windt, W., Aelterman, P., and Verstraete, W. (2005). Bioreductive deposition of paslladium (0) nanoparticles on *Shewanella oneidensis* with catalytic activity towards reductive dechlorination of polychlorinated biphenyls. *Environmental Microbiology*, 7(3), 314–325.

Wong, K. K., Douglas, T., Gider, S., Awschalom, D. D., and Mann, S. (1998). Biomimetic synthesis and characterization of magnetic proteins (magnetoferritin). *Chemistry of Materials*, 10(1), 279–285.

Xiao, Y., Ju, H., and Chen, H. (1999). Hydrogen peroxide sensor based on horseradish peroxidase-labeled Au colloids immobilized on gold electrode surface by cysteamine monolayer. *Analytica Chimica Acta*, 391, 73–82.

Yu, J. X., and Li, T. H. (2011). Distinct biological effects of different nanoparticles commonly used in cosmetics and medicine coatings. *Cell & Bioscience*, 1, 19.

Zang, W., Chen, Z., Liu, H., Zhang, L., Gaoa, P., and Li, D. (2011). Biosynthesis and structural characteristics of selenium nanoparticles by *Pseudomonas alcaliphila. Colloids and Surfaces B: Biointerfaces*, 88, 196–201.

5 A Way Forward With Nano-Antimicrobials as Food Safety and Preservation Concern

A Look at the Ongoing Trends

Ajay Singh,
Mata Gujri College, India
Rohit Sangwan,
University of Potsdam, Germany
Pradyuman Kumar,
Department of Food Engineering and Technology, SLIET, India
Ramandeep Kaur
Punjab Agriculture University, India

CONTENTS

5.1 INTRODUCTION

Microbial contamination is the recent challenge encountered by food industries from a preservation perspective and has attracted much attention by economists, policy-makers, and food researchers too (Baranwal et al., 2018). Despite recent technological advancement, many validations to date are still lying in infancy stages, hence the need to explore more at a fast pace. Presently available conventional methods to avert these issues are not sufficient enough; therefore, better enhancement with upgraded technologies is continuously in need to exploit and secure humankind's food availability and preservation as well (Bajpai et al., 2018). Nanotechnology is the field that involves the control and application of technology at the nanoscale. These nanomaterials behave as a whole to carry out all the functioning for what we want to exploit; hence, nanomaterial almost resembles similar a big molecule (Baranwal et al., 2018). A wide variety of materials exist today that considered as nanostructured materials (NSMs), but the term *NSM* validates only those materials which belong to a 1- to 100-nm range. NSMs may exhibit large particle size (>100 nm) when they combine with other materials (like polymers, biomolecules) to form composite NSMs. The classification of NSMs is divided into three categories, which are also separated into sub-categories. Three main categories are inorganic NSMs, organic NSMs, and nanocomposites (Dastjerdi and Montazer, 2010). Biochemical mechanisms are involved in the degradation of microbial enzymes, alterations in cytoplasmic membranes of bacteria, and changes in membrane integrity due to the release of metal ions which cause antimicrobial activities. Metal nanoparticles (NPs) differentiate their response to the bacterial species based on the presence of the peptidoglycan layer. The passage of NPs into cell structures is generally divided into two types—open or non-specific transport and specific transport (Chudobova et al., 2015). Problems in antimicrobials are always noticed with less stability and fast degradability. The encapsulation of food antimicrobials agents with nanocarriers systems increases the high surface area–to–volume ratio and results in the increment of the concentration of antimicrobials at the specific site of the occurrence of microorganisms and allows them to reach at the target site without getting depleted (Blanco-Padilla et al., 2014).

5.2 CLASSIFICATION OF NSMs

NSM shave widespread domains for the types of nanomaterials and different synthetics pathways that followed to provide an antimicrobial effect at the nanoscale (Bhushan, 2010). They are also showing a huge variation in their shapes and structures, such as crystals, wire, sheets, and capsules. NSMs are mainly classified into three major categories (Figure 5.1). Inorganic NSMs include different categories for the material used and various forms, and some of them have sub-categories too. Nanocrystals are showing the proper alignment of atoms in the single or polycrystalline configuration in one dimension (<100 nm). Nanowires are wire-like structures, having a diameter of few nanometres, and nanoshells are the type of NPs that have spherical shapes with a dielectric core, which is covered by the thin metallic shell. Two sub-categories of NPs are metal and metal oxide NPs with a dimension of less than 100 nm. Carbon nanotubes and quantum dots have specific cylindrical carbon structures and semiconducting properties, respectively. Organic NSMs consist of

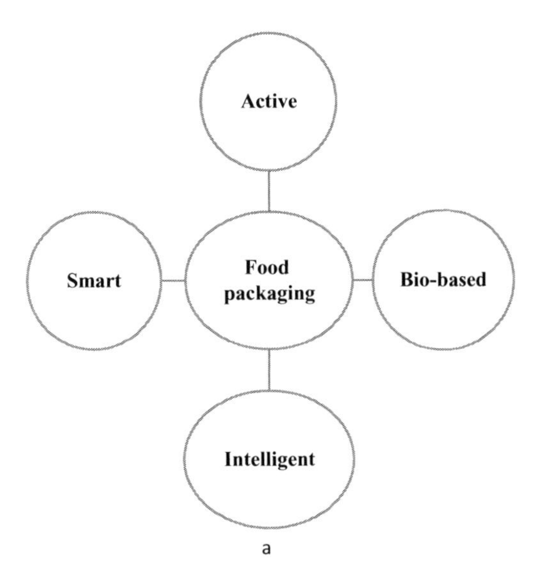

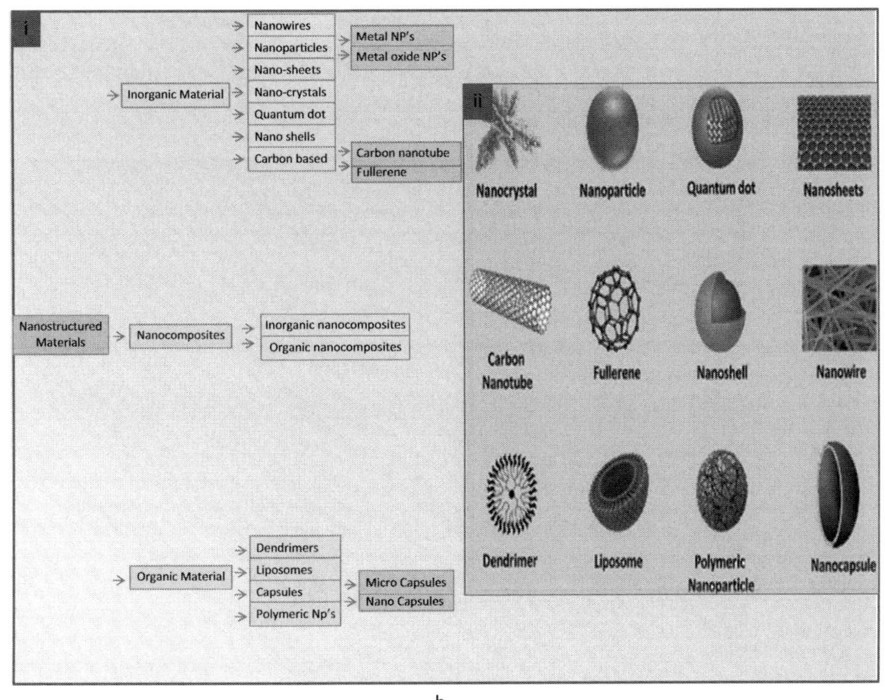

FIGURE 5.1 (a) Classification of advanced food packaging in food industry; (b) classification of nanomaterials in three categories (left) and structure of some nanomaterials (right) (Barnwal et al., 2016).

four different types—nano/microcapsules, dendrimers, polymeric NPs, and liposomes. These are composed of more complex three-dimensional structures in comparison to inorganic NSMs. The last major category is of nanocomposites, which are divided into organic and inorganic composites (Dastjerdi and Montazer, 2010).

5.2.1 Current Scenario and Ongoing Trends

Many of the NSMs described earlier have importance in various sectors like agricultural and industrial processes, food preservation, water treatment, antibiotics development, and consumer products (mentioned in Table 5.1). Mechanisms followed by them mostly have common targets that affecting cell membrane by the release of metal ion of Ag+ and Au+ in metal oxide NPs. In some cases, carbon-based structures in nanotubes, graphene, and fullerene cause the generation of reactive oxygen species (ROS) and produce reactive nascent oxygen (Dakal et al., 2016). Although in some different nanoscale materials, the mechanism of antimicrobial activity and how the cell wall is changing in its integrity are still not clear (Kong et al., 2016). In addition to the taking part directly in the mechanism, a few organic NSMs are used as carriers of inorganic NPs and antimicrobials to provide encapsulation to them for delivering efficiently to their respected targets.

TABLE 5.1
Different Nanostructured Materials (NSMs) and Their Dimensions, Targeted Microbial Strains, Effects, and Applications

NSMs	Dimensions(nm)	Microbes Tested	Effects	Application
AgNPs	20–30	*E. coli* and *S. aureus*	Diminished growth of bacterial species was evident	Water filters, storage containers, and food packaging
AuNPs	7.7–3.9	*B. subtilis*, *K. mobilis*, and *E. coli*	Antibacterial action	Antibiotic drug delivery system
ZnONP	25 and 40	*S. aureus*, *S. marcescens*, and *P. mirabilis*	Prominent inhibition of the bacterial strains	Antimicrobial creams, lotions, and ointments
Graphene oxide- chitosan (CS-GO) Nanocomposites	Not reported	*E. coli* and *B. subtilis*	Efficient bacterial inactivation	Food packaging
CuONP	20–95	*S. epidermidis*, and *E. coli*	Suppress the growth of tested bacteria	Next generation antibiotics and biosensing
Dendrimer	Not reported	*S. aureus*, and *P. aeruginosa*	Enhanced inhibition of microbes	Antibacterial agent
Nanofibres	200–550	*E. coli* and *P. aeruginosa*	Detectable antibacterial effect	Water treatment, medical and health care products

5.3 BIOCHEMICAL AND CELLULAR MECHANISMS

The inhibition of microbial growth by affecting cell entities and the inside environment is also a complex process; understanding the background and foreground for reactions to take place is necessary. This part of the chapter more focused on different ways to synthesize NPs and the effects of metal NPs with different nonmetals.

5.3.1 Synthesis of Nanoparticles

In general, it follows a common procedure for all NPs that have a combination of some metal salts, mostly with capping ligands; the next step of altering the pH or by adding some reducing agent will synthesize NPs. Nowadays, the production of magnetic NPs is coming in trend because of their utility in magnetic resonance imaging, treatment of wastewater, and ongoing researches in biomedicine and biotechnology (Lu et al., 2007). Biocompatibility is the main issue and is the main concern for consideration while working on them. The production of different NPs has not been executed through a single protocol approach. Varied methods applications such as thermal decomposition for preparing small-sized NPs, aqueous synthesis of magnetic NPs by hydrothermal reduction, and microemulsion or microwave synthesis are among the list. The preparation of colloidal NPs, Fe_3O_4 stabilized by citric acid in a single-step process. Hyperthermia treatment is done with magnetic NPs functionalized by citric acid (Nigam et al., 2011).

5.3.2 Mechanism—Nanoparticle Conjugation With Chitosan

Infections caused by pathogenic bacterial strains such as *Escherichia coli*, *Streptococcus*, and *Staphylococci* have started showing resistance towards antibiotic drugs, and there is a necessity for introducing a new method to resolve this problem (Chudobova et al., 2014). Studies on the interactions between metal particles and cell structures confirm their suitability as tools for protection against infectious bacterial strain. Two alternative pathways by which metal NPs are transported to the cell membranes are discovered (Nies and Silver, 1995). The first one is the non-specific transporters, also known as "open gates", provide rapid transport through the cell membrane, and the chemiosmotic gradient allows the transport of heavy metals (Nieboer and Richardson, 1980; Schreurs and Rosenberg, 1982). The second way is used less often; when required, it is more specific, slower, and consumes more energy (Nies, 1999). Chitosan is a positively charged polysaccharide biopolymer composed of glucosamine having β-1-4-glycosidic linkage (Bonilla et al., 2013). Chitosan can be degraded by enzymes that cause nontoxicity and antimicrobial activity (Andrews et al., 2002). Previous research acknowledged the interaction between negatively charged microbial cell membranes having an NH_3^+ group at the monomer of glucosamine tends to leakage of intracellular constituents of cells (Helander et al., 2001).

An interaction is reported in the chitosan–silver NPs composite with molecular iodine. The expression of green fluorescent protein in *E. coli* has been used to suggest the mechanism of the process. Significant results of enhanced bactericidal activity of the nanocomposite are found in the presence of iodine. Measurements from flow cytometry and transmission electron microscopy reveal the attachment of bacteria to the composite and damage of the cell wall of bacteria, where iodine-treated

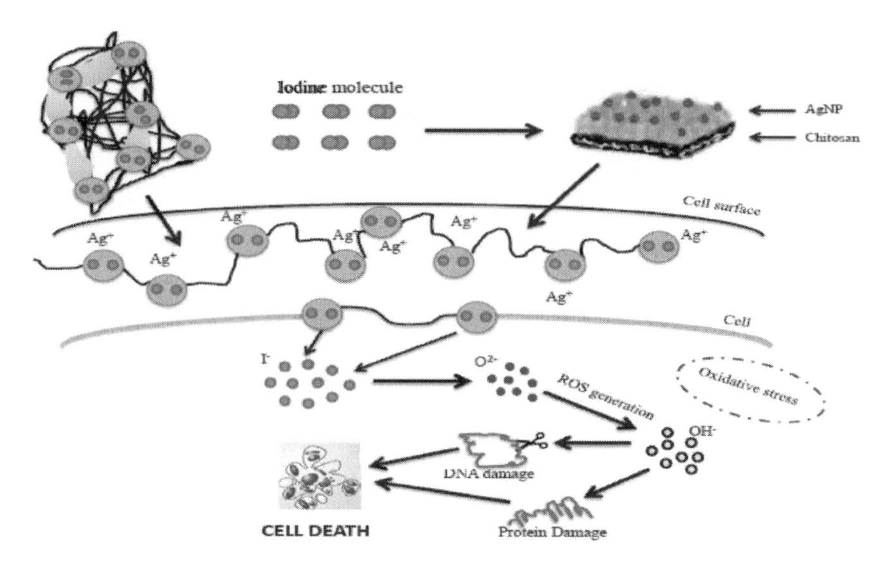

FIGURE 5.2 Mechanism of antibacterial activity of iodinated chitosan-silver nanoparticles composite (Banerjee et al., 2010; Chudobova et al., 2015).

composite are present (Banerjee et al., 2010). In addition to this, a combination of iodine and a nanocomposite also allows us to generate oxidative stress in the cytoplasm of bacterial cells by recruiting ROS. The chitosan-silver NPs composite, in addition to the molecular iodine generated iodine atoms(I^-) and ROS(O^{2-}), enters into the cytoplasm and generates oxidative stress facilitates the process of DNA damage, and protein damage leads to the process of cell death (Figure 5.2).

5.4 NANOCARRIERS FOR ANTIMICROBIALS

Nano-encapsulation prevents the degradation of bioactive compounds, such as carbohydrates, proteins, vitamins, and antioxidants (Sekhon, 2010). Such functional foods can be valued for a long time through preservation. Nano-encapsulation, in brief, means to the nanoscale application of molecules coated within the layers, viz. coating and films. In the present era, nano-based smart packaging has been developed with a better mechanical strength of the material, barrier properties, nanosensing for foreign material, and pathogen detection by antimicrobial films (Mihindukulasuriya and Lim, 2014). Nanocarriers are based on sustained-release properties, better pharmacokinetic profile, and biocompatibility with tissue and cells.

5.4.1 Nano-Encapsulated Systems

Nanoemulsions are the most common encapsulated systems configured with stable colloidal systems of size less than or equal to 100 nm. Prepared by dispersing one miscible liquid in another immiscible liquid with the help of suitable emulsifiers, nanoemulsions are not the same as microemulsions because of optical transparency and more shelf stability while the size distribution of droplets remains the same. The preparation of nanoemulsions depends on the functionality and desired structure perform by using two types of methods: high-energy and low-energy methods. High-energy methods are ultrasonication, microfluidization, and high-pressure homogenization. Solvent diffusion is used as a lower energy method (Burguera and Burguera, 2012). Emulsions are single emulsions (oil in water), multiple emulsions (oil–water–oil or water–oil–water), and multilayer emulsion, whereby oil droplets are surrounded by a different nano-size layer of polyelectrolytes (Weiss et al., 2006). In oil-in-water nanoemulsion delivery of encapsulated poorly water-soluble food antimicrobials stabilize the active compound. NPs are known as nanospheres and nanocapsules. Nanocapsules comprise a vesicular system consisting of actives in an inner liquid core and nanospheres; a polymer matrix here activates present at the surface of a sphere or covered within the particle (Ahsan et al., 2002). Performances are measured during delivery and depend on the physical state, size, charge, and morphology of NPs (Chen et al., 2006). Some methods that are suitable for producing NPs having food applications are polymerization, electro-spraying, salting out, nanoprecipitation, and emulsification. Nanoliposomes are structurally spherical, with water-soluble polar heads directed to the inside and outside of the cell membrane, and hydrophobic phospholipids tails are associated with the bilayer. These structures can release and deliver amphiphilic, water, and lipid-soluble materials. Many natural sources, such as soy, milk, and eggs, contain phospholipids and are used in the production of nanoliposomes to provide better biological activity (Mozafar et al., 2008). Actions involved in the interaction between liposomes and cells are by fusion with the cell membrane, endocytosis, and adsorption of liposomes onto the cell surface (Torchilin, 2005).

Nanofibres are produced by electrospinning methods, and morphologically, these are ultra-thin structures with dimensions less than 100 nm. In the electrospinning method, continuous polymer fibres are produced through the external electric field imposed on a polymeric solution. Electrospun fibres are important for various food applications, like additive delivery systems and edible films, because of their high surface area–to–volume ratio. Properties like biodegradability, antibacterial properties, and hydrophobicity allow nanofibres to use in food, textile, cosmetics, and pharmaceutical industry (Blanco-Padilla et al., 2014).

5.4.2 Production Technology in Use

Production of nanocapsules is done by top-down and bottom-up techniques (Singh, 2018). Techniques include in the top-down approach are solvent extraction, high-pressure homogenization, and emulsification. In the emulsification process, two immiscible liquids are allowed to mix by using a surfactant (Solans et al., 2005). High-pressure homogenization is used for high-quality drug particle production at the industrial level. Solvent extraction performed in a solvent media which reduces the risk of instability in physico-chemical properties and contamination by the pathogen. Nanoprecipitation, coacervation, and supercritical fluid are general bottom-up approaches. Nanoprecipitation is the process of solvent displacement involving precipitation of an organic solution polymer and the aqueous solution is diffused by an organic solvent. This type of solvent displacement results in the formation of nanospheres and nanocapsules. In supercritical fluid precipitation, the solubilization of bioactive material and polymers is performed in a supercritical fluid, which passes through a nozzle. Evaporation of this fluid done by spraying methods results in

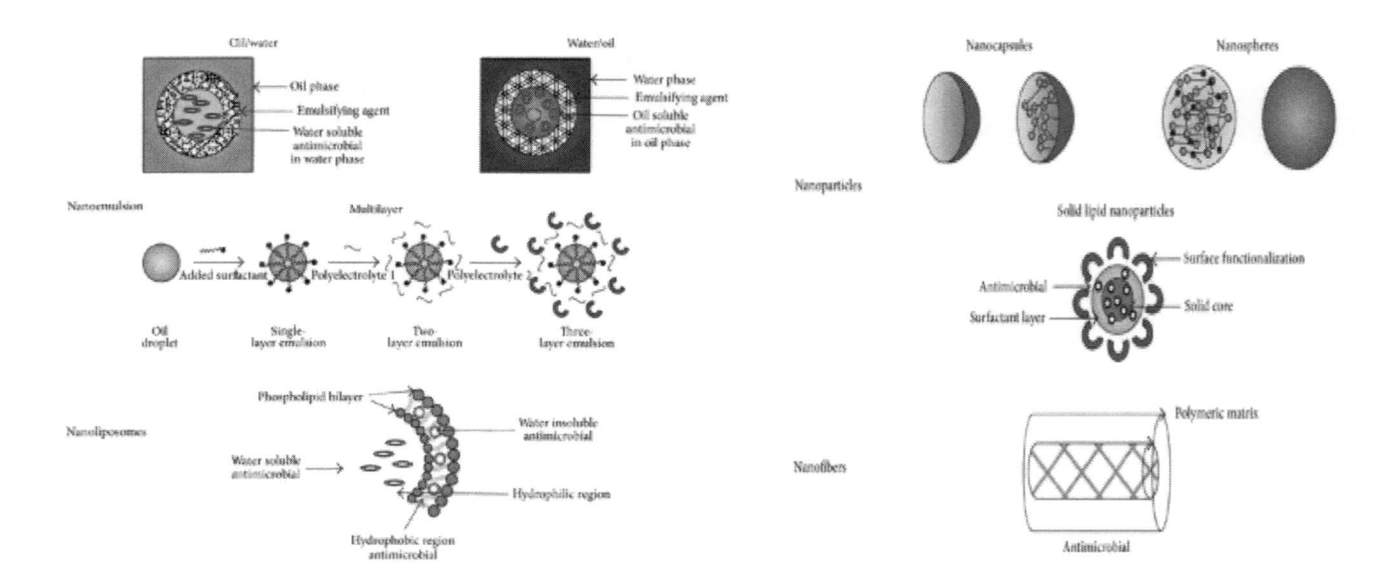

FIGURE 5.3 Nanoscale delivery systems showing entrapment of food antimicrobials (Blanco-Padilla et al., 2014).

precipitation solute particles (Reis et al., 2006). Phase separation of polyelectrolyte mixture from a solution is performed by the coacervation technique.

5.4.3 Nanocarrier Systems

These carrier systems can be lipid-, carbohydrate-, and protein-based. Lipid-based nanosystems are more efficient towards encapsulation and bear less toxicity (Sekhon, 2010). It is the main application for the delivery and protection of foods and nutraceuticals. Nanoemulsions, nanocochleate, nanoliposomes, and solid-lipid NPs are some examples of lipid-based nanosystems. Carbohydrate-based systems require chemical and heat treatments that do not allow to scaling the process to a very extent. Carbohydrates are polysaccharides that have a huge molecular weight and a complex structure and can entrap small compounds suitable for delivery systems. Many popular polysaccharides such as starch (most abundant in plants), cellulose esters (enteric and non-enteric cellulose), pectin (linear anionic polysaccharide), guar gum (water-soluble polysaccharide), alginate (linear polysaccharide from brown sea algae), dextran (bacterial polysaccharide of glucan, and, most common, chitosan (cationic, biocompatible, linear, and biodegradable polymer) are widely used as carbohydrate-based nanosystems (Fathi et al., 2014).

5.5 NANOANTIMICROBIALS FOOD APPLICATIONS (COMMERCIAL VIEW)

Nanotechnology offers several applications to food and agricultural industries in the delivery of agrochemicals, the advanced strategy of detecting animal and plant microbes, food supplements and additives to increase bioavailability, targeted delivery of nutraceuticals by nanoencapsulation, the selective removal and binding of pathogens from food by NPs, and temperature and moisture monitoring by biodegradable nanosensors (Chaudhry et al., 2008; Sekhon, 2014; Chawla et al., 2019). Current limited applications in the food industry are due to regulatory and toxicological issues. Major applications of nanoencapsulation are in the pharmaceutical industries and in its role in antimicrobial active food packaging, opening its path to the food industry. Specificity in the type of nanomaterials is used to develop certain types of desired properties into food products (Sandoval, 2009; Chawla et al., 2018). Many applications entering field instrumentation technology include designing nanosensors to determine soil conditions, crop health, and plant, and animal pathogens. Nanocapsules are efficiently used to improve the bioavailability of nutraceuticals and to deliver vaccines in the medical sector. The targeted delivery of DNA to plants through genetic engineering techniques uses NPs, and the attachment of fluorescent NPs to the antibodies can detect chemical and foodborne pathogens (Ravichandran, 2010). Nanoclays, nanofilms, and nanotubes are preventing spoilage and oxygen absorption as barrier materials and gelation agents. Nanoemulsions provide better dispersion of nutrients and antifungal surface coatings are provided by silver, zinc, and magnesium NPs. Furthermore, silicate NPs have applications towards the production of heat-resistant films which are stronger and lighter compared to other conventionally produced films. The delivery process of nutrients is improved specifically by nanocochleate without affecting colour and taste of food and for better drug delivery carrier nanocrystal composite made of cellulose are used (Gallocchio et al., 2015).

5.6 IMPROVEMENT OF NANOANTIMICROBIALS

Mechanisms behind the action of antimicrobials are still indistinct and unclear. Suggestions vary from the development of oxidative stress by the generation of free radicals to reports declining the association of antimicrobial effect with metabolic regulatory mechanism (Dakal et al., 2016; Chawla et al., 2020). In the future, further research work can be done to determine the exact pathways for NSMs' action on microbes. The depletion of metal NPs in the downstream process starts accumulating at different levels of an ecosystem and are able to cause toxicity in flora and fauna.

The development of better technology and improvement in the synthesis of NPs and nano-encapsulation lead to less retention of nanomaterials. The future aim of nano-antimicrobials can be easily stated by ensuring fewer effects on the host organism and more antimicrobial activity towards microbes. The research interest of food scientists is now shifting towards the transformation of the future of food packaging by using carbon nanotubes to detect food spoilage by microbes and toxicity levels of proteins. To hold to big promises for the food safety and health risks of consumers, the detection of analytical methods and better characterization of nanomaterials in toxicological data are required. For scaling up the production of nanocarriers in the food industry, a concern of research should be in the direction to design new methods of scaling and cost minimization to identify low-cost ingredients. More advancement in nanomaterials such as carbon nanotubes and quantum dots is required for quantification of microbes and their toxin level to address food safety issues (Ibarra and Chen, 2016). Big data generation by new engineering applications that increasingly affect the new field of research and application may encourage advancement in food safety. (King et al., 2017). Nano-antimicrobials needs more research work to gain more knowledge about the exact biochemical mechanisms of their action inside cell systems and more improvement in the developing technology for detection of nanosystems and nanomaterials to fulfil food safety requirements.

REFERENCES

Ahsan, F., Rivas, I. P., Khan, M.A., and Torres Suarez, A.I. (2002). Targeting to macrophages: Role of physicochemical properties of particulate carriers—liposomes and microspheres—on the phagocytosis by macrophages. *Journal of Controlled Release*, 79(1–3), 29–40.

Alonso, D., Gimeno, M., Sepúlveda-Sánchez, J.D. and Shirai, K. (2010). Chitosan-based microcapsules containing grapefruit seed extract grafted onto cellulose fibres by a non-toxic procedure. *Carbohydrate Research*, 345(6), 854–859.

Andrews, L.S., Keys, A.M., Martin, R.L., Grodner, R., and Park, D.L. (2002). Chlorine dioxide wash of shrimp and crawfish an alternative to aqueous chlorine. *Food Microbiology*, 19(4), 261–267. DOI 10.1006/yfmic.493.

Bajpai, V.K., Kamle, M., Shukla, S., Mahato, D.K., Chandra, P., Hwang, S.K., ... Han, Y.K. (2018). Prospects of using nanotechnology for food preservation, safety, and security. *Journal of Food and Drug Analysis*, 26(4), 1201–1214. https://doi.org/10.1016/j.jfda.2018.06.011.

Banerjee, M., Mallick, S., Paul, A., Chattopadhyay, A., and Ghosh, S.S. (2010). Heightened reactive oxygen species generation in the antimicrobial activity of a three-component iodinated chitosan-silver nanoparticle composite. *Langmuir*, 26(8), 5901–5908. DOI 10.1021/la9038528.

Baranwal, A., Mahato, K., Srivastava, A., Maurya, P. K., and Chandra, P. (2016). Phytofabricated metallic nanoparticles and their clinical applications. *RSC Advances*, 6(107), 105996–106010.

Baranwal, A., Srivastava, A., Kumar, P., Bajpai, V. K., Maurya, P. K., and Chandra, P. (2018). Prospects of nanostructure materials and their composites as antimicrobial agents. *Frontiers in Microbiology*, 9(March). https://doi.org/10.3389/fmicb.2018.00422.

Bhushan, B. (2010). *Springer Handbook of Nanotechnology*. Berlin; Germany: Springer Science & Business Media.

Blanco-Padilla, A., Soto, K.M., Hernández Iturriaga, M., and Mendoza, S. (2014). Food antimicrobials nanocarriers. *The Scientific World Journal*, https://doi.org/10.1155/2014/837215.

Bonilla, J., Atares, L., Vargas, M., and Chiralt, A. (2013). Properties of wheat starch film-forming dispersions and films as affected by chitosan addition. *Journal of Food Engineering*, 114(3), 303–312. DOI 10.1016/j.jfoodeng.2012.08.005.

Chaudhry, Q., Scotter, M., Blackburn, J., Ross, B., Boxall, A., Castle, L., Aitken, R., and Watkins, R. (2008). Applications and implications of nanotechnologies for the food sector. *Food Additives & Contaminants: Part A: Chemistry, Analysis, Control, Exposure & Risk Assessment,* 25(3), 241–258.

Chawla, P., Kaushik, R., Shiva Swaraj, V.J., and Kumar, N. (2018). Organophosphorus pesticides residues in food and their colorimetric detection. *Environmental Nanotechnology, Monitoring and Management*, 10, 292–307.

Chawla, P., Kumar, N., Kaushik, R., and Dhull, S.B. (2019). Synthesis, characterization and cellular mineral absorption of gum arabic stabilized nanoemulsion of *Rhododendron arboreum* flower extract. *Journal of Food Science and Technology*, 56(12), 5194–5203.

Chawla, P., Kumar, N., Bains, A., Dhull, S.B., Kumar, M., Kaushik, R., and Punia, S. (2020). Gum Arabic capped copper nanoparticles: Characterization and their utilization. *International Journal of Biological Macromolecules*, 146, 232–242.

Chen, H., Weiss, J., and Shahidi, F. (2006). "Nanotechnology in nutraceuticals and functional foods," *Food Technology*, 60(3), 30–36.

Chudobova, D., Cihalova, K., Kopel, P., Ruttkay-nedecky, B., Kizek, R., and Adam, V. (2015). *Antimikrobiálnínanomateriály v potravinářství. Kvasny Prumysl*, 2015(2), 51–56.

Chudobova, D., Dostalova, S., Blazkova, I., et al. (2014). Effect of ampicillin, streptomycin, penicillin and tetracycline on metal resistant and non-resistant *Staphylococcus aureus*. *International Journal of Environmental Research and Public Health*, 11(3), 3233–3255.

Dakal, T.C., Kumar, A., Majumdar, R.S., and Yadav, V. (2016). Mechanistic basis of antimicrobial actions of silver nanoparticles. *Frontiers in Microbiology*, 1831. DOI 10.3389/fmicb.2016.01831.

Dastjerdi, R. and Montazer, M. (2010). A review on the application of inorganic nano-structured materials in the modification of textiles: Focus on anti-microbial properties. *Colloids and Surfaces B: Biointerfaces,* 79, 5–18. DOI 10.1016/j.Colsurfb.2010.03.029.

Fathi, M., Martín, A., and McClements, D.J. (2014). Nanoencapsulation of food ingredients using carbohydrate based delivery system. *Trends in Food Science and Technology*, 39(1), 18–39.

Gallocchio, F., Belluco, S., and Ricci, A. (2015). Nanotechnology and food: brief overview of the current scenario. *Procedia Food Science*, 5, 85–88. https://doi.org/10.1016/j.profoo.2015.09.022.

Inbaraj, BS, and Chen, BH. (2016). Nanomaterial-based sensors for of foodborne bacterial pathogens and toxins as well as pork adulteration in meat products. *Journal of Food and Drug Analysis,* 24, 15e28.

King, T., Cole, M., Farber, J.M., Eisenbrand, G., Zabaras, D., Fox, E.M., et al. (2017). Food safety for food security: Relationship between global megatrends and developments in food safety. *Trends in Food Science and Technology,* 68, 160e75.

Kong, M., Chen, X. G., Liu, C. S., Liu, C. G., Meng, X. H., and Yu, L. J. (2008). Antibacterial mechanism of chitosan microspheres in a solid dispersing system against *E. coli. Colloids and Surfaces B: Biointerfaces,* 65, 197–202. DOI 10.1016/j.colsurfb.2008.04.003.

Kong, Y. L., Gupta, M. K., Johnson, B. N., and McAlpine, M. C. (2016). 3D printed bionic nanodevices. *Nano Today*, 11(3), 330–350.

Lu, A. H., Salabas, E. L., and Schuth, F. (2007). Magnetic nanoparticles: Synthesis, protection, functionalization, and application. *Angewandte Chemie International Edition*, 46(8), 1222–1244. DOI 10.1002/anie.200602866.

Mihindukulasuriya, S. D. F. and Lim, L. T. (2014). Nanotechnology development in food packaging: A review. *Trends in Food Science and Technology*, 40, 149–167.

Nieboer, E. and Richardson, D. H. S. (1980). The replacement of the non-descript term heavy-metals by a biologically and chemically significant classification of metal-ions. *Environmental Pollution Series B Chemical and Physical*, 1(1), 3–26. DOI 10.1016/0143-148x(80)90017-8.

Nies, D. H. (1999). Microbial heavy-metal resistance. *Applied Microbiology and Biotechnology*, 51(6), 730–750.

Nies, D. H., and Silver, S. (1995). Ion efflux systems involved in bacterial metal resistances. *Journal of Industrial Microbiology*, 14(2), 186–199.

Nigam, S., Barick, K. C., and Bahadur, D. (2011). Development of citrate -stabilized Fe_3O_4 nanoparticles: Conjugation and release of doxorubicin for therapeutic applications. *Journal of Magnetism and Magnetic Materials*, 323(2), 237–243. DOI: 10.1016/j.jmmm.2010.09.009.

Ravichandran, R. (2010). Nanotechnology applications in food and food processing: Innovative green approaches, opportunities and uncertainties for global market. *International Journal of Green Nanotechnology: Physics and Chemistrym*, 1(2), P72–P96, https://doi.org/10.1080/19430871003684440.

Reis, C. P., Neufeld, R. J., Ribeiro, A. J. and Veiga, F. (2006). Nanoencapsulation I. Methods for preparation of drug-loaded polymeric nanoparticles. *Nanomedicine*, 2(1), 8–21.

Sandoval, B. (2009). Perspectives on FDA's regulation of nanotech: Emerging challenges and potential solutions. *Comprehensive Reviews in Food Science and Food Safety,* 8, 375e93.

Schreurs, W. J. A. and Rosenberg, H. (1982). Effect of silver ions on transport and retention of phosphate by *Escherichia coli. Journal of Bacteriology*, 152(1), 7–13.

Sekhon, B. (2010). Food nanotechnology–An overview. *Nanotechnol, Sci Appl*, 3, 1–15.

Sekhon, B. S. (2014). Nanotechnology in agri-food production: An overview. *Nanotechnol, Science and Applications*, 7, 31–53.

Singh, P. (2018). Nanotechnology in food preservation. *Journal of Food Science Research*, *9*(2), 441–447. https://doi.org/10.15740/has/fsrj/9.2/441-447.

Solans, C., Izquierdo, P. J., Nolla, J., Azemar, N. and García-Celma, M.J. (2005). Nano-emulsions. *Current Opinion in Colloid & Interface Science*, 10(3–4), 102–110.

Torchilin, V. P. (2005). Recent advances with liposomes as pharmaceutical carriers, *Nature Reviews Drug Discovery*, 4(2), 145–160.

Weiss, J., Takhistov, P., and McClements, D.J. (2006a). Functional materials in food nanotechnology. *Journal of Food Science*, 71(9), R107–R116.

Weiss, J., Takhistov, P. and McClements, J. (2006). Functional materials in food nanotechnology. *Journal of Food Science*, 71, R107–R116. 10.1111/j.1750-3841.2006.00195.x.

Zuidam, N. J. and Shimoni, E. (2010). Overview of Microencapsulates for use in food products or processes and methods to make them. In *Encapsulation technologies for active food ingredients and food processing* (pp. 3–29). Springer, New York, NY.

6 Biogenic Nanoparticles

A New Paradigm for Treating Infectious Diseases in the Era of Antibiotic Resistance

Kanika Dulta, Kiran Thakur, Parveen Chauhan, and P.K. Chauhan
Shoolini University, India

CONTENTS

6.1 INTRODUCTION

The first organism to start life on earth was bacteria; because of this, they have become extremely adapted over time. The most notable discovery of humankind in the medical world is the discovery of antibiotics in the 20th century (WHO, 2017). It all started with the discovery of Salvarsan, the medicine capable of treating infectious disease—syphilis—without any toxic effects. But the finding of penicillin in 1928 by Alexander Fleming was serendipity; this started the vigorous research on antibiotics, leading the world towards its "golden age" between the 1950s and 1960s (WHO, 2017). More than 20 new types of antibiotics were yielded between 1930 and 1962. However, because of the increasing number and evolution of new resistant bacteria, the pharmaceutical industry faces difficulty to discover new molecules with antibacterial properties (Coates et al., 2011; Aslam et al., 2018). In medical history, the discovery of antibiotics is seen as its greatest accomplishment. In nearly all medical procedures, like surgeries, organ transplantation, and infection treatment, antibiotics are considered the most important tool. Sadly, the unchecked or, to a great extent, their misuse has caused the development of antibiotic-resistant bacterial species. This rapid growth in bacterial resistance to antibiotics is now a threat to the remedial accomplishment (World Health Organization, 2014).

Excluding all the negative social and economic effects, they also hold the danger of the spread of epidemic infections (Baluja et al., 2018). Antimicrobial resistance (AMR), announced as the "biggest threats to global health" by the World Health Organization (Davis et al., 2018). According to the European Union, 25,000 people die yearly because of AMR. Worldwide around 7 lakhs deaths per year were documented due to the multidrug-resistant (MDR) microorganisms (Sharma et al., 2016; Betts et al., 2018). The most established antibiotic mechanisms identified in bacteria are (1) reduction in absorption of antimicrobial drugs and/or increased outflow of drugs, (2) alterations of antibiotic target, (3) development of drug degrading/altering enzymes in microorganisms, and (4) formation of a biofilm layer around the bacteria which avoids its direct contact with antibiotics (Blecher et al., 2011a). These contingencies, in the end, result in either a lower retention of drugs in microbial cells or a short intracellular residence of drugs; because of this, the therapeutic levels of drugs cannot be easily reached (Huh and Kwon, 2011). As a consequence of the greater amount and repeated use of drugs, unfavorable effects can happen in human beings and animals.

Pathogenic microorganisms are nearly immune to all types of antibiotics currently being used. Over the last few decades, there have been no reports on the development of any new antibiotic class. Additionally, antibiotic development and commercialization are time-and money-consuming processes, which include the discovery of new antibiotics, multiple clinical trials, and licensing (Bartlett et al., 2013). It is also a well-known fact that bacterial resistance can come into view very quickly to any new antibiotics, resulting in a decrease in antibiotic use and a regression in sales. Failing in the development of antibiotics can increase the rate of mortality due to surgical infections after chemotherapy or organ transplants (Adeniji, 2018). Thus, there is an immediate need for pioneering new drugs to fight these problems. As a result, scientists have started showing interest in obtaining a rapid diagnostic and targeted therapy by either avoiding or amending the use of typical antibiotics. Hence, the need to think of a more efficient and less/non-toxic treatment was felt, which led to advances in biotechnology and nanotechnology.

Since the last century, nanotechnology has been a well-known field of research. *Nanotechnology* was introduced by Richard P. Feynman (Nobel laureate) in his famous 1959 lecture "There's Plenty of Room at the Bottom"; many revolutionary developments have been achieved in nanotechnology. Nanoparticles (NPs) are a wide class of materials produced at nanoscale having dimensions less than 100 nm (Laurent et al., 2010; Chawla et al., 2020) and can be categorized as zero-dimension (0D), one-dimension (1D), two-dimension (2D), or three-dimension (3D) based on overall shape (Tiwari et al., 2012).

About NP synthesis, the term *biogenic* covers different methods and practices, such as reducing dissolved metals to NPs using cell extracts of plants and using microorganisms and their natural abilities to synthesize NPs (Chawla et al., 2019; Agarwal et al., 2020). For studying biogenic NP synthesis and their industrial applications, bacteria represent the most optimistic field due to their easy genetic manipulation and growth.

At present, metallic nanomaterials have become extremely large and rapidly emerging substances of science areas. The increased attention of nanomaterial construction, especially NPs, is because of their captivating properties revealed by their size, large surface area, and exceptional surface activity displaying outstanding catalytic, electrical, and optical properties (Singh et al., 2020). Because of this, metallic NPs have a variety of applications in research methodology and advance micro- and nanotechnologies (Pathania et al., 2018). They are good heterogeneous catalysts, widely used in electronics, thin-film fabrication technology and in microelectronic device manufacturing (Popa et al., 2007). Controlled NP synthesis is achieved using a combination of modern and conventional colloid techniques. Metallic NPs have a strong absorption spectrum in the visible range because of consistent oscillations of electrons on the particulate surface (known as surface plasmon resonance [SPR]). The SPR spectrum range of metallic NPs has many applications in biotechnology and attracts many researchers towards it (Amendola and Meneghetti, 2009; Abdel et al., 2012).

Heavy metals like copper, silver, and gold are known for their application in infectious diseases since ancient times (Dhull et al., 2019). Out of all these metals, silver is more frequently used because of its known medicinal properties as an antimicrobial. The biggest public health care concern is antibiotic-resistant organisms, and against this, silver shows potential activities (Prabhu and Poulose, 2012). Recently, silver (AgNPs) have replaced metallic silver. In agriculture, healthcare, and industry, metal NPs are frequently used (Chawla et al., 2018). However, in many studies, AgNPs were reported as a major antimicrobial agent, which can fight against multiple infectious ailments and could be utilized as novel nanomedicine (Rai et al., 2009; Murphy et al., 2015). Also, other metallic NPs, along with noble metal oxide and metal NPs, have been studied and examined for their properties like antimicrobial drug carriers or as antimicrobial potential (Brandelli, 2012; Rai et al., 2015).

In this chapter, the authors have discussed the antimicrobial potential of metal NPs against different bacteria. Specific instances of using metal oxide NPs, gold, and silver are highlighted and the potential mechanisms of action on microbial cells are described.

6.2 METAL NANOPARTICLES—OVERVIEW

Due to distinctive properties, structures, and extraordinary characteristics exhibited by NPs, metal oxides are extensively studied class of inorganic solids. Transition metal oxides exhibit several industrial applications. Nano-powders, nanotubes, and NPs hold remarkable applications in household uses, bioremediation, medicine, and industries. To explore morphology and scale-dependent applications, 1D nanostructures of metal oxides are ideal systems. This is the main area of focus in nanotechnology and nanoscience due to the easy availability and presence of metal oxides in a variety of shapes, structures, composition, and physical and chemical properties (Zhai et al., 2009).

6.2.1 Synthesis of Nanoparticles

For the synthesis of NPs, numerous methods can be employed and these are widely distributed into two major categories, that is (1) the top-down approach and (2) the bottom-up approach (Wang and Xia 2004; Iravani 2011). These methods are further categorized into various subcategories based on the adopted protocols, operations, and reaction conditions. To date, in chemical and physical fields, several methods are utilized for NPs synthesis, but these are expensive and involve the use of toxic chemicals, which is why the use of biological synthesis is preferred more. Plant, fungal, and bacterial extracts are also used for the synthesis because these kinds of methods are cost-effective, very reliable, and non-toxic (Singh et al., 2015). Out of all the suggested methods for synthesis of NPs, two methods, that is chemical reduction and biological synthesis, were widely accepted because of their benefits in controlling the size and morphology of NPs perfectly.

Top-down and bottom-up pathways are the most commonly used processes in the synthesis of NPs. In the top-down pathway, with the help of various lithographic approaches like chemical etching and ball milling/chemical milling, the material in bulk is converted into smaller nanoscale particles. The biggest limitation of this procedure is that it adds imperfections in the structure's surface, and NPs are greatly dependent on their surface structure (Gudikandula and Maringanti, 2016). However, in the bottom-up pathway, oxidation and bio-reduction procedures are used to prepare NPs from smaller striating materials. The possibility of having any flaw is reduced, as in this procedure, atoms aggregate to form nuclei range at the nanoscale. In the biological procedure, stabilizing and capping mediators (flavonoids, phytochemicals like phenolics, co-factors, and terpenoids) that provide greater stability are used (Korbekandi et al., 2009; Thakkar et al., 2010). In more advanced studies considering the factors like ease of synthesis, cost-efficacy, stable NPs, the function of microbes in green synthesis are also considered (Durán et al., 2007; Huang et al., 2007). From simple prokaryotic bacterial cells to compound eukaryotes like angiosperms, a wide range of life forms are used in the biogenic synthesis (Ahmad et al., 2003).

6.2.2 Methods Commonly Used for Nanoparticles Synthesis

6.2.2.1 Synthesis of Nanoparticles by the Physical Method

NPs synthesis by physical methods includes sonochemistry, laser ablation, radiolysis, and ultraviolet (UV) irradiation, among others. During the physical synthesis process, the vaporization of metal atoms takes place, followed by condensation on a variety of supports, which result in the rearrangement of the metallic atoms and finally aggregated as a small group of metallic NPs (Hurst et al., 2006). These methods required extremely refined instruments, chemicals, and radiative heating, as well as high power expenditure, which lead to a high functioning price.

6.2.2.2 Synthesis of Nanoparticles by the Chemical Method

NP synthesis is a declination of metallic ions in the solution using chemicals. In the circumstances of the reaction mixture, metal ions may help in the process of nucleation or aggregation to the appearance of a small group of metals. The chemicals generally used as reducing agents are sodium borohydride, hydrogen, and hydrazine (Egorova and Revina, 2000). Stabilizing agents like natural or synthetic polymers, such as natural rubber, cellulose, co-polymers micelles, and chitosan, are also utilized. These chemicals are hydrophobic so that they need the count of several organic solvents such as dimethyl, ethane, toluene, formaldehyde, and chloroform. These chemicals are poisonous and are non-biodegradable, which limits the manufacture extent. Some of the poisonous chemicals may also pollute the surface of NPs and make them inappropriate for certain biomedical applications (Patel et al., 2015). In this circumstance to eradicate all these drawbacks of chemical and physical methods and researchers are focusing on the substitute process of synthesis of metal NPs (Figure 6.1).

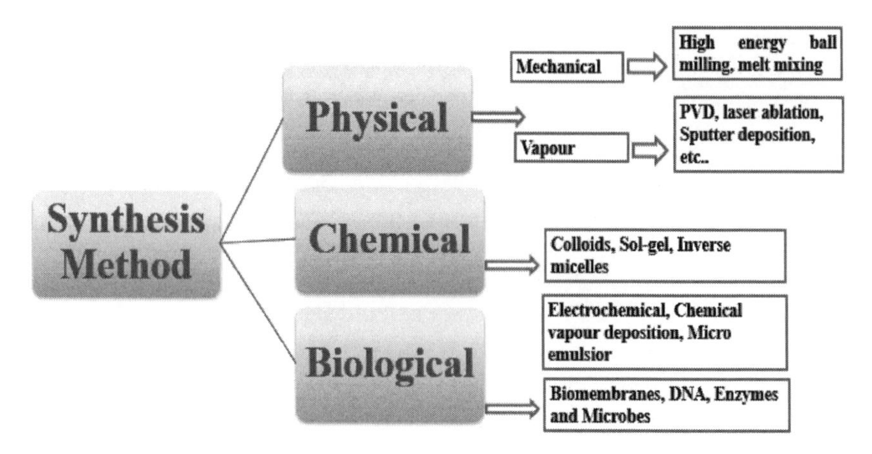

FIGURE 6.1 Different methods utilized for the metal-based nanoparticles synthesis.

6.2.2.3 Synthesis of Nanoparticles by the Biological Method

In recent years, the biogenic synthesis process of metallic NPs has concerned substantial attention. The biogenic NP synthesis process is achieved by using microorganisms and plants (Mukunthan and Balaji, 2012). Biosynthesis can offer NPs of a better-defined size and morphology compared to some of the other physico-chemical methods of manufacturing (Narayan and Sakthivel, 2008). It has been found that the microbial-based synthesis process is readily eco-friendly and well suited to the use of products for pharmacological applications, but synthesis by the use of microorganisms is often costlier than production with plant materials. The major assistance of the plant-based synthesis approach over the physical and chemical process is that plant-based synthesis is cheaper, eco-friendly, and easily extend the process for the synthesis of NPs in large-scale, and there is no need to use pressure, high temperature, and toxic chemicals (Shankar et al., 2004).

A lot of research has been done on the biological synthesis of metal NPs employing microorganisms such as fungi, bacteria, plants, and algae (Figure 6.2). This is due to their reducing or antioxidant potential that is dependable for the reduction of metal NPs. It is found that microbe-mediated synthesis is not valid for large-scale manufacturing, because it needs high septic conditions and special preservation so that the synthesis of NPs from the plant source is more valuable over microbes due to the simple scale-up process and no extra requirement of maintaining cell culture (Dhuper et al., 2012). Employing plant extract for NP synthesis also reduces the additional requirement of microorganism isolation and preparation of culture medium which increases the cost-competitive capability over the synthesis of the NP by microbes. Plant-mediated production is a one-step procedure towards synthesis, whereas microbes during the course of time may lose their ability to produce NPs by mutation; thus, research on plants is expanding fast. Several synthesis processes have been developed, including the chemical decline of metal ions in aqueous solutions with or without stabilizing agents, thermal decayed in organic solutions, and so on.

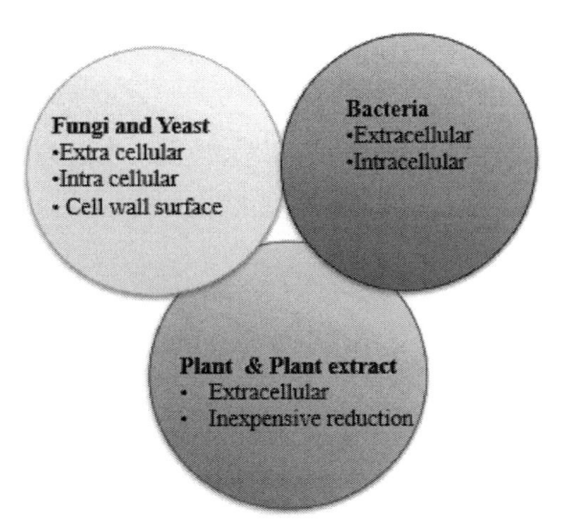

FIGURE 6.2 Biological method for the synthesis of metal-based nanoparticles.

6.2.2.3.1 Synthesis of Nanoparticles From Fungi

Fungi have secondary metabolites and active biomolecules that are very necessary for the NPs synthesis. Fungal species, for example. *F. oxysporum*, exude polymers, proteins, and enzymes that help in the production of metal NPs (Riddin et al., 2006). These components improve the yield and constancy of NPs. In other studies, it was found that several fungal species can synthesize NPs using extracellular amino acid residues. For example, the yeast surface contains aspartic acid and glutamic acid that reduces silver ions into silver metal in the occurrence of a sufficient quantity of light (Nam et al., 2008). Ahmad et al. (2003) observed that fungal species like *F. oxysporum* has reductase enzyme in the cytosol which reduces silver ions into silver metal in the existence of nicotinamide adenine dinucleotide ($NADH^+$), a reducing component (Ahmad et al., 2003). Phytochelatins are a group of compounds that are found mostly in fungus and have a high capability of reducing silver ion into silver metal (Lee and Jun, 2019). Sanghi and Verma (2009) utilized supernatant obtained from fungal species *Coriolus versicolor* to make AgNPs. In this study, Fourier-transform infrared (FTIR) data established the presence of hydroxyl groups in the mycelium of a fungus which contributes electrons to silver ion and condensed it into silver to form AgNPs. It also confirmed the presence of aromatic and aliphatic amines and some proteins in the extract of fungus that act as capping agents to stabilize the formed AgNPs. Moreover, it was confirmed that these compounds stabilized the silver metal by joining with protein through amide bonds. Tan et al. (2002) also reported the participation of SH group-containing protein from the fungal extract in the capping and stabilization of AgNPs. Das et al. (2009) had used mycelia of *R. oryzae* for the synthesis of nano conjugate of gold NPs (AuNPs) through in situ reductions of chloroauric acid ($HAuCl_4$) in an acidic medium (pH 3). *Verticillium* fungus is also a good intermediary for the synthesis of AgNPs. Recently, it is reported that biomass of fungi intracellularly synthesized NPs on exposure to silver nitrate ($AgNO_3$) in an acidic medium (pH 5.5–6) (Mukherjee et al., 2001b).

6.2.2.3.2 Synthesis of NPs From Bacteria

The synthesis of NPs from bacteria includes two approaches: intracellular and extracellular approaches. Extracellular NPs synthesis has more benefits than intracellular synthesis because it is less time-consuming and it does not require any downstream process for setting of NPs by microorganisms (Singh et al., 2016). Reductase enzymes occur inside the cell of bacteria that speed up the decline of metal ions into metal NPs. *D. radiodurans* bacteria have vast antioxidant properties and are extremely resistant to rays and oxidative stress (Li et al., 2016). It makes it positive for use in the green synthesis of AuNPs by ionic appearance. The fictitious AuNPs were constant for a longer time and showed superior antimicrobial potential. Kunoh et al. (2017) used a bacterial strain, *Leptothrix*, for the production of AuNPs by sinking gold salt in an aqueous standard (Kunoh et al., 2017). The utilization of guanine residues of RNA molecules and 2-deoxy guanosine was done for reducing the gold salt.

6.2.2.3.3 Synthesis of NPs by Plant Source

The synthesis of NPs by plant source is extra capable in conditions of obtaining an advanced production than the microorganism. Flora encloses several metabolites and biochemicals (e.g., polyphenols) that can be used as reducing agents in biogenic NPs synthesis. NPs synthesized by a plant source are eco-friendly (avoid the use of poisonous substances) and cost effective. The stability of these NPs is higher compared to those which are synthesized from microorganisms (Singh et al., 2016). Synthesis of NPs from plant sources can be categorized into three groups such as intracellular, extracellular, and using phytochemicals. The extracellular way is utilized when a plant extract is used as the raw material for the synthesis of NPs. Synthesis of NPs from an intracellular manner occurred inside the plant tissue cells by the exploitation of cellular enzymes. Post synthesis, the NPs are improved by rupture the plant cell wall. The occurrence of a higher quantity of phytochemicals in plant extract is linked to the synthesis of NPs in high yield (Mohammadinejad et al., 2019). Synthesis of NPs by phytochemicals is not a general method as it requires information of the exact phytochemical present, which is needed for the stabilized NPs synthesis.

6.3 MECHANISM OF SYNTHESIS

The precise mechanism for the synthesis of NPs using biological agents has not been discovered yet. It has been found that as different biological agents react differently with metal ions so some different biomolecules are responsible for NP synthesis. The mechanism for intra-and extracellular NPs synthesis of various biological agents is also distinct. The microorganisms' cell wall plays a significant role in the intracellular NPs synthesis. The negatively charged cell wall interacts electrostatically with the positively charged metal ions. The enzymes present in the cell wall reduce the metal ions to NPs and eventually disperse the smaller NPs through the cell wall.

A mechanism for intracellular synthesis of NPs using *Verticillium* sp. was reported by Mukherjee et al. (2001a). This mechanism of synthesis is divided into bio-reduction, trapping, and synthesis. The fungal cell surface interacts electrostatically as it comes in contact with metal ions and traps the ions. The metal ions are then reduced by the enzymes found in the cell wall. Finally, the synthesis of NPs and aggregation of particles takes place. In the case of *Lactobacillus* sp., it was observed that during the initial phase of NP synthesis, the nucleation of clusters of metal ions occurs, and therefore, there is an electrostatic interaction between the bacterial cell and metal clusters which ultimately leads to the nanoclusters formation (Nair and Pradeep 2002).

Finally, nanoclusters of smaller size diffused through the bacterial cell wall. In actinomycetes, metal ions are reduced on the surface of mycelia along with cytoplasmic membrane which contributes to the formation of NPs (Ahmad et al., 2003b). Sintubin et al. (2009) demonstrated the mechanism for the synthesis of AgNPs using lactic acid bacteria. According to them, as pH increases, more competition occurs between protons and metal ions for negatively charged binding sites.

The mechanism for the extracellular synthesis of NPs using microbes is found to be essentially mediated by nitrate reductase. The fungi secreted enzyme nitrate reductase helps in the bio-reduction of metal ions and NPs synthesis.

Several researchers supported nitrate reductase for the extracellular synthesis of NPs (Durán et al., 2005; Kumar et al., 2007b; He et al., 2007; Gade et al., 2008; Ingle et al., 2008). Durán et al. (2005) conducted the nitrate reductase assay test through the reaction of nitrite with 2,3-diaminonaphthalene. The emission spectrum showed two major fluorescence intensity peaks at 405 and 490 nm, about the maximum nitrite and 2,3-diaminonapthotriazole (DAN), respectively. With the addition of 0.1% KNO_3 solution, the size of these two bands was found to be increased, indicating the existence of nitrate reductase. It was therefore concluded that the reductase enzyme is responsible for the reduction of Ag^+ ions and the subsequent production of AgNPs.

In the analysis, Ingle et al. (2008) further confirmed the results of Durán et al. (2005), using commercially available nitrate reductase disk; the colored the disk turned reddish from white when confronted with fungal filtrate, signifying the existence of nitrate reductase. Thus, in the case of fungi, it may be concluded that the NADH-dependent reductase enzyme is concerned with the reduction of Ag^+ to Ag^0.

In the case of AuNPs using *Rhodopseudomonas capsulate*, the same mechanism was also reported (He et al., 2007). The *R. capsulata* bacterium is responsible for the secretion of NADH and NADH-dependent co-factor enzymes. The bio-reduction of gold ions was found to be triggered by the electron transfer from the NADH via NADH-dependent reductase as an electron carrier. The gold ions then receive electrons and are reduced to Au (0) and hence result in the AuNPs formation.

Nangia et al. (2009) suggested the bacteria *Stenotrophomonas maltophilia* for synthesizing AuNPs. The authors demonstrated that AuNPs should be biosynthesized and stabilized through charge capping in *S. maltophilia* which included the NADPH-dependent reductase enzyme that converts Au^{3+} to Au^0 through enzymatic metal reduction phase of an electron shuttle. The synthesis of cadmium sulfide (CdS) NPs using *Schizosaccharomyces pombe* was found to be based on a response to stress protein (Kowshik et al., 2003). Phytochelatin enzyme is activated upon exposure of cadmium to *S. pombe* and synthesizes phytochelatins.

The cytoplasmic cadmium is chelated to the phytochelatin–Cd complex. A vacuolar protein-type adenosine triphosphate–binding cassette then transports phytochelatin–Cd complex across the vacuolar membrane. A high molecular-weight phytochelatin –CdS^{-2} complex or CdS nanocrystal sulfide is applied to the matrix inside the vacuole sulfide.

Mukherjee et al. (2008) reported a possible mechanism on the encapsulation of NPs using the Raman spectrum. The Raman spectra indicated that C=O bonds of carboxylate ions and Ag–N bonds from the free amine groups lie perpendicular to the nanosilver surface and get directly associated with the capping of AgNPs. Broadened symmetric and asymmetric bands of carbon dioxide (CO_2) were observed, which resulted due to the distortion of the respective bonds and encapsulation of silver NPs. The absence of a peak at Ag–S vibration indicated that disulfide linkages do not play any role in stabilization. Thus, the authors conclude that some of the peptide linkages from amino acids undergo hydrolysis to synthesize free carboxylate ions and free amino groups which possibly act in the encapsulation of silver NPs.

6.4 CHARACTERIZATION

6.4.1 Ultraviolet-Visible Absorption Spectroscopy

Absorbance spectroscopy is used for the determination of a solution optical properties. A light is transmitted through the sample solution and the quantity of absorbed light is measured. The wavelength is varied, and at each wavelength, the absorbance is measured. The absorbance can be used to calculate the solution concentration using the Beer–Lambert law. Subbaiya et al. (2014) reported that when a 1 mM silver nitrate solution was treated with *Nerium obander* plant extract, the optical measurement of UV-visible spectrophotometer has a specific absorbance peak, like 410 nm.

6.4.2 X-Ray Diffraction Analysis

X-ray diffraction (XRD) is a conventional technique for the determination of crystallographic structure and morphology. There is an increase or decrease in intensity with the amount of constituent. This technique is used to establish the metallic nature of particles and gives information on the translational symmetry, size, and shape of the unit cell from peak positions and information on electron density in the unit cell, namely where the atoms are located from peak intensities. XRD analysis with various NPs has been studied by various research workers to find the high crystallinity of the prepared sample (YelilArasi et al., 2012).

6.4.3 Fourier-Transform Infrared Spectroscopy

Fourier-transform infrared (FTIR) spectroscopy is used to investigate the vibrational properties of amino acids and co-factors that are prone to slight structural changes (Rana et al., 2018). On one hand, the lack of precision of this technique helps one to specifically investigate the vibrational properties of almost all co-factors, side chains of amino acids, and water molecules. On the other hand, we can use FTIR difference spectroscopy caused by a reaction to identify vibrations corresponding to the single chemical groups involved in a particular reaction. Different strategies are used to identify the infrared (IR) signatures of each residue of interest in the resulting FTIR difference spectra. Using FTIR spectroscopy, the green synthesized AgNP was analysed using different leaf extracts and showed characteristic peaks (Murugan and Kumara, 2014).

6.4.4 Microscopic Techniques

These techniques are mainly used for morphological NP studies, namely transmission electron microscopy (TEM) and scanning electron microscopy (SEM). These techniques are mainly used for morphological properties of NPs, which were, more or less, uniform in shape and size (Shobha et al., 2014).

6.4.5 Transmission Electron Microscopy

TEM is a method of microscopy in which an electron beam is transmitted through an ultra-thin layer, communicating with the material as it passes. An image is created by the interaction of the transmitted electrons through the specimen; the image is magnified and centered on an imaging system, such as a fluorescent screen, a photographic film plate, or a sensor such as a Charged-Couple Device (CCD) camera can detect it. TEM constitutes a significant tool of research in several scientific areas, both in physical and biological sciences. TEM is used in cancer research, virology, materials science, environmental research, nanotechnology, and semiconductor research.

6.4.6 Scanning Electron Microscopy

Scanning electron microscope research characterization is used to determine the scale, shape, and morphologies of the NPs. SEM provides high-resolution images of the target surface of a sample. The scanning electron microscope operates the same way as an optical microscope, but it detects photons rather than the electron scattered from the sample. Because electrons can be accelerated by an electric potential, the wavelength can be made shorter than the one of photons. That allows the SEM to magnify images up to 200,000 times and measurements of particle size, characterization of sputter-coated sample, and sensitivity up to 1 nm (Umer et al., 2012).

6.4.7 Energy-Dispersive X-Ray

Energy-dispersive X-ray spectroscopy is an analytical method used to analyse an atomic sample or to classify a chemical. This is one of the type variants of X-ray fluorescence spectroscopy that focuses on examining a sample through interactions between matter and electromagnetic radiation, measuring the X-rays emitted by the matter in the reaction of hitting the charged particles. Its characterization capabilities are largely due to the fundamental principle that each element has a unique atomic structure that allowing X-rays characteristic of the atomic structure of an element to be distinguished uniquely from each other.

6.4.8 Thermo Gravimetric Analysis/Differential Thermal Analyser

Thermo gravimetric analysis is a thermal analysis technique which, in a controlled environment, measures the weight change in a material as a function of temperature and time.

This is very useful for investigating a material's thermal stability or investigating its behavior in various atmospheres (e.g. inert or oxidizing). It is suitable for use in all types of solid materials, including organic or inorganic materials.

6.5 APPLICATIONS OF NANOPARTICLES

The use of NPs in food, cancer treatment, drug and gene delivery, and diagnostic tools, among other areas, has been widely studied in the past few years. Nowadays, it has become a topic of interest among many scientists not only due to their wide applications but also due to their small size and large surface-to-volume ratio, which differentiate them from larger particles of bulk material (Mittal et al., 2014). Also, in the field of biomedicine, NPs have gained popularity in recent times. In the diagnostic and therapeutic fields of medicine, NPs have significant benefits (Navalakhe and Nandedkar, 2007).

6.5.1 Application of Nanoparticles in Drug Delivery

NPs involved in drug delivery entrap drugs either to enhance delivery or uptake by the target cells and a reduction in the toxicity of the free drug to non-target organs (Wim, 2008).

6.5.2 Applications of Nanoparticles in Food

Food that uses nanotechnology as methods or adds NPs while cultivation, production, processing, or packaging is known as nano food. Nano food can be developed for many reasons such as for food safety betterment, improvement in flavours and nutrition value, decreasing manufacturing, and consumer costs. Also, they provide other benefits, including health-promoting additives, increasing shelf life, and the addition of new flavours. The applications of nanotechnology are widely increasing, covering all areas from the very first step, agriculture to the ultimate final product processing. It is also used in improving nutrient bioavailability.

6.5.3 Application of Nanoparticles in Gene Delivery

To get the desirable protein introduced in the selected host cell, the gene is introduced with the help of the gene delivery technique. Nowadays, there are different types of preliminary gene delivery systems, mainly employing viral vectors like retroviruses and adenoviruses, nucleic acid transfection, and nucleic acid electroporation. (Kami et al., 2011).

6.5.4 Application of Nanoparticles in Cancer Treatment

Considering the need of finding a cancer treatment many studies on NPs are under investigation currently. Some metals like gold, silver, and some magnetic oxides (mainly Fe_3O_4) are topics of interest. At present, a wide range of NP systems is studied for their biomedical properties, including cancer therapeutics, which include valuable metals (mostly silver and gold systems) and some magnetic oxides (in particular magnetite Fe_3O_4), and they have received much interest and considered as natural NPs. For cancer treatment, the unique up conversion process of upconversion nanoparticles (UCNPs) may be utilized to activate photosensitive therapeutic agents (Cheng et al., 2013).

6.5.5 Other Applications of Nanoparticles

In recent years, NPs are involved with new applications in areas like information and communication technology, power engineering, industrial engineering, environmental engineering, chemical industry, medicine, in pharmaceuticals and cosmetics, and so on. Nanoscale materials have been around us for decades and are used in many daily used things like window glass, sunglasses, car bumpers, paints, and so on; however, a few are newly discovered and are used in making sunscreens and cosmetics, textiles, coatings, sports goods, explosives, propellants, and pyrotechnics, or some of their applications are under investigation (e.g. in batteries, fuel cells, solar cells, electronic storage media, light sources, display technologies, bio detectors, and bioanalysis, medical implants, new organs, and drug delivery systems). In conclusion, the discovery of new NPs and their extraordinary applications are increasing every day.

6.5.6 Applications of Nanoparticles as Antimicrobial Agents

The antimicrobial benefits provided by NPs not only solve the problem of resistance mechanisms but also include the enzyme inactivation, reduce the permeability of the cell, modify target sites/enzymes, overexpress efflux pumps to increase efflux and to prevent from the antibacterial activity of antimicrobial agents (Mulvey and Simor, 2009; Baptista et al., 2018). However, NPs conjugated antibiotics indicate cooperative effects against bacteria, inhibit the formation of biofilms, and are used to fight against MDR organisms (Pelgrift and Friedman 2013a; Baptista et al., 2018). Many distinguishing features of NPs make them alternatives to traditional antibiotics. First, the large surface area–to–volume ratio of NPs increases the contact area with target organisms. NPs can act as nanoscale molecules, communicating with bacterial cells, regulating cell membrane penetration, and interfering with molecular pathways (Rai et al., 2012; Dakal et al., 2016; Durán et al., 2016; Hemeg, 2017). Second, NPs may improve the inhibitory effects of antibiotics. Saha et al. (2007) showed that AuNPs paired with ampicillin, streptomycin, or kanamycin could decrease the minimum inhibitory concentrations (MICs) of the antibiotic equivalents against both gram-negative and gram-positive bacteria. Similarly, Gupta et al. (2017) showed a mutually stimulating effect of functioned AuNPs and fluoroquinol one antibiotics for the treatment of MDR *E. coli* infections. However, during the preparation of NPs and their interaction with bacteria, different antibacterial properties of NPs, such as size, shape, physiochemical properties, chemical modification, solvents, and environmental factors, play an important role (Beyth et al., 2015). At last, to overcome antibiotic resistance, combinations of antibiotics and NPs provide complex antimicrobial mechanisms (Huh and Kwon, 2011). According to Gupta et al. (2017), the use of AuNPs and fluoroquinolone antibiotics for the treatment of MDR *E. coli* bacterial strains is a synergetic way.

6.6 ANTIMICROBIAL ACTIVITY OF NANOPARTICLES

In the field of scientific publication and research, the popularity of NPs application is growing day by day. The biogenically synthesized NPs known as the green generation of NPs have led to sustainable developments in diagnostic medicine and clinical applications (Stiufiuc et al., 2013; Kargara et al., 2015; Jana and Pal, 2015). NPs are considered as a better solution than antibiotics because they have better outcomes with minimal side effects and major effects on bacterial MDR. The green synthesized metal or metal oxide NPs have the potential to become antibacterial, antifungal, and antiparasitic medicines (Acharyulu et al., 2014; Gopinath et al., 2014; Naika et al., 2015; Awwad et al., 2015; Hussain et al., 2015). Multiple in vivo and in vitro studies have analysed the toxicity of metal NPs. It is observed in several studies that toxicity caused by metal-based NPs can affect the biological behavior of the organ, tissue, cellular, subcellular, and protein levels. Easy penetration of the NPs in the cell is due to its unique size, which leads to many adverse effects. Due to their physicochemical and biological properties, it is well accepted that metal-based NPs are a better option as therapeutic agents and antimicrobials (Camporotondia et al., 2013; Kamaraj et al., 2014; Samavati and Ismail, 2017). NPs have extremely large biocidal effects (Jayaseelan et al., 2012; Stiufiuc et al., 2013) and are stated captivating in a scientific field, particularly for the creation of a new class of antimicrobials (Periasamy et al., 2012). However, the antimicrobial action of AgNPs is of a wide spectrum. The metabolically and morphologically different microorganisms emerge to join with multidimensional mechanisms of NPs to interact with microbes (Kim et al., 2007). The unique level of biocidal effects and unique mechanisms of impairment is achieved because of their size, shape, structure, and mode of interaction with the microbial surface. Shape, size, the concentration of NPs, and microorganisms influence the bactericidal effects of NPs (Liu, Dai et al., 2010; Mirzajani et al., 2011). NPs, being smaller in size, appear to have a high likelihood of penetrating deeper into the bacterial cell wall and come in contact, with the membrane adding to the cell membrane damage. Indisputably, the bactericidal brilliance is prevailing in the smaller size of NPs than larger size with a positive zeta potential. In connection with the bacterial cell, the electrostatic forces are created by the NPs having zeta potentials. It is the potential difference between the scattering media and the stationary layer of fluid connected to the scattered particle. Bacterial cell wall promotes attraction between two particles leading to the penetration in the bacterial membrane because of its negative charge.

6.6.1 Antimicrobial Mechanism of Metallic Nanoparticles

Since around 4000 BCE, metals have been known to have antibacterial properties such as, silver chalices were used for drinking water (Alexander, 2009; Ivask et al., 2014). Recent researches on nanophysics attract scientists to understand various metallic NPs antibacterial properties (Grass et al., 2011; Kailasa et al., 2019). Antibacterial mechanism of metallic NPs is still a topic of debate, and three main processes are presumed for this, which include the generation of reactive oxidative species (ROS), the releasing process of ions, and, finally, the interaction of NPs with the cell membrane (Figure 6.3). Comparing metallic NPs with their salts, they have

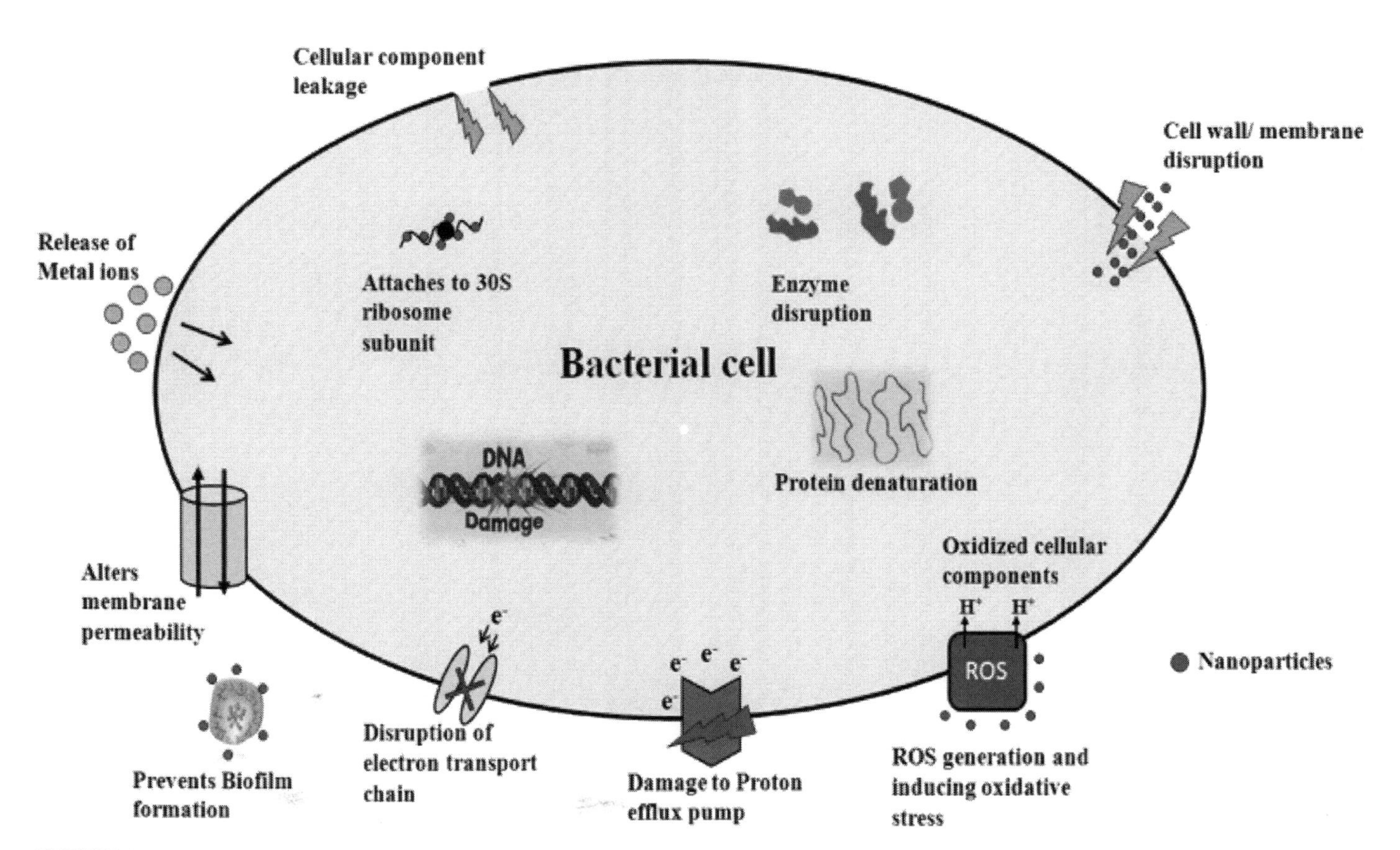

FIGURE 6.3 Mechanisms of antimicrobial activity of the metal NPs.

better potential to fight against bacterial infections (Zhang et al., 2010; Pelgrift et al., 2013; Dizaj et al., 2014; Arias and Murray, 2015). The antibacterial mechanism is mostly influenced by the size of NPs (Raghupathi et al., 2011; Guzman et al., 2012b; Azam et al., 2012; Kim et al., 2012; You et al., 2012; Ali et al., 2016).

6.6.1.1 Entering the Cell

Attachment of the nanometer range metallic ions with the cell through transmembrane protein is the initial step of the antibacterial mechanism. The next step is to create structural changes in the cell membrane and obstructing its transport channels (Dutta et al., 2012); this whole process depends on the size. After this, NPs may be incorporated, which will generate ionization inside the cell and cause destruction to the intracellular structures leading to cell death (Figure 6.3; Dizaj et al., 2014).

6.6.1.2 Reactive Oxidative Species Generation

For NPs antibacterial effectiveness, the formation of ROS plays a critical role (Figure 6.2). ROS contains ephemeral oxidants, like superoxide radicals (O^{-2}), hydrogen peroxide (H_2O_2), hydroxyl radicals (OH^{-1}) and singlet oxygen (O^{-2}; Raffi et al., 2008; Baek and An, 2011). ROS can result in damage to the peptidoglycans of cell membranes, DNA, mRNA, proteins, and ribosomes because of its high reactivity (Pelgrift et al., 2013). Translation, transcription, the electron transport chain, and enzymatic activity are also inhibited by ROS (Raffi et al., 2008). The generation of ROS is used as a main mechanism of toxicity by a few metal oxide NPs (Nel et al., 2006; Dutta et al., 2012).

6.6.1.3 Protein Inactivation and DNA Destruction

To deactivate the function of enzymes, metal atoms attach with thiol groups of enzymes. It is observed that the bonding of hydrogen between two antiparallel strands of DNA is also disturbed by metal ion attachment, which ultimately destructs the molecule of DNA (Figure 6.2). The true metal ions also show a tendency to attach to DNA, after they get inside the cell, but it is still under investigation (Jung et al., 2008).

6.7 FACTORS AFFECTING THE ANTIMICROBIAL ACTIVITY OF NANOPARTICLES' CONCENTRATION AND SIZE

In NPs, the role of size and concentration has been analysed. At a low concentration of 0.01 ppm, silver NPs with different sizes has been investigated for activity (Poole 2002; Kim et al., 2007). To kill and destroy bacteria, the smallest AgNPs were more effective in comparison to larger ones. The small-sized NPs discharged more silver cations because of their high surface-to-volume ratio; therefore, they are more effective in comparison to large-sized NPs (Torres et al., 2013).

6.7.1 Chemical Composition

The base of NPs is their chemical composition, which determines the differences in their activities. The NPs are observed to produce ROS (titanium oxide [TiO_2], silicon oxide [SiO_2], and zinc oxide [ZnO_2]) against *B. subtilis* and *E. coli*. The ascending order of the biocidal activity of these compounds was SiO_2 to TiO_2 to ZnO. By 10-ppm concentration of ZnONPs the growth of *B. subtilis* was 90% inhibited, while 1000 and 2000 ppm of TiO_2 and SiO_2, respectively, arerequired to inhibit the development of *B. subtilis* to 90%. Although, the inhibition effect of NPs on *E. coli* was partly at 10 ppm of ZnONPs and 500 ppm of TiO_2NPs and SiO_2NPs (Huang et al., 2008). Additionally, it was explained that light or dark conditions do not affect the bactericidal activity, but the growth inhibition includes mechanism-based ROS production.

6.7.2 The Shape of Nanoparticles

Studies showed that various shapes (elongated rod, truncated triangular, and spherical) of AgNPs have different biocidal activity intensity. The difference in the colony-forming unit count of *E. coli* depends on the media inoculated with different shapes of NPs. The morphology of NPs seems to stimulate the activity of NPs. The shape-determined activity was dependent on the term of facets. The spherical NPs primarily had 100 facets, rod-shaped NPs had 111 facets on the side surface and 100 on the end, and truncated triangular NPs have 111 facets on top basal planes. The 111 facets are of high atom density that aid the antibacterial reactivity of NPs (Liu et al., 2010).

6.7.3 Target Microorganisms

It is reported in many studies that NPs showed greater biocidal activity against gram-negative rod-shaped bacteria as compared to gram-positive bacteria. *E. coli* and *S. aureus* were used to analyse the impact of AgNPs and results showed way more activity against *E. coli* (MIC 3.3-3.6) than *S. aureus* (MIC more than 33 nm). The higher concentration of peptidoglycans is present in the cell wall of gram-positive bacteria (*S. aureus*) which depicts the difference in the results (Wang et al., 2010). Huang et al. revealed the activity of ZnONPs against both the gram-positive (*S. aureus*) and gram-negative (*E. coli*) bacteria (Huang et al., 2008). However, in another study, ZnONPs were found to be most effective against *S. aureus* than *E. coli* and *P. aeruginosa* (Mao et al., 2004).

6.7.4 Photoactivation

TiO_2NPs showed good activity against *E. coli*, which noticeably increases with UV radiations. Additionally, it was observed that TiO_2NPs without photoactivation have negligible activity, that is it shows about 20 % of growth inhibition of *S. aureus*. On the other side, ZnONPs show escalating activity after photoactivation by UV-visual radiation (Shankar et al., 2003; Philip 2009). ROS production can be enhanced with photoactivation with blue light, which further enhances the activities of these NPs (Lipovsky et al., 2011). However, it is also a well-known fact that no surviving microorganisms have shown any effect because of bacterial incubation with NPs in the dark (Huang et al., 2008).

6.8 BIOGENIC METALLIC NANOPARTICLES FOR ANTIBACTERIAL APPLICATIONS

Silver, copper oxide, gold, titanium oxide, magnesium oxide, and zinc oxide are the important metal NPs used as antimicrobial agents because their potent antibacterial and antifungal effects are well known (Beyth et al., 2015; Zhang, 2015). Recently, several investigations reported that different metal oxide NPs showed biocidal action against several pathogenic bacteria. The antimicrobial potential of NPs is well known to be the purpose of the surface area in contact with the microbes (Ravishankar and Jamuna, 2011; Franci et al., 2015; Chiriac et al., 2016). The bactericidal action of metal NPs has been accredited to their smaller size and high surface to volume ratio, which enables them to interact closely with the outer membranes of microbes and is not only because of the discharge of metal ions in the solution (Ruparelia et al., 2008). Specific metal ions are necessary for the development of all organisms, and their privation can cause damage to the structure of plasma membranes and nucleic acids. However, the excess of these ions or the presence of other unnecessary ions can be fatal to the cells, which was often caused by protein dysfunction, oxidative stress, or membrane damage (Lemire et al., 2013). Some metal NPs are effective antimicrobial agents against several pathogenic microbes. This antimicrobial activity is responsible for water treatment, biomedical and surgical devices, synthetic textiles, and food processing or packaging (Ravishankar and Jamuna, 2011).

6.8.1 Silver Nanoparticles

AgNPs have been used against several pathogenic bacteria due to their relatively low cytotoxicity. Several studies reported that AgNPs have significant antibacterial activity, but their mechanism of action is not fully understood (Ravishankar and Jamuna, 2011; Beyth et al., 2015). However, a few studies described the mechanism of action against bacteria, and it is assumed that the high affinity of silver toward phosphorus and sulfur is the key element for the antimicrobial effect (Ravishankar and Jamuna, 2011; Maiti et al., 2014; Franci et al., 2015). The silver ions have an affinity for nitrogen and sulfur and can disrupt or inhibit the structure of proteins by binding to amino and thiol groups, which affects the viability of bacterial cells (Beyth et al., 2015). Also, AgNPs released several silver ions that can interact with phosphorus moieties in a nucleic acid (DNA), leading to inactivation of the replication process, or they can react with sulfur-containing amino acids, resulting in the inhibition of protein or enzyme functions (León-Silva et al., 2016). Another mechanism reported that ROS are induced by AgNPs, forming free radicals with a bactericidal action membrane (Franci et al., 2015). ROS causes damage to the intracellular structures, particularly mitochondria and the nucleus; besides ribosome denaturation, they also result in protein synthesis inhibition (León-Silva et al., 2016). Earlier studies have shown that the bactericidal action of AgNPs is dependent on their shape, size, and zeta potential. Lesser dimensions (<30 nm) have a superior capability to penetrate microbes, and the electrostatic force produced when NPs with a positive zeta potential encounter the negative surface charge of the organism. This promotes a closer communication between the two entities and probably the penetration through the cell membranes (Franci et al., 2015).

AgNPs have significant fungicidal and bactericidal activities that have been employed in the fabric industry, paints, food, and packaging industries (Suresh et al., 2012). Extensive green synthesis of AgNPs of various sizes and shapes from bacteria, plants, fungi, and yeast has been studied widely (Singh et al., 2016). The basic antibacterial action mechanism of AgNPs has been reported due to the release of silver ions or the intracellular deposition of NPs (Kim et al., 2016; Hoseinnejad et al., 2018). The detailed mechanism involves cell membrane damage, inhibition of mRNA synthesis, disruption of energy metabolism, and the generation of oxidative stress due to ROS formation. AgNPs released silver ions that have been found to interact with sulfur- and phosphorus-containing amino acids (extrinsic and intrinsic proteins) in the cell wall and plasma membrane of microbes (Hindi et al., 2009). The initial interaction of silver ions with the cell surface of microbes starts with the attraction of positively charged silver with negatively charged microbial cells, thus leading to the development of several pores in the cell membrane and outflow of intracellular materials. This process results in an electrochemical instability in the cells and permits the silver ions to pass through the cell membrane into the cytoplasm of cells to interact with the intracellular content which ultimately results in permanent cell damage (Dakal et al., 2016).

Silver ions also inhibit the structure of enzymes that are vital for adenosine triphosphate (ATP) synthesis, inhibit respiratory enzymes leading to the synthesis of ROS, damage genetic material, and destabilize and disrupt the cell membrane. NPs, due to their small size with more surface area, have a greater capability to cross the cell membrane (Lara et al., 2010). NPs are more active against gram-negative bacteria in comparison to gram-positive bacteria having a thicker peptidoglycan layer. A higher content of peptidoglycan in the cell wall of gram-positive bacteria gives more resistance against the NPs' penetration. Many reports of antimicrobial action of AgNPs have linked their toxicity with the shape and size of the particles (Raza et al., 2016). The NPs with the larger surface area have been reported to release silver ions at a higher rate, which is a major factor for high antibacterial activity (Tang and Zheng, 2018). Antibacterial action of AgNPs has been studied against MDR bacteria such as *P. aeruginosa*, *E. coli*, *S. aureus*, *S. pyogenes*, and *K. pneumonia* (Gopinath et al., 2015; Jinu et al., 2017). This bactericidal effect is attributed to the inhibition of cell membrane, protein, and nucleic acid synthesis.

Furthermore, AgNPs have also been shown to improve the antimicrobial activity of antibiotics, such as penicillin-G, amoxicillin, clindamycin, vancomycin, and particularly erythromycin, against *E. coli* and *S. aureus* (Shahverdi et al., 2007). Also, silver carbene complexes encapsulated in NPs are efficient against MDR bacteria, which includes MRSA, *A. baumannii*, *B. cepacia*, *P. aeruginosa*, and *K. pneumoniae* (Leid et al., 2012). The strong bactericidal action of AgNPs against the MDR bacteria is mostly due to the disruption of microbial cells. Despite having multiple mechanisms for antimicrobial effects, the latest investigation describing a pretreatment of microbial cells with a sub-lethal concentration of AgNPs showed minor membrane damage, lower levels of intracellular ROS, and the greater amount of intracellular ATP when bacterial cells were further exposed to the antibiotic ampicillin. This concluded that the pretreatment of cells with sub-lethal concentrations of AgNPs leads to lifelong responses that increase the antibiotic stress resistance in microbes at numerous levels (Kaweeteerawat et al., 2017).

6.8.2 Gold Nanoparticles

AuNPs possess advanced stability when it is in contact with organic fluid, and they are considered harmless to human cells. Only AuNPs are considered as biologically static and have an absence of antimicrobial properties but can be customized to have a chemical or photothermal functionality (Zhang, 2015). AuNPs, when joined with photosensitizers, can kill cancer cells and various pathogens resistant to many drugs (León-Silva et al., 2016). AuNPs can soak up near-infrared (NIR) radiation, so they can transmit this energy into the adjacent environment as heat, and once close to bacterial cells, the heat produced by NIR radiation cause permanent cell damage (Norman et al., 2008; Beyth et al., 2015). A lot of studies confirmed the large-scale AuNP antibacterial activity linked to antibiotics, for instance, vancomycin, cefaclor ampicillin, and the antibacterial enzyme lysozyme against gram-negative, as well as gram-positive, bacteria (Zhang, 2015; Payne et al., 2016). In the case of cefaclor, the AuNPs causes porous cell walls, resulting in the cell contents leakage, leading to cell death (Ravishankar and Jamuna, 2011). One of the main broadly studied biogenic NPs is gold NPs (AuNPs). Mostly, the shape of AuNPs is spherical (Aromal et al., 2012), hexagonal (Sheny et al., 2012), and triangular (Suman et al., 2014). Rod-shaped NPs are also observed in different studies. AuNPs are synthesized using the entire plant or by a mixture of different components, used as reducing agents. The morphology of synthesized NPs is defined by the extracts nature which exhibits bio-reductant properties.

Galaxaura-synthesized AuNPs are one of the crucial examples where a broad variety of shapes (rod, spherical, hexagonal, triangular) and size (4–77 nm) of NPs were formed (Abdel-Raouf et al., 2017). One more vital finding as a result of pH on AuNPs size was observed with a core size of 6 nm and 18 nm which were obtained at pH 9 and pH 2, respectively, from the extract of mango peel (Yang et al., 2014). AuNPs are documented for their biocompatibility to microbial cells with no bactericidal or bacteriostatic activity. However, antibiotics incorporated AuNPs are exposed to have a powerful bactericidal effect against the drug-resistant bacteria. The ampicillin leap AuNPs have been demonstrated to spoil ampicillin-resistant bacteria, which includes MRSA, *Enterobacter aerogenes*, *P. aeruginosa*, and *E. coli* K-12 sub strain DH5-alpha by numerous mechanisms. AuNPs-AMP (Anti-Microbial Peptide) can overcome bacterial articulation at higher concentrations of beta-lactamase, furthermore, a transmembrane pump, catalyzingdrug efflux from the bacterial cell is also inhibited by AuNP-AMP (Brown et al., 2012).

6.8.3 Copper Oxide Nanoparticles

Copper oxide is cost-effective and can mix with polymers easily. It is moderately stable in terms of physical and chemical properties (Gawande et al., 2016). The precise bactericidal results for CuNPs are still unknown. CuONPs activity is known to be dependent on bacterial species, and its antibacterial effect also takes place by membrane damage and ROS production (Lemire et al., 2013; Beyth et al., 2015), which revealed that the loss of viability of bacterial cell has been related employing the uptake of copper ions and enlarged construction of reactive oxygen species. Other authors observed that the surfaces of copper do not make major oxidative DNA damage in vivo and various other proof recommend that DNA damage is not the main reason for copper-mediated surface killing (Warnes et al., 2012).

DNA and RNA released from dead cells are slowly corrupted at a rate that enhances employing the copper content of the surface alloy. In this context, Beyth et al. (2015) reported that because of the superior compatibility of amines and carboxyl groups on cell surfaces of the CuNPs of copper have a superb bactericidal activity against *B. subtilis* and *B. anthracis*. CuONPs gained vital significance owing to their value in antimicrobial capacity, cosmetics, pharmaceutical industry, farming, transport, and power (El-Batal et al., 2017). It is fairly easy to manufacture CuONPs using chemical methods, but it has numerous disadvantages, such as high toxicity, low potency, atmosphere surliness, and high cost. The synthesis of CuONPs of different biogenic sources, for example, polysaccharides such as chitosan, alginate, pectin, bacteria, and leaf extracts are preferred. Unlike silver, gold, and other NPs, it has been a challenge to make stable CuONPs due to their proneness to oxidation when subjected to an aqueous medium (Singh et al., 2013). While there are a few reports on the synthesis of CuONPs under static circumstances by salts of copper, there is extremely partial information that recommends the production of metallic CuONPs in non-static circumstances (El-Batal et al., 2018). The biogenic synthesis of CuONPs is moderately new, and behaviour is looking into to compose it with an eco-friendly approach. The method subsequently shows CuONPs have prophylactic activity and therefore have an electrostatic attraction between Cu^{+2} and plasma membrane which helps in disrupting the membrane sheath and destroy the cells (Hoshino et al., 1999). Upon being occupied up by the cell, Cu^{+2} ions are energetically much simpler to pass through a lipid bilayer, leading to ROS production, protein oxidation, and lipid peroxidation (Bogdanovic et al., 2014). CuONPs have shown a high bactericidal effect for both gram-positive and gram-negative bacteria (Bogdanovic et al., 2014; DeAlba-Montero et al., 2017). The wide-range antimicrobial capacity of CuONPs recommended the probable utilization in the treatment of wound healing (Hostynek and Maibach, 2003; Borkow and Gabbay, 2004).

6.8.4 Zinc Oxide Nanoparticles

ZnONPs have been proved to be toxic for gram-positive and gram-negative bacteria, as well as for the spores, which are resistant to elevated pressure and temperature (Ma et al., 2013). Alternatively, zinc oxidepossesses the minimal outcome of body cells, which makes it secure for utilization as the antimicrobial mediator (Espitia et al., 2012). A certain survey presented the antimicrobial capacity of ZnONPs is based on the selective size and concentration of the particles (Ma et al., 2013). Along with this, the mechanism accounted for the antibacterial capacity of ZnONPs, and the production of hydrogen peroxide (H_2O_2) from the zinc oxide surface is considered more efficient for the inhibition of growth of bacteria. The inhibitory outcome of hydrogen by alive cells stimulates the catalase enzyme; it stimulates the dissolution of H_2O_2 to H_2O and O_2. It also attracts electrons from bacterial cell walls and damages the basic molecular structure of the lipid layer and cell proteins on the cell surface. The cell walls are oxidized and destroyed or break completely. Synthesis of ZnONPs can be done using diverse sources (Madhumitha et al., 2016). They are harmless with high-quality photocatalysis and high lucidity. Synthesis of ZnONPs can also be made from different plant parts, such as bark, flowers, leaves, rhizomes, roots, and fruits (Ahmed et al., 2017). ZnONPs demonstrate a probable antibacterial capacity (Bhuyan et al., 2015) and a good deprivation of the photo and have applications in the medicine delivery (Ali et al., 2016) and anticancer properties (Vimala et al., 2014). ZnONPs are broadly experienced metallic NPs for antimicrobials. The broad-spectrum gram-positive and gram-negative bacteria, for example *L. monocytogenes, S. aureus*, and *E. coli*, has verified sensitivity against ZnONPs (Jones et al., 2008; Liu et al., 2009). The management of ZnONPs microbial cells undergoes ROS, membrane destruction of reducing sugars, lipid peroxidation, reducing sugars membrane, proteins, cell viability, and DNA (Tiwari et al., 2018). ZnONPs generate ROS for instance hydrogen peroxide and superoxide anion in the cells (Kumar et al., 2011; Horie et al., 2012). ROS cause membrane leakage of nucleic acids and proteins by enhancing lipid peroxidation. Zn^{+2} ions released by NPs also injure the plasma membrane and cooperate with intracellular machinery (McDevitt et al., 2011; Li et al., 2011). ZnONPs was shown to slow down the growth of carbapenem-resistant *A. baumannii* by producing ROS and causing damage of membrane, signifying that ZnONPs residential as an option to carbapenems (Beta-lactam; Tiwari et al., 2018).

6.8.5 Titanium Dioxide Nanoparticles

TiO_2NPs have drawn considerable interest because they show unique and enhanced properties compared with their counterparts of bulk material. These NPs demonstrate quantum size effects, in which materialistic effects such as physical and chemical properties depend heavily on particle size (Othman et al., 2014). NPs made from (TiO_2NPs) are photocatalytic. TiO_2NPs decompose organic compounds when exposed to non-lethal UV light at a wavelength of less than 385 nm by the formation and continuous release of superoxide ions and hydroxyl radicals (Beyth et al., 2015). The TiO_2NPs of a nanometerscale can further boost titanium oxide antimicrobial activity (Othman et al., 2014). TiO_2NPs possess interesting catalytic, optical, dielectric, and antibacterial properties and, thus, can be used in various industries (Vamathevan et al., 2002), sensors (Varghese et al., 2003), biosensors (Zhou et al., 2005), solar cells (Zhang et al., 2014), and medical diagnostics as image-contrast agents (Kirillin et al., 2009).

TiO_2NPs with various morphologies such as nanotubes and nanorods are usually synthesized using various stabilizing and reducing agents (Chen and Mao, 2007). Hydrothermal preparation is an alternative solution because of its low efficacy and easiness (Mali et al., 2011), but green paths are necessary to be built to provide a consistent source in adequate quantities with no harmful environmental effects. TiO_2NPs are synthesized from plants (Sankar et al., 2015) and fungi also (Rajakumar et al., 2012; Hunagund et al., 2016). TiO_2NPs often demonstrate antimicrobial effects through multiple mechanisms indicating that the possibility of microbial cells developing resistance to these NPs is extremely low. TiO_2NPs have been stated to have a bactericidal effect against *S. aureus*, *Escherichia coli*, *Pseudomonas aeruginosa*, and *E. faecium* (Foster et al., 2011; Li et al., 2012). TiO_2NPs destroy microorganisms by producing ROS with the exposure of UV radiation (Li et al., 2012). The produced ROS interfere with the oxidative phosphorylation of the cell membrane, which results in cell death. A recent study suggested that exposing cells to TiO_2 photocatalysis

TABLE 6.1
Nanoparticle Activity Against Pathogens and Mode of Action

Type of Nanoparticles (NPs)	Targeted Bacteria and Antibiotic Resistance	Antibacterial Mechanisms	References
Silver (Ag) NP	*Escherichia coli*, MRSA, vancomycin-resistant *Enterococcus* (VRE), extended-spectrum beta-lactamase (ESBL)-producing organisms, MDR *Pseudomonas aeruginosa*, *Klebsiella pneumoniae*, carbapenem-and polymyxin B-resistant *A. baumannii*, carbapenem-resistant *P. aeruginosa, Staphylococcus epidermidis*, and carbapenem-resistant *Enterobacteriaceae* (CRE)	Reactive oxygen species (ROS) generation, lipid peroxidation, inhibition of cytochromes in the electron transport chain, bacterial membrane disintegration, inhibition of cell wall synthesis, increase in membrane permeability, dissipation of proton gradient resulting in lysis, adhesion to cell surface causing lipid and protein damage, ribosome destabilization, intercalation between DNA bases	Dizaj et al. (2014); Cavassin et al. (2015); Rudramurthy et al. (2016); Hemeg (2017); Zaidi et al. (2017)
Gold (Au) NP	Methicillin-resistant *Staphylococcus aureus* (MRSA)	Loss of membrane potential, disruption of the respiratory chain, reduced ATPase activity, decline in tRNA binding to ribosome subunit, bacterial membrane disruption, generation of holes in the cell wall	Chen et al. (2014); Dizaj et al. (2014); Hemeg (2017); Zaidi et al. (2017)
Copper (Cu) NP	MDR *E. coli*, *A. baumannii*	Dissipation of cell membrane, ROS generation, lipid peroxidation, DNA degradation and protein oxidation	Chatterjee et al. (2014); Dizaj et al. (2014); Cavassin et al. (2015); Hemeg (2017); Zaidi et al. (2017)
Zinc oxide (ZnO) NP	*E. coli*, *Klebsiella oxytoca, K. pneumoniae*, MRSA, *Enterobacteraerogenes* ESBL-producing *E. coli*, *K. pneumoniae*	ROS production, disruption of membrane, adsorption to cell surface, and lipid and protein damage	Cavassin et al. (2015); Rudramurthy et al. (2016); Hemeg (2017)
Titanium dioxide (TiO2)	*E. coli*, *S. aureus*, *P. aeruginosa and Enterococcus faecium*	ROS generation, adsorption to the cell surface	Rudramurthy et al. (2016); Hemeg (2017)
Magnesium oxide (MgO) NP	*E. coli, S. aureus*	ROS generation, lipid peroxidation, electrostatic interaction, alkaline effect	Rudramurthy et al. (2016)

would rapidly inactivate the regulatory signalling stage, efficiently decreases the coenzyme-independent respiratory chains. They decrease the ability to absorb and hold iron and phosphorous and lowers the ability of biosynthesis and degrade heme (Fe-S cluster) groups (Foster et al., 2011).

6.8.6 Magnesium Oxide Nanoparticles

In the human body, after potassium, magnesium is the second common intracellular cation and has a total 25% mineral. As compared with magnesium oxide (MgO) NPs, TiO_2NPs, AgNPs, and CuNPs can be synthesized from economical and available precursors. The main function of MgONPs in bacteria is to induce ROS and inhibit essential microbial enzymes. The alkaline effect of MgONPs is considered as another primary aspect in the antibacterial activity (Li et al., 2012). According to Tang and Lv (2014), MgO can incapacitate the microorganisms by generating ROS; however, the mechanism of adsorption and direct particle penetration of the cell membrane may be other potential forms of pathogen inactivation, such alternatives must be considered for further aspects. In contrast, Hossain et al. (2014) explored the virulence of three separate MgONPs in *E. coli* and documented the absence of ROS formation for two MgONPs. The authors proposed that proteomics results clearly showed the absence of oxidative stress and indicated that the primary mechanism of cell death is associated with damage to the cell membrane which does not appear to be related to lipid peroxidation. Like other NPs, the MgONPs also contain the ROS, which is the leading mechanism behind their antimicrobial activity (Hossain et al., 2014). MgONPs physically interact with the surface of the cell and disrupt the membrane integrity, leading to membrane leakage (Blecher et al., 2011a). Additionally, they damage the cells by irreversible intracellular biomolecular oxidation. Other research, however, showed that MgONPs exhibit greater antibacterial activity without ROS and lipid peroxidation. The authors recommended that the antibacterial activity of MgONPs is related to the interface of NPs with the plasma membrane of microbes, pH change, and release of Mg^{+2} ions (Leung et al., 2014). The antimicrobial action of MgONPs has also been demonstrated because of the adsorption of halogen molecules on the surface of MgO (Blecher et al., 2011a). Over the ancient years, NPs, especially nano-silver coating, have served as an antibacterial measure in dental implants, bone prostheses, and surgical instruments and as a coating for wound dressing to fight the microorganisms in lesions (Correa et al., 2015; Burdusel et al., 2018). These NPs target the bacterial cells and disturb the crucial function of cell membranes such as membrane respiration and membrane permeability (Dakal et al., 2016; Slavin et al., 2017). They also react with intracellular components such as nucleic acids and polypeptides inhibit the divisions of cells and transfer of genes (Guzman et al., 2012a; Azam et al., 2012). Many reports show the bactericidal properties of numerous NPs, mainly zinc, magnesium, gold, copper, titanium, and silver (Vimbela et al., 2017; Hoseinnejad et al., 2018).

The mode of action of NPs and antibiotics seems identical in the instance of involvement of proteins, RNA, and DNA, as well as membrane disruption (Correa et al., 2015). However, these metallic NPs exhibit antimicrobial activity through multiple mechanisms, which decrease the possibility of resistance in microorganisms to grow against them (Slavin et al., 2017). Microbial cells will need to acquire various simultaneous gene mutations to establish resistance against these NPs, which is rare. Also, the composite of these NPs by a green synthesis mode will result in the formation of polysaccharides, small biologically active compounds, and proteins. Their interface with the NPs, thus further enhance their antimicrobial action against the MDR microorganisms. In this chapter, we have addressed some metallic NPs which are synthesized by green method(s) and their mechanism of action in contrast to several pathogenic microorganisms.

6.9 CONCLUDING REMARKS AND FUTURE PERSPECTIVES

In the end, the authors would like to conclude that microorganisms are generally able to establish antibiotic resistance due to deprived diagnostics, overdose, and incapability of drugs. Microorganisms caused infections are a grave global healthcare issue. Biogenic metallic NPs were designed to tackle these problems and demonstrated a good efficacy against various pathogens, either individually or in combination with antibiotics. However, there are some significant details, which should be considered to use such NPs for therapeutic applications. NPs are active in targeting and distribution in the body if utilized as a drug carrier for the cure of site-specific infections. Due to the antimicrobial functions of metallic NPs, they can be suitable for textiles, food processing, agriculture (nano-fertilizers and nano-pesticides), cosmetics, water treatment, washing machines, computer keyboards, and self-cleaning coatings on mobile phones. For these purposes, however, the biogenic NPs have not yet been commercialized. The main challenge for biogenic NPs is to strike the right balance among applicability, scalability, and cost of development. Therefore, a lot of work will be needed in this respect to concentrate on economic ways of developing biogenic NPs which will make them easily accessible for all kinds of future applications related to either antimicrobial era or another.

REFERENCES

Abdel, H.M. Mady, M.M. and Ghannam, M.M. (2012). Physical properties of different gold nanoparticles: Ultraviolet-visible and fluorescence measurements. *Journal of Nanomedicine and Nanotechnology*. 3, 3.

Abdel-Raouf, N. Al-Enazi, N.M. and Ibraheem, I.B.M. (2017). Green biosynthesis of gold nanoparticles using *Galaxaura elongata* and characterization of their antibacterial activity. *Arabian Journal of Chemistry*. 10, 3029–3039.

Acharyulu, N.P. Dubey, R.S. Swaminadham, V. Kalyani, R.L. Kollu, P. and Pammi, S.V.N. (2014). Green synthesis of CuO nanoparticles using *Phyllanthus amarus* leaf extract and their antibacterial activity against multidrug resistance bacteria. *International Journal of Engineering Research and Technology*. 3, 639–641.

Adeniji, F. (2018). Global analysis of strategies to tackle antimicrobial resistance. *International Journal of Pharmacy Practice*. 26, 85–89.

Agarwal, A., Pathera, A.K., Kaushik, R., Kumar, N. Dhull, S. B., Arora, S. and Chawla, P. ((2020)). Succinylation of milk proteins: Influence on micronutrient binding and functional indices. *Trends in Food Science and Technology*. 97, 254–264.

Ahmad, A. Mukherjee, P. Senapati, S. et al. (2003). Extracellular biosynthesis of silver nanoparticles using the fungus *Fusarium oxysporum*. *Colloids and Surfaces B*. 28, 313–318.

Ahmad, A. Senapati, S. Khan, M.I. et al. (2003). Intracellular synthesis of gold nanoparticles by a novel alkalotolerant actinomycete, *Rhodococcus* species. *Nanotechnology*. 14, 824–828.

Ahmed, S. Annu Chaudhry, S.A. and Ikram, S. (2017). A review on biogenic synthesis of ZnO nanoparticles using plant extracts and microbes: A prospect towards green chemistry. *The Journal of Photochemistry and Photobiology B: Biology*. 166, 272–284.

Alexander, J.W. (2009). History of the medical use of silver. *Surgical Infections* 10, 289–292.

Ali, K. Dwivedi, S. Azam, A. Saquib, Q. Al-Said, M.S. Alkhedhairy, A.A. and Musarrat, J. (2016). Aloe vera extract functionalized zinc oxide nanoparticles as nanoantibiotics against multi-drug resistant clinical bacterial isolates. *Journal Colloid Interface Sciences*. 472, 145–156.

Amendola, V. and Meneghetti, M. (2009). Size evaluation of gold nanoparticles by UV-vis spectroscopy. *Journal of Physical Chemistry C*. 113, 4277–4285.

Arias, C.A. and Murray, B.E. (2015). A new antibiotic and the evolution of resistance. *New England Journal of Medicine*. 372, 1168–1170.

Aromal, S.A. Vidhu, V.K. and Philip, D. (2012). Green synthesis of well-dispersed gold nanoparticles using *Macrotyloma uniflorum*. *Spectrochimica Acta Part A: Molecular and Biomolecular Spectroscopy*. 85, 99–104.

Aslam, B. Wang, W. Arshad, M.I. Khurshid, M. Muzammil, S. Rasool, M.H. and Salamat, M.K.F. (2018). Antibiotic resistance: A rundown of a global crisis. *Infection and Drug Resistance*. 11, 1645–1658.

Awwad, A.M. Albiss, B.A. and Salem, N.M. (2015). Antibacterial activity of synthesized copper oxide nanoparticles using *Malva sylvestris* leaf extract. *Sikkim Manipal University Medical Journal*. 2, 91–101.

Azam, A. Ahmed, A.S. Oves, M. Khan, M.S. and Memic, A. (2012). Size-dependent antimicrobial properties of CuO nanoparticles against Gram-positive and -negative bacterial strains. *International Journal of Nanomedicine*. 7, 3527–3535.

Baek, Y.W. and An, Y.J. (2011). Microbial toxicity of metal oxide nanoparticles (CuO, NiO, ZnO, and Sb_2O_3) to *Escherichia coli, Bacillus subtilis*, and *Streptococcus aureus*. *Science of the Total Environment*. 409, 1603–1608.

Baluja, Z. Nabi, N. and Ray, A. (2018). Challenges in antimicrobial resistance: An update. *EC Pharmacology and Toxicology*. 6, 865–877.

Baptista, P. V. Mccusker, M. P. Carvalho, A., Ferreira, D. A. Mohan, N. M. Martins, M. et al. ((2018)). Nano-strategies to fight multidrug resistant bacteria—"a battle of the titans". *Frontiers in Microbiology* 9, 1441. doi: 10.3389/fmicb.2018.01441

Bartlett, J.G. Gilbert, D.N. and Spielberg, B. (2013). Seven ways to preserve the miracle of antibiotics. *Public Infectious Diseases Society*. 56, 1445–1450.

Betts, J.W. Hornsey, M. and La Ragione, R.M. (2018). Novel antibacterials: Alternatives to traditional antibiotics. *Advances in Microbial Physiology*. 73, 123–169.

Beyth, N. Houri, Y. Domb, A. Khan, W. and Hazan, R. (2015). Alternative antimicrobial approach: Nanoantimicrobial materials. *Evidence-Based Complementary and Alternative Medicine*. 24, 60–62.

Bhuyan, T. Mishra, K. Khanuja, M. Prasad, R. and Varma, A. (2015). Biosynthesis of zinc oxide nanoparticles from *Azadirachta indica* for antibacterial and photocatalytic applications. *Materials Science in Semiconductor Processing*. 32, 55–61.

Blecher, K. Nasir, A. and Friedman, A. (2011a). The growing role of nanotechnology in combating infectious disease. *Virulence*. 2, 395–401.

Blecher, K. Nasir, A., and Friedman, A. (2011b). 1. *Virulence,* 2(5), 395–401. doi: 10.4161/viru.2.5.17035.

Bogdanovic, U. Lazic, V. Vodnik, V. Budimir, M. Markovi, C. Z. and Dimitrijevi, C. S. (2014). Copper nanoparticles with high antimicrobial activity. *Materials Letters*. 128, 75–78.

Borkow, G. and Gabbay, J. (2004). Putting copper into action: Copper-impregnated products with potent biocidal activities. *The Society for Experimental Biology*. 18, 1728–1730.

Brandelli, A. (2012). Nanostructures as promising tools for delivery of antimicrobial peptides. *Mini Reviews in Medicinal Chemistry.* 12, 731–741.

Brown, A.N. Smith, K. Samuels, T.A. Lu, J. Obare, S.O. and Scott, M.E. (2012). Nanoparticles functionalized with ampicillin destroy multiple-antibiotic-resistant isolates of *Pseudomonas aeruginosa* and *Enterobacter aerogenes* and methicillin-resistant *Staphylococcus aureus*. *Applied Environmental Microbiology.* 78, 2768–2774.

Burdusel, A.C. Gherasim, O. Grumezescu, A.M. Mogoanta, L. Ficai, A. and Andronescu, E. (2018). Biomedical applications of silver nanoparticles: An up-to-date overview. *Nanomaterials.* 8, 681.

Camporotondia, D.E. Fogliaa, M.L. Alvareza, G.S. Meberta, A.M. and Diaza, L.E. (2013). Antimicrobial properties of silica modified nanoparticles. *Microbial Pathogens and Strategies for Combating Them: Science, Technology and Education.* 2013, 283–290.

Cavassin, E. D. De Figueiredo, L. F. Otoch, J. P. et al. (2015). Comparison of methods to detect the *in vitro* activity of silver nanoparticles (AgNP) against multidrug resistant bacteria. *Journal of Nanobiotechnology.* 13, 64.

Chatterjee, Arijit and Chakraborty, Ruchira and Basu, Tarakdas. (2014). Mechanism of antibacterial activity of copper nanoparticles. *Nanotechnology.* 25, 135101. 10.1088/0957-4484/25/13/135101.

Chawla, P. Kaushik, R. Shiva Swaraj, V. J. and Kumar, N. (2018). Organophosphorus pesticides residues in food and their colorimetric detection. *Environmental Nanotechnology, Monitoring and Management.* 10, 292–307.

Chawla, P. Kumar, N. Kaushik, R. and Dhull, S.B. (2019). Synthesis, characterization and cellular mineral absorption of gum arabic stabilized nanoemulsion of *Rhododendron arboreum* flower extract. *Journal of Food Science and Technology.* 56, 5194–5203.

Chawla, P. Kumar, N. Kumar, M. Kaushik, R. and Punia, S. (2020). Gum Arabic capped copper nanoparticles: Characterization and their utilization. *International Journal of Biological Macromolecules.* 146, 232–242.

Chen, C. W. Hsu, C. Y. Lai, S. M. Syu, W. J. Wang, T. Y. and Lai, P. S. (2014). Metal nanobullets for multidrug resistant bacteria and biofilms. *Advanced Drug Delivery Reviews* 78, 88–104. doi: 10.1016/j.addr.2014.08.004

Chen, X. and Mao, S.S. (2007). Titanium dioxide nanomaterials: Synthesis, properties, modifications, and applications. *Chemical Reviews.* 107, 2891–2959.

Cheng, L. Wang, C. and Liu, Z. (2013). Upconversion nanoparticles and their composite nanostructures for biomedical imaging and cancer therapy. *Nanoscale.* 5, 23–37.

Chiriac, V. Stratulat, D.N. and Calin, G. (2016). Antimicrobial property of zinc-based nanoparticles. *IOP Conference Series: Materials Science and Engineering.* 133, 012055.

Coates, A.R. Halls, G. and Hu, Y. (2011). Novel classes of antibiotics or more of the same. *Brazilian Journal of Pharmacognosy.* 163, 184–194.

Correa, J.M. Mori, M. Sanches, H. L. Da Cruz, A. D. Poiate, E. Jr and Poiate, I. A. (2015). Silver nanoparticles in dental biomaterials. *International Journal of Biomaterials.* 2015, 485–275. 10.1155/2015/485275.

Dakal, T.C. Kumar, A. Majumdar, R.S. and Yadav, V. (2016). Mechanistic basis of antimicrobial actions of silver nanoparticles. *Frontiers in Microbiology.* 7, 1831.

Das, S.K. Das, A.R. and Guha, A.K. (2009). Gold nanoparticles: Microbial synthesis and application in water hygiene management. *Langmuir.* 25, 8192–8199.

Davis, M. Whittaker, A. Lindgren, M. Djerf-Pierre, M. Manderson, L. and Flowers, P. (2018). Understanding media publics and the antimicrobial resistance crisis. *Global Public Health.* 13, 1158–1168.

DeAlba-Montero, I. Guajardo-Pacheco, J. Morales-Sanchez, E. et al. (2017). Antimicrobial properties of copper nanoparticles and amino acid chelated copper nanoparticles produced by using a soya extract. *Bioinorganic Chemistry Applied.* 10, 649–618.

Dhuper, S. Panda, D. and Nayak, P.L. (2012). Green synthesis and characterization of zero valent iron nanoparticles from the leaf extract of *Mangifera indica*. *Nano Trends Journal of Nanotechnology and Its Applications*. 13, 16–22.

Dizaj, SM. Lotfipour, F. Barzegar-Jalali, M. Zarrintan, M.H. and Adibkia, K. (2014). Antimicrobial activity of the metals and metal oxide nanoparticles. *Materials Science and Engineering C*. 44, 278–284.

Dhull, S.B. Anju, M. Punia, S. Kaushik, R. and Chawla, P. (2019). Application of Gum Arabic in nanoemulsion for safe conveyance of bioactive components. In: Prasad R., Kumar V., Kumar M., and Choudhary D. (eds) *Nanobiotechnology in Bioformulations. Nanotechnology in the Life Science*. Springer, Cham. Chapter 3, pp. 85–98. DOI: 10.1007/978-3-030-17061-5-3.

Durán, N. Durán, M. De Jesus, M. Seabra, A. Favaro, W. and Nakazato, G. (2016). Silver nanoparticles: A new view on mechanistic aspects on antimicrobial activity. *Nanomedicine* 12, 789–799. DOI: 10.1016/j.nano.2015.11.016

Durán, N. Marcato, P. D. Alves, O. L. DeSouza, G. and Esposito, E. (2005). Mechanistic aspects of biosynthesis of silver nanoparticles by several *Fusarium oxysporum* strains. *Journal of Nanobiotechnology*. 3, 1–8.

Durán, N. Marcato, P.D. De, S. et al. (2007). Mechanistic aspects of biosynthesis of silver nanoparticles by several *Fusarium oxysporum* strains. *Journal of Biomedical Nanotechnology*. 3, 203–208.

Dutta, R.K. Nenavathu, B.P. Gangishetty, M.K. and Reddy, A.R. (2012). Studies on antibacterial activity of ZnO nanoparticles by ROS induced lipid peroxidation. *Colloids Surfaces B Bio Interfaces*. 94, 143–150.

Egorova, E.M. and Revina, A.A. (2000). Synthesis of metallic nanoparticles in reverse micelles in the presence of quercetin. *Colloids and Surface. A*. 168, 87–96.

El-Batal, A.I. Al-Hazmi, N.E. Mosallam, F.M. and El-Sayyad, G.S. (2018). Biogenic synthesis of copper nanoparticles by natural polysaccharides and *Pleurotus ostreatus* fermented fenugreek using gamma rays with antioxidant and antimicrobial potential towards some wound pathogens. *Microbiology Pathogens*. 118, 159–169.

El-Batal, A.I. El-Sayyad, G.S. El-Ghamery, A. and Gobara, M. (2017). Response surface methodology optimization of melanin production by *Streptomyces cyaneus* and synthesis of copper oxide nanoparticles using gamma radiation. *Journal of Clusters Science*. 28, 1083–1112.

Espitia, P.J.P. Soares, N.F.F. Coimbra J.S.R. de Andrade, N.J. and Cruz, R.S. (2012). Zinc oxide nanoparticles: Synthesis, antimicrobial activity and food packaging applications. *Food Bioprocess Technology*. 5, 1447–1464.

Foster, H.A. Ditta, I.B. Varghese, S. and Steele, A. (2011). Photocatalytic disinfection using titanium dioxide: Spectrum and mechanism of antimicrobial activity. *Applied Microbiology and Biotechnology*. 90, 1847–1868.

Franci, G. Falanga, A. and Galdiero, S. (2015). Silver nanoparticles as potential antibacterial agents. *Molecules*. 20, 8856–8874.

Gade, A.K. Bonde, P. Ingle, A.P. Marcato, P.D. Durán, N. and Rai, M.K. (2008). Exploitation of *Aspergillus niger* for synthesis of silver nanoparticles. *Journal of Biobased and Material Bioenergy*. 2, 243–247.

Gawande, M.B. Goswami, A. Felpin, F.X. et al. (2016). Cu and Cu-based nanoparticles: Synthesis and applications in catalysis. *Chemical Reviews*. 116, 3722–3811.

Gopinath, P.M. Narchonai, G. Dhanasekaran, D. Ranjani, A. and Thajuddin, N. (2015). Mycosynthesis, characterization and antibacterial properties of AgNPs against multidrug resistant (MDR) bacterial pathogens of female infertility cases. *Asian Journal Pharmaceutical Sciences*. 10, 138–145.

Gopinath, K. Shanmugam, V.K. Gowri, S. Senthil, V. Kumaresan, S. and Arumugam A. (2014). Antibacterial activity of ruthenium nanoparticles synthesized using *Gloriosa superba* L. leaf extract. *Journal of Nanostructures in Chemistry*. 4, 83.

Grass, G. Rensing, C. and Solioz, M. (2011). Metallic copper as an antimicrobial surface. *Applied and Environmental Microbiology.* 77, 1541–1548.

Gudikandula, K. and Maringanti, S.C. (2016). Synthesis of silver nanoparticles by chemical and biological methods and their antimicrobial properties. *Journal of Experimental Nanoscience.* 11, 714–721.

Gupta, A. Saleh N.M. Das, R. Landis, R. Bigdeli, A. Motamedchaboki, K. et al. ((2017)). Synergistic antimicrobial therapy using nanoparticles and antibiotics for the treatment of multidrug-resistant bacterial infection. *Nano Futures.* 1, 015004. DOI: 10.1088/2399-1984/aa69f.

Guzman, M. Dille, J. and Godet, S. (2012a). Synthesis and antibacterial activity of silver nanoparticles against Gram-positive and Gram-negative bacteria. *Nanomedicine Nanotechnology, Biology Medicine.* 8, 37–45.

Guzman, M. Dille, J. and Godet, S. (2012b). Synthesis and antibacterial activity of silver nanoparticles against gram-positive and gram-negative bacteria. *Nanomedicine: Nanotechnology, Biology and Medicine.* 8(1), 37–45. DOI: 10.1016/j.nano.2011.05.007.

He, S. Guo, Z. Zhang, Y. Zhang, S. Wang, J. and Gu, N. (2007). Biosynthesis of gold nanoparticles using the bacteria *Rhodopseudomonas capsulata. Material Letters.* 61, 3984–3987.

Hemeg, H.A. (2017). Nanomaterials for alternative antibacterial therapy. *International Journal of Nanomedicne* 12, 8211–8225.

Hindi, K.M. Ditto, A.J. Panzner, M.J. et al. (2009). The antimicrobial efficacy of sustained release silver-carbene complex-loaded L-tyrosine polyphosphate nanoparticles: Characterization, *in vitro* and *in vivo* studies. *Biomaterials.* 30, 3771–3779.

Horie, M. Fujita, K. Kato, H. et al. (2012). Association of the physical and chemical properties and the cytotoxicity of metal oxide nanoparticles: Metal ion release, adsorption ability and specific surface area. *Metallurgy Integrations and Biomaterials Sciences.* 4, 350–360.

Hoseinnejad, M. Jafari, S.M. and Katouzian, I. (2018). Inorganic and metal nanoparticles and their antimicrobial activity in food packaging applications. *Critical Reviews in Microbiology.* 44, 161–181.

Hoshino, N. Kimura, T. Yamaji, A. and Ando, T. (1999). Damage to the cytoplasmic membrane of *Escherichia coli* by catechin-copper (II) complexes. *Free Radical Biology and Medicine.* 27, 1245–1250.

Hossain, F. Perales-Perez, O.J. Hwang, S. and Roman, F. (2014). Antimicrobial nanomaterials as water disinfectant: Applications, limitations and future perspectives. *Science Total Environment.* 466, 1047–1059.

Hostynek, J.J. and Maibach, H.I. (2003). Copper hypersensitivity: Dermatologic aspects—An overview. *Review Environmental Health.* 18, 153–183.

Huang, J. Li, Q. and Sun, D. (2007). Biosynthesis of silver and gold Nanoparticles by novels sundried *Cinnamomum camphora* leaf. *Nanotechnology.* 18, 104.

Huang, X. Zheng, D. and Yan, G. (2008). Toxicological effect of ZnO Nanoparticles based on bacteria. *Langmuir.* 24, 4140–4144.

Huh, A.J. and Kwon, Y.J. (2011). Nanoantibiotics: A new paradigm for treating infectious diseases using nanomaterials in the antibiotic's resistant era. *Journal of Controlled Release Society.* 156, 128–145.

Hunagund, S.M. Desai, V.R. Kadadevarmath, J.S. Barretto, D.A. Vootla, S. and Sidarai, A.H. (2016). Biogenic and chemogenic synthesis of TiO_2 NPs via hydrothermal route and their antibacterial activities. *Advances.* 6, 97438–97444.

Hurst, S.J. Lytton-Jean, A.K.R. and Mirkin, C.A. (2006). Maximizing DNA loading on a range of gold nanoparticle size. *Analytical Chemistry.* 78, 8313–8318.

Hussain, I. Singh, N.B. Singh, A. Singh, H. and Singh, S.C. (2015). Green synthesis of nanoparticles and its potential application. *Biotechnology Letters.* 15, 2026–2027.

Ingle, A. Gade, A. Pierrat, S. Sonnichsen, C. and Rai, M.K. (2008). Mycosynthesis of silver nanoparticles using the fungus *Fusarium acuminatum* and its activity against some human pathogenic bacteria. *Current Nanoscience.* 4, 141–144.

Iravani, S. (2011). Green synthesis of metal nanoparticles using plants. *Green Chemistry*. 13, 2638.

Ivask, A. ElBadawy, A. Kaweeteerawat, C. et al. (2014). Toxicity mechanisms in *Escherichia coli* vary for silver nanoparticles and differ from ionic silver. *American Chemical Society Nano*. 8, 374–386.

Jana, S. and Pal, T. (2015). Synthesis, characterization and catalytic application of silver nanoshell coated functionalized polystyrene beads. *Journal of Nanoscience Nanotechnology*. 7, 2151–2156.

Jayaseelan, C. Rahuman, A.A. Kirthi, A.V. et al. (2012). Novel microbial route to synthesize ZnO nanoparticles using *Aeromonas hydrophily* and their activity against pathogenic bacteria and fungi. *Spectrochimica Acta A: Molecular and Biomolecular Spectroscopy*. 90, 78–84.

Jinu, U. Jayalakshmi, N. SujimaAnbu, A. Mahendran, D. Sahi, S. and Venkatachalam, P. (2017). Biofabrication of cubic phase silver nanoparticles loaded with phytochemicals from *Solanum nigrum* leaf extracts for potential antibacterial, antibiofilm and antioxidant activities against MDR human pathogens. *Journal of Clusters Science*. 28, 489–505.

Jones, N. Ray, B. Ranjit, K.T. and Manna, A.C. (2008). Antibacterial activity of ZnO nanoparticle suspensions on a broad spectrum of microorganisms. *Microbiology Letters*. 279, 71–76.

Jung, W.K. Koo, H.C. Kim, K.W. Shin, S. Kim, S.H. and Park, Y.H. (2008). Antibacterial activity and mechanism of action of the silver ion in *Staphylococcus aureus* and *Escherichia coli*. *Applied of Environmental Microbiology*. 74, 2171–2178.

Kailasa, S.K. Park, T.J. Rohit, J.V. and Koduru, J.R. (2019). Antimicrobial activity of silver nanoparticles. *Nanoparticles in Pharmacotherapy*. 461–484. 10.1016/B978-0-12-816504-1.00009-0.

Kamaraj, P. Vennila, R. Arthanareeswari, M. and Devikala, S. (2014). Biological activities of tin oxide nanoparticles Synthesized using plant extract. *World Journal Pharmacy and Pharmaceutical Sciences*. 3, 382–388.

Kami, D. Takeda, S. and Itakura, Y. (2011). Application of magnetic nanoparticles to gene delivery. *International Journal of Molecular Science*. 12, 3705–3722.

Kargara, H. Ghasemi, F. and Darroudid, M. (2015). Bioorganic polymer-based synthesis of cerium oxide nanoparticles and their cell viability assays. *Ceramics International*. 41, 1589–1594.

Kaweeteerawat, C. Na Ubol, P. Sangmuang, S. Aueviriyavit, S. and Maniratanachote, R. (2017). Mechanisms of antibiotic resistance in bacteria mediated by silver nanoparticles. *Journal of Toxicology Environmental Health*. 80, 1276–1289.

Kim, J.S. Kuk, E. Yu, K.N. et al. (2007). Antimicrobial effects of silver nanoparticles. *Nanomedicine*. 3, 95–101.

Kim, T. Braun, G.B. She, Z.G. Hussain, S. Ruoslahti, E. and Sailor, M.J. (2016). Composite porous silicon-silver nanoparticles as theragnostic antibacterial agents. *Applied Material and Interfaces*. 8, 30449–30457.

Kim, T.H. Kim, M. Park, H.S. Shin, U.S. Gong, M.S. and Kim, H.W. (2012). Size-dependent cellular toxicity of silver nanoparticles. *Journal of Biomedical Material Research*. 100, 1033–1045.

Kirillin, M. Shirmanova, M. Sirotkina, M. Bugrova, M. Khlebtsov, B. and Zagaynova, E. (2009). Contrasting properties of gold nanoshells and titanium dioxide nanoparticles for optical coherence tomography imaging of skin: Monte Carlo simulations and in vivo study. *Journal of Biomedicine Optimization*. 14, 021017.

Korbekandi, H. Iravani, S. and Abbasi, S. (2009). Production of nanoparticles using organisims. *Critical Reviews of Biotechnology*. 29, 279–306.

Kowshik, M. Ashataputre, S. Kharrazi, S. et al. (2003). Extracellular synthesis of silver nanoparticles by a silver-tolerant yeast strain MKY3. *Nanotechnology*. 14, 95–100.

Kumar, S.A. Abyaneh, M.K. Gosavi, S.W. et al. (2007). Nitrate reductase-mediated synthesis of silver nanoparticles from $AgNO_3$. *Biotechnology Letters.* 29, 439–445.
Kumar, A.S. Ansary, A.A. Ahmad, A. and Khan, M.I. (2007). Extracellular biosynthesis of CdSe quantum dots by the fungus, *Fusarium oxysporum. Journal of Biomedical and Nanotechnology.* 3, 190–194.
Kumar, A. Pandey, A.K. Singh, S.S. Shanker, R. and Dhawan, A. (2011). Engineered ZnO and TiO (2) nanoparticles induce oxidative stress and DNA damage leading to reduced viability of *Escherichia coli. Free Radicals Biology Medicine.* 51, 1872–1881.
Kunoh, T. Takeda, M. Matsumoto, S. Suzuki, I. Takano, M. Kunoh, H. and Takada, J. (2017). Green synthesis of gold nanoparticles coupled with nucleic acid oxidation. *ACS Sustainable Chemistry and Engineering.* 6, 364–373.
Lara, H.H. Ayala-Núñez, N.V. IxtepanTurrent, L.D.C. and Rodríguez Padilla, C. (2010). Bactericidal effect of silver nanoparticles against multidrug-resistant bacteria. *World Journal of Microbiology and Biotechnology.* 26, 615–621.
Laurent, S. Forge, D. Port, M. Roch, A. Robic, C. Vander Elst, L. and Muller, R.N. (2010). Magnetic iron oxide nanoparticles: Synthesis, stabilization, vectorization, physicochemical characterizations, and biological applications. *Chemical Review.* 110, 2574.
Lee, S.H. and Jun, B.H. (2019). Silver nanoparticles: Synthesis and application for nanomedicine. *International of Journal of Molecular Science.* 20, 865.
Leid, J.G. Ditto, A.J. and Knapp, A. (2012). *In vitro* antimicrobial studies of silver carbene complexes: Activity of free and nanoparticle carbene formulations against clinical isolates of pathogenic bacteria. *Journal of Antimicrobial Chemotherapy.* 67, 138–148.
Lemire, J.A. Harrison, J.J. and Turner, R.J. (2013). Antimicrobial activity of metals: Mechanisms, molecular targets and applications. *Nature Reviews Microbiology.* 11, 371–384.
León-Silva, S. Fernández-Luqueño, F. and López-Valdez, F. (2016). Silver nanoparticles (AgNP) in the environment: A review of potential risks on human and environmental health. *Water Air Soil Pollution.* 227, 306.
Leung, Y.H. Shen, Z. and Gethings, L.A. (2014). Mechanisms of antibacterial activity of MgO: Non-ROS mediated toxicity of MgO nanoparticles towards *Escherichia coli. Small.* 10, 1171–1183.
Li, J. Li, Q. Ma, X. et al. (2016). Biosynthesis of gold nanoparticles by the extreme bacterium *Deinococcus radiodurans* and an evaluation of their antibacterial properties. *International Journal of Nanomedicine.* 11, 5931.
Li, M. Zhu, L. and Lin, D. (2011). Toxicity of ZnO nanoparticles to *Escherichia coli*: Mechanism and the influence of medium components. *Environmental Science and Technology.* 45, 1977–1983.
Li, Y. Zhang, W. Niu, J. and Chen, Y. (2012). Mechanism of photogenerated reactive oxygen species and correlation with the antibacterial properties of engineered metal-oxide nanoparticles. *Nano.* 6, 5164–5173.
Lipovsky, A. Gesanken, A. Nitzan, Y. and Lubart, R. (2011). Enhanced inactivation of bacteria by metal-oxide nanoparticles combined with visible light irradiation. *Lasers in Surgery and Medicine.* 43, 236–240.
Liu, H.L. Dai, S.A. Fu, K.Y. and Hsu, S.H. (2010). Antibacterial properties of silver: Nanoparticles in three different sizes and their nanocomposites with a new waterborne polyurethane. *International Journal of Nanomedicine.* 5, 1017–1028.
Liu, P. Duan, W. Wang, Q. and Li, X. (2010). The damage of outer membrane of *Escherichia coli* in the presence of TiO_2 combined with UV light. *Colloids Surface B Biointerfaces.* 78, 171–176.
Liu, Y. He, L. Mustapha, A. Li, H. Hu, Z.Q. and Lin, M. (2009). Antibacterial activities of zinc oxide nanoparticles against *Escherichia coli* O157: H7. *Journal of Applied Microbiology.* 107, 1193–1201.

Ma, H. Williams, P.L. and Diamond SA. (2013). Ecotoxicity of manufactured ZnO nanoparticles - a review. *Environmental Pollution.* 172, 76–85.
Madhumitha, G. Elango, G. and Roopan, S.M. (2016). Biotechnological aspects of ZnO nanoparticles: Overview on synthesis and its applications. *Applied Microbiology and Biotechnology.* 100, 571–581.
Maiti, S. Krishnan, D. Barman, G. Ghosh, S.K. and Laha, J.K. (2014). Antimicrobial activities of silver nanoparticles synthesized from *Lycopersicon esculentum* extract. *Journal of Analytical Science and Technology.* 5, 40.
Mali, S.S. Betty, C.A. Bhosale, P.N. and Patil, P.S. (2011). Hydrothermal synthesis of rutile TiO_2 with hierarchical microspheres and their characterization. *Crystal Engineering Communication.* 13, 6349–6351.
Mao, R.Y.C. Gao, X. Burt, J.L. Belcher, A.M. Georgiou, G. and Lverson, B.L. (2004). Bacterial biosynthesis of cadmium sulfide nanocrystals. *Chemical Biology.* 11, 1553–1559.
McDevitt, C.A. Ogunniyi, A.D. Valkov, E. et al. (2011). A molecular mechanism for bacterial susceptibility to zinc. *Pathology.* 7 e1002357.
Mirzajani, F. Ghassempour, A. Aliahmadi, A. and Esmaeili, M.A. (2011). Antibacterial effect of silver nanoparticles on *Staphylococcus aureus. Research in Microbiology.* 162, 542–549.
Mittal, J. Batra, A. Singh, A., and Sharma, M.M. (2014). Phytofabrication of nanoparticles through plant as nanofactories. *Advanced Natural Sciences Nanosciences and Nanotechnology.* 5, 10.
Mohammadinejad, R. Shavandi, A. Raie, D.S. et al. (2019). Plant molecular farming: Production of metallic nanoparticles and therapeutic proteins using green factories. *Green Chemistry.* 21, 1845–1865.
Mukherjee, P. Ahmad, A. Mandal, D. et al. (2001a). Bioreduction of $AuCl_4$ ions by them fungus *Verticillium* sp and surface trapping of the gold nanoparticles formed. *Angewante Chemie International Edition.* 40, 3585–3588.
Mukherjee, P. Ahmad, A., Mandal, D.S. et al. (2001b). Fungus-mediated synthesis of silver nanoparticles and their immobilization in the mycelial matrix: A novel biological approach to nanoparticle synthesis. *Nano Letters.* 1, 515–519.
Mukherjee, P. Roy, M. Mandal, B.P. et al. (2008). Green synthesis of highly stabilized nanocrystalline silver particles by a non-pathogenic and agriculturally important fungus *T. asperellum. Nanotechnology.* 19, 103–110.
Mukunthan, K.S. and Balaji, S. (2012). Cashew apple juice (*Anacardium occidentale* L.) speeds up the synthesis of silver nanoparticles. *International Journal of Green Nanotechnology.* 4, 71–79.
Mulvey, M. R. and Simor, A. E. (2009). Antimicrobial resistance in hospitals: How concerned should we be? *CMAJ* 180, 408–415. DOI: 10.1503/cmaj.080239.
Murphy, M. Ting, K. Zhang, X. Soo, C. and Zheng, Z. (2015). Current development of silver nanoparticle preparation, investigation, and application in the field of medicine. *Journal of Nanomaterial.* (2015). 1–12. 10.1155/2015/696918.
Murugan, A. and Kumara, K. (2014). Biosynthesis and characterization of silver nanoparticles using the aqueous extract of *vitex negundo. linn. World Journal of Pharmaceutical Science.* 3, 1385–1393.
Naika, H.R. Lingaraju, K. Manjunath, K.D. et al. (2015). Green synthesis of CuO nanoparticles using *Gloriosa superba* L. extract and their antibacterial activity. *Journal of Taibah University Sciences.* 9, 7–12.
Nair, B. and Pradeep, T. (2002). Coalescence of nanoclusters and formation of submicron crystallites assisted by *Lactobacillus* strains. *Crystal Growth Design.* 2, 293–298.
Nam, K.T. Lee, Y.J. Krauland, E.M. Kottmann, S.T. and Belcher, A.M. (2008). Peptide-mediated reduction of silver ions on engineered biological scaffolds. *ACS Nano.* 2, 1480–1486.

Nangia, Y. Wangoo, N. Goyal, N. Shekhawat, G. and Suri C.R. (2009). A novel bacterial isolate *Stenotrophomonas maltophilia* as living factory for synthesis of gold nanoparticles. *Microbial Cell Factories.* 8, 39. 10.1186/1475-2859-8-39.

Navalakhe R.M. and Nandedkar, T.D. (2007). Application of nanotechnology in biomedicine. *Indian Journal of Experimental Biology.* 45, 160–165.

Nel, A. Xia, T. Madler, L. and Li, N. (2006). Toxic potential of materials at the nano level. *Science.* 311, 622–627.

Norman, R.S. Stone, J.W. Gole, A. Murphy, C.J. and Sabo-Attwood, T.L. (2008). Targeted photothermal lysis of the pathogenic bacteria, *Pseudomonas aeruginosa*, with gold nanorods. *Nano Letters.* 8, 302–306.

Othman, S.H. Salam, N.R.A. Zainal, N. Basha, R.K. and Talib, R.A. (2014). Antimicrobial activity of TiO_2 nanoparticle-coated film for potential food packaging applications. *International Journal of Photoenergy.* (2014) 1–6 10.1155/2014/945930.

Patel, P. Agarwal, P. Kanawaria, S. Kachhwaha, S. and Kothari, S.L. (2015). Plant-based synthesis of silver nanoparticles and their characterization. *Nanotechnol. Plant Science.* pp. 271–288. Springer, Berlin.

Pathania, R. Khan, H. Kaushik, R. and Khan, M.A. (2018). Essential oil nanoemulsions and their antimicrobial and food applications. *Current Research Nutrition and Food Science.* 6, 1–16.

Payne, J. Waghwani, H. Connor, M. Hamilton, W. Dowling, S. Moolani, H. Chavda, F. Badwaik, V. Lawrenz, M. and Dakshinamurthy, R. (2016). Novel synthesis of kanamycin conjugated gold nanoparticles with potent antibeacterial activity. *Frontiers in Microbiology.* 7. 10.3389/fmicb.2016.00607.

Pelgrift, R.Y. and Friedman, A.J. (2013a). Nanotechnology as a therapeutic tool to combat microbial resistance. *Advances in Drug Delivery Review.* 65, 1803–1815.

Pelgrift, R. and Friedman, A. J (2013b). Nanotechnology as a therapeutic tool to combat microbial resistance. *Advanced Drug Delivery Reviews.* 65, 1803–1815. doi: 10.1016/j.addr.2013.07.011

Periasamy, S. Joo, H.S. Duong, A.C. et al. (2012). How *Staphylococcus aureus* biofilms develop their characteristic structure. *Proceedings of the National Academy of Sciences.* 109, 1281–1286.

Philip, D. (2009). Biosynthesis of Au, Ag and Au-Ag nanoparticles using edible mushroom extract. *Spectrochimica Acta A.* 73, 374–381.

Poole, K. (2002). Mechanisms of bacterial biocide and antibiotic resistance. *Journal of Applied Microbiology.* 92, 55S–64S.

Popa, M., Pradell, T., Crespo, D., and Calder, J.M. (2007). Stable silver colloidal dispersions using short chain polyethylene glycol. *Colloids and Surfaces A: Physicochemical and Engineering Aspects.* 303, 184–190.

Prabhu, S. and Poulose, E.K. (2012). Silver nanoparticles: Mechanism of antimicrobial action, synthesis, medical applications, and toxicity effects. *International Nano Letters.* 2, 32.

Raffi, M. Hussain, F. Bhatti, T.M. Akhter, J.I. Hameed, A. and Hasan, M.M. (2008). Antibacterial characterization of silver nanoparticles against *E. coli* ATCC-15224. *Journal of Material Science and Technology.* 24, 192–196.

Raghupathi, K.R. Koodali, R.T. and Manna, A.C. (2011). Size-dependent bacterial growth inhibition and mechanism of antibacterial activity of zinc oxide nanoparticles. *Langmuir.* 27, 4020–4028.

Rai, M. Yadav, A and Gade, A. (2009). Silver nanoparticles as a new generation of antimicrobials. *Biotechnology Advances.* 27, 76–83.

Rai, M.K., Deshmukh, S.D., Ingle, A.P., and Gade, A.K. (2012). Silver nanoparticles: The powerful nanoweapon against multidrug-resistant bacteria. *Journal of Applied Microbiology.* 112, 841–852. DOI: 10.1111/j.1365-2672.2012.05253.x.

Rai, M. Ingle, A.P. Gupta, I. and Brandelli, A. (2015). Bioactivity of noble metal nanoparticles decorated with biopolymers and their application in drug delivery. *International Journal of Pharmacy*. 496, 159–172.

Rajakumar, G. Rahuman, A.A. Rooan, S.M. Khanna, V.G. Elango, G. Kamaraj, C. Zahir, A.A. and Velayutham, K. (2012). Fungus-mediated biosynthesis and characterization of TiO_2 nanoparticles and their activity against pathogenic bacteria. *Spectrochimica Acta A: Molecular Biomolecules Spectroscopy*. 91, 23–29.

Rana, B. Kaushik, R. Kaushal, K. Kaushal, A. Gupta, S. Upadhay, N. Rani, P. and Sharma, P. (2018). Application of biosensors for determination of physicochemical properties of zinc fortified milk. *Food Bioscience*. 21, 117–124.

Ravishankar, R.V. and Jamuna, B.A. (2011). Nanoparticles and their potential application as antimicrobials. Science against microbial pathogens, communicating current research and technological advances. *Badajoz: Formatex*. 197–209.

Raza, M.A. Kanwal, Z. Rauf, A. Sabri, A.N. Riaz, S. and Naseem, S. (2016). Size- and shape-dependent antibacterial studies of silver nanoparticles synthesized by wet chemical routes. *Nanomaterials*. 6, 74.

Riddin, T. Gericke, M. and Whiteley, C. (2006). Analysis of the inter-and extracellular formation of platinum nanoparticles by *Fusarium oxysporum* f. sp. *lycopersici* using response surface methodology. *Nanotechnology*. 17, 3482.

Rudramurthy, G.R. Swamy, M.K. Sinniah, U.R. and Ghasemzadeh, A. (2016). Nanoparticles: Alternatives against drug-resistant pathogenic microbes. *Molecules*. 21, 836. 10.3390/molecules21070836.

Ruparelia, J.P. Chatterjee, A.K. Duttagupta, S.P. and Mukherji, S. (2008). Strain specificity in antimicrobial activity of silver and copper nanoparticles. *Acta Biomaterials*. 4, 707–716.

Saha, B. Bhattacharya, J. Mukherjee, A. Ghosh, A. Santra, C. Dasgupta, A. K. et al. (2007). In vitro structural and functional evaluation of gold nanoparticles conjugated antibiotics. *Nanoscale Research Letters*. 2, 614–622. doi: 10.1007/ s11671-007-9104-2

Samavati, A. and Ismail, A.F. (2017). Antibacterial properties of copper substituted cobalt ferrite nanoparticles synthesized by coprecipitation method. *Particuology*. 30, 158–163.

Sanghi, R. and Verma, P. (2009). Biomimetic synthesis and characterisation of protein capped silver nanoparticles. *Bioresources Technology*. 100, 501–504.

Sankar, R. Rizwana, K. Shivashangari, K.S. and Ravikumar, V. (2015). Ultra-rapid photocatalytic activity of Azadirachtaindica engineered colloidal titanium dioxide nanoparticles. *Applied Nanoscience*. 5, 731–736.

Shahverdi, A.R. Fakhimi, A. Shahverdi, H.R. and Minaian, S. (2007). Synthesis and effect of silver nanoparticles on the antibacterial activity of different antibiotics against *Staphylococcus aureus* and *Escherichia coli*. *Nanomedicine: Nanotechnology, Biology and Medicine*. 3, 168–171.

Shankar, S.S. Ahmad, A. Pasricha, R. and Sastry, M. (2003). Bioreduction of chloroaurate ions by geranium leaves and its endophytic fungus yields gold nanoparticles of different shapes. *Journal of Materials Chemistry*. 13, 1822–1826.

Shankar, S.S. Rai, A. Ahmad, A. and Sastry, M. (2004). Rapid synthesis of Au, Ag, and bimetallic Au core–Ag shell nanoparticles using Neem (*Azadirachta indica*) leaf broth. *Journal of Colloid Interface Sciences*. 275, 496–502.

Sharma, S. Kaushik, R. Sharma, P. Sharma, R. Thapa A. and Indumathi, K.P. (2016). Antimicrobial activity of herbs against *Yersinia enterocolitica*. *The Annals of the University Dunarea de Jos of Galati - Food Technology*. 40, 119–134.

Sheny, D.S. Mathew, J and Philip, D. (2012). Synthesis characterization and catalytic action of hexagonal gold nanoparticles using essential oils extracted from *Anacardium occidentale*. Sp. *Spectrochimica Acta Part A: Molecular and Biomolecular Spectroscopy*. 97, 306–310.

Shobha, G. Vinutha, M and Ananda, S. (2014). Biological synthesis of copper nanoparticles and its impact. *International Journal of Pharmaceutical Sciences and Research Invention*, 3(8), 6, 28, 38.

Singh, A. Singh, N.B. Hussain, I. Singh, H. and Singh, S.C. (2015). Plant nanoparticle interaction: An approach to improve agricultural practices and plant productivity. *International Journal of Pharmaceuticals Sciences*. 4, 25–40.

Singh, B.P. Jena, B.K. Bhattacharjee, S. and Besra, L. (2013). Development of oxidation and corrosion resistance hydrophobic graphene oxide-polymer composite coating on copper. *Surface Coating Technology*. 232, 475–481.

Singh, P. Kim, Y.J. Zhang, D. and Yang, D.C. (2016). Biological synthesis of nanoparticles from plants and microorganisms. *Trends in Biotechnology*. 34, 588–599.

Singh, R. Thakur, P. Thakur, A. Kumar, H. Chawla, P. Jigneshkumar V.R. Kaushik, R. and Kumar, N. (2020). Colorimetric sensing approaches of surface modified gold and silver nanoparticles for detection of residual pesticides: A Review. *International Journal of Environmental Analytical Chemistry*. DOI: 10.1080/03067319.2020.1715382

Sintubin, L. Windt, W.E. Dick, J. Mast, J. Ha, D.V. Verstarete, W. and Boon, N. (2009). Lactic acid bacteria as reducing and capping agent for the fast and efficient production of silver nanoparticles. *Applied Microbiology and Biotechnology*. 87, 741–749.

Slavin, Y.N. Asnis, J. Hafeli, U.O. and Bach, H. (2017). Metal nanoparticles: Understanding the mechanisms behind antibacterial activity. *Journal of Nanobiotechnology*. 15, 1–20.

Stiufiuc, R. Iacovita, C. and Lucaciu, C.M. et al. (2013). SERS- active silver colloids prepared by reduction of silver nitrate with short-chain polyethylene glycol. *Nanoscale Research Letters*. 8, 47.

Subbaiya, R. Shiyamala, M. Revathi, K. Pushpalatha, R. and Masilamani Selvam, M. (2014). Biological synthesis of silver nanoparticles from *Nerium oleander* and its antibacterial and antioxidant property. *International Journal of Current Microbiology and Applied Science*. 3, 83–87.

Suman, T.Y. Rajasree, S.R. Ramkumar, R. Rajthilak, C. and Perumal, P. (2014). The Green synthesis of gold nanoparticles using an aqueous root extract of *Morinda citrifolia* L. *Spectrochimica Acta Part A: Molecular and Biomolecular Spectroscopy*. 118, 11–16.

Suresh, A.K. Pelletier, D.A. Wang, W. Morrell-Falvey, J.L. Gu, B. and Doktycz, M.J. (2012). Cytotoxicity induced by engineered silver nanocrystallites is dependent on surface coatings and cell types. *Journal of Surfaces and Colloids*. 28, 2727–2735.

Tan, Y. Wang, Y. Jiang, L. and Zhu, D. (2002). Thiosalicylic acid-functionalized silver nanoparticles synthesized in one-phase system. *Journal of Colloid Interface and Science*. 249, 336–345.

Tang, S. and Zheng, J. (2018). Antibacterial activity of silver nanoparticles: Structural effects. *Advanced Healthcare Materials*. 7, 1701503. 10.1002/adhm.201701503.

Tang, Z.X. and Lv, B.F. (2014). MgO nanoparticles as antibacterial agent: Preparation and activity. *Brazilian Journal of Chemical Engineering*. 31, 591–601.

Thakkar, K.N. Mhatre, S.S. and Parikh, R.Y. (2010). Biological synthesis of metallic nanoparticles. *Nanomedicine*. 6, 257–262.

Tiwari, V. Mishra, N. Gadani, K. Solanki, P.S. Shah, N.A. and Tiwari, M. (2018). Mechanism of anti-bacterial activity of zinc oxide nanoparticle against carbapenem-resistant *Acinetobacter baumannii*. *Frontiers Microbiology*. 9, 1218.

Tiwari, J.N. Tiwari, R.N. and Kim, K.S. (2012). Zero-dimensional, one dimensional, two-dimensional and three-dimensional nanostructured materials for advanced electrochemical energy devices. *Progress Material Science*. 57, 24–803.

Torres, L.A. Gmez, T.J.R. Padron, G.H. Santana, F.B. Hernandez, J.F. and Castano, V.M. (2013). Silver nanoprisms and nanospheres for prosthetic biomaterials. Agricultural waste mango peel extract and it's *in vitro* cytotoxic effect on two normal cells. *Materials Letters*. 134, 67–70.

Umer, A. Naveed, S. and Ramzan, N. (2012). Selection of a suitable method for the synthesis of copper nanoparticles. *Nano*. 7, 18.

Vamathevan, V. Amal, R. Beydoun, D. Low, G. and McEvoy, S. (2002). Photocatalytic oxidation of organics in water using pure and silver-modified titanium dioxide particles. *Journal of Photochemistry and Photobiology A: Chemistry*. 148, 233–245.

Varghese, O.K. Gong, D. Paulose, M. Ong, K.G. and Grimes, C.A. (2003). Hydrogen sensing using titania nanotubes. *Sensors Actuators B*. 93, 338–344.

Vimala, K. Sundarraj, S. Paulpandi, M. Vengatesan, S. and Kannan, S. (2014). Green synthesized doxorubicin loaded zinc oxide nanoparticles regulates the Bax and Bcl-2 expression in breast and colon carcinoma. *Process Biochemistry*. 49, 160–172.

Vimbela, G.V. Ngo, S.M. Fraze, C. Yang, L. and Stout, D.A. (2017). Antibacterial properties and toxicity from metallic nanomaterials. *International Journal of Nanomedicine*. 12, 3941–3965.

Wang, Y. and Xia, Y. (2004). Bottom-up and top-down approaches to the synthesis of monodispersed spherical colloids of low melting-point metals. *Nano Letters*. 4, 2047–2050.

Wang, Z. Lee, Y.H. Wu, B. et al. (2010). Antimicrobial activities of aerosolized transition metal oxide Nanoparticles. *Chemosphere*. 80, 525–528.

Warnes, S.L. Caves, V. and Keevil, C.W. (2012). Mechanism of copper surface toxicity in *Escherichia coli* O157: H7 and *Salmonella* involves immediate membrane depolarization followed by slower rate of DNA destruction which differs from that observed for Gram-positive bacteria. *Environmental Microbiology*. 14, 1730–1743.

WHO. (2017). *Antibacterial Agents in Clinical Development: An Analysis of the Antibacterial Clinical Development Pipeline, Including Tuberculosis; Organ*. GWH (Ed.) WHO/EMP/IAU/2017.122017; WHO: Geneva, Switzerland.

Wim, H. (2008). Drug delivery and nanoparticles: Applications and hazards. *International Journal Nanomedicine*. 3, 133–149.

World Health Organization. (2014). *Antimicrobial Resistance: Global Report on Surveillance*, World Health Organization.

Yang, N. WeiHong, L. and Hao, L. (2014). Biosynthesis of Au nanoparticles using agricultural waste mango peel extract and it's in vitro cytotoxic effect on two normal cells. *Material Letters*. 134, 67–70.

YelilArasi, A. Hema, M. Tamilselvi, P. and Anbarasan, R. (2012). Synthesis and characterization of SiO_2 nanoparticles by sol-gel process. *Indian Journal of Science*. 1, 6–10.

You, C. Han, C. Wang, X. et al. (2012). The progress of silver nanoparticles in the antibacterial mechanism, clinical application and cytotoxicity. *Molecular Biology Reports*. 39, 9193–9201.

Zaidi, S. Misba, L. and Khan, A.U. (2017). Nano-therapeutics: A revolution in infection control in post antibiotic era. *Nanomedicine*. 13, 2281s–22301.

Zhai, T. Fang, X. Liao, M. Xu X. Zeng, H. Yoshio, B. and Golberg, D. (2009). A Comprehensive review of one-dimensional metal-oxide nanostructure photodetectors. *Sensors, 9(8)*, 6504.

Zhang, J. Li, S. Ding, H. et al. (2014). Transfer and assembly of large area TiO_2 nanotube arrays onto conductive glass for dye sensitized solar cells. *Journal of Power Sources*. 247, 807–812.

Zhang, L. Jiang, Y. Ding, Y. Daskalakis, N. Jeuken, L. and Povey, A.J. (2010). Mechanistic investigation into antibacterial behaviour of suspensions of ZnO nanoparticles against *E. coli*. *Journal of Nanoparticle Research*. 12, 1625–1636.

Zhang, X. (2015). Gold nanoparticles: Recent advances in the biomedical applications. *Cell Biochemistry and Biophysics*. 72, 771–775.

Zhou, H. Gan, X. Wang, J. Zhu, X. and Li, G. (2005). Hemoglobin-based hydrogen peroxide biosensor tuned by the photovoltaic effect of nano titanium dioxide. *Analytical Chemistry*. 77, 6102–6104.

7 Nanoparticles and Antibiotic Drug Composite

A Novel Approach Towards Antimicrobial Activity

Aarti Bains
Chandigarh Group of Colleges Landran, India
Dipsha Narang and
Maharaja Lakshman Sen Memorial College, India
Prince Chawla
Lovely Professional University, India
Sanju Bala Dhull
Chaudhary Devi Lal University, India

CONTENTS

7.1 INTRODUCTION

Nanotechnology is a technique that plays an important role in the field of modern science. Nanotechnology is applied in the manipulation of nanoparticles (NPs) which are associated with the provocative phenomenon in living beings and, therefore, have proved to be most effective in scientific knowledge (Tibbals, 2017). NPs, being small in size, ranging from 10 to 100 nm, have unique chemical, physical, and biological properties and, therefore, have been studied extensively over the last decade (Soppimath et al., 2001; Dowling et al., 2004). NPs are highly persistent and synthesized through different processes; therefore, they can be used in the food and agriculture industries and the pharmaceutical industry (Rani et al., 2017; Chawla et al., 2019, 2020). NPs of varied sizes and properties are used in nanotechnologies in medical science for the production of safe and effective production of the drugs (Bamrungsap et al., 2012). In the present era, due to the continuous usage of antibiotics, all bacteria, whether gram-positive and gram-negative, are gaining resistance against them. The World Health Organization (WHO), in its report, also mentioned antibiotic resistance as the most important health threat issue of the 21st century (Munita and Arias, 2016). The resistance of bacteria against antibiotics is, therefore, a topic of concern, and therefore, some relevant factors need to be discussed to understand the mechanism behind it. In the environment, there are present natural antibiotic agents, and to survive against these antimicrobial molecules, the co-resident microorganisms have developed a resistance mechanism to overcome their effect. As a result, the microorganism is considered intrinsically resistant to these naturally occurring antimicrobial agents (Munita and Arias, 2016). Another factor responsible for resistance is the development of acquired resistance resulting from the acquisition of external genetic determinants of resistance through horizontal gene transfer and chromosomal mutation of genes which are associated with the mechanism of action in antimicrobial compounds. Another mechanism for resistance is vertical gene transfer and the efflux pump (Hastings et al., 2004). To survive against antibiotics, the bacteria also develop an enzyme known as β-lactamase that provides resistance against β-lactam antibiotics, which are derivatives of penicillin carbapenems and cephalosporin. To overcome the resistance effect, different approaches for the synthesis of highly effective new antibiotics, as well as a change the structure of β-lactam by adding moieties, are done (Butler et al., 2017). The procedure for developing highly effective antibiotics and changing the moiety of antibiotics is highly time-consuming and expensive. Therefore, NP metal ions conjugated with small fragments of DNA, proteins, and drugs are employed for the delivery application (Kraft et al., 2014). The NPs have a large volume–surface area ratio and are compatible with antibiotics; hence, they do not affect their activity (Wang et al., 2017). NPs in conjugation with antibiotics work together and enhance the effectiveness, and therefore, the results are advanced antimicrobial agents (Gao et al., 2018). The NP and antibiotic composite work together in such a manner that if bacteria are resistant to one of them, another can inhibit the growth of bacteria in a different manner (Zheng et al., 2018). The approach for conjugating NPs with antibiotics requires a high quantity of antibiotics and requires multiple steps for the synthesis of NPs that include surface modification and the development of binding sites for antibiotics to attach to the surface of the NPs (Miller et al., 2015). Among these approaches, the surface binding of antibiotics with NPs to inhibit the growth of pathogenic microorganisms is most widely used in pharmaceutical industries (Ahmed et al., 2016). The NP used for coating is required to have excellent antimicrobial activity and should have low toxicity (Jamuna-Thevi et al., 2011). Therefore, in the present chapter antimicrobial resistance, mechanism, and synergistic effect of NPs and antibiotics, along with their mode of action against pathogenic bacteria, are discussed.

7.2 ANTIBACTERIAL RESISTANCE AGAINST ANTIBIOTICS

The resistance of bacteria against antibacterial agents is the ability of this organism against the antagonizing effect of antibacterial agents on reproduction. The development of the antimicrobial resistance against the antibiotic results due to the indiscriminate and continuous use of antibiotics (Gottlieb and Nimmo, 2011). Therefore, this resistance has become a major public health issue across the world. To survive, microorganisms develop four types of resistance mechanisms, which are discussed next (see Figure 7.1).

7.2.1 Intrinsic Resistance/Natural Resistance

This type of resistance is intrinsic and acquired naturally by microorganisms. The microorganism on its cell wall does not include the target sites for the attachment of antibiotics; therefore, antibiotics are not able to reach the targets (Aghdam et al., 2016).

7.2.2 Acquired Resistance

The bacteria become resistant to antibiotics by acquiring changes in the genetic level. It may be chromosomally acquired and plasmidly acquired or an extrachromosomal

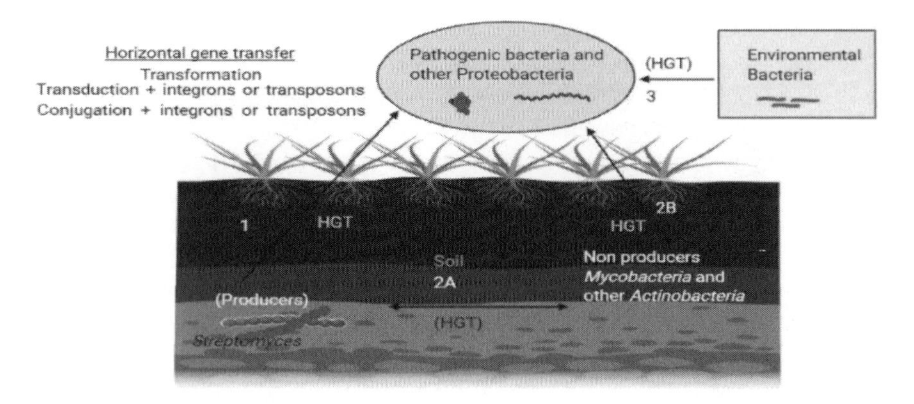

FIGURE 7.1 Schematic representation of source and reservoir of antimicrobial compounds resistant genes in nature and different pathway of their spread.

resistance. In chromosomally acquired resistance, spontaneous mutation in the bacterial chromosomes occurs. These mutations result due to chemical and physical factors (ultraviolet) which cause changes in the target sites present on the cell wall of bacteria or reduce the permeability of antibiotics. In extrachromosomal resistance, bacterial extrachromosomal genetic elements are transferred by either plasmid, transposons, and intergons. Among these, plasmids are large DNA fragments that independently replicate from chromosomes. The plasmid consists of genes that generate enzymes, resulting in the inactivation of antibiotics. These resistance genes and another genetic material of plasmids responsible for the resistance of bacteria are transferred through transduction by a bacteriophage, transformation by competence factor, conjugation by sex pili between two bacteria, and the transposition mechanism (Munita and Arias, 2016).

7.2.3 Cross-Resistance

This type of resistance can be chromosomal acquired or extrachromosomal acquired. This type of condition usually occurs when bacteria are resistant to two or more antibiotics having a similar mechanism and are structurally similar to each other. Sometimes this condition can be seen in unrelated drug groups (Maier et al., 2018).

7.2.4 Multidrug Resistance

Multidrug-resistant (MDR) bacteria are the microorganisms that show resistance to three or more antibiotic classes. The condition of MDR occurs due to the continuous and inappropriate use of drugs. MDR may occur due to the accumulation of multiple-resistance R-plasmid genes that code for the resistance to a single drug and may be due to an increase in the expression of genes that code for multidrug efflux pumps, the inactivation of enzymes, and structural changes in target sites present on the cell wall or membrane of the bacteria (Pelgrift and Friedman, 2013).

7.3 MECHANISM OF RESISTANCE TO ANTIBIOTICS

Bacteria, to survive, has developed a sophisticated mechanism of resistance against antimicrobial agents. Bacterial cells achieve resistance against antimicrobial agents through multiple biochemical pathways that include gene mutation encoding the target site of antimicrobial, efflux overexpression that results in the exclusion of antibiotics outside the cell, and target sites protection through proteins, as shown in Figure 7.2 (a) and (b) (Singh et al., 2012). All bacteria, gram-positive or

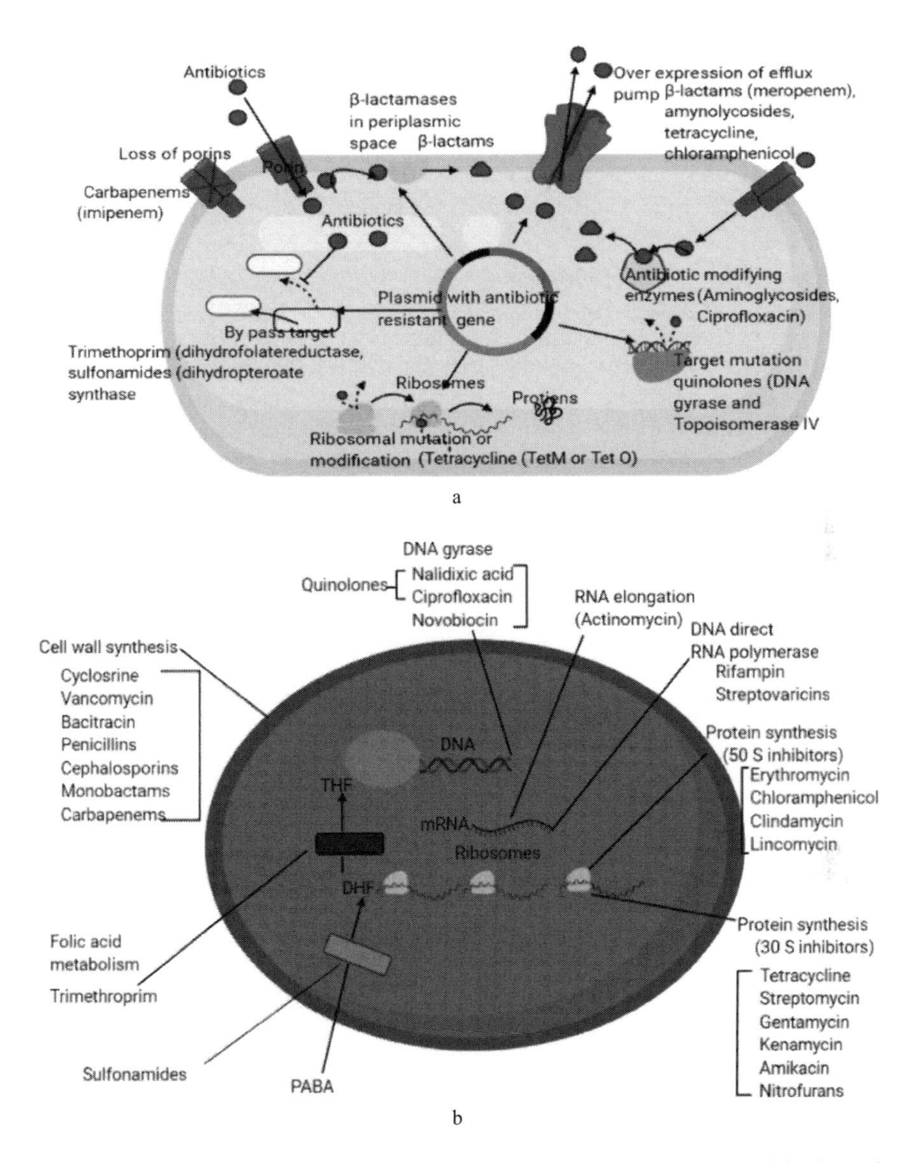

FIGURE 7.2 Mechanism of resistance of (a) gram-negative and (b) gram-positive bacteria against antibiotics.

gram-negative, evolve the mechanism of resistance against antimicrobial agents differently. Gram-positive bacteria achieve resistance against β-lactams by modifying the binding target sites present on the outer wall of the cell whereas gram-negative bacteria produce β-lactamase enzymes. The difference in the mechanism is due to the cell wall composition of both gram-positive and gram-negative bacteria. In gram-negative bacteria, β-lactams require target proteins known as porins to mediate their diffusion to reach Penicillin–binding proteins (PBPs) present in the inner membrane; therefore, bacterial cells produce enzymes known as β-lactamase that control access of β-lactams molecules to the periplasmic space and destroy molecules (Van Grieken et al., 2010). Gram-positive bacteria lack such compartmentalization, but in certain cases, the production of β-lactamase enzymes is also seen as staphylococcal or penicillinase. To provide a detailed classification of resistant mechanism against antibiotics the biochemical route is categorized into following steps that involve modification of antimicrobial agent molecules, resistance by efflux pump, and prevention of antimicrobial compounds reaching the antibiotic target sites (Munita and Arias, 2016).

7.3.1 Modification or Degradation of Antibiotic Molecules

Bacterial strains produce specific enzymes that catalyse and add specific chemical moieties to the antimicrobial compounds resulting in the inactivation of these compounds. These enzymes may destroy the molecules by itself or interrupt the antibiotics to interact with the target sites present on the membrane of the microorganisms (Hurdle et al., 2011). All these mechanisms result in the modification or destruction of antibiotics and therefore have proved to be a common strategy used to render the effectiveness of antibiotics.

7.3.2 Chemical Modification/Alteration

The gram-positive and gram-negative bacteria acquired resistance against antibiotics by introducing chemical changes with the help of enzymes. The enzymes modified the antibiotics chemically and exerted their mechanism of action at the ribosomal level to inhibit protein synthesis. These enzymes perform identical biochemical reactions in antibiotic-resistant bacterial strains and were first identified in the producer *Streptomyces* (Peterson and Kaur, 2018). Several modifying-enzyme-producing microorganisms were also identified and studied to describe frequent biochemical reactions. These biochemical reactions include (a) acetylation of aminoglycosides, chloramphenicol, and streptogramins; (b) phosphorylation of aminoglycosides and chloramphenicol; and (c) adenylation of aminoglycosides and lincosamides. The enzymes not only affect the biochemical reaction but also cause steric hindrance so that antimicrobial compounds become least active for their target sites (Munita and Arias, 2016). AAC (N-acetyl transferases), APH (O-phosphotransferases), and ANT (O-adenyltransferases) are three aminoglycoside-modification enzymes providing resistance to microorganisms against antibiotics. These enzymes act on aminoglycoside molecules and covalently modify their hydroxyl or amino groups (Ramirez and Tolmasky, 2010). In addition to aminoglycoside-modified enzymes, bacteria also produce modifying enzymes for some other classes of antibiotics. These include the bleomycin family, tallysomycin, phleomycin, and zorbamycin. The bacteria produce specific enzymes against these group of antibiotics that results in acetylation of metal free-form and inhibit the formation of the metal-binding domain of these antibiotics. Chloramphenicol is another group of antibiotics that can be modified chemically through enzymes known as chloramphenicol acetyltransferase (CATs) The CATs acetylate the antibiotic chloramphenicol and results in the modification of their chemical structure (Biswas et al., 2012).

7.3.3 Destruction in the Chemical Structure of Antibiotics

The β-lactam antibiotics resistance is conferred by the β-lactamase enzyme. These enzymes are hydrolyzing enzymes destroying the amide bond present in the β-lactam ring, which renders the effect of antimicrobial agents (Bonomo, 2017). In 1940, before the introduction of antibiotic penicillin β-lactamases enzymes were described by scientists. These enzymes are divided into four classes (A, B, C, D) based on the amino acid sequence in Amber classification (King, 2016) and into four categories having several subgroups based on Bush–Jacoby classification. Bush–Jacoby classification is mainly described their biochemical function and substrate specificity. Amber classification among two of them is considered as the backbone for the discussion of different types of β-lactamases.

Class A β-lactamases shares a property with class C and D, having serine residue in their catalytic site. Spectrum activity of class A enzymes includes monobactams instead of cephamucins and clavulanic acid inhibits most of the enzymes of this class. Class A enzymes constitute an extensive range of proteins each having distinguished catalytic properties. TEM-1 and SHV-1 (Penicillinases) hydrolyze penicillin, CTX-M (ESBLs), and carbapenemases (*Klebsiella pneumoniae* carbapenemase) are highly prevalent in gram-negative bacteria and have a high clinical impact (Docquier and Mangani, 2018). ESBLs (CTX-M) are commonly found in *Klebsiella pneumonia*, *Escherichia coli*, and other *Enterobacteriaceae* and are plasmid-encoded. Insertion sequences (ISEcp1) and transposable elements Tn402-like transposons are in association with genes that encode CTM-X enzymes. Conjugated plasmids caught insertion sequences (ISEcp1) and transposable elements Tn402-like transposons and serve them as a vehicle for dissemination (Roy and Partridge, 2017). CTM-X enzymes are the most prevalent ESBLs throughout the world and are responsible for cephalosporin resistance in *K. pneumonia* and *E. coli* (Irenge et al., 2019). Class A carbapenemases have been described into five classes out of which three are chromosomal encoded that include imipenem-hydrolyzing enzyme (IMI), *S. marcescens* enzyme (SME), and not-metalloenzyme carbapenemase (NMC), and two are plasmid-encoded that include Klabseilla pneumonia Carbapenemase (KPC) and Guiana Extended-Spectrum (GES) (Escandón-Vargas et al., 2017). KPC enzymes are predominant in *K. pneumonia* but are found in *E. coli*, *Enterobacter* spp., *Proteus mirabilis*, and *Salmonella* spp. These enzymes are also reported in non-lactose fermenting bacteria *Pseudomonas aeruginosa*.

Class B enzymes were discovered in chromosomal genes of nonpathogenic bacteria over 50 years ago. Later in the 1990s, these enzymes were isolated in pathogenic strains of *Pseudomonas aeruginosa*, *Enterobacter* spp., and *Acinetobacter* spp. These enzymes utilize zinc metal ions as a cofactor and are therefore known as metallo-β-lactamases. The enzymes along with zinc as co-factor destroy the chemical structure of the β-lactam by the nucleophilic attack (Munita and Arias, 2016). These enzymes show similarity to class A carbapenemases and are active against several β-lactams but their mode of action can be inhibited by EDTA an ion-chelating agent. There are about ten types of metallo-β-lactamases and among them, the enzymes that help in providing resistance against antibiotics compounds belong to four families Active-on-imipenem (IMP), Verona integrin-mediated mediated metallo- β –lactamase (VIM), Sao Paulo Metalo- β-lactamase (SPM), and New Delhi metallo- β- lactamase (NDM).

Class C β-lactamases are resistant to all cephalosporins and penicillins. These enzymes are resistant to clavulanic acid and are not hydrolyzed by aztreonam. Cephalosporinase AmpC is one of the most important class C enzymes, which are generally chromosomal encoded while the *bla* AmpC gene associated with these enzymes is also found in plasmids. These enzymes are generally reported in *Enterobacter cloacae*, *E. aerogenes*, *Citrobacter freundii*, *S. marcescens*, *Morganella morganii*, and *Pseudomonas aeruginosa* (Bush, 2010). The expression of AmpC enzymes under the control of complex regulatory mechanism and are highly inducible. AmpR is the transcriptional regulator that bounds to the peptidoglycan precursor (UDP-MurNAc pentapeptides) in the absence of β-lactam and, therefore, not interact with its cognate promoter. The absence of interaction of regulator with promoter suppresses the process of transcription of *bla* AMPc genes. β-lactams, when present, cause alteration in cell wall homeostasis; therefore, peptidoglycan by-products (anhydro-muropeptides) accumulate and start competing with peptidoglycan precursor (UDP-MurNAc pentapeptides) for the AmpR binding sites. The AmpR, therefore, gets released and interacts with the *bla* AMPc promoter and activates the transcription of *bla* AMPc genes (Chakraborty, 2016). Another mechanism involves the over-expression of AmpC by cytosolic amidase called AmpD. AmpD reprocesses muropeptides and decrease the concentration of peptidoglycan precursor (tri-, tetra-, and pentapeptides), and also prevents the displacement of these precursors from AmpR. The bacteria constitutively produce AMPc in large amounts and cause mutation in AmpD which affects the efficacy of cephalosporin (Berrazeg et al., 2015).

Class D β-lactamases consist of an extensive range of enzymes that have a property to hydrolyze oxacillin and are poorly inhibited by clavulanic acid; due to this, this class initially differentiated from class A penicillinases. OXA enzymes variants of this class can degrade third-generation cephalosporins and carbapenems.

7.3.4 Efflux Pumps

Multidrug efflux pumps confer resistance to different antibiotics. These are the transporters present in all organisms including prokaryotes, eukaryotes, and human cells (Blanco et al., 2016). The efflux transport system may be substrate-specific or may have broad specificity. The substrate-specific specificity is for a particular antibiotic such as tet gene determinants specific for tetracycline antibiotic and *mef* genes specific for macrolides in pneumococci and broad specificity is found in bacteria that are MDR. The efflux pump gene can be chromosomally encoded, or these genes can be located in MGEs. The efflux pump, which is chromosomally encoded, explains the

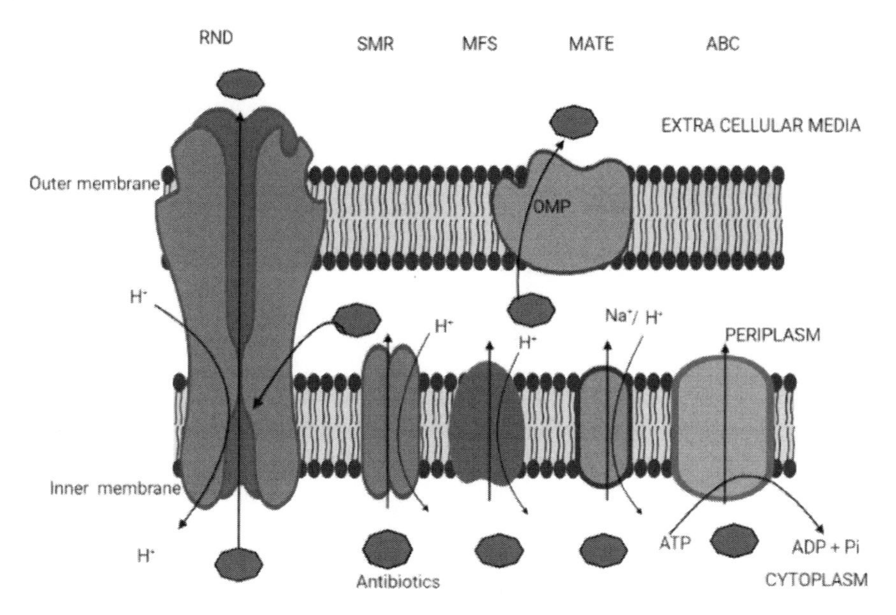

FIGURE 7.3 Bacterial efflux system. Schematic description of families of efflux transporters: RND (resistance nodulation division family), SMR (small multidrug resistance family), MFS (major facilitator superfamily), MATE (multidrug and toxic compound extrusion), ABC (adenosine triphosphate [ATP]-binding cassette superfamily), and OMP (outermembrane protein).

inheritance resistance of bacterial species to specific antibiotics. In bacteria, efflux transporters have five major families (Figure 7.3):

1. RND (Resistance nodulation division family),
2. SMR (Small multidrug resistance family),
3. MFS (Major facilitator superfamily),
4. MATE (Multidrug and toxic compound extrusion),
5. ABC (Adenosine triphosphate [ATP]–binding cassette superfamily (Misra and Bavro, 2009)

These families can be distinguished based on their structural confirmation, the source of energy they are utilizing, and their ability to extrude the substrate and bacteria in which they are distributed. Each family can be defined by similarity to sequence, specificity to the substrate, components (multiple or single), transmembrane spanning regions, and source of energy. The ABC family export substrate by utilizing energy from ATP hydrolysis; other families use proton motive force for the exportation of the substrate (Zhang et al., 2016). Among these five families, the RND superfamily is found only in gram-negative bacteria, and the remaining four are common in both gram-positive and gram-negative bacteria (Weadge et al., 2005). The MFS family of efflux pump groups is the most relevant among four in gram-positive bacteria; the most studied members of this family are NorA and PmrA from *S. aureus* and *Streptococcus pneumoniae* (Handzlik et al., 2013).

7.3.4.1 Efflux Pump and Antibiotic Resistance

In 1980, McMurry and colleagues conferred the resistance of tetracycline in *E. coli* for the first time. In efflux pump mediate tetracycline resistance Tet efflux pumps belong to the MFS family and extrude tetracycline using proton motive force as a source of energy (Kumar et al., 2013). Tet efflux pump, Tet (K), and Tet (L) are preferentially found in gram-negative bacteria and affect tetracycline and doxycycline. In addition to tetracycline-specific efflux transport system, there is the present broad specific efflux transport system which is chromosomall encoded and confers resistance against multiple drugs, thus known as a multidrug efflux pump. These MDR efflux pumps belong to the RND family and provide intrinsic resistance to bacteria against multiple antimicrobials (Mykytczuk, 2003). Efflux pumps of the RND family have tripartite structures and allow selective communication of cytoplasm with the external environment. RND efflux pumps act as proton antiporters and transport a large number of substrates, therefore conferring the resistance to tetracycline, chloramphenicol, and a few β-lactams. In addition to extruding of these antimicrobial, these pumps can extrude toxic compounds like cationic dyes, disinfectants, and bile salts (Blanco et al., 2016). AcrAB-TolC in *E. coli* is an example of an efflux pump that belongs to the RND family and is among one the most studied efflux pumps. The AcrAB-TolC consists of a transporter protein (AcrB) and linker protein (AcrA) located in the inner membrane and periplasmic space and protein channel (TolC) located in the outer membrane of the bacteria (Du et al., 2014). The transport protein AcrB consists of binding pockets that have different substrate preferences; therefore, a series of conformational changes occur when the substrate moves out of the cell. The linker protein AcrA present in the inner membrane and periplasmic membrane helps in the interaction of substrate when extruding via periplasmic space and protein channel TolC (Pos, 2009). Resistance to macrolide by bacteria is another example of efflux-mediated resistance which was first described in gram-positive bacteria *S. aureus* (Handzlik et al., 2013). The efflux pump responsible for the resistance of macrolide is encoded by *mef* A and *mef* E genes. These two *mef* genes extrude the antibiotics belongs to the macrolide class. Transposon Tn1207 located in chromosomes carried *mef* A genes and *mef* E genes are carried by the macrolide efflux genetic assembly element of DNA, also known as MEGA element fragments. These fragments are inserted in different regions of bacterial chromosomes (Santoro et al., 2019). The ABC transporter family consists of MsrA and MsrC efflux pumps, which also result in resistance of macrolide in gram-positive bacteria. MsrA is a plasmid-encoded efflux pump in *S. epidermidis*, and MsrC is a chromosomal encoded efflux pump in *E. faecalis*. MsrC efflux pump produces the least resistance to macrolides and streptogramin B. In *E. faecalis*, the *Lsa* efflux pump is studied which is chromosomally encoded and is responsible for intrinsic resistance of this bacteria to antibiotics streptogramin and lincosamides.

7.3.5 Antimicrobial Resistance by Target Site Protection

Bacteria in their chromosomes consist of genetic determinants that code proteins to mediate the protection of target sites. The genes which are responsible for the target protection mechanism are carried by carrier proteins called MGEs. Tetracycline, fluoroquinolones, and fusidic acid are the example of antibiotics which are affected by the mechanism of protection of target genes (Munita, 2016). Plasmid-mediated tetracycline resistance determinants Tet (M) and Tet (O) are widely found in different pathogenic bacterial strains. These components function similarly to the elongation factor (EF-G and EF Tu) used in the synthesis of protein and belong to GTPases translation factor superfamily (Peterson and Kaur, 2018). Both resistant determinants act with ribosomes inside the cell and replace the exact target position of tetracycline on ribosomes in a Guanosine 5'-Triphosphate (GTP) dependent manner. The resistant determinant Tet M directly replace the tetracycline from its target site and interacts itself with domain IV of 16S rRNA and tetracycline-binding site. This interaction results in the release of tetracycline from the binding site and alteration of the ribosome confirmation, thus preventing the rebinding of the antibiotic to the target site (Foster, 2017). Tet O, instead of directly replacing the tetracycline from its binding site, first competes with it for ribosomal space and alters the shape of the binding site of the antibiotic and then displaces it and resumes the process of protein synthesis (Kashinskaya, 2017). Bacteria consist of plasmid-mediated fluoroquinolones resistant determinant Qnr belongs to pentapeptide repeat family. These target determinants act as DNA homolog and likely compete for the DNA binding site of DNA gyrase and topoisomerase IV. These results decrease in the interaction of DNA gyrase with DNA and therefore quinolone is unable to form a stable, lethal DNA quinolone complex (Hooper and Jacoby, 2015). Bacteria, therefore, follow the different pathways to develop resistance against antimicrobial agents. To overcome this problem, there is a need for developing new antimicrobial compounds effective against the resistant mechanism of the bacteria. NPs are the most common inorganic compounds that have a potential remedy against conventional antibiotic resistance; therefore, they can be used as a potential approach against antibiotic-resistant microorganisms (Wang et al., 2017). NPs use the mechanism of action different from the mechanism followed by antibiotics and target several biomolecules that adversely affect the growth of resistant strains (Slavin et al., 2017). Nanomaterials including NPs of metal and metal oxide are used against MDR bacteria, and these NPs are synthesized from gold, silver, titanium, copper, zinc, and aluminium metals and metal oxides (Niño-Martínez et al., 2019). Metal-based NPs have magnetic and electrochemical properties; therefore, they have also been investigated to identify the pathogenic bacterial strains (Yuan et al., 2018). NPs possess unique physical and chemical properties due to their high surface area and density. The synthesis of NPs along with antibiotics can prove to be an eco-friendly method for their production which can be non-toxic and effective against pathogenic bacteria (Sasidharan and Pottail, 2020).

7.4 MECHANISM OF ACTION OF NANOPARTICLES AGAINST PATHOGENIC MICROORGANISMS

Metal NPs are considered as active broad-spectrum antimicrobial agents. These metal ions are supposed to form powerful coordination links with biomolecules nitrogen, oxygen, sulfur atoms and have an affinity to amines, phosphates, protein

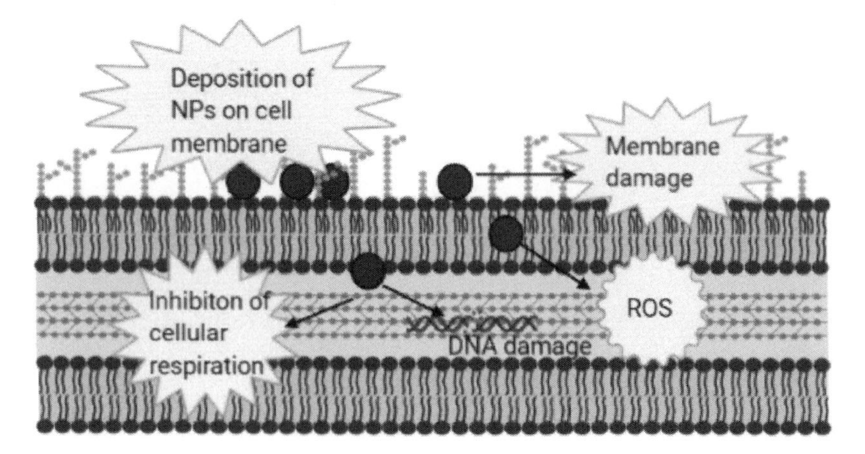

FIGURE 7.4 Toxic effect of nanoparticles on bacterial cell.

thiols, and DNA. The metal ions bind with these biomolecules and influence their functionality, which can be adopted for antimicrobial applications (Hasanzadeh and Shadjou, 2016). Metal-based NPs such as gold, copper, zinc oxide, and titanium oxide generate reactive oxygen species (ROS), which are known to be the main antimicrobial mechanism. ROS are a normal product of cells with respiratory activity and can induce oxidative stress; this destroys cell membranes, genetic material, and cell organelles. (see Figure 7.4).

7.4.1 Mechanism of Action of Silver Nanoparticles Against Pathogenic Microorganisms

The mechanism behind the antimicrobial activity of silver NPs is not known; however, different theories have been proposed on the microbicidal effect of silver NPs against pathogenic bacteria (Pareek et al., 2018). Silver NPs (AgNPs) subsequently penetrate the bacterial cell wall by anchoring it and cause structural changes in cell membranes, including changes in cell membrane permeability, and cell death. When the NPs come in contact with the cell surface, there occurs pit formation where the NPs accumulate (Huang et al., 2016). Another mechanism for bacterial cell death is the formation of free radicals by AgNPs. Electron resonance spectroscopy studies suggested that when AgNPs come in contact with bacteria there occurs the formation of silver ion free radicals. These silver ion free radicals damage the cell membrane make it porous to reach inside the cell and interact with thiol groups of vital enzymes and deactivate them. (Negm et al., 2016). The inactivation of the enzyme by silver ions results in the production of reactive oxygen species. Due to the production of ROS, these silver ions start to attack the cell by itself. Silver being a soft acid tend to react with base, and cells are composed of sulfur and phosphorus which are soft bases. The reaction takes place on cells due to the action of NPs which subsequently lead to cell death. AgNPs also act on DNA as it consists of sulfur and phosphorus as major components therefore result in DNA damage which leads to cell death

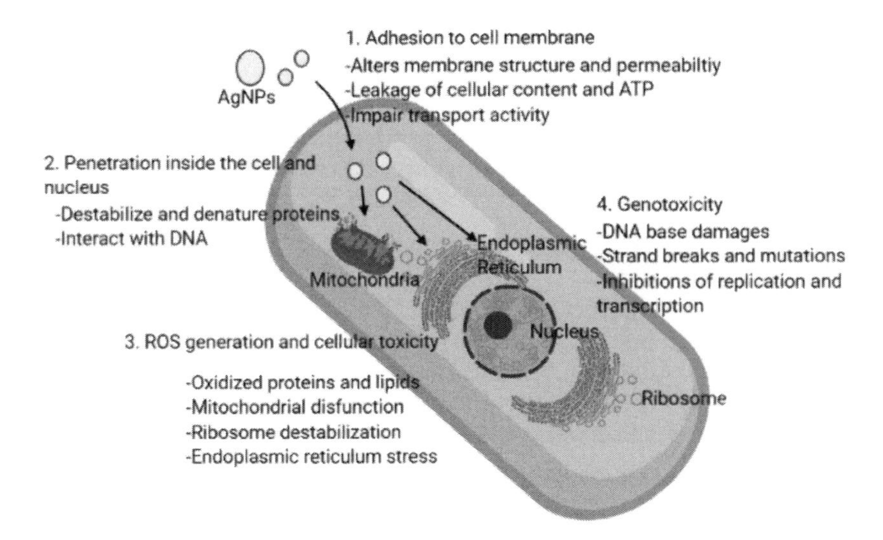

FIGURE 7.5 Cytotoxic mechanism of silver nanoparticles.

(Abdel-Hafez et al., 2016). The interaction of AgNPs with DNA alters its replication and thus terminates the microorganism. The NPs cause dephosphorylation of peptide substrate present on tyrosine residues of gram-negative bacteria and inhibit the signal transduction which results in inhibition of bacterial growth (Dakal et al., 2016) (Figure 7.5).

7.4.2 Mechanism of Action of Zinc Oxide Nanoparticles Against Pathogenic Microorganisms

NPs perform a series of mechanisms to act as an antibacterial agent. The mechanism involves loss of integrity of the cell membrane due to the disruption of a phospholipid bilayer and ROS, inducing oxidative stress. The ROS alter the process of protein synthesis and DNA replication (Das et al., 2017). Also, zinc oxide (ZnO) can generate active targeting potential in the cell membrane which results in disruption of the integrity of the cell membrane and can generate reactive oxygen species that aids in protein and DNA denaturation. The mechanism of action of ZnONPs against pathogenic bacteria, therefore, follow different steps as explained in the following sections.

7.4.2.1 Interaction of Zinc Oxide Particles With Bacterial Cell

The membrane structure of both gram-positive and gram-negative bacteria is different; therefore, the mechanism of attachment of NPs to their surface and their transport inside the cell is different (Mei et al., 2013). Gram-positive bacteria consist of a thick layer of peptidoglycan having teichoic and lipoteichoic acid present in it that act as chelating therefore provide rigidity to the cell. They transport ZnONPs inside the cell by chelating them into Zn^{2+} ions. In gram-negative bacteria, there are present ion channels known as porin proteins on the peptidoglycan layer of the cell wall.

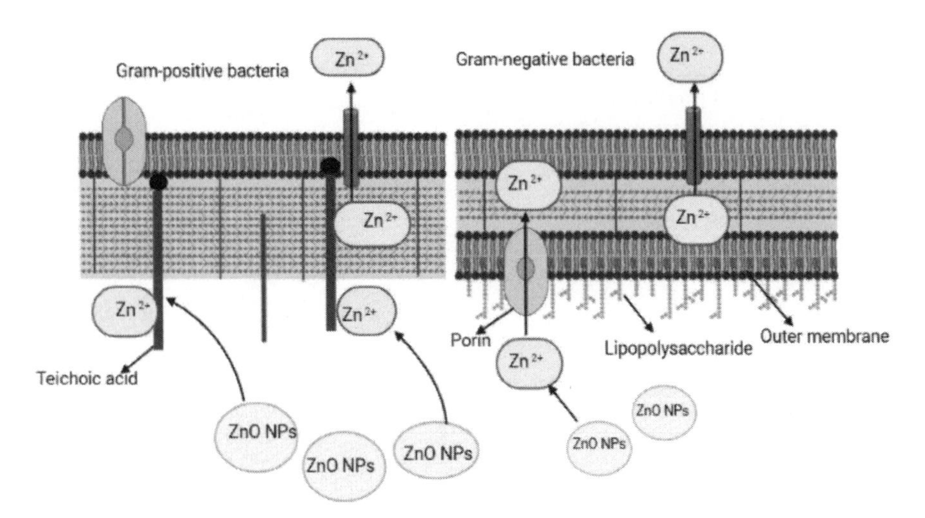

FIGURE 7.6 Schematic representation of the interaction of zinc oxide nano particles with gram-positive and gram-negative bacteria.

These ion channels facilitate the movement of NPs inside the cell by passive diffusion (Wang et al., 2017). The porin proteins also help in nutrient uptake, recognition, and cell-to-cell interactions. The mechanism of transportation of ZnONPs and its transportation inside the cell are represented in Figure 7.6. The diffusion of ZnONPs inside the cell internalized by endocytosis and nonspecific uptake depends on the shape and size of the NPs. The uptake of NPs within the cell is increased by surface functionalization. Maltodextrin, a major source of glucose, is diffused by porin proteins called LamB from outside of the membrane to the periplasmic space. Maltodextrin, when it binds to a maltodextrin binding protein, get transferred to the intracellular matrix. When the surface of NPs is conjugated with maltodextrin the internalization of NPs increased several-fold. ZnONPs, when dissolved in an aqueous medium, release Zn^{2+} ions on change of surface property (Bacchetta et al., 2014). Another common way of NP attachment to the cell surface is the electrostatic attraction of Zn^{2+} ion towards the negatively charged membrane. There occur local dissolution of attached ZnONPs due to which the concentration of Zn^{2+} in bacterial cytoplasm increases which results in membrane leakage and loss of proton motive force (Wang et al., 2017). Gram-negative bacteria exhibit less resistance on interaction with NPs due to the presence of a thin layer of peptidoglycan as compared to gram-positive bacteria. Thus, during the interaction of NPs with bacteria structural composition, the thickness of the cell wall plays an important role (Panariti et al., 2012).

7.4.2.2 Disruption of Cell Membrane

Cell membrane leakage and protrusions are considered one of the most effective reasons for the inhibitory effect of NPs on bacteria (Figure 7.7). When ZnONPs get internalized the cell, there occurs a loss of the phospholipid bilayer, and

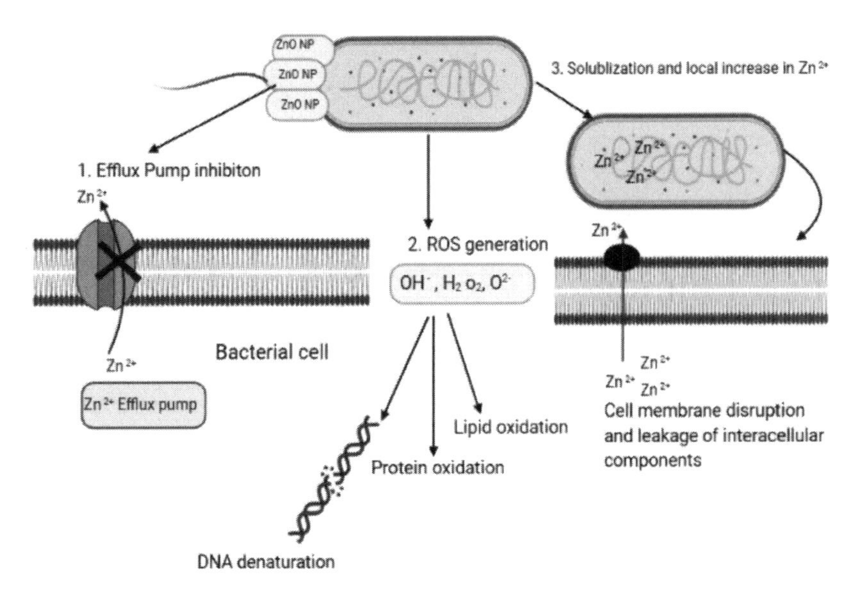

FIGURE 7.7 Mechanism of antibacterial activity shown by zinc oxide nanoparticles.

intracellular components like lipopolysaccharides and ATP leak out, which leads to cell death (Mahmoudi et al., 2011). The resting potential of the cell membrane is altered on inclusion and attachment of NPs; this induces depolarization of the cell membrane by blocking K^+ ion channel in the cell membrane. Damage to the membrane by ZnONPs can be estimated by measuring the charge of bacterial surface (Espitia et al., 2012). When proteins are exposed to ZnONPs, their expression gets regulated leads to lipid peroxidation (Roy et al., 2014). The NPs interact with proteins present on the cell membrane and inactivate them; this decreases the permeability of the membrane and causes cell death. NPs, depending on their physical parameters, can form a protein corona and interact with a thiol group, resulting in the unfolding and denaturation of proteins thus leading to cell death (Abdelhamid and Wu, 2015).

7.4.2.3 Release of Zn^{+2} from Zinc Oxide Complex Into Bacteria Cell

Enzymes play an important role in carrying out basic metabolic pathways essential for life. To induce bacterial cell death, the inhibition of enzyme activity inside the cell is considered an important mechanism (Wang et al., 2017). The high equilibrium solubility of Zn^{+2} ions released from ZnONPs depends on the physical properties. Zn^{+2} acts by inducing a conformational change in the enzymes or by distorting the active sites of enzymes which results in competitive or non-competitive reversible inhibition. The ions can inhibit the enzymes like alkaline phosphatase, DNA, and RNA polymerase by interacting with the sulfhydryl group of these enzymes. Zinc act as competitive inhibitors and inhibits the binding of aspartate and magnesium (Agarwal et al., 2018). Zinc ions impair the respiratory cycle of bacteria, which results in cell death.

7.4.2.4 Generation of Reactive Oxygen Species

NPs induce cell toxicity by generating free radicals, and these free radicals could be superoxide ions, hydroxyl ions, peroxide, and singlet oxygen. These free radicals are mainly generated on exposure to UV rays (He et al., 2014). Hydroxyl ions and superoxide ions bear negative charges; therefore, they fail to enter the bacterial cell membrane whereas peroxide ions have a positive charge and can penetrate the cell membrane and induce death of the cell (Slavin et al., 2017). When ZnONPs anneal with oxygen there occur holes on the surface of the cell membrane which results in increase surface area. Due to an increase in the surface area, there occur more absorption and diffusion of oxygen on the surface, hence the increase in reactive oxygen species production. Oxidative damage, when persisting for a long time, damages DNA, proteins, and lipids. This finally results in cell death.

7.4.2.5 Oxidative Stress Independent of Reactive Oxygen Species

In oxidative stress, independent of ROS, NPs transfer electrons to the cell membrane and cause disruption of it.

7.5 NANOPARTICLES AS THE CARRIER FOR ANTIBIOTIC DRUGS

Most of the NPs are microbicidal to a wide range of bacteria, but due to alteration in genes, the bacteria there also may be a rapid evolution of resistance in NPs (Fernando et al., 2018). To control the resistance, NPs pack a variety of the antimicrobial agents responsible for resistance therefore to overcome the bacteria require multiple simultaneous mutations in the genes. To combat resistance functionalized NPs using as antibiotics is a promising nano-platform that may also reduce the dose and toxicity of the drugs (Mishra et al., 2017). NPs, when conjugated with antibiotics, help deliver these antibiotics to the infected site and, therefore, overcome the resistance and reduce their hazardous effect on normal cells. Synergistic antimicrobial effect of NPs and antibiotics has been absorbed against bacteria, including both gram-positive and gram-negative, at extremely low concentrations (Masadeh et al., 2015). NPs when conjugated with antibiotics like penicillin G, amoxicillin, vancomycin, clindamycin, and erythromycin boost their effects against MDR bacteria (Hasani et al., 2019). MRSA, causing irreversible membrane damage, is affected by NPs conjugated with quercetin and acetylcholine; silver NPs, along with resveratrol nanocarriers, also showed potent activity against all bacteria (AlMatar et al., 2018). Therefore, it has been observed that antibiotic functionalized NPs can promote effective results against pathogenic bacteria which proved to be a reversal of antimicrobial resistance. The synergistic study of NPs along with antibiotics against different microorganisms is listed in Table 7.1.

TABLE 7.1
Effect of Nano Particles in Conjugation with Antibiotics Against Different Pathogenic Microorganisms

Nanoparticles	Antibiotics	Microorganisms	Reference
Silver	Chloramphenicol, Rifampicin, Vancomycin, and Ampicillin	*Salmonella typhi, Enterococcus faecium, Pseudomonas aeruginosa, Acinetobacter baumannii, Enterobacteraerogenes, Escherichia coli, Streptococcus mutans, Staphylococcus aureus*	Hwang et al. (2012); Brown et al. (2012); Thomas et al. (2014); Franci et al. (2015); Wan et al. (2016)
Gold	Ampicillin, Vancomycin, Kanamycin	Methicillin resistant *S. aureus, Pseudomona aeruginosa, E. aerogenes, E. coli, Streptococcus bovis, Staphylococcus epidermidis, Yersinia pestis*	Gu et al. (2003); Brown et al. (2012); Hur and Park, (2016);
Zinc oxide	Streptomycin, gentamycin, neomycin, Ciprofloxacin, ceftazidime, Ciprofloxacin, erythromycin, methicillin, vancomycin	*S. aureus, Micrococcus luteus, E. coli, P. aeruginosa, MDR A. baumannii, E. faecalis, E. faecium*	Iram et al. (2016); Balaure et al. (2019)

7.6 CONCLUSION

Microorganisms have evolved innate immune response strategies and therefore are dodging the process of examination. To fight against the infection caused by microorganisms, antibiotics have become the mainstay, but infection resistance to these antibiotics is alarmingly on the rise. The continuous use of antibiotics is the major reason for the resistance of microorganisms as it allows them to transform their genotype, resulting in the emergence of multidrug-resistant strains. The development of resistant strains emphasizes the need to develop effective therapeutic options. NPs, due to their biocidal and immunopotentiating properties, are considered an alternative to antibiotics. NPs in conjugation with existing antibiotics for bacterial infections result in lowering of uptake dosage and therefore minimize the toxicity and reduce the probability of development of resistance. The synergistic approach of NPs with antibiotics may serve as a complement to the existing therapies and may help to control the serious bacterial infections.

REFERENCES

Abdel-Hafez, S. I., Nafady, N. A., Abdel-Rahim, I. R., Shaltout, A. M., Daròs, J. A., and Mohamed, M. A. (2016). Assessment of protein silver nanoparticles toxicity against pathogenic *Alternaria solani*. *3 Biotech*, *6*(2), 199.

Abdelhamid, H. N., and Wu, H. F. (2015). Proteomics analysis of the mode of antibacterial action of nanoparticles and their interactions with proteins. *TrAC Trends in Analytical Chemistry*, *65*, 30–46.

Agarwal, A., Virk, G., Ong, C., and Du Plessis, S. S. (2014). Effect of oxidative stress on male reproduction. *The World Journal of Men's Health*, *32*(1), 1–17.

Aghdam, E. M., Hejazi, M. S., and Barzegar, A. (2016). Riboswitches: From living biosensors to novel targets of antibiotics. *Gene*, *592*(2), 244–259.

Ahmed, K. B. A., Raman, T., and Veerappan, A. (2016). Future prospects of antibacterial metal nanoparticles as enzyme inhibitor. *Materials Science and Engineering: C*, *68*, 939–947.

AlMatar, M., Makky, E. A., Var, I., and Koksal, F. (2018). The role of nanoparticles in the inhibition of multidrug-resistant bacteria and biofilms. *Current Drug Delivery*, *15*(4), 470–484.

Bacchetta, R., Moschini, E., Santo, N., Fascio, U., Del Giacco, L., Freddi, S., …, and Mantecca, P. (2014). Evidence and uptake routes for Zinc oxide nanoparticles through the gastrointestinal barrier in *Xenopus laevis*. *Nanotoxicology*, *8*(7), 728–744.

Balaure, P. C., Holban, A. M., Grumezescu, A. M., Mogoşanu, G. D., Bălşeanu, T. A., Stan, M. S., …, and Mogoantă, L. (2019). In vitro and in vivo studies of novel fabricated bioactive dressings based on collagen and zinc oxide 3D scaffolds. *International Journal of Pharmaceutics*, *557*, 199–207.

Bamrungsap, S., Zhao, Z., Chen, T., Wang, L., Li, C., Fu, T., and Tan, W. (2012). Nanotechnology in therapeutics: A focus on nanoparticles as a drug delivery system. *Nanomedicine*, *7*(8), 1253–1271.

Berrazeg, M., Jeannot, K., Enguéné, V. Y. N., Broutin, I., Loeffert, S., Fournier, D., and Plésiat, P. (2015). Mutations in β-lactamase AmpC increase resistance of Pseudomonas aeruginosa isolates to antipseudomonal cephalosporins. *Antimicrobial Agents and Chemotherapy*, *59*(10), 6248–6255.

Biswas, T., Houghton, J. L., Garneau-Tsodikova, S., and Tsodikov, O. V. (2012). The structural basis for substrate versatility of chloramphenicol acetyltransferase CATI. *Protein Science*, *21*(4), 520–530.

Blanco, P., Hernando-Amado, S., Reales-Calderon, J. A., Corona, F., Lira, F., Alcalde-Rico, M., …, and Martinez, J. L. (2016). Bacterial multidrug efflux pumps: Much more than antibiotic resistance determinants. Microorganisms, *4*(1), 14.

Bonomo, R. A. (2017). β-Lactamases: A focus on current challenges. *Cold Spring Harbor Perspectives in Medicine*, *7*(1), a025239.

Brown, A. N., Smith, K., Samuels, T. A., Lu, J., Obare, S. O., and Scott, M. E. (2012). Nanoparticles functionalized with ampicillin destroy multiple-antibiotic-resistant isolates of Pseudomonas aeruginosa and Enterobacter aerogenes and methicillin-resistant Staphylococcus aureus. *Applied and Environmental Microbiology*, *78*(8), 2768–2774.

Bush, K. (2010). Bench-to-bedside review: The role of β-lactamases in antibiotic-resistant Gram-negative infections. *Critical Care*, *14*(3), 224.

Butler, M. S., Blaskovich, M. A., and Cooper, M. A. (2017). Antibiotics in the clinical pipeline at the end of 2015. *The Journal of Antibiotics*, *70*(1), 3–24.

Chakraborty, A. K. (2016). Multi-drug resistant genes in bacteria and 21st Century problems associated with antibiotic therapy. *BioTechnology: An Indian Journal*, *12*(12), 114.

Chawla, P., Kumar, N., Bains, A., Dhull, S. B., Kumar, M., Kaushik, R., and Punia, S. (2020). Gum arabic capped copper nanoparticles: Synthesis, characterization, and applications. *International Journal of Biological Macromolecules*, *146*, 232–242.

Chawla, P., Kumar, N., Kaushik, R., and Dhull, S. B. (2019). Synthesis, characterization and cellular mineral absorption of nanoemulsions of Rhododendron arboreum flower extracts stabilized with gum arabic. *Journal of Food Science and Technology*, *56*(12), 5194–5203.

Dakal, T. C., Kumar, A., Majumdar, R. S., and Yadav, V. (2016). Mechanistic basis of antimicrobial actions of silver nanoparticles. *Frontiers in Microbiology*, *7*, 1831.

Das, B., Dash, S. K., Mandal, D., Ghosh, T., Chattopadhyay, S., Tripathy, S., …, and Roy, S. (2017). Green synthesized silver nanoparticles destroy multidrug resistant bacteria via reactive oxygen species mediated membrane damage. *Arabian Journal of Chemistry*, *10*(6), 862–876.

Docquier, J. D., and Mangani, S. (2018). An update on β-lactamase inhibitor discovery and development. *Drug Resistance Updates*, *36*, 13–29.

Dowling, A., Clift, R., Grobert, N., Hutton, D., Oliver, R., O'neill, O., …, and Ryan, J. (2004). Nanoscience and nanotechnologies: Opportunities and uncertainties. *The Royal Society & The Royal Academy of Engineering*, *46*, 618–618.

Du, D., Wang, Z., James, N. R., Voss, J. E., Klimont, E., Ohene-Agyei, T., …, and Luisi, B. F. (2014). Structure of the AcrAB–TolC multidrug efflux pump. *Nature*, *509*(7501), 512–515.

Durán, N., Durán, M., De Jesus, M. B., Seabra, A. B., Fávaro, W. J., and Nakazato, G. (2016). Silver nanoparticles: A new view on mechanistic aspects on antimicrobial activity. *Nanomedicine: Nanotechnology, Biology and Medicine*, *12*(3), 789–799.

Escandón-Vargas, K., Reyes, S., Gutiérrez, S., and Villegas, M. V. (2017). The epidemiology of carbapenemases in Latin America and the Caribbean. *Expert Review of Anti-Infective Therapy*, *15*(3), 277–297.

Espitia, P. J. P., Soares, N. D. F. F., dos Reis Coimbra, J. S., de Andrade, N. J., Cruz, R. S., and Medeiros, E. A. A. (2012). Zinc oxide nanoparticles: Synthesis, antimicrobial activity and food packaging applications. *Food and Bioprocess Technology*, *5*(5), 1447–1464.

Fernando, S. S. N., Gunasekara, T. D. C. P., and Holton, J. (2018). Antimicrobial Nanoparticles: Applications and mechanisms of action. *Srilanka Journal of Infectious Disease, 8*(1) 2–11.

Foster, T. J. (2017). Antibiotic resistance in *Staphylococcus aureus*. Current status and future prospects. *FEMS Microbiology Reviews*, *41*(3), 430–449.

Franci, G., Falanga, A., Galdiero, S., Palomba, L., Rai, M., Morelli, G., and Galdiero, M. (2015). Silver nanoparticles as potential antibacterial agents. *Molecules*, *20*(5), 8856–8874.

Gao, W., Chen, Y., Zhang, Y., Zhang, Q., and Zhang, L. (2018). Nanoparticle-based local antimicrobial drug delivery. *Advanced Drug Delivery Reviews*, *127*, 46–57.

Goswami, N., Zheng, K., and Xie, J. (2014). Bio-NCs–the marriage of ultrasmall metal nanoclusters with biomolecules. *Nanoscale*, *6*(22), 13328–13347.

Gottlieb, T., and Nimmo, G. R. (2011). Antibiotic resistance is an emerging threat to public health: An urgent call to action at the Antimicrobial Resistance Summit 2011. *Medical Journal of Australia*, *194*(6), 281–283.

Gu, H., Ho, P. L., Tong, E., Wang, L., and Xu, B. (2003). Presenting vancomycin on nanoparticles to enhance antimicrobial activities. *Nano Letters*, *3*(9), 1261–1263.

Guo, D., Zhu, L., Huang, Z., Zhou, H., Ge, Y., Ma, W., …, and Zhao, Y. (2013). Anti-leukemia activity of PVP-coated silver nanoparticles via generation of reactive oxygen species and release of silver ions. *Biomaterials*, *34*(32), 7884–7894.

Handzlik, J., Matys, A., and Kieć-Kononowicz, K. (2013). Recent advances in multi-drug resistance (MDR) efflux pump inhibitors of Gram-positive bacteria S. aureus. *Antibiotics*, *2*(1), 28–45.

Hasani, A., Madhi, M., Gholizadeh, P., Mojarrad, J. S., Rezaee, M. A., Zarrini, G., and Kafil, H. S. (2019). Metal nanoparticles and consequences on multi-drug resistant bacteria: Reviving their role. *SN Applied Sciences*, *1*(4), 360.

Hasanzadeh, M., and Shadjou, N. (2016). Pharmacogenomic study using bio-and nanobioelectrochemistry: Drug–DNA interaction. *Materials Science and Engineering: C*, *61*, 1002–1017.

Hastings, P. J., Rosenberg, S. M., and Slack, A. (2004). Antibiotic-induced lateral transfer of antibiotic resistance. *Trends in Microbiology*, *12*(9), 401–404.

He, W., Liu, Y., Wamer, W. G., and Yin, J. J. (2014). Electron spin resonance spectroscopy for the study of nanomaterial-mediated generation of reactive oxygen species. *Journal of Food and Drug Analysis*, *22*(1), 49–63.

Hooper, D. C., and Jacoby, G. A. (2015). Mechanisms of drug resistance: Quinolone resistance. *Annals of the New York Academy of Sciences*, *1354*(1), 12.

Huang, L., Zhao, S., Wang, Z., Wu, J., Wang, J., and Wang, S. (2016). In situ immobilization of silver nanoparticles for improving permeability, antifouling and anti-bacterial properties of ultrafiltration membrane. *Journal of Membrane Science*, *499*, 269–281.

Hur, Y. E., and Park, Y. (2016). Vancomycin-functionalized gold and silver nanoparticles as an antibacterial nanoplatform against methicillin-resistant Staphylococcus aureus. *Journal of Nanoscience and Nanotechnology*, *16*(6), 6393–6399.

Hurdle, J. G., O'neill, A. J., Chopra, I., and Lee, R. E. (2011). Targeting bacterial membrane function: An underexploited mechanism for treating persistent infections. *Nature Reviews Microbiology*, *9*(1), 62–75.

Hwang, I. S., Hwang, J. H., Choi, H., Kim, K. J., and Lee, D. G. (2012). Synergistic effects between silver nanoparticles and antibiotics and the mechanisms involved. *Journal of Medical Microbiology*, *61*(12), 1719–1726.

Iram, S., Khan, J. A., Aman, N., Nadhman, A., Zulfiqar, Z., and Yameen, M. A. (2016). Enhancing the anti-enterococci activity of different antibiotics by combining with metal oxide nanoparticles. *Jundishapur Journal of Microbiology*, *9*(3), e31302.

Irenge, L. M., Ambroise, J., Bearzatto, B., Durant, J. F., Chirimwami, R. B., and Gala, J. L. (2019). Whole-genome sequences of multidrug-resistant Escherichia coli in South-Kivu Province, Democratic Republic of Congo: Characterization of phylogenomic changes, virulence and resistance genes. *BMC Infectious Diseases*, *19*(1), 137.

Jamuna-Thevi, K., Bakar, S. A., Ibrahim, S., Shahab, N., and Toff, M. R. M. (2011). Quantification of silver ion release, in vitro cytotoxicity and antibacterial properties of nanostuctured Ag doped TiO_2 coatings on stainless steel deposited by RF magnetron sputtering. *Vacuum*, *86*(3), 235–241.

Kashinskaya, D. (2017). *Structural analysis and investigation of the Staphylococcus aureus ribosome and potential anticancer drugs.* (Doctoral dissertation, Strasbourg).

King, A. M. (2016). *Discovery and characterization of novel beta-lactamase inhibitors* (Doctoral dissertation), Hamilton, Ontario, Canada.

Kraft, J. C., Freeling, J. P., Wang, Z., and Ho, R. J. (2014). Emerging research and clinical development trends of liposome and lipid nanoparticle drug delivery systems. *Journal of Pharmaceutical Sciences*, *103*(1), 29–52.

Kumar, S., Floyd, J. T., He, G., and Varela, M. F. (2013). Bacterial antimicrobial efflux pumps of the MFS and MATE transporter families: A review. *Recent Research and Development Antimicrobial Agents and Chemotherapy*, *7*, 1–21.

Mahmoudi, M., Azadmanesh, K., Shokrgozar, M. A., Journeay, W. S., and Laurent, S. (2011). Effect of nanoparticles on the cell life cycle. *Chemical Reviews*, *111*(5), 3407–3432.

Maier, L., Pruteanu, M., Kuhn, M., Zeller, G., Telzerow, A., Anderson, E. E., …, and Patil, K. R. (2018). Extensive impact of non-antibiotic drugs on human gut bacteria. *Nature*, *555*(7698), 623–628.

Masadeh, M. M., Karasneh, G. A., Al-Akhras, M. A., Albiss, B. A., Aljarah, K. M., Al-Azzam, S. I., and Alzoubi, K. H. (2015). Cerium oxide and iron oxide nanoparticles abolish the antibacterial activity of ciprofloxacin against gram positive and gram negative biofilm bacteria. *Cytotechnology*, *67*(3), 427–435.

Mei, L., Lu, Z., Zhang, W., Wu, Z., Zhang, X., Wang, Y., …, and Jia, Y. (2013). Bioconjugated nanoparticles for attachment and penetration into pathogenic bacteria. *Biomaterials*, *34*(38), 10328–10337.

Miller, K. P., Wang, L., Benicewicz, B. C., and Decho, A. W. (2015). Inorganic nanoparticles engineered to attack bacteria. *Chemical Society Reviews*, *44*(21), 7787–7807.

Mishra, P. K., Mishra, H., Ekielski, A., Talegaonkar, S., and Vaidya, B. (2017). Zinc oxide nanoparticles: A promising nanomaterial for biomedical applications. *Drug Discovery Today*, *22*(12), 1825–1834.

Misra, R., and Bavro, V. N. (2009). Assembly and transport mechanism of tripartite drug efflux systems. *Biochimica et Biophysica Acta (BBA)-Proteins and Proteomics*, *1794*(5), 817–825.

Munita, J. M., and Arias, C. A. (2016). Mechanisms of antibiotic resistance. *Virulence Mechanisms of Bacterial Pathogens*, *22*, 481–511.

Mykytczuk, O. (2003). *Characterization of antimicrobial resistance mechanisms in multi-drug resistant Salmonella and Shigella strains from Brazil and Colombia*. Master dissertation, Brazil.

Negm, N. A., Kana, M. T. A., Abd-Elaal, A. A., and Elwahy, A. H. (2016). Fluorescein dye derivatives and their nanohybrids: Synthesis, characterization and antimicrobial activity. *Journal of Photochemistry and Photobiology B: Biology*, *162*, 421–433.

Niño-Martínez, N., Salas Orozco, M. F., Martínez-Castañón, G. A., Torres Méndez, F., and Ruiz, F. (2019). Molecular mechanisms of bacterial resistance to metal and metal oxide nanoparticles. *International Journal of Molecular Sciences*, *20*(11), 2808.

Panariti, A., Miserocchi, G., and Rivolta, I. (2012). The effect of nanoparticle uptake on cellular behavior: Disrupting or enabling functions? *Nanotechnology, Science and Applications*, *5*, 87.

Pareek, V., Gupta, R., and Panwar, J. (2018). Do physico-chemical properties of silver nanoparticles decide their interaction with biological media and bactericidal action? A review. *Materials Science and Engineering: C*, *90*, 739–749.

Pelgrift, R. Y., and Friedman, A. J. (2013). Nanotechnology as a therapeutic tool to combat microbial resistance. *Advanced Drug Delivery Reviews*, *65*(13–14), 1803–1815.

Peterson, E., and Kaur, P. (2018). Antibiotic resistance mechanisms in bacteria: Relationships between resistance determinants of antibiotic producers, environmental bacteria, and clinical pathogens. *Frontiers in Microbiology*, *9*, 2928.

Pos, K. M. (2009). Drug transport mechanism of the AcrB efflux pump. *Biochimica et Biophysica Acta (BBA)-Proteins and Proteomics*, *1794*(5), 782–793.

Ramirez, M. S., and Tolmasky, M. E. (2010). Aminoglycoside modifying enzymes. *Drug Resistance Updates*, *13*(6), 151–171.

Rani, M., Shanker, U., and Jassal, V. (2017). Recent strategies for removal and degradation of persistent & toxic organochlorine pesticides using nanoparticles: A review. *Journal of Environmental Management*, *190*, 208–222.

Roy, P. H., and Partridge, S. R. (2017). Genetic Mechanisms of Transfer of Drug Resistance. In *Antimicrobial Drug Resistance* (pp. 61–76). Springer, Cham, Douglas L. Mayer, Jack D. Sobel, Marc Ouellette, Keith S. Kaye, Dror Marchaim.

Roy, R., Singh, S. K., Chauhan, L. K. S., Das, M., Tripathi, A., and Dwivedi, P. D. (2014). Zinc oxide nanoparticles induce apoptosis by enhancement of autophagy via PI3K/Akt/mTOR inhibition. *Toxicology Letters*, *227*(1), 29–40.

Santoro, F., Iannelli, F., and Pozzi, G. (2019). Genomics and genetics of Streptococcus pneumoniae. *Gram-Positive Pathogens*, *7*, 344–361.

Sasidharan, S., and Pottail, L. (2020). Antimicrobial activity of metal and non-metallic nanoparticles from Cyperus rotundus root extract on infectious disease causing pathogens. *Journal of Plant Biochemistry and Biotechnology*, *29*(1), 134–143.

Singh, R., Swick, M. C., Ledesma, K. R., Yang, Z., Hu, M., Zechiedrich, L., and Tam, V. H. (2012). Temporal interplay between efflux pumps and target mutations in development of antibiotic resistance in Escherichia coli. *Antimicrobial Agents and Chemotherapy*, *56*(4), 1680–1685.

Slavin, Y. N., Asnis, J., Häfeli, U. O., and Bach, H. (2017). Metal nanoparticles: Understanding the mechanisms behind antibacterial activity. *Journal of Nanobiotechnology*, *15*(1), 65.

Soppimath, K. S., Aminabhavi, T. M., Kulkarni, A. R., and Rudzinski, W. E. (2001). Biodegradable polymeric nanoparticles as drug delivery devices. *Journal of Controlled Release*, *70*(1–2), 1–20.

Thomas, R., Nair, A. P., Soumya, K. R., Mathew, J., and Radhakrishnan, E. K. (2014). Antibacterial activity and synergistic effect of biosynthesized AgNPs with antibiotics against multidrug-resistant biofilm-forming coagulase-negative staphylococci isolated from clinical samples. *Applied Biochemistry and Biotechnology*, *173*(2), 449–460.

Tibbals, H. F. (2017). *Medical nanotechnology and nanomedicine*. CRC Press, New York.

Van Grieken, R., Marugán, J., Pablos, C., Furones, L., and López, A. (2010). Comparison between the photocatalytic inactivation of Gram-positive E. faecalis and Gram-negative *E. coli* faecal contamination indicator microorganisms. *Applied Catalysis B: Environmental*, *100*(1–2), 212–220.

Wan, G., Ruan, L., Yin, Y., Yang, T., Ge, M., and Cheng, X. (2016). Effects of silver nanoparticles in combination with antibiotics on the resistant bacteria Acinetobacter baumannii. *International Journal of Nanomedicine*, *11*, 3789.

Wang, L., Hu, C., and Shao, L. (2017). The antimicrobial activity of nanoparticles: Present situation and prospects for the future. *International Journal of Nanomedicine*, *12*, 1227.

Weadge, J. T., Pfeffer, J. M., and Clarke, A. J. (2005). Identification of a new family of enzymes with potential O-acetylpeptidoglycan esterase activity in both Gram-positive and Gram-negative bacteria. *BMC Microbiology*, *5*(1), 49.

Yuan, P., Ding, X., Yang, Y. Y., and Xu, Q. H. (2018). Metal nanoparticles for diagnosis and therapy of bacterial infection. *Advanced Healthcare Materials*, *7*(13), 1701392.

Zhang, X. C., Han, L., and Zhao, Y. (2016). Thermodynamics of ABC transporters. *Protein & Cell*, *7*(1), 17–27.

8 Nanoemulsions of Plant-Based Bioactive Compounds

Synthesis, Properties, and Applications

Naresh Butani
Shree Ramkrishna Institute of Computer Education and Applied Sciences, India

Megha D. Bhatt
G. B. Pant University of Agriculture & Technology, Pantnagar, India

Preeti Parmar
Navsari Agricultural University, India

Jaydip Jobanputra
Bhagwan Mahavir College of Science and Technology, India

Anoop K. Dobriyal
HNB Garhwal Central University, India

Deepesh Bhatt
Shree Ramkrishna Institute of Computer Education and Applied Sciences, India

CONTENTS

8.1 INTRODUCTION

Measures to enhance food safety, nutritional enrichment, and increased shelf life have always been an area of prime focus that holds immense scope for improving overall livelihood. Food preservation methods mostly utilize freezing, dehydration, varying pH, temperature, and using antimicrobial and bioactive compounds which contain the growth of pathogenic and spoilage microorganisms responsible for food-borne illness (Guzey and McClements, 2006). In this era of synthetic food additives, demands for products formulated using natural ingredients has propelled the food companies for exploring better preservation techniques for increased food safety, without causing any significant nutritional and organoleptic depletion in terms of food quality. Plants are reported to be a rich source of bioactive compounds capable of exhibiting antibacterial, antifungal, insecticidal, and antioxidant properties (Dhull et al., 2016; Kaur et al., 2018). Being safe and economical, plant-derived phytochemicals have recently received considerable attention in the area of food preservation mostly because of their innate antimicrobial properties (Khameneh et al., 2019; Dhull et al., 2020a). Nanoemulsions prepared using plant-derived compounds not only show increased shelf life but also display high efficacy against potentially pathogenic microbes in a natural way (Chawla et al., 2019; Panghal et al., 2019; Dhull et al., 2019a). However, the major limiting factor for these potentially safe plant-based bioactive compounds is their sparing solubility in the aqueous phase. At this minimal concentration, these bioactive compounds are unable to suppress the growth of a pathogenic culture which minimizes their large-scale commercial usage. To overcome this issue, several studies have conceptualized the encapsulation of these antimicrobial and bioactive compounds for generating food-grade emulsions which holds better physicochemical stabilization (Donsì, 2018). The word *emulsion* is derived from *emulgeo*, which means "to milk" which to date proves to be the best example of a natural emulsion. In an emulsion, the liquid droplets are dispersed in a liquid that can be broadly categorized as water-in-oil (W/O)-, oil-in-water (O/W)-, and oil-in-oil (O/O)-type emulsions. Another vital component termed as emulsifier also plays a significant role in dispersing two immiscible liquids named as the dispersed phase and the continuous phase. Emulsions can be prepared using artificial synthetic surfactants, namely polysorbates or natural amphiphilic biopolymers, mostly proteins that form globules. The sizes of these globules range from approximately 0.25 to 25 μM in diameter (Joyce and Nagarajan, 2017). The emulsion is termed as a heterogeneous system that is based on the type of emulsifier used which is known to alter the properties of the generated system. These can be classified broadly into microemulsions, nanoemulsions, and macroemulsions. Microemulsions possess a radius of 25 nm or less which are formed spontaneously, in which oil droplets get encapsulated in small surfactant micelles (McClements and Rao, 2011). Being spontaneously generated, these are known to be thermodynamically stable. Compared to microemulsions, the other two, namely nanoemulsions and macroemulsions, are thermodynamically unstable, and hence, a moderate amount of surfactant seems to be a prerequisite in these systems. Biopolymeric emulsifiers augment the formation of these emulsions but, being thermodynamically unstable, require minimal external energy which can either be obtained from certain types of mechanical devices or may be a form of stored chemical energy inside the system. However, macroemulsions generally appear opaque and are turbid because of their droplet size, having a radius larger than 100 nm. In contrast, nanoemulsions usually are relatively more transparent or somewhat translucent because of their smaller droplet size, usually smaller than 100 nm (McClements, 2012). This reduction in the droplet size, a consequence of significant steric stabilization between droplets, is considered to be the major reason for the higher stability of nanoemulsions as smaller size helps in minimizing the breakdown processes, like creaming, sedimentation, flocculation, Ostwald ripening, coalescence and phase inversion which are discussed in detail in a later part of the chapter (Weiss et al., 2009). Among other factors are surfactant type, surfactant concentration, mixing ratio of oil to surfactant, and sonication time, which helps to prevent the breakdown process mainly the shear-induced coalescence. It does so by minimizing the surface tension between the oil–water interface, which leads to a significant influence on droplet size and stability in the generated nanoemulsion. Nanoemulsion preparation can be broadly classified into two categories, namely high-energy emulsification and low-energy emulsification. The former method utilizes mechanical devices to generate strong disruptive forces that break the oil and water phase whereas the latter method is based on the formation of small droplets by utilizing the stored internal energy for obtaining nanoemulsions. Alternatively, nanoemulsions can be stabilized by optimizing parameters like temperature and pH and varying the hydrophilic–lipophilic balance (HLB) of the system. The importance of nanoemulsions can be justified by analysing their prime applicability in industries like the food, agrochemical, cosmetics, oil, and pharmaceutical sectors. The product range includes dairy products, bakery products, candy products, meat products, pesticides, oils, lotions, latex emulsions, creams, ointments, anaesthetics, hair and skin conditioning products, and many more. This chapter broadly summarizes most of the plant-derived compounds, their classification, and traditional usage while the major emphasis will be on recent advancements made in the area of plant-derived nanoemulsions by highlighting their medicinal, pharmaceutical, and food-based applications. Major case studies for nanoemulsion preparation of main polyphenols, flavonoids, phenolic acids, lignans, catechins, curcumin, essential oils like carvacrol, cinnamaldehyde, thymol, cymene, eugenol, terpenes; their preparation methods; and potential applications in terms of food safety are discussed.

8.2 MAJOR COMPONENTS OF NANOEMULSIONS SYSTEM

The major components involved in the nanoemulsion formulation could be summarized as in the following sections.

8.2.1 Oil Phase and Aqueous Phase

The nanoemulsion system is composed mainly of the oil phase and aqueous phase but also entails additional components for increasing stability and solubility. The oil phase comprises hydrophobic components that can be assembled into smaller droplets in presence of oils and must possess properties like polarity, density, refractive index, and, most important, low viscosity with minimal interfacial tension (Shahavi et al., 2019). These properties are often helpful in reducing the size due to lesser energy requirements. The oil phase includes components such as triglyceride and essential oils, flavour oils, oil-soluble vitamins, colours, flavours, preservatives, nutraceuticals, and pharmaceuticals. However, digestibility and bioavailability of the oil also play another vital role in generating physiological responses for the used nanoemulsion (Punia et al., 2019). Apart from showing excellent formulating properties, these types of emulsions lack stability as being susceptible to phenomena like coalescence or Ostwald ripening. Therefore, to avoid these phenomena several "ripening inhibitors", namely triglycerides such as medium- or long-chain triglycerides and sometimes vegetable oils such as corn, sunflower, and sesame oils, are frequently used (Pavoni et al., 2020). Whereas the aqueous phase comprises mostly of polar solvents mainly water for the formation of nanoemulsions. Apart from this, several other polar co-solvents, proteins, carbohydrates, surfactants, minerals, flavours, preservatives, vitamins, certain acids, and bases are also frequently used. The use of these components in aqueous phase has a prominent role in controlling the physicochemical properties, like ionic strength, polarity, density, rheology, refractive index, interfacial tension, and phase behaviour, of these nanoemulsions. However, several studies report further decrease in size of the oil droplets by the addition of viscosity enhancers before homogenization, which results in higher disruptive shear stresses built inside the homogenizer (Qian and McClements, 2011). The addition of viscosity enhancers was also found to augment its long-term stability as the addition slowed down the gravitational separation, thus reducing the process of droplet collisions. Apart from this, other degradative processes, like lipid peroxidation, is also vulnerable, which again can be prevented by the addition of water-soluble antioxidants in the aqueous phase (McClements and Decker, 2000). Thus, the aqueous phase composition plays a vital role in controlling the stability and several functional attributes of synthesized nanoemulsions.

8.2.2 Stabilizers

Being thermodynamically unstable the formation of small droplets requires special attention to prevent the process of aggregation during and after the homogenization process. The addition of stabilizers is often required to avoid these issues, namely flocculation, coalescence, Ostwald ripening, and gravitational separation, that occur frequently during the synthesis of nanoemulsions. We briefly describe some of the widely used stabilizers, namely weighting agents, texture modifiers, and ripening inhibitors (Salvia-trujillo et al., 2017).

8.2.2.1 Weighting Agents

A substance that is added to the oil droplets present in the dispersed phase of O/W nanoemulsion to equalize the density of surrounding continuous aqueous phase which helps in reducing the driving force tendency for gravitational separation, thus preventing the creaming or sedimentation process. Brominated vegetable oil, ester gum, damar gum, and sucrose acetate isobutyrate are some of the most commonly utilized weighting agents in many food and beverage industries that depend on O/W nanoemulsions (McClements and Rao, 2011). The major principle behind using these weighting agents in the presence of triglyceride oils and flavour as major components of nanoemulsion which possess a lower density as compared to water. Therefore, including the appropriate weighting agent increases their density until it becomes equal to that of the aqueous phase, thus obstructing the creaming and sedimentation process (McClements and Jafari, 2018).

8.2.2.2 Texture Modifier

Nanoemulsions often require a texture modifier to change its rheological properties to improve its stability towards gravitational separation which may also be categorized as a thickening or a gelling agent (McClements, 2015). These texture modifiers when incorporated into food formulations as thickening agents, mainly water-soluble polysaccharides, like xanthan gum, can increase the viscosity of the solution. This is due to their ability to modify the flow profile of fluid, which typically consists of soluble polymers broadly termed as hydrocolloids (Dhull et al., 2020b). Additionally, these gelling agents under their property to form physical or chemical cross-links with its neighbours can provide solid-like characteristics to the solution. Moreover, few polysaccharides bearing amphiphilic properties can also be treated as emulsifiers which are likely to be a derivative of a modified polysaccharide, such as gum Arabic, and a modified form of starch, cellulose, alginate, pectins, and cellulose. The presence of nonpolar groups or amino acid components, which are attached to the hydrophilic polysaccharide backbone, are responsible for the surface activity of these polysaccharide derivatives (Dickinson, 2009). Furthermore, these texture modifiers are known to inhibit the droplet movement, thus retarding the gravitational separation which helps in imparting stability to nanoemulsions (McClements and Jafari, 2018).

However playing a pivotal role in providing stability, these hydrocolloids, when supplemented in nanoemulsions, are sometimes also known to interfere with the gastrointestinal fate of nanoemulsion, as these may modify the nutritional properties of encapsulated lipophilic active ingredient (Salvia-trujillo et al., 2017).

8.2.2.3 A Ripening Inhibitor

In a nanoemulsion, the mean size of existing droplets presents in the dispersed phase starts increasing with time which is frequently termed as Ostwald ripening. This is because of the gradual diffusion of oil molecules from smaller to larger droplets which occurs between intervening fluid. A substance added to retard the growth of existing droplets present in the dispersed phase is termed as a ripening inhibitor. The absence of a ripening inhibitor results in an increased droplet size consequently leading to accelerated creaming which results in an unstable nanoemulsion. The properties of ripening inhibitors may be typically high hydrophobicity and very low solubility in water. Generally, food-grade vegetable oils are an optimal choice as a ripening inhibitor. These properties are mostly possessed by long-chain triglycerides such as oil from corn, palm, rapeseed, or sunflower (McClements and Rao, 2011).

8.2.3 Emulsifiers/Surfactant

Emulsifiers are surface-active amphiphilic molecules capable of adsorbing to the droplet surfaces and stabilizing both hydrophilic and lipophilic parts in their structure (Dhull et al., 2020b). Hydrophilic part shows an affinity towards nonpolar media such as oil while the lipophilic part displays affinity towards polar media, mostly water. HLB is an accepted parameter for gauging the differential affinity of an emulsifier towards water and oil phases. It represents the absolute proportionate of hydrophilic to the lipophilic fraction in a molecule. A higher HLB number of an emulsifier (greater than 10) indicates an increased affinity for water, whereas a lower HLB number (less than 10) indicates a higher relative affinity towards oil (Hasenhuettl and Hartel, 2008). As the emulsifier is known to promote the disruption of droplets and prevents droplet aggregation, this disruption of droplets is a result of reduced interfacial tension generated while using high-energy methods. Similarly, low-energy methods also create low interfacial tension while using emulsifiers; however, they generate very small droplets that are created in a particular environment like varying solution conditions. Likewise, surfactants are also known to produce thermodynamically stable nanoemulsions via promoting low interfacial tension. Small molecule surfactants, phospholipids, proteins, and polysaccharides role as surfactants are widely accepted in food emulsion and nanoemulsion stabilization due to their ability to adsorb at the oil–water interface (Salvia-trujillo et al., 2017). In a nanoemulsion, issues with the destabilization process, namely Ostwald ripening, can be prevented if surfactants are added together with oil and other components in a predefined ratio. Due to steric interactions, there is an increase in the repulsive maximum because of an additional adsorbed layer of emulsifier present on the droplet, which helps stabilize the emulsion agent from generating flocculation and coalescence. However, microemulsions are more prone to coalescence as compared to nanoemulsions because microemulsions bear a smaller droplet size as compared to nanoemulsions which prevent the reversible aggregation of droplets. Notably, the stability of a nanoemulsion against environmental stress like pH, ionic strength, temperature fluctuations, and increased storage is mostly determined by the type of emulsifier used. It is of extreme importance to use proteins and polysaccharides as emulsifiers and surfactants at higher concentrations because they have the advantage of being natural ingredients and thus are considered benign and are generally recognized as safe (GRAS). Small molecule surfactants are mostly classified according to their electrical characteristics like bearing ionic, non-ionic, and zwitterionic properties (McClements, 2015). Examples of ionic surfactants bearing a negative charge are citrem, datem, and SLS whereas lauric arginate is positively charged, all these are considered as food-grade components. The second category of surfactants for generating nanoemulsion are non-ionic surfactants, which are mostly favoured surfactants because of their relatively lower toxicity and irritability, and these can generate nanoemulsion utilizing both high and low energy approaches. Examples of non-ionic surfactants include sugar ester surfactants mainly sucrose monopalmitate, sorbitan monooleate, polyoxyethylene and ether (POE) and ethoxylated sorbitan esters (McClements and Rao, 2011). The third and final category is zwitterionic surfactants include oppositely charged ionizable groups present on the same molecule which is responsible for imparting a negative, positive, or neutral charge to the nanoemulsion solution which can deviate with the final pH of the solution. An example of zwitterionic surfactants includes lecithin, which is a phospholipid considered safe for consumption in food items (Hoeller et al., 2009).

8.3 DESTABILIZATION OF NANOEMULSION SYSTEM

All the previously mentioned components are mostly considered to provide stability to a nanoemulsion system; however, being a metastable system, it may break down over an extended period through several physicochemical mechanisms like gravitational separation which specifically includes creaming and sedimentation, coalescence, flocculation, chemical degradation, and Ostwald ripening (McClements and Rao, 2011). All mentioned destabilization mechanisms are discussed serially in the next section.

8.3.1 Gravitational Separation

The most common form for the destabilization of nanoemulsions is gravitational separation, either in the form of creaming or in sedimentation that varies based on relative densities of dispersed and the continuous phases. Droplets having a lower density starts to move upwards due to a relatively lower density than its liquid surrounding them, which is termed as creaming. Conversely, when the droplets due to a higher density than the surrounding liquid start moving downwards are termed as sedimentation. This is the major reason that conventional O/W emulsions are more prone to creaming as oil has a lower density than surrounding liquid water therefore creaming is more prominent in liquid edible oil, and due to similar reasons, the W/O emulsion is more prone to sedimentation. In some cases, due to the presence of crystalline lipids the density of lipids increases in an O/W emulsion, the increased density due to lipid crystallization. Decreasing the droplet size, adding weighting agents, and adding thickening agents, for increasing the viscosity of the aqueous phase, are some possible ways to reduce the gravitational separation (McClements and Jafari, 2018).

8.3.2 Flocculation and Coalescence

These are other issues with droplet formation that occurs due to colloidal interactions among the droplets. In flocculation, mostly the clusters are formed as two or more droplets interact together due to the presence of attractive forces that acts between the droplets. However, coalescence is the formation of a large droplet which is a result of the merging of multiple small droplets that aggregate together. The main reason for coalescence is the attractive interactions, caused due to van der Waals forces, hydrophobic forces, and depletion forces, between the smaller droplets that bring them close enough to form a larger aggregate. These attractive interactions can be negated by factors termed as repulsive interactions that include electrostatic and steric interactions. To avoid coalescence the repulsive interactions must be kept greater than the attractive interactions (McClements and Rao, 2011).

8.3.3 Ostwald Ripening

It is a process where the formation of larger droplets starts from shrinking of smaller droplets. The process where the mean size of existing droplets present in the dispersed phase starts increasing with time which is frequently termed as Ostwald ripening (Kabalnov, 2001). The solubility of the dispersed phase is more in smaller droplets when compared to bigger droplets as smaller droplets that possess higher curvature whereas bigger droplets possess smaller curvature. This difference of curvature leads to gradual diffusion of oil molecules resulting in the formation gradient which leads to the growth in the size of droplets between the intervening fluid (McClements and Jafari, 2018).

8.4 NANOEMULSIONS PREPARATION METHODS

Preparation of nanoemulsions often requires a huge amount of energy as being non-equilibrium systems; therefore, their preparation either requires a large amount of energy or the use of surfactants and most of the time both. Therefore, in this section, a major focus is on the most utilized energetic methods for nanoemulsion preparation. Briefly, these are categorized as high-energy emulsification and low-energy emulsification methods. Both methods hold equal weightage in generating small droplets; however, high-energy methods are preferred in the food industry as lower levels of an emulsifier is required and the use of a natural emulsifier is also feasible using this approach. Additionally, it can also be easily customized with huge mechanical devices for large-scale production (Salvia-trujillo et al., 2017). In the next section, a detailed description of high- and low-energy methods is specified; we initially elaborate on various high-energy methods followed by low-energy methods.

8.4.1 High-Energy-Based Methods

High-energy methods require the application of high disruptive forces with a mechanical device that is capable of causing the breakdown of oil droplets and dispersing them into the water phase. High-energy methods involve high-pressure homogenization (HPH), ultrasonic homogenization (USH), microfluidization, high-pressure microfluidic homogenization (HPMH), and colloid mills (Salvia-trujillo et al., 2017; McClements and Jafari, 2018; Espitia et al., 2019).

8.4.1.1 High-Pressure Valve Homogenizers

HPH is the most common and widely used method for generating small-sized droplets in the food and pharmaceutical industries due to its versatility and feasibility. Rather than generating emulsions, this method is more effective in reducing the size of droplets already generated during coarse emulsion from the two distinct liquids. This type of homogenizer uses a piston pump that pulls the coarse emulsion, generated using a high-shear mixer; pushes that into a chamber; and then propels it through a narrow valve located at the end of the chamber. This constant flow through the valve is maintained by uninterrupted forward and backward strokes of the piston that ultimately generate intense turbulence and hydraulic shear force through which macroscale droplets get broken down into relatively smaller droplets. A conventional HPH system operates in a pressure range not exceeding 150 MPa however, few high-capacity ultra-HPH may function beyond maximum operating pressures, which could go up to 350 to 400 MPa, resulting in the generation of ultra-small droplets. Droplet diameters ranging as small as 1 nm are reported to be produced using this method (Donsì et al., 2012; Qadir et al., 2016; Horison et al., 2019). Further reduction in the droplet size by increased homogenization is not recommended for this type of method as a further increase in pressure may lead to coalescence and ultimately leads to increased-sized droplets. The major disadvantage of this technique is the consumption of high energy, which causes increased emulsion temperature during the process. Nevertheless, this technique may further be utilized as hot and cold HPH types, in which the cold HPH technique is highly favored and mostly utilized, especially for temperature-sensitive compounds. A melted lipid phase is used to disperse the compound in both techniques, but in the hot HPH technique, a hot surfactant is used to disperse the mixture above its boiling point using a high-speed stirrer. In contrast, in the cold HPH technique, prior grounding and cooling of active compound and lipid phase are required into a cold surfactant solution which results in the formation of a cold pre-suspension (Azmi et al., 2019).

8.4.1.2 Ultrasonic Homogenization Nanoemulsions

USH employs making use of sonotrode also termed as a sonicator probe which generates mechanical ultrasound vibrations when it comes into contact with the liquid causing cavitation to occur. Cavitation is defined as the formation and collapse of vapour cavities present in the liquid. This probe, when inserted into the coarse emulsion, results in subsiding and breaking down the microdroplets near the inserted probe with the help of ultrasound energy generated from the sonotrode. USH requires a two-step mechanism; initially, the interfacial waves are generated under the acoustic field which breaks the dispersed phase resulting in a continuous phase. Next, acoustic cavitation aids in collapses these droplets into smaller droplets via pressure fluctuations. Finally, the formation of nanoemulsion droplets occurs which is the overall result of interaction among droplet breakup and droplet coalescence. Factors influencing the efficiency of USH are the duration of treatment and the power, frequency, and amplitude of the ultrasound waves. Mostly medium-scale homogenizers are utilized in those laboratories where the droplet size requirement of nanoemulsion is approximately 0.2 μm. These ultrasonic homogenizers have very simple operational protocols and are energy-efficient and affordable. Additionally, this process requires low emulsifier content and displays excellent dispersion stability. Being aseptic, these can be safely introduced into the coarse emulsion, and therefore, the risk of microbial contaminants entering the processing stage is further reduced. Sonication can be performed either in batch mode or in continuous mode, which solely depends on the design of the operating chamber. However, reports exist that highlight continuous mode to generate broader particle size distribution as compared to batch mode; the reason attributed for this is the non-homogeneous treatment of the fluids in the flow chamber. However, nanoemulsions generation through the sonication method holds several potential drawbacks, one of them being a high shear rate which may instigate an increase in temperature of the emulsion, that may go up to 80°C. This may induce abnormal detrimental effects on the heat-sensitive compounds mainly lipids. Apart from this, the sonication probe being metal derivatives may result in metal leaching into the emulsion because of cavitational abrasion. This could be attributed to additional limitations for limiting its vast commercial application (Salvia-trujillo et al., 2017).

8.4.1.3 Microfluidization

A microfluidization technique utilizes a device called a "microfluidizer" which looks similar in design to a HPH. However, compared to HPH the designing of channels for making emulsion flow within the homogenizer is extremely complicated. This method uses an interaction chamber that uses high pressure to force the product through a narrow orifice, resulting in droplet disruption. The interaction chamber from the inner side consists of small channels, termed as microchannels, that utilize a positive displacement pump having a high pressure of about 500 to 20,000 psi. As pre-emulsification is not required in this technique because the dispersed phase is directly injected into the continuous phase, by making use of microchannels, therefore, this technique is also known as the "direct" emulsification technique. Initially, the course emulsion is prepared by mixing aqueous phase and oily phase which is then further processed in the microfluidizer channels which works on the principle of bifurcating the flow of emulsion into two separate streams. This is followed by merging these separate channels in the interacting chamber which results in generating of intense disruptive forces as the fast-moving streams superimpose on each other leading to an efficient droplet disruption that further generates nanoemulsion of the submicron range (Salvia-trujillo et al., 2017; Pathania et al., 2018).

8.4.1.3.1 High-Pressure Microfluidic Homogenization

This method is a step ahead of the microfluidization technique. Both HPH and HPMH use a positive displacement pump that generates high pressure between

30 MPa and 120 MPa to create nanoemulsions, but still, the specific design for both differ considerably. However, HPMH is extremely useful in generating customizable-size, tailored nanoemulsions at a large scale. Several reports highlight the usefulness of HPMH to be more effective in the formation of nanoemulsions when compared to HPH. In similar research using the HPMH method, authors had claimed to generate stable nanoemulsions having a particle size ranging to 275.5 nm, a zeta-potential of −36.2 mV, and a viscosity of 446 cP (Liu et al., 2019).

8.4.1.4 Colloid Mills

This method makes use of a machine suitable for homogenizing mixtures to reduce the droplet sizes for preparing coarse emulsion. Colloid mills are also termed stirred mills and bear a high-speed rotor that can generate hydraulic shear by running up to 18,000 rpm twists, thereby disrupting the dispersed phase. The high-speed rotor generates high hydraulic shear to reduce the droplet size up to 1 µm; however, it can never achieve further size reduction and reach nanoscale size. Therefore, this process can be used as pre-preparative process for generation of micro-nanoemulsions. However, under certain circumstances, droplets as small as 10 nm can be generated by this method. Reports emphasize that lowering the processing temperature beyond the solidification temperature of the dispersed phase augments the process of droplet reduction, which can be justified as grinding rather than fluid disruption (Espitia et al., 2019).

8.4.2 Low-Energy-Based Methods

Low-energy methods mostly rely on generating tiny oil droplets in a mixed oil–water surfactant whenever in the solution any intrinsic physicochemical properties or environmental conditions are altered. Low-energy approaches are known to be more efficient in producing a smaller droplet size when compared to high-energy approaches, but their usage is restricted with limited types of oils and emulsifiers. Likewise, the use of proteins or polysaccharides is still restricted as emulsifiers using low-energy approaches. Ensuring higher concentrations of synthetic surfactants only makes possible the use of protein and polysaccharides to form nanoemulsions, which still limits employing these methods for food applications (McClements and Rao, 2011). Nevertheless, low-energy techniques may prove beneficial for certain commercial applications as it omits the use of complicated designer equipment, a prerequisite in high-energy homogenization that incurs high cost. Low-energy approaches include spontaneous emulsion (SE), phase inversion composition (PIC), and phase inversion temperature (PIT). We will initiate this by briefly elaborating every approach with their merits and demerits in the next section.

8.4.2.1 Spontaneous Emulsion

This method, also termed emulsification by solvent diffusion (ESD), is an affordable approach used to obtain nano-sized emulsions. Formation of an oil–water nanoemulsion is said to be very spontaneous in which the addition of surfactants holds the most prominent role. Technically, the process of synthesizing nanoemulsion through SE method includes an organic phase in which a hydrophilic surfactant and oil is added into an aqueous water phase in which major fraction can be a cosurfactant, namely ethanol, acetone, and propylene glycol. The addition of a hydrophilic surfactant and oil into an aqueous phase and co-surfactant is the key because other solvent usage is usually problematic in the food industry. Notably, increased levels of surfactants and co-surfactants are not permitted in the food and pharmaceutical industries due to regulations and high cost. Stages of generating SEs mostly follow mixing lipid phase which consists of oil having a lipophilic surfactant and a water-miscible solvent with a hydrophilic surfactant. As soon as the lipid phase is injected in the aqueous phase, which is kept under continuous magnetic stirring, the emulsion formation is initiated. Finally, the aqueous phase is removed through evaporation under reduced pressure, leading to the formation of smaller droplets in the dispersed phase enclosed with a surfactant present in the continuous phase. As surfactants hold a higher affinity towards oil or the aqueous phase, smaller droplets coated with a surfactant are generated which is a result of turbulence created at the interface of the dispersed and continuous phases. Furthermore, to minimize the droplet size in emulsions, the turbulence generated at the interface must be supplemented with co-surfactants such as ethanol, acetone, and propylene glycol. SE methods prove beneficial for the encapsulation of bioactive compounds, used immensely in the food and pharmaceutical industries, as compared to other methods as it helps to impart protection to bioactive compounds from ill effects like increased temperature and pressure which are inadvertently utilized in high-energy methods. Various groups have reported the synthesis of nanoemulsions of reduced size using the SE method in which a reduced droplet size of 50 nm was reported by (Barzegar et al., 2018). Similarly, particle sizes as low as 10 nm, 10 to 30 nm, and 50 to 500 nm were reported by Zhao et al. (2018) by applying the same surfactant level. Another study revealed an average droplet diameter of 13 to 14 nm prepared via SE method in which the authors highlighted that exposure to extreme temperature, mainly 4°C and 45°C, helped in exceeding the storage period up to 8 months. Furthermore, nanoemulsions with a diameter 109 to 139 nm, with a net negative surface zeta-potential between –28.5 mV and –35.8 mV, and with a spherical structure were successfully generated by Wang et al. (2015).

8.4.2.2 Phase Inversion Composition

PIC is also termed emulsion inversion point (EIP) and was initially reported by Shinoda and Saito (1968). It involves the gradual transition in the spontaneous curvature and thus the composition at a particular temperature at which a few components of the continuous phase are added to the components of the dispersed phase, leading to phase inversion. This method implies the generation of O/W nanoemulsions from their W/O analogs by slowly altering the water volume fraction at a specified temperature by changing the curvature of natural emulsifier. A steady increase in the water volume fraction results in phase inversion to occurs at a specific level, consequently leading to the emergence of a bi-continuous phase. Probably, this bi-continuous intermediate structure supports the trapping of one phase into the other, mainly the oil phase into the water phase, leading to the formation of droplets in droplets and finally generating oil–water nanoemulsions. Reports propose that highly stable nanoemulsions having a droplet size of 51 nm can be generated using this method by keeping temperatures as low as 70°C. Advantages of using this technique include relatively low cost and a simple apparatus. Despite the mentioned advantages, the time required for completion of this process is greater than the time required in the SE technique. A major reason being the smaller driving forces used in this method. Commercial utility of this method was explored by successfully producing nanoemulsions using the EIP method and fusing them with pineapple ice cream which allowed the existing artificial yellow dye to be replaced. The average droplet size produced was about 29.55 to 37.12 nm (Borrin et al., 2016).

8.4.2.3 Phase Inversion Temperature

PIT mostly relies on the transitional inversion, which is initiated by an imbalance prompted by the improper proportion of hydrophile to lipophile ratio in the system. The PIT method is based on the deliberate fluctuations with temperature-based alterations in the dispersity of polyoxyethylene-type non-ionic surfactants. Applying this method technically means the spontaneous generation of nanoparticles via alteration in the time- and temperature-related conditions of the components present in the system. Rapid temperature fluctuations are known to minimize the coalescence which promotes formation of stable nanoemulsion. At a low temperature, the temperature-sensitive surfactants tend to become increasingly polar, thus forming a positive spontaneous curvature at the droplet interface resulting in oil–water microemulsions. In contrast, at a higher temperature, the surfactants have the property of being increasingly lipophilic due to the dehydration of the polyoxyethylene chains, thereby forming a curvature having negative surfactant layer at the droplet interface. However, at room temperature, the combination of oil, water, and nonionic surfactant exhibits a similar affinity towards the oil, as well as the water, phase, consequently, generating a net-zero value for the spontaneous surfactant layer curvature existing at the droplet interface. Major advantages of employing this method include minimal energy requirement for generating high shear force. Several studies have indicated the successful generation of nanoemulsions using the PIT method. According to a study by Chuesiang et al. (2018), nanoemulsions were successfully prepared by heating a mixture of oil, water, and surfactant above the PIT which was followed by gradual cooling and continuous stirring, resulting in a nanoemulsion with a particle size of 101 nm. Another study investigating nanoemulsion synthesis using the PIT method resulted in the formation of particles whose size was 100 nm which were found stable up to 15 days (Su and Zhong, 2016). However, a major drawback pertaining to this system is its inability to function with ionic surfactants as in ionic surfactants, the temperature effect will not be able to modify the spontaneous curvature of this system (Liu et al., 2019). The next section mostly focuses on all pertinent and recent case studies of plant-derived bioactive compounds used as nanoemulsions with major emphasis on their commercial applications and limitations (Figure 8.1).

8.5 PLANT-DERIVED BIOACTIVE COMPOUND–BASED NANOEMULSIONS AND THEIR APPLICATIONS IN FOOD INDUSTRIES

Plant-derived phytochemicals are produced as secondary metabolites in plants and are categorized depending on their active ingredient component present (Cowan, 1999; Dhull et al., 2020c). These bioactive compounds are usually non-nutritive but confer

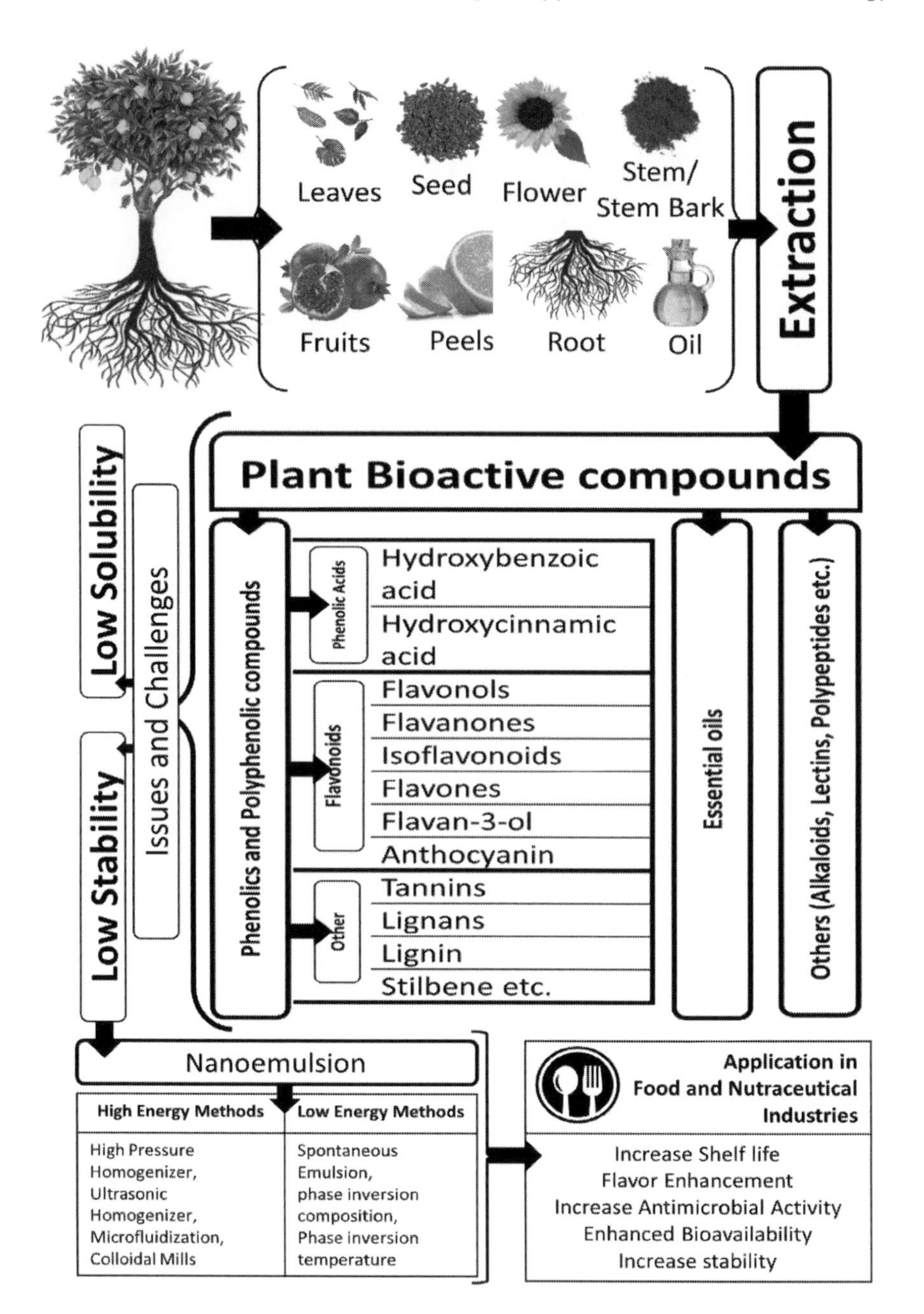

FIGURE 8.1 Major applications of plant bioactive nanoemulsions in food industries.

organoleptic properties and are, therefore, popularly termed "phytochemicals", being from plant origin. Plants can synthesize limitless compounds with wide array of beneficial advantages in terms of antioxidant, anti-inflammatory, antibacterial, antiviral, and antifungal activities (Dhull et al., 2019b). The use of these compounds in the food industry is still challenging as being sparingly soluble in water which minimizes their stability and shelf life and thus hampers their immense usage in food preparations. Various attempts to remove these undesirable properties is provided by exploring their nanosized structures, mainly by generating their nanoemulsion, which are considered safe for utilization in food formulations. As these nanoemulsions offer added advantages over conventional emulsions, they have become a preferred choice in food industries. The production and accumulation of these bioactive compounds are observed in various plant organs, namely roots, seeds, leaves, and fruits. Phenolic compounds are considered to be the most effective class of phytochemicals where major components are phenols or their oxygen-substituted derivatives (Geissman, 1963). It includes simple phenolic compounds, phenolic acids, flavonoids, quinones, tannins, and coumarins (Cowan, 1999). Apart from this, another essential class termed as essential oils, have been used for food preservation from ancient times. Antimicrobial activity of essential oils has been well summarized by (Donsì, 2018). In addition, alkaloids, lectins, and plant polypeptides are also known to possess antimicrobial activity (Cowan, 1999; Hintz et al., 2015). Specific antimicrobial properties of any particular plant species are generally attributed to the presence of these bioactive molecules. Their antimicrobial activity depends on the concentration, composition, structure, and presence of functional groups (Juneja and Sofos, 2017). In the next section, we briefly summarize various types of phytochemicals as per their classification and their traditional beneficial applications and attempt to list their potential applications as nanoemulsion, including major methods for their synthesis.

8.5.1 Polyphenolic Compounds

Regardless of diversity in the chemical structures, phenolic compounds are usually termed "polyphenols" mainly because of their natural occurrence as conjugates of at least two phenolic rings linked with mono- and polysaccharides and as esters and methyl esters. Phenolic compounds are large group of secondary metabolites in plants, which are characterized by the presence of substituted phenolic ring. They are ubiquitously distributed in the plant kingdom and are the most abundant secondary metabolite in plants. Phenolic compounds play key role in providing defense against insects and plant pathogens. They are helpful in growth and reproduction and impart colour and flavour to fruits and vegetables. The benefits of phenolics include their antimicrobial, antioxidant, anti-inflammatory, antiviral, and anti-allergenic properties, among others, which have been reviewed in detail by Cowan (1999). Polyphenolic compounds can be broadly grouped into four major categories: flavonoids, phenolic acids, hydroxycinnamic acids, and lignans (Maqsood et al., 2013).

8.5.1.1 Flavonoids

From plants, more than 5,000 structurally different flavonoids have been isolated and identified. These compounds are ubiquitously distributed in the plant kingdom,

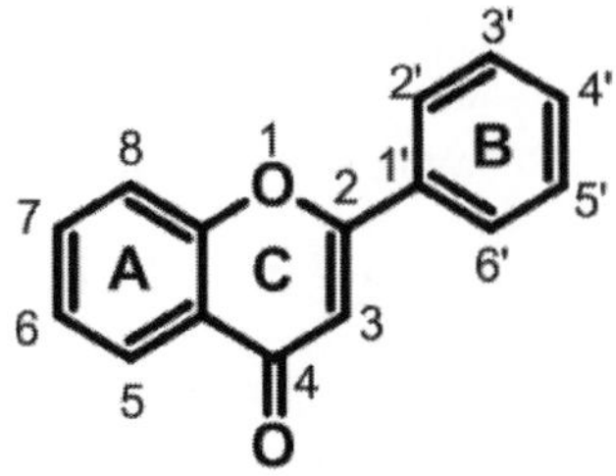

FIGURE 8.2 General flavonoid backbone.

mainly in photosynthesizing plant cells. Flavonoids are responsible for producing various attractive colours of flowers, fruit, and leaves. Flavonoids are known to exhibit antimicrobial, anti-inflammatory, oestrogenic, enzyme inhibitory, anti-diabetic, anti-allergic, antioxidant, and cytotoxic anti-tumourous activity (Cushnie and Lamb, 2005). The basic structure of flavonoid compounds is the flavane nucleus, 15-carbon (C6–C3–C6) skeleton, which consists of two benzene rings linked through a heterocyclic pyrane ring (Figure 8.2). Based on various substitution pattern, flavonoids can be categorized in six major subgroups, viz. flavonols, flavanones, is of lavonoids, flavones, flavan-3-ol, and anthocyanins (Cushnie and Lamb, 2005; Hossain et al., 2016).

8.5.1.1.1 Flavonols

Flavonols are known as 3-hydroxyflavones since they have a hydroxyl group attached to position 3 of the flavones. Flavonols are extensively concentrated in the outer parts (skin of fruits) as well as in the leaves of the higher plants. In higher plants, nearly 450 different flavone aglycones have been documented so far. The major dietetical flavonols include quercetin, kaempferol, myricetin, isorhamnetin, and fisetin. The recent work done utilizing them for nanoemulsion preparation and its method of preparation is briefly summarized.

Quercetin is biologically significant flavonoids in human dietary nutrition in plentiful amount and forms the skeletons of many other flavonoids, such as hesperidin, naringenin, and rutin. It is found in various fruits and vegetables such as apples, grapes, red wine, cherries, cranberries, red onions, green leafy vegetables, broccoli, pepper, coriander, buckwheat citrus fruits, tea, and blueberries. Quercetin shows numerous health benefits, including lowering of blood pressure, anti-tumour, cardiovascular protection, an immunomodulatory effect, an antiviral property, gastroprotective effects, and antimicrobial activity (Hossain et al., 2016). Quercetin also demonstrates high free radical scavenging activities towards hydroxyl ions, peroxyl ions, and superoxide anions (Karadag et al., 2013). Antibacterial properties exhibited by quercetin majorly employs inhibitory activity on FabZ in *Helicobacter pylori*, the inhibition of d-alanine:d-alanine ligase in *H. pylori* and *Escherichia coli*, the inhibition of DNA gyrase in *E. coli*, the inhibition of motility due to a disturbed proton motive force, a disturbance of membrane potential, and moderate inhibition of efflux pump in various bacteria (Cushnie and Lamb, 2005; Khameneh et al., 2019). Reports also indicated its antiviral activity due to inhibition of viral integrase and reverse transcriptase in HIV (Cowan, 1999). Moreover, its low solubility in water and gastric juice limit its bioavailability when taken orally. These solubility issues were addressed by utilizing HPH, using a response surface methodology, for the generation of a quercetin-based nanoemulsion with a food-grade emulsifier, were found beneficial for oral uptake (Karadag et al., 2013). As per the available literature, quercetin-based emulsion displayed increased solubility that was found to be 40 times more in medium-chain triglyceride at room temperature. Furthermore, oral pseudo-organogel-based emulsion systems for quercetin had been developed for improved bio accessibility. As compared to quercetin-loaded oil mixture, the bio-accessibility of these pseudo-organogel-based emulsions was found to be enhanced significantly. This may open new avenues for exploring their potential applications in the food, cosmetic, and pharmaceutical industries (Xu, 2014). Another study demonstrated that nanoemulsion of quercetin developed by two-step homogenization process can further be incorporated into edible gelatin films. These films exhibited antimicrobial activity against food-borne pathogens without any reduction in antioxidant activities. These polyphenol-enriched gelatin films, being of plant origin and safe, can be used as active packaging material for reducing the total microbial load of fresh meat and may further extend its shelf life by reducing the risk of foodborne illness (Khan et al., 2020).

Another example of flavonoids is rutin (also known as rutoside, quercetin-3-O-rutinoside, and sophorin), which is a common dietary flavonoid that prevalently exists in the plant kingdom. Rutin exhibits a wider range of pharmacological properties which can exploited as human nutritive medicine. It shows antimicrobial, anti-fungal, and anti-allergic properties and hence is commonly utilized as vital preventive medicine. The high radical scavenging activity of rutin makes it a suitable candidate that could be utilized as a good source of antioxidants, and due to the several applications, rutin is now categorized among important nutraceutical food product. Rutin shows antibacterial activity against a permeable *E. coli* strain (a strain into which the envA1 allele had been incorporated). Rutin was also shown to inhibit topoisomerase IV-dependent decatenation activity and induce an emergency cellular against extensive DNA damage in the repair pathway SOS response of the *E. coli* strain, thus leading to growth inhibition (Bernard et al., 1997). Rutin alone or in combination with quercetin, is reported to inhibit the growth of *E. coli* at 20°C without causing cellular death but was able to display more efficient antibacterial activity at 4°C (Rodriguez Vaquero et al., 2010; Cetin-Karaca and Newman, 2015). Although it shows excellent functional properties, its application in food industries is still hindered due to its poor solubility, low absorption, and poor bioavailability (Dammak et al., 2017; Bazana et al., 2019). To overcome this, rutin O/W nanoemulsions were prepared using a rotor-stator homogenization, using lecithin and chitosan as stabilizers, which remarkably improved their rheological properties, heat stability, kinetic degradation, and release properties when compared to rutin solution (Dammak and do Amaral Sobral, 2018). Rutin was reported as a principle component of *Physalis peruviana calyx* extract for the preparation of a nanoemulsion in which antioxidant stability of the prepared nanoemulsins was compared with free extracts under different storage conditions (7 and 25°C) and with absence of light for an extended time of 4 months. These observed results demonstrated that nanoemulsions were able to retain the antioxidant capacity during preservation, thus resulting in reuse of the extract and thereby avoiding wastage. These findings are of great interest for food industries for the development of new food products with enhanced shelf life (Bazana et al., 2019). Similar studies demonstrate that rutin-loaded O/W nanoemulsions incorporated in edible gelatin-based films initiated interactions with the gelatin film matrix, thus inducing a change in the properties of films. Reports exist to ascertain that rutin-loaded nanoemulsions were able to significantly improve the mechanical properties of gelatin films via increasing the films' resistance towards mechanical stress. Thus, the application of nanoemulsions can be extended as food additives carriers, which allows incorporating a variety of lipophilic bioactive compounds, such as rutin, into the hydrophilic matrices of biopolymers namely gelatins, for the fabrication of films due to their enhanced biodegradability (Dammak et al., 2017).

In addition, nanoemulsions of other flavonoids, for example kaempferol, isorhamnetin, and fisetin, are also reported for their beneficial medicinal activities, namely anti-inflammatory, antioxidant, anti-diabetic, and anti-cancerous properties. Kaempferol possesses free radical scavenging activity; anti-aging, anti-inflammation, and anti-diabetic property; and inflammatory pathways suppression and was known to exhibit antimicrobial, antiviral, anti-tumour, and anti-cancer activities. Myricetin, found in tea, wine, berries, fruits, and vegetables, shows notable antibacterial activity through inhibition of efflux pump (Khameneh et al., 2019). The scarce solubility of kaempferol was found similar to quercetin and myricetin in the oil phase and requires the addition of amphiphilic molecules in the lipophilic core for the generation of nanoemulsions. Nanoemulsions prepared by the mentioned method resulted in improved thermal and enhanced photo stability (Donsì, 2018).

8.5.1.1.2 Flavanones

The flavanones (also called dihydroxyflavones) are structurally different from the rest of flavonoids as they lack a double bond between C-2 and C-3 positions in the C-ring of the flavonoid skeleton, which gives them the presence of a chiral centre at C-2 position. Flavanones are extensively distributed in about 42 larger plant families, especially in *Compositae*, *Leguminosae*, and *Rutaceae*. Flavanones are abundant in citrus fruits and have been reported to exhibit antioxidant, anti-diabetic, lipid-lowering, anti-atherogenic, and anti-inflammatory activities. The main aglycones are naringenin in grapefruit, hesperetin in oranges, and eriodictyol in lemons. Hesperetin, naringenin, and naringin are reported as antibacterial agents, as they inhibited the growth of methicillin-resistant *Staphylococcus aureus* and S*treptococci* by alteration of membrane fluidity (Kumar and Pandey, 2013). Due to their high pharmaceutical values, these compounds are mainly exploited for their medicinal purposes. In a related study, a nanoemulsion was prepared from hesperidin by using high shear–high pressure homogenization bonding technology and surfactant comparison study was carried out (Liao et al., 2020). In another study, pickering nanoemulsions of hesperidin were prepared by homogenization and were applied in a gelatin-based active film, where the encapsulated hesperidin showed a significant antioxidant activity. Hesperidin nanoemulsion was found to improve the emulsifying properties of chitosan nanoparticles in the composite active film used in food packaging (Dammak et al., 2019). Similarly, another compound taxifolinis categorized under flavanones, which contains basic 3-hydroxyflavanone structure and is naturally found in onion, citrus fruits, wood of larch, rench maritime bark, tamarind seeds, and milk thistle. Taxifolin is known to exhibit antibacterial, antifungal, anti-inflammatory, antithrombotic, antioxidative, and anti-cancer properties and is utilized as a food supplement. In an associated study, it was found that a nanoemulsion preparation of taxifolin, if consumed by athletes during exercise, can significantly enhance the recovery period from exhaustion to normalcy. O/W nanoemulsions of taxifolin were produced by ultrasonication, which interestingly resulted in minimizing the risk of oxidative stress and provided homeostasis regulation in humans (Kalinina et al., 2019). Another example of flavanones is silibinin (silybin), which is a natural flavanone compound isolated from milk thistle of *Silybum marianum*, an important plant. It is structurally similar to quercetin and kaempferol and had historically been used in Chinese medicine; it demonstrates hepato-protective and anti-cancer effects (Li, 2013). Nanoemulsions of silybin from silymarin extract were prepared through a two-stage HPH. Physically stable nanoemulsions were prepared in which antioxidant properties during storage were retained for relatively longer time. This approach of nanoemulsion formation utilizing high medical-value natural plant products represents a suitable tool for the development of next-generation functional foods and pharmaceutical products with enhanced bio-safety (Calligaris et al., 2015).

8.5.1.1.3 Isoflavonoids

Isoflavonoid, nonsteroidal plant compounds, are a broad group of polyphenolic compounds which include structurally similar groups like isoflavones, isoflavonones, and isoflavans. Isoflavones, often found as glycosides, include daidzin, genistin, biochanin A, and formononetin. Isoflavones are mainly found in *Leguminosae*, where soybean is considered to be the richest source. Other sources include kidney, navy, fava, pinto, red, kudzu, lupine, and mung beans and chickpeas, split peas, peanuts, sunflower seeds, and walnut. Genistein is reported to possess anti-diabetic, anti-estrogenic, anti-cancer, antioxidant, and anti-helminthic activity (Bultosa, 2016). Genistein exhibits an increased anti-cancer activity and is widely exploited for pharmaceutical purpose rather than food. According to a study, nanoemulsions of genistein were found to be useful in cancer therapy. These prepared nanoemulsions, when incorporated in a small pouch/tablet and placed inside the cheek to let the active ingredient get absorbed directly via oral mucosa, were found to be extremely effective for maintenance therapy of oral cavity and oropharyngeal cancers (Gavin et al., 2015). Similarly, hydrogel products containing nanoemulsions of isoflavin aglycone rich fraction, containing high amounts of daidzein, genistein, and glycitein (isolated from soybean) were also reported to be effective for dermal applications (Nemitz et al., 2019). However, very limited studies have been found on the application of nanoemulsions of isoflavonoids in food.

8.5.1.1.4 Flavones

Flavones are a subclass of flavonoids characterized by a non-saturated 3-C chain and have a double bond between C-2 and C-3, like flavonols, with which they differ by the absence of hydroxyl in the 3-position, found mainly in celery, parsley, and many related herbs. The major dietary flavones include apigenin, baicalein, and luteolin. Apigenin is found in several plants and vegetables, such as parsley, chamomile, celery, olive, pigeon peas, and chamomile. As per several studies, apigenin possesses antiviral, antibacterial, antioxidant, and strong anti-inflammatory activities and blood pressure reduction properties. It is reported to inhibit the growth of *H. pylori* and *E. coli* by inhibiting d-alanine:d-alanine ligase (Wu et al., 2008). To overcome the issue of poor water solubility, apigenin-loaded W/O/W multiple nanoemulsions were prepared and passed through simulated digestion model; physical properties and digestibility at each stage were recorded which demonstrated in vitro and in vivo bioavailability enhancement in the small intestine (Kim et al., 2016). For this, O/W emulsions loaded with apigenin was prepared using HPH. The resulting apigenin nanoemulsion, as nutraceutical formulation fortified with hydrophobic flavones, was retained even after storage of 30 days (Abcha et al., 2019). Another flavone, Baicalein (5,6,7-trihydroxyflavone), present in roots of *Scutellaria baicalensis*, *S. lateriflora*, and *Oroxylum indicum*, is reported to have various pharmacological activities, such as *Candida albicans*–mediated biofilm inhibition, anti-cancer, antioxidant, anti-allergic, antiviral, and anti-inflammatory activities. It is insoluble in an acidic medium; however, it is said to be highly soluble in alkaline medium, although it is highly unstable. An O/W emulsion, with average size about 300 nm, of baicalein was prepared using HPH, which significantly enhanced its bioavailability and stability. The formulation and stability were also recorded; therefore, it could be utilized for oral delivery of poorly water-soluble compounds at the industrial level (Treesuwan et al., 2013; Yin et al., 2017).

Polymethoxy flavone, or PMF, is a general term for flavones bearing two or more methoxy groups on their basic flavane skeleton with a carbonyl group at the C-4 position. PMFs are present in ample amount in citrus fruit peels and are commonly used as a traditional Chinese medicine to relieve stomach upset and cough. PMFs have been of particular interest due to their proven broad-spectrum biological activity, including anti-inflammatory, anti-carcinogenic, and anti-atherogenic properties (Li et al., 2009). An efficient nanoemulsion-based oral delivery system, with capacity to carry almost 10 times higher concentration required for biological actions, was designed for PMFs to facilitate its application in nutraceutical and pharmaceutical products (Li et al., 2012). Tangeretin (pentamethoxyflavone) functions as a potential chemo-preventive agent as it possesses anti-inflammatory, anti-proliferative, anti-obesity, and anti-diabetic effects and anti-carcinogenic activities. Its low bioavailability when injected orally is due to its poor water solubility. However, the bioavailability of oral delivery of synthesized tangeretin nanoemulsion, prepared by HPH, was significantly enhanced against colorectal cancer (Ting et al., 2015).

8.5.1.1.5 Flavan-3-ol

Flavans-3-ol (also called flavanols) are present in fruits and beverages like tea, chocolate, grapes, lychees, apples, blueberries, gooseberries and red wine. The absence of double bonds between C-2 and C-3 and absence of carbonyl functional group C-4 in ring C of flavanol result in the generation of two chiral centres at C-2 and C-3, thus making four possible diastereoisomers. Due to presence of hydroxyl unit at position 3 of the heterocyclic C-ring they are called flavan-3-ols. The most common flavanols found in fruits and cocoa are catechin and epicatechin, while in grapes, tea, and seeds of certain leguminous plants, the main flavanols present are epicatechingallate, gallocatechin, epigallocatechin, and epigallocatechin gallate (EGCG). Among them, EGCG is the abundant polyphenol as it composes approximately 50% of total catechins in dried green tea leaves. The antioxidant action of catechin, namely anti-hypertensive, anti-inflammatory, antiviral, anti-proliferative, anti-thrombogenic, and anti-hyperlipidemic, is well established by various *in-vitro*, *in-vivo*, and physical methods (Ziaullah et al., 2015; Tsai and Chen, 2016). The highest free radical scavenging activity of EGCG among all types of tea catechins makes it a highly potent molecule, but its larger molecular size and number of hydrogen bonds result in poor bioavailability of EGCG. To enhance the bio-availability of green tea extract, W/O nanoemulsions were prepared using soy, peanut, sunflower, and corn oils that displayed the increased oxidative stability. Thus, the synthesized nanoemulsions could be used as a promising vehicle for delivering green tea bioactive compounds in nutraceutical applications (Puligundla et al., 2017). Another study showed the potential of nanoemulsion for development of edible film (Nunes et al., 2020) for β-carotene, which is chemically unstable which hinders its application as a nutraceutical ingredient in foods. The incorporation of (−)-epigallocatechin-3-gallate (EGCG) and alpha-lactalbumin was carried out in the continuous phase to evaluate their efficiency to inhibit β-carotene degradation in nanoemulsions. EGCG was found effective in protecting β-carotene in various emulsion systems without negatively impacting lipid oxidation, signifying its use to support the incorporation of β-carotene into food emulsions (Liu et al., 2016). The study concluded that catechin-containing nanoemulsions could be safely applied in the field of nutraceuticals for enhancing their shelf life (Gadkari and Balaraman, 2015). Recent attempts to replace the petroleum-based packaging, the development of natural polymer–based edible film, such as gelatin, is underway.

8.5.1.1.6 Anthocyanins and Anthocyanidins

Anthocyanins, the group of water-soluble pigments found in vegetables like radish and red/purple cabbage, fruits like berries and blackcurrants, flower petals (red rose, blue chicory, and purple passionflower), and certain varieties of grains. These pigments are glycosylated anthocyanidins which contain natural flavylium salts as basic structure termed as anthocyanins. Considerable attention has been paid to anthocyanins because of their potential health benefits of being bactericidal, antiviral, hepatoprotective, and anti-inflammatory and having antioxidant, anti-carcinogenic, anti-obesity, and anti-diabetic effects as well as preventive effects on cardiovascular and degenerative diseases (Ziaullah et al., 2015; Rabelo et al., 2018). Anthocyanin shows antimicrobial effects by various possible mechanisms including, cytoplasmic membrane destabilization, permeabilization of plasma membrane, extracellular microbial enzymes inhibition, inhibition of anti-adherence of bacteria to epithelial cells, direct actions on microbial metabolism, and deprivation of the substrates required for microbial growth (Cisowska et al., 2011; Nile and Park, 2014). Anthocyanins are vulnerable to high pH, light, heat, and oxygen during processing and storage. Due to anthocyanins large particle size and low zeta potential, conventional techniques like micro-encapsulation have failed to offer any stability. Therefore, a nanoemulsion of *Myrciaria cauliflora* peel extract containing 0.8% anthocyanins and 2.56% of total flavonoids was prepared by low-energy methods, thus resulting in reduced economic cost of the process which makes it a feasible option for industrial application (Garcia et al., 2019). In a similar study, anthocyanins and total phenolics were extracted from blueberry pomace, and a food-grade double-nanoemulsion system was successfully developed by using HPH. The findings were useful for designing a co-encapsulation-based system with a high proportion of anthocyanins for the intended application of blueberry pomace in functional foods and other products with enhanced health benefits (Bamba et al., 2018). To overcome the low stability of anthocyanins, a food-grade W/O nanoemulsion for açaí berry, a type of anthocyanin, was generated by using HPH which resulted in enhanced stability even after long storage. As the açaí berry carries a great nutritional profile; its astringency and nutty taste somehow confine its consumption by certain populations. This promising approach of W/O nanoemulsion synthesis may be utilized as a possible application for preparing food product such as sport drinks and pharmaceutical sectors by masking the possible off-taste of this fruit (Rabelo et al., 2018).

8.5.1.2 Phenolic Acids

Phenolic acids (phenol carboxylic acids) are organic compounds that contain a phenolic ring and an organic carboxylic acid functional group. Phenolic acids, the most widely distributed plant non-flavonoid phenolic compounds, are chemical derivatives of benzoic acid or cinnamic acid (Dhull et al., 2019b, 2020c). They are distributed broadly in fruits like blueberries, kiwi fruit, plums, cherries, and apples, vegetables, and cereal grains. They exert excellent antioxidant activity; their salts are used mainly for antifungal and broad-spectrum antibacterial properties, which make them desirable candidates for the pharmaceutical, cosmetics, and food industries (Dhull and Sandhu, 2018). Phenolic acids are mainly categorized into two major subgroups, namely hydroxybenzoic acid and hydroxycinnamic acid. The four commonly found hydroxybenzoic acids are p-hydroxybenzoic, protocatechuic, vanillic, and syringic acids. The four most common hydroxycinnamic acids are ferulic, caffeic, p-coumaric, and sinapic acids. The antimicrobial action of hydroxybenzoic acids followed the same diffusion pattern like other weak organic acids; it was passed across the membrane by un-dissociated acid exert antimicrobial activity by acidification of cytoplasm, leading to cell death (Almajano et al., 2007). It seems that usually hydroxycinnamic acid (p-coumaric, caffeic, and ferulic acids)–based natural products are capable of exerting antimicrobial action by interfering with membrane integrity; the more lipophilic nature of p-coumaric makes it most powerful among all (Campos et al., 2009).

To further explore their role as nanoemulsions, a novel W/O/W multiphase nanoemulsion of hydrophilic arbutin and hydrophobic coumaric acid was developed by using hydrocolloids. Increased stability of these compounds was recorded even under unfavourable conditions. The outcome of this study suggests that these emulsions could act as a novel delivery method to co-deliver hydrophilic arbutin and hydrophobic coumaric acid for increasing the bio-accessibility of these nutraceutically important phenolic compounds (Huang et al., 2019).

In a related study, nanoemulsions loaded with jaboticaba extract (*Plinia peruviana*) were found having remarkable antioxidant properties and were able to hold a significant concentration of phenolics, flavonoids, and ellagic acid, thus proving their utility for pharmaceutical and cosmetic applications (Mazzarino et al., 2018).

Generally, ferulic acid (4-hydroxy-3-methoxycinnamic acid), exhibits very low solubility in the aqueous phase, thereby limiting its application in pharmaceutical and food products. In this regard, a stable nanoemulsion of ferulic acid, by applying the spontaneous emulsification method, was found stable throughout 12 weeks of storage period at 4°C (Ebrahimi et al., 2013).

Similarly, W/O/W multiple nanoemulsions using gallic acid in the internal aqueous phase were generated where more than 50% antioxidant activity was retained after storage of 28 days (Martins et al., 2020).

Three different acids, viz. vanillic, caffeic, and syringic acid, were individually tested for their ability to provide antioxidant activity for the formulation of virgin olive oil nanoemulsion. These acids were incorporated in the aqueous phase to evaluate their effect on emulsion properties and oxidation stability at various aqueous phase ratios. The crucial role of aqueous phase ratio was elucidated, and caffeic acid was found to be most effective in providing stability against oxidation. Additionally, all the tested phenolic acids were able to reduce the air/water surface tension, thus facilitating an increased emulsion formulation (Katsouli et al., 2017).

In another study, stable trans-cinnamic acid (trans-CA) nanoemulsions were fabricated utilizing low-energy SE methods. It demonstrated enhanced bacteriostatic and bactericidal activity against S. *aureus*, *S. typhimurium*, and *Pseudomonas aeruginosa* at subcellular size as compared to the pure chemical compound. This was found to be very effective in the preservation of freshly chopped lettuce through microbial inhibition in freshly dissected fruits and vegetables (Letsididi et al., 2018).

Golden spice, turmeric, is another prominent example of polyphenol that is highly utilized since the ancient time of Ayurveda and traditional Chinese medicine contains curcumin. Curcumin (and its two related compounds, the curcuminoids: demethoxycurcumin and bis-demethoxycurcumin) is derived from the rhizomes of turmeric plant and is widely known for its antioxidant, antimicrobial, anti-cancer, anti-inflammatory, anti-Alzheimer's, psychotic and neurotic activities, anti-diabetic, and wound-healing properties (Jiang et al., 2020). It exhibits bactericidal activity by damaging the cell membranes of gram-positive and gram-negative bacteria (Khameneh et al., 2019). However, low water solubility, degradation of structure, and poor bioavailability limit its application. Recently, oil-based nanoemulsion fabrication methods for curcumin have been deeply reviewed, which are categorized as low- and high-energy emulsification methods (Jiang et al., 2020). Studies highlight that using an oil–water–oil-based phase inversion method can generate curcumin-loaded nanoemulsions, and by optimizing parameters like mechanical stirring, aqueous flow pumping, this could be scalable up to industrial level. Results indicated that after storage of 60 days, 70% of curcumin was retained in emulsion when compared to other systems (Borrin et al., 2016). Apart from these two, a different nanocarrier system has also been designed which prevents degradation during processing and storage; both systems use same level of surfactant concentration. The release of entrapped curcumin was sustained and both the systems were found to have potential in inhibiting lipid oxidation (Chuacharoen and Sabliov, 2019). Similarly, curcumin nanoemulsions were produced with various surfactant concentrations using HPH, and were applied to a commercial milk system. The nanoemulsion was found stable for a month at room temperature, retaining effective radical scavenging activity, and was also able to inhibit the process of lipid oxidation in milk. In a similar study, preparation of the curcumin nanoemulsion with milk protein named sodium caseinate was carried out for incorporating it in ice cream. The nanoemulsion particle size was found to be influenced by heating, pH, and ionic strength; however, their release kinetics in simulated gastrointestinal digestion suggested their stability against pepsin digestion. The formulation was successfully applied to ice cream, and sensory attributes were evaluated; interestingly, no significant difference was reported as compared to control. The outcomes advocate that ice cream can be an appropriate dairy product for the enhanced delivery of lipophilic bioactive components like curcumin which can be potentially used for therapeutic purposes (Kumar et al., 2016).

Curcumin nanoemulsion, with casein and soy polysaccharide compacted complex, was prepared at an optimized condition and stored for 500 days at lower temperature 4°C. Sustained release was also obtained in simulated gastric and intestinal fluid with rapid and effective absorption in mice model. This study suggests that casein and soy polysaccharide complex nanoemulsion to be an appropriate system for oral delivery of lipophilic nutrients and drugs (Xu et al., 2017). In another study, a uniform curcuminoids-loaded nanoemulsion, for intended oral application, was also prepared with coconut oil by PIT method. These properties make curcumin nanoemulsion as suitable systems for food and beverage industry (Joung et al., 2016).

8.5.1.3 Tannins

"Tannins" are an important group of complex polymeric phenolic substances originally used in the tanning of leather and animal hides or precipitating gelatin from solutions, a property termed astringency. They are present in almost every part of plant, namely bark, wood, leaves, fruits, and roots, and their occurrence is reported in a wide range of plant families (Dhull et al., 2016, 2020a). Tannins exhibit toxic effect on filamentous fungi, yeasts, some viruses, and bacteria. Their antimicrobial action is attributed to their inhibitory activity against DNA gyrase and the inhibition of other enzymes (Cowan, 1999; Khameneh et al., 2019). Tannin has gained a great deal of attention in recent years due to its excellent antioxidant properties. Tannins are divided into the following three broad classes (Maqsood et al., 2013):

1. Hydrolyzable tannins (gallotannins: gallic acid, quinic acid, tannic acid, ellagitannins: ellagic acid, castalagin, vescalagin and hydrolysable tannin oligomer: agrimoniin, rugosinD)
2. Condensed tannins also referred to as proanthocyanidins (oligomers of flavan-3-ols: catechin, epicatechin; flavan-3,4-diol: leucoanthocyanidin)
3. Complex tannins (stenophyllanin A, acutissimin B, mongolicain A, stenophynin A, etc.).

Tannic acid, an important gallotannin belonging to the hydrolysable class of tannin, is a specific tannin that formally contains 10 galloyl (3,4,5-trihydroxyphenyl) units surrounded with glucose. Tannic acid does not contain carboxyl groups, but its weakly acidic nature is because of the multiple phenolic hydroxyls (Ribeiro et al., 2018). It exhibits a diverse range of biological activities, including antioxidant, antibacterial, and antiviral activity. Due to its capacity to interact with proteins, polysaccharides, alkaloids, and metal ions, it is used to enhance emulsifying properties and antioxidant activity. Role of tannic acid as nanoemulsion stabilizer has also been explored by O/W emulsions using a high-pressure microfluidizer by using polyphenol-polysaccharide (tannic acid and β-glucan) complexes. The results suggest the use of these complex is limited in the range of pH and temperature but definitely increase the application of β-glucan as a functional ingredient in foods (Li et al., 2019a). Studies indicated that alginate-based edible films when loaded with tannic acid and quercetien were able to enhance the shelf-life of food items. These alginate-based edible biofilms were analysed on the basis of different parameters, viz. thickness of film, tensile strength, percentage elongation at break, light transmission, film opacity, water vapour permeability, water sorption kinetics, and antioxidant and antimicrobial activity after 11 days storage at 4°C. As per the authors, pork patties wrapped in tannic acid nanosolution–incorporated films showed lower microbial counts (total plate count, yeast and mould count, and psychrophilic count) as compared to control films (Rao, 2020).

The effect of tannic acid and two other plant-based emulsifiers, namely quillaja saponin (QS) and gum arabic (GA), on a nanoemulsion loaded with β-carotene was also evaluated in a simulated gastrointestinal tract. Emulsifier type and tannic acid addition had no significant effect on β-carotene bio-accessibility but the addition of tannic acid could efficiently inhibited temperature-induced β-carotene degradation which is reported to increase at a higher temperature. These outcomes may facilitate the design of more effective nutraceutical-loaded functional foods and beverages (Li et al., 2019b).

Proanthocyanidins, the condensed tannins, are oligomeric and polymeric products of the flavonoid biosynthetic pathway and are present in flowers, olives, onions, green and black tea, broccoli, nuts, dark chocolate, cocoa, red wine, bark, and seeds of various plants. Berries and fruits like lingonberry, banana, pomegranate, cranberry, oranges, grapefruit, black elderberry, black chokeberry, black currant, blueberry are the greatest sources of proanthocyanidins. They exert antioxidant, cardio-protective, neuro-protective, immunomodulatory, lipid-lowering and anti-obesity, anti-cancer, and antimicrobial activity (Rauf et al., 2019).

In a comparative study, both condensed and hydrolysable tannins were used to generate microcapsules and nanoemulsions and were fabricated by ultrasonic irradiation. The generated nanoemulsions displayed increased stability at slightly alkaline conditions (pH > 8) and were partially disassembled even under acidic conditions (pH < 7). This pH stability of nanoemulsion can be exploited in pH-triggered controlled release nanoemulsions, which can further be explored for food, pharmaceutical, or biomedical applications (Bartzoka et al., 2017).

Since a limited range of emulsifiers can be used in food and beverage products to get the required functional performance, due to consumer health concerns, in attempt to find suitable food-grade emulsifiers, complex of proteins with other food grade natural products are great deal of concern in recent days. In a recent study, the lotus seedpod proanthocyanidin (LSPC) when incorporated in whey protein, was able to stabilize the β-carotene-loaded nanoemulsions at various pH and also exhibited better antioxidant activity. Studies suggest that LSPC-whey protein complexes can be used as effective emulsifiers and be useful for developing more efficacious functional foods and beverages for nutraceutical-loaded nanoemulsions (Chen et al., 2020).

8.5.1.4 Lignans

Lignans dimers of phenylpropanoid (C6–C3) units joined by the central carbons of their side chains. They are fibre-associated compounds widely dispersed throughout the plant kingdom and found in human common diet, including grains, nuts, seeds, vegetables, and drinks such as tea, coffee, or wine. Grain sources of lignans include oil seeds (flax, rapeseed, and sesame), whole-grain cereals (wheat, oats, rye, barley, and millets), and legumes (soybean). Flaxseed carries the highest concentration of dietary lignans, as secoisolariciresinol diglucoside. Sesamin, matairesinol, pinoresinol, and lariciresinol are other dietary lignans. Lignan shows various biological activities like anti-cancer, antioxidant, anti-mitotic, anti-hypertensive, antiviral, estrogenic, and insecticidal properties.

Multiple attempts to synthesize nanoemulsion using flaxseed oil, which is a good source of α-linolenic acid (ALA), is reported to inhibit cardiovascular disease, non-alcoholic fatty liver disease, insulin resistance, type 2 diabetes, and neurodegenerative disorders. However, a higher concentration of ALA, having a larger surface area of nanoemulsified flaxseed oil, is prone to rapid lipid oxidation in the presence of heat, light and oxygen, which limits their incorporation into functional foods and beverages (Arab-Tehrany et al., 2012). Limited researchers have evaluated flaxseed lignan extract and secoisolariciresinol, secoisolariciresinol diglucoside, and p-coumaric acid for their antioxidant/pro-oxidant effect in nanoemulsions, and interestingly, these were found to improve the stability of flaxseed oil nanoemulsion, suggesting its application in function for food industries (Cheng et al., 2019). In a similar study, impacts of sesame lignans, natural sesamol (SOH), and sesamin on the stability of flaxseed O/W emulsion were assessed. SOH exhibited higher antioxidant activities, an improved physicochemical property, and a stronger interfacial barrier in flaxseed nanoemulsion (Wang et al., 2019).

8.5.1.5 Stilbene

The name, stilbene (1,2-diphenylethylene), was originally derived from the Greek word *stilbos*, which means "shining". Stilbenoids are hydroxylated derivatives of stilbene, with C_6-C_2-C_6 structure. Stilbenoids are characterized by two phenyl groups linked by a transethane bond. The most famous stilbenoid, resveratrol, has several hydroxyl groups attached to its phenyls. It is present in strongly pigmented vegetables; fruits like some varieties of grapes, berries, and peanuts; and red wine. It is known to provide several health benefits like antioxidant, anti-carcinogenic, chemo-preventive, and cardio-protective effects. Low water solubility accounts for its poor bio-availability. By applying HPH, an O/W nanoemulsion system was developed and tested for oral delivery of resveratrol for its effect on cell viability, cell permeability, and sustained release. Studies revealed a better delivery system with enhanced stability and efficient bioavailability, which can further be explored in nutraceuticals and functional food systems (Sessa et al., 2014). In another study, nanoemulsion was fabricated by the SE method, using grape seed oil and orange oil together in the oil phase. The low-energy nanoemulsion method was successful in the development of stable delivery system (Davidov-pardo and Julian, 2015). Interestingly, in another study, resveratrol nanoemulsion, in tuna fish oil, stabilized by octenyl succinic and hybrid modified starch (OSA-MS) Hi-cap 100, was generated by using a high-energy method and was successfully recommended for application in dairy and beverage products (Shehzad et al., 2020). Other fabrication methods, effect of various oil phase, effect of droplet size, and effect of emulsifiers have been recently deeply reviewed by Choi and McClements (2020).

8.5.2 Essential Oils

Essential oils (EOs), quintaessentia (as they carry fragrance), are mixtures of lipophilic, volatile, aromatic compounds and are insoluble in water but soluble in organic solvents. EOs are a complex mixture of low molecular weight compounds like alkaloids, phenols, and terpenes. Essential oils are considered as naturally occurring terpenic mixtures, including carbohydrates, phenols, alcohols, ethers, aldehydes and ketones that are known to be responsible for biological activity. The general chemical formula for terpenes is $C_{10}H_{16}$, they occur in various forms like diterpenes, triterpenes, and tetraterpenes, hemiterpenes, and sesquiterpenes. When the compounds carry additional elements, usually oxygen, they are termed terpenoids (Cowan, 1999; Pavoni et al., 2020). EOs are derived from spices, herbs, and medicinal plant parts like flowers, buds, seeds, leaves, twigs, bark, wood, fruit, and roots by way of steam or dry distillation, expression, or solvent extraction (Punia et al., 2019). EOs are widely known in traditional medicine. EOs have been explored as flavouring additives, food preservative, fragrance agent, antioxidants, anti-allergic, anti-lice, anti-dandruff, anti-cancer, and anti-inflammatory. Not all, but many EOs exert strong antibacterial, antiviral, and antifungal activities, encouraging their application also as natural antimicrobials in food and beverage products (Friedman et al., 2002). The antimicrobial activity of essential oils can be explained by their chemical composition (usually up to 85% major components and rest are other trace components). The major antimicrobial components can be grouped by their chemical similarities. For example, aldehydes, phenols, and terpene alcohols are considered as the most active antimicrobial components. In cinnamaldehyde, antimicrobial activity is due to its major component aldehyde. Phenols are aromatic compounds, and thymol, carvacrol, and eugenol found in thyme, oregano, and clove oils, respectively, are well known for their excellent antimicrobial activities. Linalool (linear monoterpene alcohol of lavender oil) and menthol (cyclic monoterpene alcohol) have a wide spectrum antimicrobial activity. The reported antibacterial mechanisms of essential oils include disturbances of the permeability of cell membranes and disruption of the electron flow, causing loss of cellular components or influx of other substances into the cell and disturbances of the pH gradient, that ultimately leads to cell death (Ferdes, 2018). Several of these essential oils are considered a GRAS category and are approved by U.S. Food and Drug Administration (Joyce and Nagarajan, 2017).

Mainly EOs and in some cases plant oils themselves are used as oil phase for making food-grade nanoemulsions rather than encapsulating EOs. Applying EOs nanoemulsion in food products is still an uphill task because a wide variety of food spoilage microorganism requires different antimicrobial compounds having broad-spectrum activity. Loss of antimicrobial activity of EOs can be due to their interaction with other food ingredients, such as lipid, proteins and minerals; adsorption on interfaces of real food; and their uneven distribution. So to apply EOs in real food systems to obtain a similar efficiency as the synthetic system, a slightly higher concentration of EO is required (Donsì and Ferrari, 2016). Major applications of EOs nanoemulsion includes prevention of spoilage to extend the shelf life of food by coating of nanoemulsions on food, by incorporating nanoemulsions in biopolymers, by direct application of nanoemulsions in food systems. Recent update of various applications has been summarized in Tables 8.1 and 8.2, whereas some miscellaneous applications are reported separately.

Penicillium italicum, a phytopathogenic fungus infecting fruits and vegetables, can cause significant loss during harvesting and transportation of fruits. A nanoemulsion of garlic EO (prepared by ultrasonic method) in comparison to garlic oil was evaluated for its fungitoxic capabilities in a model citrus fruit system, where the synthesized nanoemulsion showed a better bioavailability (Long et al., 2020). Similarly, augmented antibacterial activity of a nanoemulsion prepared from EOs of lime was reported against *E. coli*, *Salmonella spp.*, and *S. aureus* (Liew et al., 2020). Antimicrobial activity is also reported for nanoemulsion prepared from Geraniol and linalool (Balta et al., 2017), 1-8-Cineol, p- Cymene, α-Pinene, and Limonene (Quatrin et al., 2017). Recently, the role of peppermint oil in improving oxidative stability and antioxidant capacity of a prepared nanoemulsion system was studied. Peppermint oil was co-encapsulated with borage seed oil (one of the richest sources of γ-linolenic acid and linoleic acid), which resulted in improved oxidative stability. This may help to attain provides novel insights regarding the development of functional foods, dietary supplements, and beverages (Rehman et al., 2020). Role of EO nanoemulsion in improving the texture of pork patties for elderly members of the population was also investigated. In the study, a canola oil nanoemulsion was supplemented in pork patties, and it was found to improve organoleptic qualities such as tenderness and juiciness. The outcome will definitely bring insights for exploring new ways towards tenderizing food products designed for elderly people in a natural manner (Lee et al., 2020). Considering an example of the malting industry, where Fusarium head blight becomes a principle safety concern as *Fusarium* mycotoxin contamination is a prominent issue in barley malting. To address this issue, the efficacy of clove oil nanoemulsions on *Fusarium* growth and mycotoxin production was investigated during malting process. Results demonstrated that clove oil nanoemulsion resulted in higher mycotoxin inhibitory activity with less impact on the final malt flavour (Wan et al., 2020). Furthermore, carvacrol nanoemulsion can be used to sanitize stainless steel surfaces as it can reduce bacterial and yeast count on a steel surface. Results depict that if combined with acidic electrolysed water, the carvacrol nanoemulsion was able to reduce the microbial load of fresh-cut vegetables significantly (Sow et al., 2017). Carvacrol nanoemulsions were also able to inactivate low levels of *S. enterica* and *E. coli* in sprouting radish seeds (Landry et al., 2015, 2016). Interestingly, direct application of EO nanoemulsion has been reported in products such as fruit juice to reduce the microbial population. The approach has been investigated by some researchers, for example eugenol-loaded antimicrobial nanoemulsion (Ghosh et al., 2014) and thyme (*Thymus vulgaris*) essential oil–based antimicrobial nanoemulsion (Patel and Ghosh, 2020). We have enlisted many applications of nanoemulsion plant bioactive compounds, but the scope is limitless for application in the food, pharmaceutical, edible vaccine, functional food, and nutraceutical industries.

TABLE 8.1
Application of EOs NE by Incorporation in Film Matrices

Essential Oil (EO)	Packaging Material	Outcome of the Study	References
Zataria multiflora	Basil seed gum-based edible film	Antimicrobial edible film Extended shelf life Enhanced mechanical strength of film Retarded release of nanoemulsion	Hashemi Gahruie et al. (2017)
Thymol	Co-emulsified gelatin and lecithin film	Effective antimicrobial activity Sustained release of EOs Applied as novel and generally recognized as safe (GRAS) biodegradable packaging materials	Li et al. (2020)
Cumin seed	Active cellulose papers	First report on the effect of active packaging containing cumin seed essential oil (CSEO) on quality attributes of meat with focus on controlled release and increased stability of CSEO Good antioxidant and antimicrobial activities Shelf-life extension	Hemmatkhah et al. (2020)
Bergamot	Whey protein isolate (WPI) films	Enhanced mechanical and water vapour permeability Batter antimicrobial and antioxidant activities	Sogut (2020)
Lemongrass	Glycerol-plasticized cassava–starch based film	Improved colorimetric attributes, thermal stability, barrier to moisture, and mechanical properties Batter antimicrobial and antioxidant activities	Mendes et al. (2020)
Thymol	Quinoa protein/chitosan edible film	Significant decrease in fungal growth Improved shelf life	Rabelo et al. (2018)

TABLE 8.2
Application of EOs NE by Coating the Food System

Essential Oil	Fabrication Method	Food System	Outcome of the Study	References
Ginger	Ultrasonic	Chicken breast fillet	The lowest colour difference The lowest cooking loss Antimicrobial effect	Noori et al. (2018)
Cinnamon	Ultrasound	Fresh strawberries	Prolonged shelf life Strongest antimicrobial activity Lower loss in physico-chemical properties	Chu et al. (2020)
Cinnamon	Phase inversion temperature	Asian seabass fillets	Prevented the alteration of colour and texture qualities Extended the shelf life Delayed the growth of bacteria	Chuesiang et al. (2020)
Ferulago angulate + chitosan	Ultrasonic	Rainbow trout fillets	Texture, colour, and overall acceptability Antibacterial activity against fish pathogens Prolonged shelf life	Shokri et al. (2020)
Star anise + polylysine + nisin	Ultrasonic	Ready-to-eat Yao meat products	Better antimicrobial effect Good retention of colour, odour, and overall acceptance Improved quality and shelf life	Liu et al. (2020)
Zataria multiflora	Ultrasonic	Beef fillet	Beef fillet was inoculated with *Escherichia coli* O157:H7 Decrease growth of *E. coli* O157:H7	Alavi et al. (2020)

Essential Oil	Fabrication Method	Food System	Outcome of the Study	References
Zataria multiflora	Ultrasonic	Trout fillet	Microbial growth inhibition Prolonged storage	Khanzadi et al. (2020)
Zataria multiflora Boiss	Ultrasonic	Turkey meat	Preservation of microbial quality Prolonged shelf life	Keykhosravy et al. (2020)
Buniumpersicum Boiss	Ultrasonic	Turkey meat	Preservation of microbial quality, Prolonged shelf life	Keykhosravy et al. (2020)
Citrus	Ultrasonic	Rainbow trout fillet	Delayed the auto-oxidation reactions of unsaturated fatty acids Extending the shelf life of fish oil	Uçar, (2020)
Lemon Grass	Phase inversion temperature	'Rocha' pears	Better firmness after storage Reduced fruit colour evolution No scald symptoms	Gago et al. (2020)
Thymol	Spontaneous	Refrigerated strawberries	Better aroma score after storage Increased shelf life No alteration in quality parameters	Robledo et al. (2018a)

8.6 CONCLUSION

Plants comprise numerous bioactive compounds broadly termed as phytochemicals, further classified as phenolics, flavonoids, essential oils, lectins, alkaloids, terpenes, and their derivatives. Their varied roles extend from plant pollination via producing attractive pigments to defend against herbivory and pathogens, which is somehow attributed to the antimicrobial properties of these active metabolites. Moreover, these compounds are also known to possess antioxidant properties and many other health benefits, making them suitable candidates for further exploration. Although there are multiple reports on traditional and medicinal values of these secondary metabolites, still a very limited literature is available regarding plant bioactive compound–based nanoemulsions, specifically in food industries. The major issues in the generation of these types of nanoemulsion are their sparingly soluble nature and their low shelf life which further limits their industrial applications. In this scenario, nanoemulsion generation seems to be suitable and a viable alternative because of their increased surface area and relatively smaller droplet size that amazingly lead to improved properties. But because they are thermodynamically unstable, an additional requirement of high energy is a prerequisite for nanoemulsion generation, which can be provided through specialized instruments generating huge shear force or, alternatively, these can be fabricated by making use of various emulsifiers. In this chapter, we have emphasized various prospects of plant-based nanoemulsions like they exhibit higher antimicrobial properties, being inexpensive and are stable and non-toxic. Therefore, it is evident that plant-derived nanoemulsions can be an alternative to chemical food preservatives as being more efficient and relatively safe. This makes them a potential candidate for suitable applications in food-based industries. In this study, we have majorly emphasized recent advancements made in the area of plant-derived nanoemulsions by highlighting their medicinal, pharmaceutical, and food-based applications.

REFERENCES

Abcha, I., Souilem, S., Neves, M. A., et al. (2019). Ethyl oleate food-grade O/W emulsions loaded with apigenin: Insights to their formulation characteristics and physico-chemical stability. *Food Research International*, 116, 953–962. doi: 10.1016/j.foodres.2018.09.032

Alavi, S. H., Khanzadi, S., Hashemi, M., et al. (2020). Effect of alginate coatings containing Zataria Multiflora Boiss essential oil in forms of coarse emulsion and Nano-emulsion on inoculated Escherichia coli O157: H7 in beef fill. *Journal of Nutrition,Fasting and Health*, 8, 90–96.

Almajano, M. P., Carbo, R., Delgado, M. E., and Gordon, M. H. (2007). Effect of pH on the antimicrobial activity and oxidative stability of oil-in-water emulsions containing caffeic acid. *Journal of Food Science*, 72, C258–C263.

Arab-Tehrany, E., Jacquot, M., Gaiani, C., et al. (2012). Beneficial effects and oxidative stability of omega-3 long-chain polyunsaturated fatty acids. *Trends in Food Science and Technology*, 25, 24–33.

Azmi, N. A. N., Elgharbawy, A. A. M., Motlagh, S. R., and Samsudin, N. (2019). Nanoemulsions: Factory for food, pharmaceutical and cosmetics. *Process Review*, 7, 1–34.

Balta, I., Brinzan, L., Stratakos, A. C., et al. (2017). Geraniol and linalool loaded nanoemulsions and their antimicrobial activity. *Bulletin of the University of Agricultural Sciences and Veterinary Medicine Cluj-Napoca Animal Science and Biotechnologies*, 74, 157–161.

Bamba, B. S. B., Shi, J., Tranchant, C. C., et al. (2018). Coencapsulation of polyphenols and anthocyanins from blueberry pomace by double emulsion stabilized by whey proteins: Effect of Homogenization parameters. *Molecules*. doi: 10.3390/molecules23102525

Bartzoka, E. D., Lange, H., Mosesso, P., and Crestini, C. (2017). Synthesis of nano- and microstructures from proanthocyanidins, tannic acid and epigallocatechin-3-: O -gallate for active delivery. *Green Chemistry*, 19, 5074–5091. doi: 10.1039/c7gc02009k

Barzegar, H., Mehrnia, M. A., Nasehi, B., and Alipour, M. (2018). Fabrication of peppermint essential oil nanoemulsions by spontaneous method: Effect of preparing conditions on droplet size. *Flavour and Fragrance Journal*, J33, 351–356.

Bazana, M. T., da Silva, S. S., Codevilla, C. F., et al. (2019). Development of nanoemulsions containing Physalis peruviana calyx extract: A study on stability and antioxidant capacity. *Food Research International*, 125, 108645. doi: 10.1016/j.foodres.2019.108645

Bernard, F.-X., Sable, S., Cameron, B., et al. (1997). Glycosylated flavones as selective inhibitors of topoisomerase IV. *Antimicrobial Agents and Chemotherapy*, 41, 992–998.

Borrin, T. R., Georges, E. L., Moraes, I. C. F., and Pinho, S. C. (2016). Curcumin-loaded nanoemulsions produced by the emulsion inversion point (EIP) method: An evaluation of process parameters and physico-chemical stability. *Journal of Food Engineering*, 169, 1–9.

Bultosa, G. 2016. *Functional foods: overview, Reference Module in Food Sciences*. Elsevier. doi: 10.1016/B978-0-08-100596-5.00071-8.

Calligaris, S., Comuzzo, P., Bot, F., et al. (2015). Nanoemulsions as delivery systems of hydrophobic silybin from silymarin extract: Effect of oil type on silybin solubility, in vitro bioaccessibility and stability. *LWT - Food Science and Technology*, 63, 77–84. doi: 10.1016/j.lwt.2015.03.091

Campos, F. M., Couto, J. A., Figueiredo, A. R., et al. (2009). Cell membrane damage induced by phenolic acids on wine lactic acid bacteria. *International Journal of Food Microbiology*, 135, 144–151.

Cetin-Karaca, H., and Newman, M. C. (2015). Antimicrobial efficacy of plant phenolic compounds against Salmonella and Escherichia Coli. *Food Bioscience*, 11, 8–16.

Chawla, P., Kumar, N., Kaushik, R., and Dhull, S. B. (2019). Synthesis, characterization and cellular mineral absorption of nanoemulsions of *Rhododendron arboreum* flower extracts stabilized with gum arabic. *Journal of Food Science Technology*, 56, 5194–5203.

Chen, Y., Zhang, R., Xie, B., et al. (2020). Lotus seedpod proanthocyanidin-whey protein complexes: Impact on physical and chemical stability of β-carotene-nanoemulsions. *Food Research International*, 127, 108738. doi: 10.1016/j.foodres.2019.108738

Cheng, C., Yu, X., McClements, D. J., et al. (2019). Effect of flaxseed polyphenols on physical stability and oxidative stability of flaxseed oil-in-water nanoemulsions. *Food Chemistry*, 301, 125207. doi: 10.1016/j.foodchem.2019.125207

Choi, S. J., and McClements, D. J. (2020). Nanoemulsions as delivery systems for lipophilic nutraceuticals: Strategies for improving their formulation, stability, functionality and bioavailability. *Food Science and Biotechnology*, 29, 149–168. doi: 10.1007/s10068-019-00731-4

Chu, Y., Gao, C. C., Liu, X., et al. (2020). Improvement of storage quality of strawberries by pullulan coatings incorporated with cinnamon essential oil nanoemulsion. *LWT - Food Science and Technology*, 122, 109054. doi: 10.1016/j.lwt.2020.109054

Chuacharoen, T., and Sabliov, C. M. (2019). Comparative effects of curcumin when delivered in a nanoemulsion or nanoparticle form for food applications: Study on stability and lipid oxidation inhibition. *LWT - Food Science and Technology*, 113, 108319. doi: 10.1016/j.lwt.2019.108319

Chuesiang, P., Sanguandeekul, R., and Siripatrawan, U. (2020). Phase inversion temperature-fabricated cinnamon oil nanoemulsion as a natural preservative for prolonging shelf-life of chilled Asian seabass (*Lates calcarifer*) fillets. *LWT - Food Science and Technology*, 125, 109122. doi: 10.1016/j.lwt.2020.109122

Chuesiang, P., Siripatrawan, U., Sanguandeekul, R., et al. (2018). Optimization of cinnamon oil nanoemulsions using phase inversion temperature method: Impact of oil phase composition and surfactant concentration. *Journal of Colloid and Interface Science*, 514, 208–216.

Cisowska, A., Wojnicz, D., and Hendrich, A. B. (2011). Anthocyanins as antimicrobial agents of natural plant origin. *Natural Product Communications*, 6, 150–156.

Cowan, M. M. (1999). Plant products as antimicrobial agents. *Clinical Microbiology Reviews*, 12, 564–582. doi: 10.1128/cmr.12.4.564

Cushnie, T. P. T., and Lamb, A. J. (2005). Antimicrobial activity of flavonoids. *The International Journal of Antimicrobial Agents*, 26, 343–356. doi: 10.1016/j.ijantimicag.2005.09.002

Dammak, I., and do Amaral Sobral, P. J. (2018). Investigation into the physicochemical stability and rheological properties of rutin emulsions stabilized by chitosan and lecithin. *Journal of Food Engineering*, 229, 12–20. doi: 10.1016/j.jfoodeng.2017.09.022

Dammak, I., de Carvalho, R. A., Trindade, C. S. F., et al. (2017). Properties of active gelatin films incorporated with rutin-loaded nanoemulsions. *International Journal of Biological Macromolecules*, 98, 39–49. doi: 10.1016/j.ijbiomac.2017.01.094

Dammak, I., Lourenço, R. V., and do Amaral Sobral, P. J. (2019). Active gelatin films incorporated with Pickering emulsions encapsulating hesperidin: Preparation and physicochemical characterization. *Journal of Food Engineering*, 240, 9–20.

Davidov-pardo, G., and Julian, D. (2015). Nutraceutical delivery systemss: Resveratrol encapsulation in grape seed oil nanoemulsions formed by spontaneous emulsification. *Food Chemistry*, 167, 205–212. doi: 10.1016/j.foodchem.2014.06.082

Dhull, S. B., Anju, M., Punia, S., et al. (2019a). Application of Gum Arabic in Nanoemulsion for Safe Conveyance of Bioactive Components. In Prasad, R., Kumar, V., Kumar, M., and Choudhary, D. (Eds), *Nanobiotechnology in Bioformulations*. Cham: Springer, pp. 85–98.

Dhull, S. B., Kaur, P., and Purewal, S. S. (2016). Phytochemical analysis, phenolic compounds, condensed tannin content and antioxidant potential in Marwa (*Origanum majorana*) seed extracts. *Resource-Efficient Technologies*, 2, 168–174.

Dhull, S. B., Kaur, M., and Sandhu, K. S. (2020a). Antioxidant characterization and in vitro DNA damage protection potential of some Indian fenugreek (*Trigonella foenum-graecum*) cultivars: Effect of solvents. *Journal of Food Science and Technology*, 57, 1–10.

Dhull, S. B., Punia, S., Kidwai, M. K., et al. (2020c). Solid-state fermentation of lentil (Lens culinaris L.) with Aspergillus awamori: Effect on phenolic compounds, mineral content, and their bioavailability. *Legume Science*, 2, e37.

Dhull, S. B., Punia, S., Sandhu, K. S., et al. (2019b). Effect of debittered fenugreek (*Trigonella foenum-graecum* L.) flour addition on physical, nutritional, antioxidant, and sensory properties of wheat flour rusk. *Legume Science*, 2, e21.

Dhull, S. B., and Sandhu, K. S. (2018). Wheat-Fenugreek composite flour noodles: Effect on functional, pasting, cooking and sensory properties. *Current Research in Nutrition and Food Science*, J6, 174–182.

Dhull, S. B., Sandhu, K. S., Punia, S., et al. (2020b). Functional, thermal and rheological behavior of fenugreek (*Trigonella foenum-graecum* L.) gums from different cultivars: A comparative study. *International Journal of Biological Macromolecules*, 159, 406–414.

Dickinson, E. (2009). Hydrocolloids as emulsifiers and emulsion stabilizers. *Food Hydrocolloids*, 23, 1473–1482.

Donsì, F. (2018). Applications of nanoemulsions in foods. *Nanoemulsions Formulation, Application and Characterisation* 349–377. doi: 10.1016/B978-0-12-811838-2.00011-4

Donsì, F., and Ferrari, G. (2016). Essential oil nanoemulsions as antimicrobial agents in food. *Journal of Biotechnology*, 233, 106–120. doi: 10.1016/j.jbiotec.2016.07.005

Donsì, F., Sessa, M., and Ferrari, G. (2012). Effect of emulsifier type and disruption chamber geometry on the fabrication of food nanoemulsions by high pressure homogenization. *Industrial & Engineering Chemistry Research*, 51, 7606–7618.

Ebrahimi, P., Ebrahim-Magham, B., Pourmorad, F., and Honary, S. (2013). Ferulic acid lecithin-based nano-emulsions prepared by using spontaneous emulsification process. *Iranian Journal of Chemistry and Chemical Engineering*, 32, 17–25.

Espitia, P. J. P., Fuenmayor, C. A., and Otoni, C. G. (2019). Nanoemulsions: Synthesis, characterization, and application in bio-based active food packaging. *Comprehensive Reviews in Food Science and Food Safety*, 18, 264–285. doi: 10.1111/1541-4337.12405

Ferdes, M. (2018). Antimicrobial compounds from plants. *Fight Antimicrob Resist*, 243–271. doi: 10.5599/obp.15.15

Friedman, M., Henika, P. R., and Mandrell, R. E. (2002). Bactericidal activities of plant essential oils and some of their isolated constituents against Campylobacter jejuni, Escherichia coli, Listeria monocytogenes, and Salmonella enterica. *Journal of Food Protection*, 65, 1545–1560.

Gadkari, P. V., and Balaraman, M. (2015). Extraction of catechins from decaffeinated green tea for development of nanoemulsion using palm oil and sunflower oil based lipid carrier systems. *Journal of Food Engineering*, 147, 14–23. doi: 10.1016/j.jfoodeng.2014.09.027

Gago, C., Antão, R., Dores, C., et al. (2020). The effect of nanocoatings enriched with essential oils on "Rocha" pear long storage. *Foods*, 9, 1–15. doi: 10.3390/foods9020240

Garcia, N. O. S., Fernandes, C. P., and da Conceição, E. C. (2019). Is it possible to obtain nanodispersions with jaboticaba peel's extract using low energy methods and absence of any high cost equipment? *Food Chemistry*, 276, 475–484. doi: 10.1016/j.foodchem.2018.10.037

Gavin, A., Pham, J. T. H., Wang, D., et al. (2015). Layered nanoemulsions as mucoadhesive buccal systems for controlled delivery of oral cancer therapeutics. *International Journal of Nanomedicine*, 10, 1569.

Geissman, T. A. (1963). *Flavonoid compounds, tannins, lignins and, related compounds. Comprehensive biochemistry*. Elsevier, pp. 213–250.

Ghosh, V., Mukherjee, A., and Chandrasekaran, N. (2014). Eugenol-loaded antimicrobial nanoemulsion preserves fruit juice against, microbial spoilage. *Colloids and Surfaces B: Biointerfaces*, 114, 392–397. doi: 10.1016/j.colsurfb.2013.10.034

Guzey, D., and McClements, D. J. (2006). Formation, stability and properties of multilayer emulsions for application in the food industry. *Advances in Colloid and Interface Science*, 128, 227–248.

Hasenhuettl, G.L, and Hartel, R. W. (2008). *Food emulsifiers and their applications*. New York: Springer.

Hashemi Gahruie, H., Ziaee, E, Eskandari, M. H., and Hosseini, S. M. H. (2017). Characterization of basil seed gum-based edible films incorporated with Zataria multiflora essential oil nanoemulsion. *Carbohydrate Polymers*, 166, 93–103. doi: 10.1016/j.carbpol.2017.02.103

Hemmatkhah, F., Zeynali, F., and Almasi, H. (2020). Encapsulated cumin seed essential oil-loaded active papers: Characterization and evaluation of the effect on quality attributes of beef hamburger. *Food and Bioprocess Technology*, 13, 533–547. doi: 10.1007/s11947-020-02418-9

Hintz, T., Matthews, K. K., and Di, R. (2015). The use of plant antimicrobial compounds for food preservation. *BioMed Research International*, 10, 77–80.

Hoeller, S., Sperger, A., and Valenta, C. (2009). Lecithin based nanoemulsions: A comparative study of the influence of non-ionic surfactants and the cationic phytosphingosine on physicochemical behaviour and skin permeation. *International Journal of Pharmaceutics*, 370, 181–186.

Horison, R., Sulaiman, F. O., Alfredo, D., and Wardana, A. A. (2019). Physical characteristics of nanoemulsion from chitosan/nutmeg seed oil and evaluation of its coating against microbial growth on strawberry. *Food Research*, 3, 821–827.

Hossain, M. K., Dayem, A. A., Han, J., et al. (2016). Molecular mechanisms of the anti-obesity and anti-diabetic properties of flavonoids. *International Journal of Molecular Sciences*. doi: 10.3390/ijms17040569

Huang, H., Belwal, T., Liu, S., et al. (2019). Novel multi-phase nano-emulsion preparation for co-loading hydrophilic arbutin and hydrophobic coumaric acid using hydrocolloids. *Food Hydrocolloids*, 93, 92–101. doi: 10.1016/j.foodhyd.2019.02.023

Jiang, T., Liao, W., and Charcosset, C. (2020). Recent advances in encapsulation of curcumin in nanoemulsions: A review of encapsulation technologies, bioaccessibility and applications. *Food Research International*, 132, 109035. doi: 10.1016/j.foodres.2020.109035

Joung, H. J., Choi, M., and Kim, J. T., et al. (2016). Development of food-grade curcumin nanoemulsion and its potential application to food beverage system: Antioxidant property and in vitro digestion. *Journal of Food Science*, 81, N745–N753.

Joyce, N. M., and Nagarajan, R. (2017). Recent research trends in fabrication and applications of plant essential oil based nanoemulsions nanomedicine & nanotechnology. *Journal of Nanomedicine & Nanotechnology*, 8, 1–10. doi: 10.4172/2157-7439.1000434

Juneja, V. K., and Sofos, J. N. (2017). *Microbial control and food preservation theory and practice*. New York: Springer Nature.

Kabalnov, A. (2001). Ostwald ripening and related phenomena. *Journal of Dispersion Science and Technology*, 22(1), 1–12.

Kabalnov, A. S., and Schchukin, E. D. (1992). Ostwald ripening theory: Applications to fluorocarbon emulsion stability. *Advances in Colloid and Interface Science*, 38, 69–97.

Kalinina, I. V., Potoroko, I. Y., Nenasheva, A. V., et al. (2019). Prospects for the application of taxifolin based nanoemulsions as a part of sport nutrition products. *Human Sport Medicine's*, 18(4), 100–107. doi: 10.14529/hsm190114

Karadag, A., Yang, X., Ozcelik, B., and Huang, Q. (2013). Optimization of preparation conditions for quercetin nanoemulsions using response surface methodology. *Journal of Agricultural and Food Chemistry*, 61, 2130–2139. doi: 10.1021/jf3040463

Katsouli, M., Polychniatou, V., and Tzia, C. (2017). Influence of surface-active phenolic acids and aqueous phase ratio on w/o nano-emulsions properties; model fitting and prediction of nano-emulsions oxidation stability. *Journal of Food Engineering*, 214, 40–46. doi: 10.1016/j.jfoodeng.2017.06.017

Kaur, P., Dhull, S. B., Sandhu, K. S., et al. (2018). Tulsi (Ocimum tenuiflorum) seeds: In vitro DNA damage protection, bioactive compounds and antioxidant potential. *Journal of Food Measurement and Characterization*, 12, 1530–1538.

Keykhosravy, K., Khanzadi, S., Hashemi, M., and Azizzadeh, M. (2020). Chitosan-loaded nanoemulsion containing Zataria Multiflora Boiss and Buniumpersicum Boiss essential oils as edible coatings: Its impact on microbial quality of turkey meat and fate of inoculated pathogens. *International Journal of Biological Macromolecules*, 150, 904–913. doi: 10.1016/j.ijbiomac.2020.02.092

Khameneh, B., Iranshahy, M., Soheili, V., et al. (2019). Review on plant antimicrobials: A mechanistic viewpoint. *Antimicrobial Resistance and Infection Control*, 8, 1–28.

Khan, M. R., Sadiq, M. B., and Mehmood, Z. (2020). Development of edible gelatin composite films enriched with polyphenol loaded nanoemulsions as chicken meat packaging material. *CyTA – Journal of Food*, 18, 137–146. doi: 10.1080/19476337.2020.1720826

Khanzadi, S., Keykhosravy, K., Hashemi, M., and Azizzadeh, M. (2020). Alginate coarse/nanoemulsions containing Zataria multiflora Boiss essential oil as edible coatings and the impact on microbial quality of trout fillet. *Aquaculture Research*, 51, 873–881. doi: 10.1111/are.14418

Kim, B. K., Cho, A. R., and Park, D. J. (2016). Enhancing oral bioavailability using preparations of apigenin-loaded W/O/W emulsions: In vitro and in vivo evaluations. *Food Chemistry*, 206, 85–91. doi: 10.1016/j.foodchem.2016.03.052

Kumar, D. D., Mann, B., and Pothuraju, R., et al. (2016). Formulation and characterization of nanoencapsulated curcumin using sodium caseinate and its incorporation in ice cream. *Food & Function*, 7, 417–424.

Kumar, S., and Pandey, A. K. (2013). Chemistry and biological activities of flavonoids: An overview. *The Scientific World*, J2013, 162750. doi: 10.1155/2013/162750

Landry, K. S., Komaiko, J., Wong, D. E., et al. (2016). Inactivation of Salmonella on sprouting seeds using a spontaneous carvacrol nanoemulsion acidified with organic acids. *Journal of Food Protection*, 79, 1115–1126. doi: 10.4315/0362-028X.JFP-15-397

Landry, K. S., Micheli, S., Julian, D., and Mclandsborough, L. (2015). Effectiveness of a spontaneous carvacrol nanoemulsion against *Salmonella enterica Enteritidis* and *Escherichia coli* O157 : H7 on contaminated broccoli and radish seeds. *Food microbiology*, 51, 10–17. doi: 10.1016/j.fm.2015.04.006

Lee, J., Kim, H., Choi, M. –J., and Cho, Y. (2020). Improved physicochemical properties of pork patty supplemented with oil-in-water nanoemulsion. *Food Science of Animal Resources*, 40, 262–273. doi: 10.5851/kosfa.2020.e11

Letsididi, K. S., Lou, Z., Letsididi, R., et al. (2018). Antimicrobial and antibiofilm effects of trans-cinnamic acid nanoemulsion and its potential application on lettuce. *LWT - Food Science and Technology*, 94, 25–32. doi: 10.1016/j.lwt.2018.04.018

Li, F. (2013). Discovery of survivin inhibitors and beyond: FL118 as a proof of concept. *International review of cell and molecular biology*. Elsevier, pp. 217–252.

Li, R., Peng, S., Zhang, R., et al. (2019a). Formation and characterization of oil-in-water emulsions stabilized by polyphenol-polysaccharide complexes: Tannic acid and β-glucan. *Food Research International*, 123, h266–275. doi: 10.1016/j.foodres.2019.05.005

Li, R., Tan, Y., Dai, T., et al. (2019b). Bioaccessibility and stability of β-carotene encapsulated in plant-based emulsions: Impact of emulsifier type and tannic acid. *Food & Function*, 10, 7239–7252. doi: 10.1039/c9fo01370a

Li, S., Pan, M. H., Lo, C. Y., et al. (2009). Chemistry and health effects of polymethoxyflavones and hydroxylated polymethoxyflavones. *The Journal of Functional Foods*, 1, 2–12. doi: 10.1016/j.jff.2008.09.003

Li, X., Yang, X., Deng, H., et al. (2020). Gelatin films incorporated with thymol nanoemulsions: Physical properties and antimicrobial activities. *International Journal of Biological Macromolecules*, 150, 161–168. doi: 10.1016/j.ijbiomac.2020.02.066

Li, Y., Zheng, J., Xiao, H., and McClements, D. J. (2012). Nanoemulsion-based delivery systems for poorly water-soluble bioactive compounds: Influence of formulation parameters on polymethoxyflavone crystallization. *Food Hydrocolloid*, 27, 517–528. doi: 10.1016/j.foodhyd.2011.08.017

Liao, Y., Zhong, L., Liu, L., et al. (2020). Comparison of surfactants at solubilizing, forming and stabilizing nanoemulsion of hesperidin. *Journal of Food Engineering*, 281, 110000. doi: 10.1016/j.jfoodeng.2020.110000

Liew, S.N., Utra, U., Alias, A.K., Tan, T.B., Tan, C.P., Yussof, N.S., (2020). Physical, morphological and antibacterial properties of lime essential oil nanoemulsions prepared via spontaneous emulsification method. *LWT - Food Science and Technology* 128, 109388. https://doi.org/10.1016/j.lwt.2020.109388

Liu, L., Gao, Y., McClements, D. J., and Decker, E. A. (2016). Role of continuous phase protein, (−)-epigallocatechin-3-gallate and carrier oil on β-carotene degradation in oil-in-water emulsions. *Food Chemistry*, 210, 242–248. doi: 10.1016/j.foodchem.2016.04.075

Liu, Q., Huang, H., Chen, H., et al. (2019). Food-grade nanoemulsions: Preparation, stability and application in encapsulation of bioactive compounds. *Molecules*, 24, 1–37.

Liu, Q., Zhang, M., Bhandari, B., et al. (2020). Effects of nanoemulsion-based active coatings with composite mixture of star anise essential oil, polylysine, and nisin on the quality and shelf life of ready-to-eat Yao meat products. *Food Control*, 107, 106771. doi: 10.1016/j.foodcont.2019.106771

Long, Y., Huang, W., Wang, Q., and Yang, G. (2020). Green synthesis of garlic oil nanoemulsion using ultrasonication technique and its mechanism of antifungal action against *Penicillium italicum. Ultrasonics Sonochemistry*, 64, 104970. doi: 10.1016/j.ultsonch.2020.104970

Maqsood, S., Benjakul, S., and Shahidi, F. (2013). Emerging role of phenolic compounds as natural food additives in fish and fish products. *Critical Reviews in Food Science and Nutrition*, 53, 162–179. doi: 10.1080/10408398.2010.518775

Martins, C., Higaki, N. T. F., Montrucchio, D. P., et al. (2020). Development of W1/O/W2 emulsion with gallic acid in the internal aqueous phase. *Food Chemistry*, 314, 126174. doi: 10.1016/j.foodchem.2020.126174

Mazzarino, L., da Silva Pitz, H., Lorenzen Voytena, A. P., et al. (2018). Jaboticaba (*Plinia peruviana*) extract nanoemulsions: Development, stability, and in vitro antioxidant activity. *Drug Development and Industrial Pharmacy*, 44, 643–651. doi: 10.1080/03639045.2017.1405976

McClements, D. J. (2012). Nanoemulsions versus microemulsions: Terminology, differences, and similarities. *Soft Matter*, 8, 1719–1729. doi: 10.1039/c2sm06903b

McClements, D. J. (2015). *Food emulsions: Principles, practices, and techniques*. CRC Press.

McClements, D. J., and Decker, E. A. (2000). Lipid oxidation in oil-in-water emulsions: Impact of molecular environment on chemical reactions in heterogeneous food systems. *Journal of Food Science*, 65, 1270–1282.

McClements, D. J., and Jafari, S. M. (2018). *General aspects of nanoemulsions and their formulation*. Elsevier Inc.

McClements, D. J., and Rao, J. (2011). Food-grade nanoemulsions: Formulation, fate, and potential toxicity food-grade nanoemulsions: Formulation, fabrication, properties, performance, biological fate, and potential toxicity. *Critical Reviews in Food Science and Nutrition*, 51, 285–330. doi: 10.1080/10408398.2011.559558

Mendes, J. F., Norcino, L. B., Martins, H. H. A., et al. (2020). Correlating emulsion characteristics with the properties of active starch films loaded with lemongrass essential oil. *Food Hydrocolloids*, 100, 105428. doi: 10.1016/j.foodhyd.2019.105428

Nemitz, M. C., von Poser, G. L., and Teixeira, H. F. (2019). In vitro skin permeation/retention of daidzein, genistein and glycitein from a soybean isoflavone rich fraction-loaded nanoemulsions and derived hydrogels. *Journal of Drug Delivery Science and Technology*, 51, 63–69.

Nile, S. H., and Park, S. W. (2014). Edible berries: Bioactive components and their effect on human health. *Nutrition*, 30, 134–144.

Noori, S., Zeynali, F., and Almasi, H. (2018). Antimicrobial and antioxidant efficiency of nanoemulsion-based edible coating containing ginger (*Zingiber officinale*) essential oil and its effect on safety and quality attributes of chicken breast fillets. *Food Control*, 84, 312–320. doi: 10.1016/j.foodcont.2017.08.015

Nunes, J. C., Melo, P. T. S., Lorevice, M. V., et al. (2020). Effect of green tea extract on gelatin-based films incorporated with lemon essential oil. *Journal of Food Science and Technology*. doi: 10.1007/s13197-020-04469-4

Panghal, A., Chhikara, N., Anshid, V., et al. (2019). Nanoemulsions: A Promising Tool for Dairy Sector. In Prasad, R., Kumar, V., Kumar, M., and Choudhary, D. (Eds) *Nanobiotechnology in Bioformulations* (pp. 99–117). Cham: Springer.

Patel, A., and Ghosh, V. (2020). Thyme (*Thymus vulgaris*) essential oil–based antimicrobial nanoemulsion formulation for fruit juice preservation. *Biotechnological applications in human health*. Singapore: Springer, pp. 107–114.

Pathania, R., Huma, K., Ravinder, K., and Mohammed Azar, K. (2018). Essential oil nanoemulsions and their antimicrobial and food applications. *Current Research in Nutrition and Food Science Current Research in Nutrition and Food Science*, 06, 626–643.

Pavoni, L., Perinelli, D. R., Bonacucina, G., and Cespi, M. (2020). An overview of micro- and nanoemulsions as vehicles for essential oils: Formulation, preparation and stability. *Nanomaterials Review*, 10, 1–24.

Puligundla, P., Mok, C., Ko, S., et al. (2017). Nanotechnological approaches to enhance the bioavailability and therapeutic efficacy of green tea polyphenols. *The Journal of Functional Foods*, 34, 139–151. doi: 10.1016/j.jff.2017.04.023

Punia, S., Sandhu, K. S., Siroha, A. K., and Dhull, S. B. (2019). Omega 3-Metabolism, absorption, bioavailability and health benefits-A review. *Pharma Nutrition*, 10, 100162.

Qadir, A., Faiyazuddin, M. D., Hussain, M. D. T., et al. (2016). Critical steps and energetics involved in a successful development of a stable nanoemulsion. *Journal of Molecular Liquids*, 214, 7–18.

Qian, C., and McClements, D. J. (2011). Formation of nanoemulsions stabilized by model food-grade emulsifiers using high-pressure homogenization: Factors affecting particle size. *Food Hydrocolloids*, 25, 1000–1008.

Quatrin, P. M., Verdi, C. M., de Souza, M. E., et al. (2017). Antimicrobial and antibiofilm activities of nanoemulsions containing *Eucalyptus globulus* oil against *Pseudomonas aeruginosa* and *Candida* spp. *Microbial Pathogenesis*, 112, 230–242. doi: 10.1016/j.micpath.2017.09.062

Rabelo, C. A. S., Taarji, N., Khalid, N., et al. (2018). Formulation and characterization of water-in-oil nanoemulsions loaded with açaí berry anthocyanins: Insights of degradation kinetics and stability evaluation of anthocyanins and nanoemulsions. *Food Research International*, 106, 542–548. doi: 10.1016/j.foodres.2018.01.017

Rao, G. (2020). *Studies on development of edible active packaging films incorporated with natural polyphenol nanoemulsions for pork patties*. Tirupati, India: Sri Venkateswara Veterinary University.

Rauf, A., Imran, M., Abu-Izneid, T., et al. (2019). Proanthocyanidins: A comprehensive review. *Biomedicine & Pharmacotherapy*, 116, 108999. doi: 10.1016/j.biopha.2019.108999

Rehman, A., Jafari, S. M., Tong, Q., et al. (2020). Role of peppermint oil in improving the oxidative stability and antioxidant capacity of borage seed oil-loaded nanoemulsions fabricated by modified starch. *International Journal of Biological Macromolecules*, 153, 697–707. doi: 10.1016/j.ijbiomac.2020.02.292

Ribeiro, M., Simões, L. C., and Simões, M. (2018). Biocides. In: Fourth, E. (Ed), *Schmidt TMBT-E of M*. Oxford: Academic Press, pp. 478–490

Robledo, N., López, L., Bunger, A., et al. (2018a). Effects of antimicrobial edible coating of thymol nanoemulsion/quinoa protein/chitosan on the safety, sensorial properties, and quality of refrigerated strawberries (Fragaria × ananassa) under commercial storage environment. *Food and Bioprocess Technology*, 11, 1566–1574. doi: 10.1007/s11947-018-2124-3

Robledo, N., Vera, P., López, L., et al. (2018b). Thymol nanoemulsions incorporated in quinoa protein/chitosan edible films; antifungal effect in cherry tomatoes. *Food Chemistry*, 246, 211–219. doi: 10.1016/j.foodchem.2017.11.032

Rodriguez Vaquero, M. J., Aredes Ferrnandez, P. A., Manca de Nadra, M. C., and Strasser de Saad, A. M. (2010). Phenolic compound combinations on Escherichia coli viability in a meat system. *Journal of Agricultural and Food Chemistry*, 58, 6048–6052.

Salvia-trujillo, L., Soliva-fortuny, R., Rojas-gra, M. A., et al. (2017). Edible nanoemulsions as carriers of active ingredients : A review. *The Annual Review of Food Science and Technology*, 8, 439–466. doi: 10.1146/annurev-food-030216-025908

Sessa, M., Luisa, M., Ferrari, G., et al. (2014). Bioavailability of encapsulated resveratrol into nanoemulsion-based delivery systems. *Food Chemistry*, 147, 42–50. doi: 10.1016/j.foodchem.2013.09.088

Shahavi, M. H., Hosseini, M., Jahanshahi, M., et al. (2019). Evaluation of critical parameters for preparation of stable clove oil nanoemulsion. *The Arabian Journal of Chemistry*, 12, 3225–3230.

Shehzad, Q., Rehman, A., Ali, A., et al. (2020). Preparation and characterization of resveratrol loaded nanoemulsions. *International Journal of Agriculture Innovations and Research*, 8, 300–310.

Shinoda, K., and Saito, H. (1968). The effect of temperature on the phase equilibria and the types of dispersions of the ternary system composed of water, cyclohexane, and nonionic surfactant. *Journal of Colloid and Interface Science*, 26, 70–74.

Shokri, S., Parastouei, K., Taghdir, M., and Abbaszadeh, S. (2020). Application an edible active coating based on chitosan- Ferulagoangulata essential oil nanoemulsion to shelf life extension of Rainbow trout fillets stored at 4 °C. *International Journal of Biological Macromolecules*, 153, 846–854. doi: 10.1016/j.ijbiomac.2020.03.080

Sogut, E. (2020). Active whey protein isolate films including bergamot oil emulsion stabilized by nanocellulose. *Food Packaging and Shelf Life*, 23, 100430. doi: 10.1016/j.fpsl.2019.100430

Sow, L. C., Tirtawinata, F., Yang, H., et al. (2017). Carvacrol nanoemulsion combined with acid electrolysed water to inactivate bacteria, yeast in vitro and native microflora on shredded cabbages. *Food Control*, 76, 88–95. doi: 10.1016/j.foodcont.2017.01.007

Su, D., and Zhong, Q. (2016). Lemon oil nanoemulsions fabricated with sodium caseinate and Tween 20 using phase inversion temperature method. *Journal of Food Engineering*, 171, 214–221.

Ting, Y., Chiou, Y. S., Pan, M. H., et al. (2015). In vitro and in vivo anti-cancer activity of tangeretin against colorectal cancer was enhanced by emulsion-based delivery system. *The Journal of Functional Foods*, 15, 264–273. doi: 10.1016/j.jff.2015.03.034

Treesuwan, W., Ichikawa, S., Wang, Z., et al. (2013). Formulation and storage stability of baicalein-loaded oil-in-water emulsions. *European Journal of Lipid Science and Technology*, 115, 1115–1122.

Tsai, Y. J., and Chen, B. H. (2016). Preparation of catechin extracts and nanoemulsions from green tea leaf waste and their inhibition effect on prostate cancer cell PC-3. *International Journal of Nanomedicine*, 11, 1907–1926. doi: 10.2147/IJN.S103759

Uçar, Y. (2020). Antioxidant effect of nanoemulsions based on citrus peel essential oils: Prevention of lipid oxidation in trout. *European Journal of Lipid Science and Technology*, 122, 1900405. doi: 10.1002/ejlt.201900405

Wan, J., Jin, Z., Zhong, S., et al. (2020). Clove oil-in-water nanoemulsion: Mitigates growth of *Fusarium graminearum* and trichothecene mycotoxin production during the malting of *Fusarium* infected barley. *Food Chemistry*, 312, 126120. doi: 10.1016/j.foodchem.2019.126120

Wang, X., Yu, K., Cheng, C., et al. (2019). Impact of sesame lignan on physical and oxidative stability of flaxseed oil-in-water emulsion. *Oil Corp Science*, 4, 254–266.

Wang, Z., Neves, M. A., Isoda, H., and Nakajima, M. (2015). Preparation and characterization of micro/nano-emulsions containing functional food components. *Japan Journal of Food Engineering*, 16, 263–276.

Weiss, J., Gaysinsky, S., Davidson, M., and McClements, J. (2009). Nanostructured encapsulation systems: Food antimicrobials. *Global issues in food science and technology*. San Diego: Elsevier, pp. 425–479.

Wu, D., Kong, Y., Han, C., et al. (2008). D-Alanine: D-alanine ligase as a new target for the flavonoids quercetin and apigenin. *The International Journal of Antimicrobial Agents*, 32, 421–426.

Xu, G., Wang, C., and Yao, P. (2017). Stable emulsion produced from casein and soy polysaccharide compacted complex for protection and oral delivery of curcumin. *Food Hydrocolloids*, 71, 108–117. doi: 10.1016/j.foodhyd.2017.05.010

Xu, X. (2014). *Improve Bioaccessibility of Quercetin Using Pseudo-Organogel Based Nanoemulsions*. Graduate School- Rutgers University, New Brunswick.

Yin, J., Xiang, C., Wang, P., Yin, Y., and Hou, Y. (2017). Biocompatible nanoemulsions based on hemp oil and less surfactants for oral delivery of baicalein with enhanced bioavailability. *International Journal of Nanomedicine*, 12, 2923–2931. doi: 10.2147/IJN.S131167

Zhao, S., Tian, G., Zhao, C., et al. (2018). The stability of three different citrus oil-in-water emulsions fabricated by spontaneous emulsification. *Food Chemistry*, 269, 577–587.

Ziaullah, R. H. P. V. (2015). Application of NMR spectroscopy in plant polyphenols associated with human health. In Ur-Rahman, A., Choudhary, M.I.B.T.-A. of NMRS (Eds.), *Applications of NMR spectroscopy*. Bentham Science Publishers, vol. 2, pp. 3–92.

9 Essential Oil Nanoemulsions

As Natural Antimicrobial Agents

Kiran Bala, Sanju Bala Dhull, and Sneh Punia
Chaudhary Devi Lal University, India
Aradhita Barmanray
Guru Jambheshwar University of Science and Technology, India

CONTENTS

9.1 INTRODUCTION

Currently, in spite of the availability of a large number of preservation techniques, food poisoning and spoilage by microorganisms are major concerns in the food sector (Ghosh et al., 2013a). Consumer needs of natural, wholesome, and safe food free from microorganisms with greater shelf life have raised the interest in the use of natural antimicrobials to preserve food. Therefore, the use of natural antimicrobial substances (bioactive compounds which hinder the growth of spoilage causing microorganisms) as food preservatives, is the latest topic of research for the food scientists and technologists. During recent years, the use of essential oils as natural antimicrobial agents in food has gained much popularity which offers a good prospectus to meet the need of natural preservatives. The exponential growth rate (5% annually) of the essential oil industry supports their widespread use for the present scenario (Pavoni et al., 2020). Essential oils, which are commonly used in the food, cosmetics, and pharmaceutical industries, are plant-derived secondary metabolites containing a mixture of volatile and non-volatile components (Punia et al., 2019). Due to the presence of biochemical compounds, these oils show antioxidant, antibacterial, antiviral, and antifungal properties (Martin-Pinero et al., 2019). These are generally recognized as safe (GRAS) for humans and animals use in form of flavouring compounds in the United States (Hyldgaard et al., 2012; Shah et al., 2012; Bhargava et al., 2015).

Owing to very high antimicrobial efficiency, essential oils have gained noteworthy consideration in the field of food preservation, and the working capacity of essential oils gets increased when encapsulated in proper delivery systems. There are many benefits of encapsulation of essential oils in nanoemulsions form, and recently, most of the essential oil nanoemulsions are prepared using synthetic surfactants. The uses of a nanometric delivery system for essential oil encapsulation have many additional benefits over conventional macroemulsions. Owing to their small droplet size, nanoemulsions possess good practical importance, as these appear transparent and are more stable as compared to conventional emulsions with respect to coalescence, cream, flocculation, and Ostwald ripening (Lu et al., 2018; Panghal et al., 2019). These are stable colloidal systems of very small-sized droplets (ranging 20–500 nm) found dispersed in a continuous phase which may be either aqueous (oil-in-water) or oily (water-in-oil), and these emulsions are stabilized by amphiphilic molecules called surfactants (Moreno-Trejo et al., 2019; Dhull et al., 2019). In nanoemulsions, the physical stability of essential oils is enhanced due to prevention of interaction with other food components, and it also overcomes dosage limitations problem with their low water solubility (Sessa et al., 2015). Additionally, the nanoemulsions increase the bioactivity of essential oils, refining the passive mechanisms of cell absorption, allowing the reduction of doses needed to show antimicrobial activity, therefore minimizing the influence on taste, flavour, and aroma (Donsì et al., 2011). This chapter mainly focuses on some important points regarding characteristics, formulation, antimicrobial efficiency, mechanism of microbial action, and safety regulations of essential oil nanoemulsions.

9.2 MAJOR CHALLENGES ASSOCIATED WITH FOOD APPLICATION OF ESSENTIAL OIL NANOEMULSION

Even though the use of essential oils emulsions as natural antimicrobial agents provides a good alternative to chemical preservatives, high volatile nature, little water solubility, and strong odour are some major limitations linked with their use in food products. In addition to aforementioned problems, incorporation of oil-based substances in aqueous food products is also a major technological issue, owing to their chemical and physical instability and lipophilic nature (Amaral and Bhargava, 2015; Chawla et al., 2019, 2020). Thermodynamically, nanoemulsions are unfavourable systems as in generating oil–water interface; positive free energy gets escorted, resulting in breakdown with time. There are some physicochemical phenomena, that is gravitational separation, flocculation, coalescence, and Ostwald ripening, associated with splitting of nanoemulsions. Some forces (inter-particle repulsive, attractive, gravitational, molecular, and flow forces) are related to these destructive processes (Karthik et al., 2017; Prakash et al., 2018). Therefore, nanoemulsions can be a better option for food use of essential oils, which is suggested by a lot of studies reported in the literature.

9.3 COMPOSITION AND CHARACTERISTICS OF ESSENTIAL OILS

Essential oils have been used as flavours and fragrances since ancient times, owing to their aromatic and volatile characteristics, and these are commonly extracted from various plant parts like leaf, seed, bark, and fruit, as well as the root, using steam distillation, hydro-distillation, dry distillation, and mechanical pressing methods (Burt, 2004; Bauer et al., 2008; Xue, 2015). These hydrophobic liquids are complex mixtures of 20 to 60 low molecular weight compounds, including terpenes (limonene, α-pinene), terpenoids (thymol, linalool), phenylpropanoids (eugenol, cinnamaldehyde), and others (allicin), depending on their chemical structure (Hyldgaard et al., 2012; Prakash et al., 2018). A lot of studies reported in the literature indicated the antimicrobial potential of essential oils. Eugenol, limonene, carvacrol, thymol, citral, anethole, carvone, chitosan, nisin, and estragole are examples of some natural food antimicrobials. Despite their amazing benefits, the use of essential oils is limited because of some properties like decomposition at elevated temperatures, highly volatile nature, and less water solubility, as well as instability at storage shelf life (Pavoni et al., 2020). Developments of suitable formulations may provide a solution for this problem, and their use can be enhanced globally. The preparation of nanoemulsions using encapsulation technique is a good approach due to easy formation, handling, and cost-effectiveness. The appropriate selection of essential oils for the incorporation into food products should be according to organoleptic characteristics of the final food products, and the amount of essential oils addition must be controlled to prevent the toxicological influences and for consumer acceptance, as well as for an economical point of view.

9.4 ANTIMICROBIAL EFFICACY OF ESSENTIAL OILS

Since ancient times, essential oils have been used as antimicrobial, anti-inflammatory, antioxidant, expectorant, digestive, and diuretic agents and as food preservatives and insecticides (Pavoni et al., 2020). The use of natural compounds in place of synthetic ones is the major demand of the present. In this regard, the use of essential oils as natural antimicrobials will be a better choice from a safety point of view. The antimicrobial property of essential oils is mainly attributed to their phenolic compounds and interaction with microbial cell membranes. Essential oils penetrate through bacterial cell membrane which is responsible for the leakage of ions and cell contents, resulting in cellular decomposition (Burt, 2004). For the addition of these oils as preservatives in foods and use against human pathogens, cytotoxic studies are very important. A literature survey suggested that essential oils are effective against a broad range of

TABLE 9.1
Examples of Food Systems Used Essential Oils as Antimicrobials

Essential Oil	Targeted Microorganism	Food	References
Oregano oil	*Listeria monocytogenes, Salmonella Typhimurium, Escherichia coli* O157:H7	Lettuce	Bhargava et al. (2015)
Carvacrol	*S. enteric, Enteritidis E. coli*	Broccoli and radish seed	Landry et al. (2015)
Carvacrol	*S. enteric, Enteritidis E. coli*	Mung bean and alfalfa seeds	Landry et al. (2014)
Mandarin oil	*L. innocua*	Green beans	Severino et al. (2014)
Mandarin oil	*L. innocua*	Green beans	Donsi et al. (2015)
Lemongrass oil	*S. typhimurium, E. coli*	Plums	Kim et al. (2013)
Cinnamaldehyde	*Lactobacillus delbrueckii, Saccharomyces cerevisiae, E. coli*	Apple and pear juice	Donsì et al. (2012)
Eugenol	*E. coli* O157:H7, *L. monocytogenes*	Fruit juice	Ghosh et al. (2014)

Source: Adapted from Amaral and Bhargavaw (2015).

disease-causing microorganisms like *L. monocytogenes*, *E. coli* O157:H7, *L. innocua*, *Bacillus cereus*, *Shigella dysenteria*, *Salmonella typhimurium*, and *Staphylococcus aureus* (Guerra-Rosas et al., 2017). Some examples of food systems which use essential oils as antimicrobial agents are listed in Table 9.1.

Lemongrass essential oils has been reported effective against various food pathogens in minimally processed foods, fruit juices, chocolate, minced meat, and fish products (Salvia-Trujillo et al., 2014a). Antimicrobial agents can also be added to the food packaging materials using nano-encapsulation mechanisms (Appendini and Hotchkiss, 2002; Blanco-Padilla et al., 2014). Essential oils face difficulty in direct application to food because the reactive components of oils deteriorate when they come into contact with the proteins, fats, and minerals of food. Additionally, antimicrobial efficacy gets decreased with uneven dispersal in real foods. Therefore, there is a need for enhancing the antimicrobial efficiency of essential oils using various encapsulation systems. The nano-encapsulation process acts as a vehicle for delivery of essential oils, and because of very small size, high tissue penetration power, and simple cellular uptake process, these nanocarriers boost the bioactivity of essential oils (Pavoni et al., 2020). Low-energy methods are most commonly employed for the preparation of antimicrobial nanoemulsions in a very easy and economical way. For understanding the concept, consider that if any producer or processor is provided with a liquid containing a mixture of antimicrobial oil and surfactant, when added to a container of water, it will start the formation of an antimicrobial nanoemulsion automatically, and furthermore, its application on the surface of commodities, either by dipping or spraying, followed by washing with water to clean residual particles (Komaiko and McClements, 2016) will be very effective process.

9.5 PREPARATION OF NANOEMULSIONS

Knowledge about the science behind the preparation of nanoemulsion is very necessary for controlling droplet size. Basically, it is a two-step process in which, first of all, macroemulsion is prepared and, in the second step, is converted to a nanoemulsion. In nanoemulsions, an oil centre remains surrounded by single or multiple layers of surfactant molecules which attain spherical shape and are arranged in such a manner that polar heads suspend toward the aqueous medium whereas a non-polar tail extends towards the lipophilic nucleus. Nanoemulsions are kinetically stable systems (aqueous and oils phases coexist) and can be stored for a long time with maximum stability at 4°C and 25°C (Nirmal et al., 2018). Essential oils have low interfacial tension, low viscosity, and high polarity, which render them suitable for the formation of nanoemulsions. With the help of suitable surfactants (surface-active particles which consist of both hydrophilic and lipophilic part in their molecular structure) and mechanical shearing, immiscible phases are mixed in, which one phase gets dispersed in another preventing coalescence (Nor Bainun et al., 2015). Food-grade nanoemulsions can be formulated using various methods based on required structure and functionality. The preparation procedure of nanoemulsions using various methods is shown in Figure 9.1.

Two techniques (high energy and low energy) are most commonly employed for nanoemulsions preparation. High-energy methods, including microfluidization, ultrasonication, and high-pressure homogenization, create strong disruptive forces to reduce the droplet size, where a slow-energy methods (solvent diffusion) are helpful in blending all emulsion constituents supporting spontaneous emulsification (Blanco-Padilla et al., 2014). Figure 9.2 is a pictorial representation of high-pressure homogenization technique used for nanoemulsion formulation.

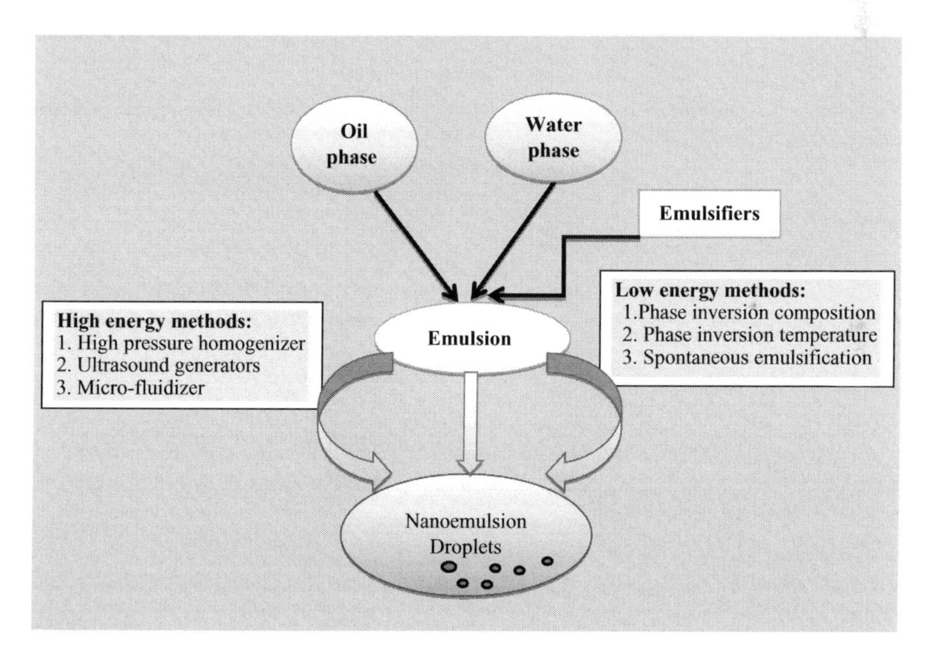

FIGURE 9.1 Schematic representation of nanoemulsion preparation using various methods.

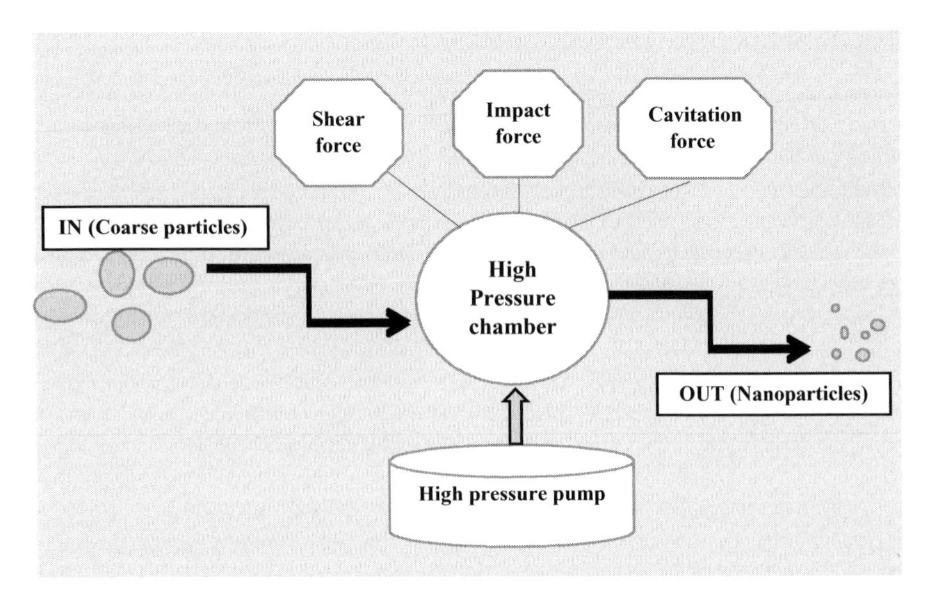

FIGURE 9.2 Preparation of nanoparticles using high pressure homogenization technique.

High-energy methods are most commonly used in the food sector, owing to their scale-up and equipment availability and highly versatile nature; also, these methods also produce nanoemulsions without including organic solvents. However, this method is not cost-effective, as the energy requirement is very high. In these processes, the oil phase is divided into uniform and smooth fine droplets using mechanical size reduction (coarse droplets converted to smaller ones). Ultrasound is a feasible technique for the formulation of lemongrass oil–alginate nanoemulsions with the desired translucency and droplet size (Salvia-Trujillo et al., 2013a). In low-energy approaches (such as spontaneous emulsification, emulsion phase inversion, and phase inversion temperature), molecular building blocks get directly added to structured systems based on thermodynamic equilibrium, and nanoemulsions are formed from intrinsic physicochemical characteristics of the oil phase and surfactants. This emulsification method is cost-effective as internal chemical energy of system is used and very simple stirring is required. In addition to this, a smaller droplet size is produced in a low-energy process as compared to a high-energy process (Solans and Solé, 2012). Low-energy methods have some drawbacks. For example the type of emulsifiers and oils used are limited, and a greater quantity of surfactants is required. The use of stabilizers, including emulsifiers (surfactants, gum arabic, soy, egg, caseinate), ripening reducing agents, texture improvers (gelatin, whey protein isolate, sucrose, high fructose corn syrup, pectin, xanthan, glycerol), and weighting agents (brominated vegetable oil, ester gums) help in the enhancement of the kinetic stability of nanoemulsions (Salem and Ezzat, 2018). Lecithins, polysorbates, sugar esters or biopolymers, animal or vegetables proteins, and natural gums are the examples of some generally used food-grade surfactants (emulsifying agents). In nanoemulsions, 5% to 10%is most commonly used concentration of surfactants. For preparation of food-grade nanoemulsions, microfluidization is most commonly employed emerging technique. In this technique, with the help of a high-pressure pump, a coarse emulsion is passed through interaction chamber which contains two flow channels arranged in such a manner so that both streams impose on each other at very high speed. It produces very high shearing action, and fine emulsion obtained after that (Salvia-Trujillo et al., 2013b). Bursting of a bubble at an oil/water interface and evaporative ripening are also recently developed technique used in nanoemulsion preparation (Gupta et al., 2016). Various techniques are available to measure the particular properties of nanoemulsions. For example droplet size is measured using dynamic light scattering (DLS) spectrophotometer, transmission electron microscopy (TEM), and scanning electron microscopy (SEM). Stability is assessed by zeta potential, and the microstructure is determined using TEM, SEM and atomic force microscopy (AFM; Silva et al., 2012).

9.5.1 Types of Delivery Systems Used for Nanoemulsions

Surface characteristics and particle size of droplets are the two major factors which decide the biological performance of nanoemulsion formulations (Nor Bainun et al., 2015). For the nano-encapsulation of active constituents of essential oils, a large number of delivery systems have been developed involving nanoemulsions, nanoparticles (nanosphere, nanocapsule, solid-liquid nanoparticle), liposomes, nanofibres, and so on. Among the encapsulations systems, emulsions are desirable delivery systems for many applications due to a very small oil droplet size. Nanoemulsions show the following advantages:

- Kinetic stability which prevent particle aggregation and gravitational separation (Guerra-Rosas et al., 2017)
- Very rapid absorption, toxicologically safe, carriers of hydrophobic components, and suitable to use in beverages owing to transparency (Blanco-Padilla et al., 2014)

Natural antimicrobial agents can be properly protected from adverse environmental and other unsuitable conditions by the nano-encapsulation process which possesses control delivery, enhancing antimicrobials concentration and passive cellular absorption procedures caused higher antimicrobial activity (Blanco-Padilla et al., 2014). Oil-in-water (O/W) nanoemulsions are commercially useful delivery systems because these have a high surface area and small size and are optically clear, as well as having a low rate of gravitational separation and flocculation characteristics (Sharif et al., 2017) which increase their functionality and interaction with biological systems. However, there are also some disadvantages of nanoemulsions like these emulsions are less stable in acidic conditions, very fast release of active constituents, and limited use of low-energy methods in the food sector.

9.5.2 Use of Nanoemulsions as Edible Coating on Foods

Since earlier times, the shelf life of food products has been increased using edible coatings as these create a barrier between the environment and food thus by forming a physical barrier between the environment and food and regulating mass transfer. Owing to a problem related to the immobilization of nano-droplets on the solid food surface, there is a need for suitable methods for incorporating active ingredients on the surface of foods. Therefore, in this regard, nanoemulsion-based edible coatings would be an effective approach. These coatings offer several extra advantages like superior film homogeneity, better antimicrobial property, less compound dosage, reduced interactions with other food components, improved stability to the compounds under stressing situations, and lower mass transfer rate through the coating (Prakash et al., 2018). The impact of a nanoemulsion-based edible coating on minimally processed apples showed positive advantages of slower respiration rate and psychrophilic bacterial growth rate as well as a higher rate of *E. coli* inactivation (Salvia-Trujillo et al., 2015). A combination of modified chitosan-based coatings having mandarin essential oils and high hydrostatic pressure applied on green beans showed significant antimicrobial synergistic influence against *L. innocua* over a long storage period (Donsi et al., 2015).

However, some controversial reports regarding antimicrobial influence of nanoemulsion-based coatings on foods are also found in the literature. Montmorillonite and ginger essential oil nanoemulsions were added to gelatin-based films, and these nanocomposite-activated films showed no antimicrobial properties but good antioxidant activity (Alexandre et al., 2016). The latest advancements in the nanoemulsion-based edible coatings possess great market potential for health-conscious consumers (Dasgupta et al., 2019). Low-fat cut cheese showed increased shelf life, quality, and nutritional characteristics when coated with a nanoemulsion-based coating prepared with oregano essential oil (2.0% w/w) and mandarin fibre (Artiga-Artigas et al., 2017). A modified chitosan-based bioactive coating having 0.05% mandarin essential oil when combined with UV-C and gamma irradiation proved to be a very effective way of reducing the growth of *Listeria innocua* on surface of green beans (Severino et al., 2014). Nanoemulsion-based sodium caseinate–edible coating containing ginger (*Zingiber officinale*) oil (6%) significantly influences the quality and shelf life of chicken breast fillets (raw poultry meat) by reducing total aerobic psychrophilic bacteria, yeasts, and moulds (Noori et al., 2018). More antimicrobial properties of cinnamaldehyde nanoemulsion can be obtained by controlling droplet size, and effective edible films can be prepared from fruit processing wastes (pectin and papaya puree) for obtaining antimicrobial packaging material (Otoni et al., 2014).

9.6 ANTIMICROBIAL POTENTIAL OF ESSENTIAL OIL NANOEMULSIONS

Antimicrobial action of essential oil nanoemulsions is stronger and wide in spectrum compared to free essential oils. A lot of studies are reported in the literature regarding the antimicrobial efficacy of essential oil nanoemulsions. The self-assemble potential of sodium caseinate was used for nanoemulsion preparation after neutralization of alkaline thyme oil and caseinate mixture (Zhang and Zhong, 2020). Encapsulated thyme oil was found more effective against *E. coli* O157: H7 and *Staphylococcus aureus* compared to free oil in milk samples, and it was therefore concluded that thyme oil nanoemulsions can be effectively used as natural preservatives. Clear and transparent nanoemulsions of lemon myrtle essential oil (extracted from *Backhousia citriodora*) prepared using ultrasonication showed very good antibacterial activity against gram-positive (*S. aureus–ATCC 33591, L. monocytogenes–ATCC 19111*) and gram-negative (*E. coli–ATCC 11775, Pseudomonas aeruginosa–ATCC 9626*) whereas anise myrtle essential oils nanoemulsions were not found effective against previously mentioned microorganisms (Nirmal et al., 2018). Antimicrobial efficiency of nanoemulsions having oregano, lemongrass, thyme, and mandarin essential oils and high methoxy pectin against *E. coli* and *L. innocua* indicated that it was dependent on type of oil instead of droplet size and gradually reduced during storage period owing to loss of volatile components (Guerra-Rosas et al., 2017). Table 9.2 compiles the information about the use of nanoemulsions having essential oils with their targeted microorganisms.

A food-grade nanoemulsion (droplet diameter, 29.6 nm) of basil oil (*Ocimum basilicum*) prepared using Tween 80 and ultrasonication changed permeability and surface features of *E. coli* in a very impressive way (Ghosh et al., 2013a). In another study, Ghosh et al. (2013b) demonstrated that cinnamon oil nanoemulsion (droplet diameter, 65 nm) prepared in a similar manner as that of basil oil, possessed bactericidal action against *Bacillus cereus* and therefore can be used for preservation of minimally processed food. Nanoemulsions containing eugenol were found effective against *S. aureus* in orange juice samples (Ghosh et al., 2014). Antimicrobial essential oil nanoemulsions of lemongrass (*Cymbopogon citratus*) prepared with low-energy methods were stabilized with chitosan and oleic acid, along with enhancement of antibacterial activity against 9 bacterial and 10 fungal strains (Bonferoni et al., 2017). Stable cinnamon bark oil nanoemulsions can be successfully formulated using a combination of Tween 80 and lauric arginate surfactants without influencing their antimicrobial potential and can be used in complex food systems like milk (Hilbig et al., 2016). In addition to this, the combined use of nanotechnology and synergistic influence of systems containing more than one antimicrobial component (D-limonene and nisin) showed very good physical stability and antimicrobial properties (Zhang et al., 2014). Zein nanoparticles encapsulated with essential oil were observed to have more water dispersion (14-fold) which increased their preservative action against human disease–causing bacteria like *E. coli* (Wu et al., 2012). A nanoemulsion of clove essential oil was also proved as effective antibacterials by Anwer et al. (2014). Likewise, thymol nanoemulsions prepared using spontaneous emulsification process were also act as antimicrobial agents against some common and foodborne pathogens and provide a different sanitizing process which can be used as agricultural sprays, aerosols, and washes (Li et al., 2017). *Trans*-cinnamaldehyde nanoemulsion (0.8 wt%) retarded the activity of *S. typhimurium* and *S. aureus* in watermelon juice (Jo et al., 2015).

Biopolymer-based (basil seed gum) edible films prepared with an added nanoemulsion of *Zataria multiflora* essential oil were found effective against gram-positive and -negative bacteria and increased the shelf life by reducing the release of volatile compounds (Gahruie et al., 2017). Anise oil (Topuz et al., 2016), *Citrus medica L.* var. sarcodactylis essential oil (Lou et al., 2017), negatively charged food-grade clove oil (Majeed et al., 2016), black cumin essential oil (Sharif et al., 2017), peppermint oil (Liang et al., 2012), tea tree oil (Li et al., 2016), *Thymus daenensis* essential oil (Moghimi et al., 2016a), sage (*Salvia officinalis*) essential oil (Moghimi et al., 2016b),

TABLE 9.2
List of Purpose of nanoemulsions Containing Essential Oil with Surfactant Used for Preparation and their Targeted Microorganisms

Essential Oil	Targeted Microorganism	Purpose	Surfactant Used	References
Eucalyptus	*Proteus mirabilis*	Pharmaceuticals	Tween 20	Saranya et al. (2012)
Clove	*Escherichia coli*, *Staphylococcus aureus*, *Salmonella typhi*, and *Pseudomonas aeruginosa*	Pharmaceuticals, foods, and disinfectants	Tween 20	Hamed et al. (2012)
Basil	*E. coli*	Foods	Tween 80	Ghosh et al. (2013)
Sunflower	*Lactobacillus delbrueckii* and *Saccharomyces cerevisiae*	Foods	Tween 20	Donsì et al. (2011)
Palm	*L. delbrueckii*, *S. cerevisiae*, and *E. coli*	Foods	Tween 20 and glycerol monooleate	Baldissera et al. (2013)
Crab wood	*Trypanosoma evansi*	Pharmaceuticals	Span 80 Tween 80	Shah et al. (2012)
Peppermint	*S. aureus* and *Listeria monocytogenes*	Foods	Triglycerol	Gaysinsky et al. (2005)
Thyme	*E. coli* and *L. monocytogenes*	Foods	Span 20	Ziani et al. (2011)
Anise	*L. monocytogenes* and *E. coli*	Food safety	Triglycerides	Orav et al. (2008)
Lemon	*L. monocytogenes* and *S. typhimurium*	Pharmaceuticals cosmetics, foods, and disinfectants	Tween 80	Kriegel et al. (2010)
Savory	*S. aureus*, *L. monocytogenes*, and *Bacillus cereus*	Food safety	Tween 80	Tozlu et al. (2011)
Clove bud Disinfectants	*B. cereus* and *L. monocytogenes*	Cosmetics and foods	Tween 20	Baldissera et al. (2013)
Tea tree and Sage	*Trichophyton rubrum.*	Pharmaceutics	Tween 80	Flores et al. (2013); Deena and Thoppil (2000)
Neem	*Vibrio vulnificus*	Pharmaceutical	Tween 20	Ghotbi et al. (2014)

Source: Adapted from Pathania et al. (2018).

oregano oil (Bhargava et al., 2015), *Pimpinella anisum* oil (Hashem et al., 2018), thyme oil (Chang et al., 2015), and citral essential oil (Lu et al., 2018) were found effective against various microorganisms studied. Therefore, it is cleared that essential oil nanoemulsions are promising tools for food preservation. In contrast to the earlier-mentioned studies, some controversial reports are also presented in the literature. In a study regarding nanoemulsions of thymol and eugenol, which were co-emulsified by lecithin and arginate, Ma et al. (2016) determined that gram-negative bacteria (*E. coli* O157:H7) was not affected by mixture of lauric arginate with lecithin in milk. However, this combination was a good idea for nano-emulsifying essential oil components for addition to foods which require much transparency.

9.6.1 Mechanism of Action of Antimicrobials

Surface charge, droplet size, and the preparation of nanoemulsions influence the movement of essential oils to the cell membrane and their interaction at different sites of microbial cell membranes. The dispersion rates of essential oil in aqueous phase get enhanced due to nanoemulsion formation which causes increases in the antimicrobial property when a minimum inhibitory concentration is above water solubility. Many authors also described some controversy about this statement. Generally, four ways of nanoemulsions interaction with microbial cell membranes are observed (Donsì and Ferrari, 2016):

- Passive transport through the cell membrane
- Fusion with the phospholipid bilayer
- Partition in the aqueous phase
- Electrostatic interaction with the cell membrane

When nanoemulsion droplets containing essential oil as an active constituent come in contact with bacterial cell wall, these get diffused into the cell wall, and lysis of the wall takes place, followed by the release of cytoplasmic cell components and death of bacteria. The influence (positive or negative) of encapsulation in nanoemulsions on antimicrobial properties of essential oils is shown in Table 9.3. Further studies are needed to fully understand the mechanism of essential oils work on microbes which will support the formation of better nanoemulsions in terms of the types of essential oil used and about target pathogen.

9.6.2 Factors Influencing Antimicrobial Efficiency of Nanoemulsions

There are a lot of factors affecting overall antimicrobial efficacy of essential oil nanoemulsions including kinds of essential oils and emulsifiers, ripening inhibitor, and method of fabrication used. All these factors may change the basic characteristics, that is droplet size, and the surface charge of nanoemulsions resulted in reduced interaction with microbial cell membranes. The type of ripening inhibitor and concentration used affect the antimicrobial activity of nanoemulsions. Higher concentrations reduce antimicrobial activity. Thyme oil nanoemulsions prepared by adding corn oil as a ripening inhibitor were found effective against spoilage yeast (*Zygosaccharomyces*

TABLE 9.3
Report Related To Mechanisms of Antimicrobial Action for Various Nanoemulsions Containing Essential Oils, With Respect to their Formulation, Droplet Size and Microbial Strain

Oil Phase	Emulsifier	Mean droplet Size (nm)	Microbial strain	Impactofnano-emulsion	Mechanism	Reference
Carvacrol/ cinnamaldehyde/ D-limonene	Tween 20–monoolein/sugar esters/lecithin/Pea proteins	170–240	*Saccharomyces cerevisiae*, *Escherichia coli*, *Lactobacillus delbrueckii*	+	Effect of formulation on the solubilization of essential oils (EOs), improving their interaction with the cell membrane Nanoemulsion droplets behave as nano tanks for sustained release of EOs.	Donsìet al. (2012)
Clove oil and canola oil	Modified starch	150	*Staphylococcus aureus*, *E. coli*, *Listeria monocytogenes*	+/–	Higher activity of EO nanoemulsion than free EO only against gram-positive due to the interaction of modified starch with their cell wall.	Majeed et al. (2016)
Lemongrass oil	Tween 80 and sodium alginate	5	*E. coli*	++	Association of the emulsifier with some constituents of the biological membrane promotes the antimicrobial activity of EO nanoemulsions.	Salvia-Trujillo et al. (2014a, b)
D-limonene	Tween 80	16	*S. cerevisiae*, *S. aureus*, *Bacillus subtilis*, *E. coli*	++	Higher antimicrobial activity of EO nanoemulsion than free EO	Zhang et al. (2014)
Peppermint oil and MCT	Modified starch	146	*L. monocytogenes*, *S. aureus*	–	Nanoemulsions (a) prolong antibacterial activities through sustained release and(b) limit the contact of EO with the membrane of bacteria, decreasing their concentration in aqueous phase	Liang et al. (2012)

Eugenol	Whey protein and maltodextrin conjugates	127	*E. coli, L. monocytogenes*	– –	Binding between emulsifier and EO reduces the resulting antimicrobial activity	
Eugenol and bean oil	Tween 80/SDS	140	*E. coli*	++	Nanoemulsion enhanced antibacterial activity of EO. Anionic surfactant (SDS) was more active than non-ionic (Tween 80)	Li et al. (2015)
Eugenol and MCT	Tween 80	80–3000	*E. coli*	– –	Smaller emulsion droplets reduce the EO antimicrobial activity due to EO accumulation at droplet interfaces.	Terjung et al. (2012)
Tea tree oil	Lecithin	175–74	*S. cerevisiae, E. coli, L. delbrueckii*	+ +	Nanoemulsions improve the dispersibility of poorly soluble EOs.	Donsì et al. (2011)
Thyme oil	Tween 80 + lecithin	143	*E. coli*	+ +	Nanoemulsions improve the access of EOs to bacterial cells and (b) their ability to disrupt the cell membrane.	Moghimi et al. (2016)
Thyme oil	Sodium caseinate and soy lecithin	82	*E. coli, S. enteritidis, L. monocytogenes*	+	Binding between emulsifier and EO promotes antimicrobial activity;(b) Nanoemulsions improve the dispersibility of poorly soluble EOs.	Xue et al. (2015)
D-limonene	Tween 80 + monoolein/modified starch	155–366	*S. cerevisiae, E. coli, L. delbrueckii*	+	Solubility of EOs is increased in the aqueous phase.	Donsì et al. (2011)

Source: Adapted from Donsì and Ferrari (2016).

bailii) but were influenced by the type and concentration of inhibitor (Chang et al., 2012). High-energy emulsification treatments like ultrasound and microfluidization can effectively produce nanoemulsion sizes ranging from 30 to 600 nm. But nanoemulsions of lemongrass oil–alginate having better functionality and more effectiveness against *E. coli* were obtained with microfluidization as compared to ultrasonication (Salvia-Trujillo et al., 2014a). The effect of emulsion (i.e. lemon myrtle oil and soybean oil) droplet size on antimicrobial efficacy against five bacteria was analysed. Reported results clarified that the antimicrobial property is mainly due to active constituents present in the emulsion and is not influenced by high surface tensions, as well as cell wall diffusion activity of nano-sized droplets (Buranasuksombat et al., 2011). The influence of delivery systems on the antimicrobial property of carvacrol, cinnamaldehyde, and limonene encapsulated in sunflower oil droplets was analysed, and on the basis of the researchers' observations, it can be concluded that the antimicrobial activity of molecules depends on their dissolution concentration in the aqueous system, which is further influenced by the selection of emulsifier (Donsì et al., 2012).

9.7 SAFETY AND REGULATORY ISSUES REGARDING NANOEMULSIONS

It is a difficult task to determine the safety of nanoscale particles (such as nanoemulsions). Therefore, very little research has been conducted regarding their toxicity and safety. Toxicological influences of nanoemulsions on biological systems depend on many factors like composition, structure (particle size and charge), and method of administration (Wani et al., 2018). Synthetic surfactants (lactic acid esters, lauric arginate, Tweens, etc.) and some other agents (weighting agents, ripening inhibitors, and solvents), when used in greater proportion, may cause the toxicity or irritation. These may cause enhanced cell permeability rates, interruption of active absorption, loose junctions, and changes in the uptake of bioactive molecules. The influence of nanoemulsions on biological systems is briefly described in Table 9.4.

Although nanotechnology is a promising tool for utilization of essential oils, a lack of scientific evidence and reliable evaluation regarding their influence on human health, as well as on the environment, enforce a limit on their broad application (Pavoni et al., 2020). On the basis of various studies reported in the literature, it is concluded that the application of nanoemulsions is safe except for situations in which very high concentrations are used. The nano-regulations are still in immature, and no global regulatory authority exists in this field yet; however, in Switzerland and the European Union, regulations have been developed and enforced (Wani et al., 2018). Therefore, there is a great need for the development of nano-regulations for the analysis and elaboration of safety measures of nano products so that additional benefits from nanotechnology can be achieved and applied for the betterment of consumer health.

9.8 CONCLUDING REMARKS

The use of nanoemulsions in the confectionery, nutraceutical, dairy, beverage, and food-packaging industries has a great potential to fulfil the consumer needs of the modern era. Recently, antimicrobial aspect of essential oils is in limelight for the preservation purpose

TABLE 9.4
Effect of Nanoemulsions on Biological Processes

Study	Bioactive Compounds	Surfactants Used	Size	Subject/Cell Line	Result	References
Hepaticgene expression and serum metabolites	Vitamin E	–	112 nm	Male Sprague Dawley rats	Alteration of energy and xenobiotic metabolism, formation of ketone bodies and elevation of stearoyl-CoA desaturase	Park et al. (2017)
Safety profiling	Olmesartan medoxomil	Solutol HS15	<50 nm	Wistar rats	No potential hematologic, biochemical, or structural toxicity	Gorain et al. (2016)
Cytotoxicity and antiangiogenic efficacy	Resveratrol	Transcutol HP	103 nm	Breast cancer cells (MCF-7). Chick embryo chorioallantoic membrane	Cytotoxicto MCF-7 cells and antiangiogenic	Pund et al. (2014)
Toxicity	*Eucalyptus staigeriana* essential oil	Tween 80	274.3 nm	Female Wistar albino rats	Ovicidaland larvicidal effects on *H. contortus*. No hematologic alterations in rats	Ribeiro et al. (2015)
Antiglioma, cytotoxicity, and genotoxicity		Tween 80	<250 nm	Peripheral human blood. Rat glioma cell line (C6)	Low toxicity against human blood cell sat 0.05 mg/mL	Mota Ferreira et al. (2016)
Dyslipidemia	Garlic oil	Tween 80	24.9 nm	Wistar rats	No hematologic or histological changes at 0.46 mL/kg dosage. Immediate mortality at 18.63 mL/kg	Ragavan et al. (2017)
Safety	Quercetin	Polyoxyethylene (20) oleylether, polyoxyethylene (3) oleylether, and cetyl trimethyl ammonium chloride		Chorioallantoic membrane	Causes hemorrhagic phenomena	Dario et al. (2016)

Source: Adapted from Wani et al. (2018).

of food commodities owing to people's preferences for natural preservatives in place of synthetic ones. Nanotechnology is an effective tool in this regard which helps in the preparation of essential oil nanoemulsions to overcome the instability problems associated with essential oils. This mild preservation method is very effective against disease-causing foodborne pathogens and offers safe and wholesome food without any change in nutritional, as well as sensory, characteristics of eatables. However, in future essential oil nanoemulsions–related challenges must be overcome for proper exploitation.

REFERENCES

Alexandre, E. M. C., Lourenço, R. V., Bittante, A. M. Q. B., Moraes, I. C. F., and do Amaral Sobral, P. J. (2016). Gelatin-based films reinforced with montmorillonite and activated with nanoemulsion of ginger essential oil for food packaging applications. *Food Packaging and Shelf Life*, 10, 87–96.

Amaral, D. M. F., and Bhargava, K. (2015). Essential oil nanoemulsions and food applications. *Advances in Food Technology and Nutritional Sciences–Open Journal*, 1(4), 84–87.

Anwer, M. K., Jamil, S., Ibnouf, E. O., and Shakeel, F. (2014). Enhanced antibacterial effects of clove essential oil by nanoemulsion. *Journal of Oleo Science*, 63(4), 347–354.

Appendini, P., and Hotchkiss, J. H. (2002). Review of antimicrobial food packaging. *Innovative Food Science and Emerging Technologies*, 3(2), 113–126.

Artiga-Artigas, M., Acevedo-Fani, A., and Martín-Belloso, O. (2017). Improving the shelf life of low-fat cut cheese using nanoemulsion-based edible coatings containing oregano essential oil and mandarin fiber. *Food Control*, 76, 1–12.

Baldissera, M. D., Da Silva, A. S., Oliveira, C. B., Zimmermann, C. E., Vaucher, R. A., Santos, R. C., Rech, V. C., Tonin, A. A., Giongo, J. L., Mattos, C. B., and Koester, L. (2013). Trypanocidal activity of the essential oils in their conventional and nanoemulsion forms: In vitro tests. *Experimental Parasitology*, 134(3), 356–3561.

Bauer, K., Garbe, D., and Surburg, H. (2008). *Common fragrance and flavor materials: Preparation, properties and uses*. John Wiley and Sons.

Bhargava, K., Conti, D. S., da Rocha, S. R., and Zhang, Y. (2015). Application of an oregano oil nanoemulsion to the control of foodborne bacteria on fresh lettuce. *Food Microbiology*, 47, 69–73.

Blanco-Padilla, A., Soto, K. M., Hernández Iturriaga, M., and Mendoza, S. (2014). Food antimicrobials nanocarriers. *The Scientific World Journal*, 2014, Article 837215.

Bonferoni, M. C., Sandri, G., Rossi, S., Usai, D., Liakos, I., Garzoni, A., and Ferrari, F. (2017). A novel ionic amphiphilic chitosan derivative as a stabilizer of nanoemulsions: Improvement of antimicrobial activity of *Cymbopogon citratus* essential oil. *Colloids and Surfaces B: Biointerfaces*, 152, 385–392.

Buranasuksombat, U., Kwon, Y. J., Turner, M., and Bhandari, B. (2011). Influence of emulsion droplet size on antimicrobial properties. *Food Science and Biotechnology*, 20(3), 793–800.

Burt, S. (2004). Essential oils: Their antibacterial properties and potential applications in foods-a review. *International Journal of Food Microbiology*, 94, 223–253.

Chang, Y., McLandsborough, L., and McClements, D. J. (2012). Physical properties and antimicrobial efficacy of thyme oil nanoemulsions: Influence of ripening inhibitors. *Journal of Agricultural and Food Chemistry*, 60(48), 12056–12063.

Chang, Y., McLandsborough, L., and McClements, D. J. (2015). Fabrication, stability and efficacy of dual-component antimicrobial nanoemulsions: Essential oil (thyme oil) and cationic surfactant (lauric arginate). *Food Chemistry*, 172, 298–304.

Chawla, P., Kumar, N., Bains, A., Dhull, S. B., Kumar, M., Kaushik, R., and Punia, S. (2020). Gum arabic capped copper nanoparticles: Synthesis, characterization, and applications. *International Journal of Biological Macromolecules*, 146, 232–242.

Chawla, P., Kumar, N., Kaushik, R., and Dhull, S. B. (2019). Synthesis, characterization and cellular mineral absorption of nanoemulsions of Rhododendron arboreum flower extracts stabilized with gum arabic. *Journal of Food Science and Technology*, 56(12), 5194–5203.

Dario, M. F., Oliveira, C. A., Cordeiro, L. R., Rosado, C., Ines de Fatima, A. M., Maçoas, E., and Velasco, M. V. R. (2016). Stability and safety of quercetin-loaded cationic nano-emulsion: In vitro and in vivo assessments. *Colloids and Surfaces A: Physicochemical and Engineering Aspects*, 506, 591–599.

Dasgupta, N., Ranjan, S., and Gandhi, M. (2019). Nanoemulsions in food: Market demand. *Environmental Chemistry Letters*, 17(2), 1003–1009.

Deena, M. J., and Thoppil, J. E. (2000). Antimicrobial activity of the essential oil of *Lantana camara*. *Fitoterapia*, 71(4), 453–455.

Dhull, S. B., Anju, M., Punia, S., Kaushik, R., and Chawla, P. (2019). Application of Gum Arabic in Nanoemulsion for Safe Conveyance of Bioactive Components. In R. Prasad, V. Kumar, M. Kumar, and D. Choudhary (Eds.), *Nanobiotechnology in Bioformulations* (pp. 85–98). Cham: Springer.

Donsì, F., Annunziata, M., Sessa, M., and Ferrari, G. (2011). Nanoencapsulation of essential oils to enhance their antimicrobial activity in foods. *LWT-Food Science and Technology*, 44(9), 1908–1914.

Donsì, F., Annunziata, M., Vincensi, M., and Ferrari, G. (2012). Design of nanoemulsion-based delivery systems of natural antimicrobials: Effect of the emulsifier. *Journal of Biotechnology*, 159(4), 342–350.

Donsì, F., and Ferrari, G. (2016). Essential oil nanoemulsions as antimicrobial agents in food. *Journal of Biotechnology*, 233, 106–120.

Donsi, F., Marchese, E., Maresca, P., Pataro, G., Vu, K. D., Salmieri, S., and Ferrari, G. (2015). Green beans preservation by combination of a modified chitosan based-coating containing nanoemulsion of mandarin essential oil with high pressure or pulsed light processing. *Postharvest Biology and Technology*, 106, 21–32.

Flores, F. C., De Lima, J. A., Ribeiro, R. F., Alves, S. H., Rolim, C. M., Beck, R. C., and Da Silva, C. B. (2013). Antifungal activity of nanocapsule suspensions containing tea tree oil on the growth of *Trichophytonrubrum*. *Mycopathologia*, 175(3–4), 281–286.

Gahruie, H. H., Ziaee, E., Eskandari, M. H., and Hosseini, S. M. H. (2017). Characterization of basil seed gum-based edible films incorporated with *Zataria multiflora* essential oil nanoemulsion. *Carbohydrate Polymers*, 166, 93–103.

Gaysinsky, S., Davidson, P. M., Bruce, B. D., and Weiss, J. (2005). Growth inhibition of *Escherichia coli* O157: H7 and *Listeria monocytogenes* by carvacrol and eugenol encapsulated in surfactant micelles. *Journal of Food Protection*, 68(12), 2559–2566.

Ghosh, V., Mukherjee, A., and Chandrasekaran, N. (2013b). Ultrasonic emulsification of food-grade nanoemulsion formulation and evaluation of its bactericidal activity. *Ultrasonics Sonochemistry*, 20(1), 338–344.

Ghosh, V., Mukherjee, A., and Chandrasekaran, N. (2014). Eugenol-loaded antimicrobial nanoemulsion preserves fruit juice against, microbial spoilage. *Colloids and Surfaces B: Biointerfaces*, 114, 392–397.

Ghosh, V., Saranya, S., Mukherjee, A., and Chandrasekaran, N. (2013a). Cinnamon oil nanoemulsion formulation by ultrasonic emulsification: Investigation of its bactericidal activity. *Journal of Nanoscience and Nanotechnology*, 13(1), 114–122.

Ghotbi, R. S., Khatibzadeh, M., and Kordbacheh, S. (2014). *Preparation of neem seed oil nanoemulsion*. In *Proceedings of the 5th International Conference on Nanotechnology: Fundamentals and Applications*, Prague, Czech Republic, Paper Aug 11 (No. 150, pp. 11–13).

Gorain, B., Choudhury, H., Tekade, R. K., Karan, S., Jaisankar, P., and Pal, T. K. (2016). Comparative biodistribution and safety profiling of *Olmesartan medoxomil* oil-in-water oral nanoemulsion. *Regulatory Toxicology and Pharmacology*, 82, 20–31.

Guerra-Rosas, M. I., Morales-Castro, J., Cubero-Márquez, M. A., Salvia-Trujillo, L., and Martín-Belloso, O. (2017). Antimicrobial activity of nanoemulsions containing essential oils and high methoxyl pectin during long-term storage. *Food Control*, 77, 131–138.

Gupta, A., Eral, H. B., Hatton, T. A., and Doyle, P. S. (2016). Nanoemulsions: Formation, properties and applications. *Soft Matter*, 12(11), 2826–2841.

Hamed, S. F., Sadek, Z., and Edris, A. (2012). Antioxidant and antimicrobial activities of clove bud essential oil and eugenol nanoparticles in alcohol-free microemulsion. *Journal of Oleo Science*, 61(11), 641–648.

Hashem, A. S., Awadalla, S. S., Zayed, G. M., Maggi, F., and Benelli, G. (2018). *Pimpinella anisum* essential oil nanoemulsions against *Tribolium castaneum*-insecticidal activity and mode of action. *Environmental Science and Pollution Research*, 25(19), 18802–18812.

Hilbig, J., Ma, Q., Davidson, P. M., Weiss, J., and Zhong, Q. (2016). Physical and antimicrobial properties of cinnamon bark oil co-nanoemulsified by lauric arginate and Tween 80. *International Journal of Food Microbiology*, 233, 52–59.

Hyldgaard, M., Mygind, T. and Meyer, R. L. (2012). Essential oils in food preservation: Mode of action, synergies, and interactions with food matrix components. *Front Microbiology*, 3, 12.

Jo, Y. J., Chun, J. Y., Kwon, Y. J., Min, S. G., Hong, G. P., and Choi, M. J. (2015). Physical and antimicrobial properties of trans-cinnamaldehyde nanoemulsions in water melon juice. *LWT-Food Science and Technology*, 60(1), 444–451.

Karthik, P., Ezhilarasi, P., and Anandharamakrishnan, C. (2017). Challenges associated in stability of food grade nanoemulsions. *Critical Reviews in Food Science and Nutrition*, 57, 1435–1450.

Kim, I. H., Lee, H., Kim, J. E, Song, K. B., Lee, Y. S., Chung, D. S., and Min, S. C. (2013). Plum coatings of lemongrass oil-incorporating carnauba wax-based nanoemulsion. *Journal of Food Science*, 78, E1551–E1559.doi: 10.1111/17503841.12244.

Komaiko, J. S., and McClements, D. J. (2016). Formation of food-grade nanoemulsions using low-energy preparation methods: A review of available methods. *Comprehensive Reviews in Food Science and Food Safety*, 15(2), 331–352.

Kriegel, C., Kit, K. M., McClements, D. J., and Weiss, J. (2010). Nanofibers as carrier systems for antimicrobial microemulsions. II. Release characteristics and antimicrobial activity. *Journal of Applied Polymer Science*, 118(5), 2859–2868.

Landry, K. S., Chang, Y., McClements, D. J., and McLandsborough, L. (2014). Effectiveness of a novel spontaneous carvacrol nanoemulsion against *Salmonella enterica Enteritidis* and *Escherichia coli* O157: H7 on contaminated mung bean and alfalfa seeds. *International Journal of Food Microbiology*, 187, 15–21. doi: 10.1016/j.ijfoodmicro.2014.06.030.

Landry, K. S., Micheli, S., McClements, D. J., and McLandsborough, L. (2015). Effectiveness of a spontaneous carvacrol nanoemulsion against *Salmonella enterica Enteritidis* and *Escherichia coli O157: H7* on contaminated broccoli and radish seeds. *Food Microbiology*, 51, 10–17. doi: 10.1016/j.fm.2015.04.006.

Li, J., Chang, J. W., Saenger, M., and Deering, A. (2017). Thymol nanoemulsions formed via spontaneous emulsification: Physical and antimicrobial properties. *Food Chemistry*, 232, 191–197.

Li, W., Chen, H., He, Z., Han, C., Liu, S., and Li, Y. (2015). Influence of surfactant and oil composition on the stability and antibacterial activity of eugenol nanoemulsions. *LWT-Food Science and Technology*, 62(1), 39–47.

Li, M., Zhu, L., Liu, B., Du, L., Jia, X., Han, L., and Jin, Y. (2016). Tea tree oil nanoemulsions for inhalation therapies of bacterial and fungal pneumonia. *Colloids and Surfaces B: Biointerfaces*, 141, 408–416.

Liang, R., Xu, S., Shoemaker, C. F., Li, Y., Zhong, F., and Huang, Q. (2012). Physical and antimicrobial properties of peppermint oil nanoemulsions. *Journal of Agricultural and Food Chemistry*, 60(30), 7548–7555.

Lou, Z., Chen, J., Yu, F., Wang, H., Kou, X., Ma, C., and Zhu, S. (2017). The antioxidant, antibacterial, antibiofilm activity of essential oil from *Citrus medica L.* var. sarcodactylis and its nanoemulsion. *LWT*, 80, 371–377.

Lu, W. C., Huang, D. W., Wang, C. C., Yeh, C. H., Tsai, J. C., Huang, Y. T., and Li, P. H. (2018). Preparation, characterization, and antimicrobial activity of nanoemulsions incorporating citral essential oil. *Journal of Food and Drug Analysis*, 26(1), 82–89.

Ma, Q., Davidson, P. M., and Zhong, Q. (2016). Nanoemulsions of thymol and eugenol co-emulsified by lauric arginate and lecithin. *Food Chemistry*, 206, 167–173.

Majeed, H., Liu, F., Hategekimana, J., Sharif, H. R., Qi, J., Ali, B., and Zhong, F. (2016). Bactericidal action mechanism of negatively charged food grade clove oil nanoemulsions. *Food Chemistry*, 197, 75–83.

Martin-Pinero, M. J., Ramirez, P., Munoz, J., and Alfaro, M. C. (2019). Development of rosemary essential oil nanoemulsions using a wheat biomass-derived. *Colloids and Surfaces B: Biointerfaces*, 173.

Moghimi, R., Aliahmadi, A., McClements, D. J., and Rafati, H. (2016b). Investigations of the effectiveness of nanoemulsions from sage oil as antibacterial agents on some food borne pathogens. *LWT-Food Science and Technology*, 71, 69–76.

Moghimi, R., Ghaderi, L., Rafati, H., Aliahmadi, A., and McClements, D. J. (2016a). Superior antibacterial activity of nanoemulsion of Thymus daenensis essential oil against E. coli. *Food Chemistry*, 194, 410–415.

Moreno-Trejo, M. B., Rodriguez-Rodríguez, A. A., Suarez-Jacobo, A., and Sánchez-Domínguez, M. (2019). Development of Nano-Emulsions of Essential Citrus Oil Stabilized with Mesquite Gum. In *Nanoemulsions-Properties, Fabrications and Applications*. IntechOpen.

Mota Ferreira, L., Gehrcke, M., Ferrari Cervi, V., Eliete Rodrigues Bitencourt, P., Ferreira da Silveira, E., Hofstatter Azambuja, J., and Rorato Sagrillo, M. (2016). Pomegranate seed oil nanoemulsions with selective antiglioma activity: Optimization and evaluation of cytotoxicity, genotoxicity and oxidative effects on mononuclear cells. *Pharmaceutical Biology*, 54(12), 2968–2977.

Nirmal, N. P., Mereddy, R., Li, L., and Sultanbawa, Y. (2018). Formulation, characterisation and antibacterial activity of lemon myrtle and anise myrtle essential oil in water nanoemulsion. *Food Chemistry*, 254, 1–7.

Noori, S., Zeynali, F., and Almasi, H. (2018). Antimicrobial and antioxidant efficiency of nanoemulsion-based edible coating containing ginger (Zingiber officinale) essential oil and its effect on safety and quality attributes of chicken breast fillets. *Food Control*, 84, 312–320.

Nor Bainun, I., Alias, N. H., and Syed-Hassan, S. S. A. (2015). Nanoemulsion: Formation, characterization, properties and applications-a review. *Advanced Materials Research*, Trans Tech Publications Ltd., 1113, 147–152.

Orav, A., Raal, A., and Arak, E. (2008). Essential oil composition of Pimpinella anisum L. fruits from various European countries. *Natural Product Research*, 22(3), 227–232.

Otoni, C. G., de Moura, M. R., Aouada, F. A., Camilloto, G. P., Cruz, R. S., Lorevice, M. V., and Mattoso, L. H. (2014). Antimicrobial and physical-mechanical properties of pectin/papaya puree/cinnamaldehyde nanoemulsion edible composite films. *Food Hydrocolloids*, 41, 188–194.

Panghal, A., Chhikara, N., Anshid, V., Charan, M. V. S., Surendran, V., Malik, A., and Dhull, S. B. (2019). Nanoemulsions: A Promising Tool for Dairy Sector. In *Nanobiotechnology in Bioformulations* (pp. 99–117). Cham: Springer.

Park, C. Y., Jang, C. H., Lee, D. Y., Cho, H. T., Kim, Y. J., Park, Y. H., and Imm, J. Y. (2017). Changes in hepatic gene expression and serum metabolites after oral administration of overdosed vitamin-E-loaded nanoemulsion in rats. *Food and Chemical Toxicology*, 109, 421–427.

Pathania, R., Kaushik, R., and Khan, M. A. (2018). Essential oil nanoemulsions and their antimicrobial and food applications. *Current Research in Nutrition and Food Science Journal*, 6(3), 626–643.

Pavoni, L., Perinelli, D. R., Bonacucina, G., Cespi, M., and Palmieri, G. F. (2020). An overview of micro-and nanoemulsions as vehicles for essential oils: Formulation, preparation and stability. *Nanomaterials*, 10(1), 135.

Prakash, A., Baskaran, R., Paramasivam, N., and Vadivel, V. (2018). Essential oil based nanoemulsions to improve the microbial quality of minimally processed fruits and vegetables: A review. *Food Research International*, 111, 509–523.

Pund, S., Thakur, R., More, U., and Joshi, A. (2014). Lipid based nanoemulsifying resveratrol for improved physicochemical characteristics, in vitro cytotoxicity and in vivo antiangiogenic efficacy. *Colloids and Surfaces B: Biointerfaces*, 120, 110–117.

Punia, S., Sandhu, K. S., Siroha, A. K., and Dhull, S. B. (2019). Omega 3-metabolism, absorption, bioavailability and health benefits-a review. *PharmaNutrition*, 10, 100–162.

Ragavan, G., Muralidaran, Y., Sridharan, B., Ganesh, R. N., and Viswanathan, P. (2017). Evaluation of garlic oil in nano-emulsified form: Optimization and its efficacy in high-fat diet induced dyslipidemia in Wistar rats. *Food and Chemical Toxicology*, 105, 203–213.

Ribeiro, W. L. C., Camurça-Vasconcelos, A. L. F., Macedo, I. T. F., dos Santos, J. M. L., de Araújo-Filho, J. V., de Carvalho Ribeiro, J., and Bevilaqua, C. M. L. (2015). In vitro effects of *Eucalyptus staigeriana* nanoemulsion on *Haemonchus contortus* and toxicity in rodents. *Veterinary Parasitology*, 212(3–4), 444–447.

Salem, M. A., and Ezzat, S. M. (2018). Nanoemulsions in Food Industry. In J. M. Milani (Ed.), *Some New Aspects of Colloidal Systems in Foods*. IntechOpen.

Salvia-Trujillo, L., Rojas-Graü, A., Soliva-Fortuny, R., and Martín-Belloso, O. (2013a). Physicochemical characterization of lemongrass essential oil–alginate nanoemulsions: Effect of ultrasound processing parameters. *Food and Bioprocess Technology*, 6(9), 2439–2446.

Salvia-Trujillo, L., Rojas-Graü, M. A., Soliva-Fortuny, R., and Martín-Belloso, O. (2013b). Effect of processing parameters on physicochemical characteristics of microfluidized lemongrass essential oil-alginate nanoemulsions. *Food Hydrocolloids*, 30(1), 401–407.

Salvia-Trujillo, L., Rojas-Graü, M. A., Soliva-Fortuny, R., and Martín-Belloso, O. (2014a). Impact of microfluidization or ultrasound processing on the antimicrobial activity against Escherichia coli of lemongrass oil-loaded nanoemulsions. *Food Control*, 37, 292–297.

Salvia-Trujillo, L., Rojas-Graü, M. A., Soliva-Fortuny, R., and Martín-Belloso, O. (2014b). Formulation of antimicrobial edible nanoemulsions with pseudo-ternary phase experimental design. *Food and Bioprocess Technology*, 7(10), 3022–3032.

Salvia-Trujillo, L., Rojas-Graü, M. A., Soliva-Fortuny, R., and Martín-Belloso, O. (2015). Use of antimicrobial nanoemulsions as edible coatings: Impact on safety and quality attributes of fresh-cut Fuji apples. *Postharvest Biology and Technology*, 105, 8–16.

Saranya, S., Chandrasekaran, N., and Mukherjee, A. M. (2012). Antibacterial activity of eucalyptus oil nanoemulsion against Proteus mirabilis. *International Journal of Pharmacy and Pharmaceutical Sciences*, 4(3), 668–671.

Sessa, M., Ferrari, G., and Donsì, F. (2015). Novel edible coating containing essential oil nanoemulsions to prolong the shelf life of vegetable products. *Chemical Engineering Transactions*, 43, 55–60.

Severino, R., Vu, K. D., Donsì, F., Salmieri, S., Ferrari, G., and Lacroix, M. (2014). Antibacterial and physical effects of modified chitosan based-coating containing nanoemulsion of mandarin essential oil and three non-thermal treatments against *Listeria innocua* in green beans. *International Journal of Food Microbiology*, 191, 82–88.

Shah, B., Davidson, P.M., and Zhong, Q. (2012). Nanocapsular Dispersion of Thymol for Enhanced dispersibility and increased antimicrobial effectiveness against *Escherichia coli* O157: H7 and *Listeria monocytogenes* in model food systems. *Applied and Environmental Microbiology*, 78, 8448–8453.

Sharif, H. R., Abbas, S., Majeed, H., Safdar, W., Shamoon, M., Khan, M. A., and Haider, J. (2017). Formulation, characterization and antimicrobial properties of black cumin essential oil nanoemulsions stabilized by OSA starch. *Journal of Food Science and Technology*, 54(10), 3358–3365.

Silva, H. D., Cerqueira, M. A., and Vicente, A. A. (2012). Nanoemulsions for food applications: Development and characterization. *Food and Bioprocess Technology*, 5, 854–867.

Solans, C., and Solé, I. (2012). Nano-emulsions: Formation by low-energy methods. *Current Opinion in Colloid & Interface Science*, 17(5), 246–254.

Terjung, N., Löffler, M., Gibis, M., Hinrichs, J., and Weiss, J. (2012). Influence of droplet size on the efficacy of oil-in-water emulsions loaded with phenolic antimicrobials. *Food & Function*, 3(3), 290–301.

Topuz, O. K., Ozvural, E. B., Zhao, Q., Huang, Q., Chikindas, M., and Golukçu, M. (2016). Physical and antimicrobial properties of anise oil loaded nanoemulsions on the survival of foodborne pathogens. *Food Chemistry*, 203, 117–123.

Tozlu, E., Cakir, A., Kordali, S., Tozlu, G., Ozer, H., and Akcin, T. A. (2011). Chemical compositions and insecticidal effects of essential oils isolated from *Achilleagypsicola, Saturejahortensis, Origanum acutidens* and *Hypericum scabrum* against broad bean weevil (Bruchusdentipes). *Scientia Horticulturae*, 130(1), 9–17.

Wani, T. A., Masoodi, F. A., Jafari, S. M., and McClements, D. J. (2018). Safety of Nanoemulsions and their Regulatory Status. In S. Jafari, and D. J. McClements (Eds.), *Nanoemulsions* (pp. 613–628). Academic Press.

Wu, Y., Luo, Y., and Wang, Q. (2012). Antioxidant and antimicrobial properties of essential oils encapsulated in zein nanoparticles prepared by liquid–liquid dispersion method. *LWT-Food Science and Technology*, 48(2), 283–290.

Xue, J. (2015). Essential oil nanoemulsions prepared with natural emulsifiers for improved food safety. PhD Thesis, University of Tennessee. https://trace.tennessee.edu/utk_graddiss/3381

Xue, J., Davidson, P. M., and Zhong, Q. (2015). Antimicrobial activity of thyme oil co-nanoemulsified with sodium caseinate and lecithin. *International Journal of Food Microbiology*, 210, 1–8.

Zhang, Y., and Zhong, Q. (2020). Physical and antimicrobial properties of neutral nanoemulsions self-assembled from alkaline thyme oil and sodium caseinate mixtures. *International Journal of Biological Macromolecules*, 148, 1046–1052.

Zhang, Z., Vriesekoop, F., Yuan, Q., and Liang, H. (2014). Effects of nisin on the antimicrobial activity of D-limonene and its nanoemulsion. *Food Chemistry*, 150, 307–312.

Ziani, K., Chang, Y., McLandsborough, L., and McClements, D. J. (2011). Influence of surfactant charge on antimicrobial efficacy of surfactant stabilized thyme oil nanoemulsions. *Journal of Agricultural and Food Chemistry*, 59(11), 6247–6255.

10 Nanoemulsions Formulated With Cinnamon Oil and Their Antimicrobial Applications

Ruhi Pathania and Bhanu Sharma
Shoolini University, India
Prince Chawla
Lovely Professional University, India
Ravinder Kaushik
Amity University, India
Mohammed Azhar Khan
Shoolini University, India

CONTENTS

10.1 INTRODUCTION

Plant-based essential oils have been utilized by various communities throughout history for homoeopathy and fragrance-based treatments. An assortment of compounds is existing in the essential oils that contrast extremely in their chemical nature (El Asbahani et al., 2015). At present, essential oil–based items are utilized in numerous industries, for example food items, pharmaceutics, cosmetics, and agricultural industries (Punia et al., 2019). Cinnamon is one of the most important essential oil that has been broadly reported (Holley and Patel, 2005).

The interest for new advances to reduce or minimize the effect of microbial pathogens from nature and to give more secure items to purchasers has been raised. Food-grade delivery systems have an important place to embed lipophilic functional components into food and beverages (Duncan, 2011). These functional components can be flavours, colours, micronutrients, antimicrobials, and so forth. Aside from their aseptic use, the utilization of these synthetic chemicals like preservatives or antimicrobials that show antimicrobial properties gives the capability to maintain the entry of microorganisms (Kanatt et al., 2012). The use of harmful chemical preservatives in food and agricultural industries has raised health concerns in consumers, thereby creating more demand for "natural" products (Troncoso et al., 2012). Consequently, there has been a huge interest in naturally produced antimicrobial agents to replace preservatives made of chemicals. Due to limited application studies on the biosafety issues of plant-based essential oil formulations, there is a need to explore their effect on animal models (Huang et al., 2010). Furthermore, the utilization of essential oil for application is difficult because of the way that these compounds are hydrophobic. Their use in agricultural and food-based industries is extremely limited as to their dissolvability in an aqueous matrix is low (Luykx et al., 2008). The problem can be overcome by formulating the essential oil into a nanoemulsion with the use of highly transparent surfactants, thereby decreasing the amount of oil required (Qian et al., 2012). Thus, the nanoemulsion approach is advantageous as it results in obtaining a homogeneous emulsion system.

10.2 NANOEMULSION

Nanotechnology in the food field leads to the arrangement of an extraordinary quantity of new products while altering the macroscale qualities of nourishment, for example surface, taste, processability, and stability (Singh, 2016; Chawla et al., 2020). Nanoemulsion technology forms an encapsulating system for functional materials; thus, degradation can be prevented, and hence, the bioavailability of the compounds is improved (Kumar and Mandal, 2018; Panghal et al., 2019). Nanoemulsions usually contain two immiscible liquids. One of these liquids is dispersed as small spherical droplets, and relying on the sort of the continuous phase, nanoemulsions are categorized as either oil-in-water or water-in-oil nanoemulsions (McClements, 2011; Dhull et al., 2019). Nanoemulsions are generally fabricated from the oil phase (organic phase), aqueous phase, surfactant, and possibly a co-surfactant (Chawla et al., 2019).

10.2.1 Organic Phase

A variety of nonpolar segments, for example glycerides, free unsaturated fats, enhance oils, fundamental oils, waxes, mineral oils, fats, oil-solvent vitamins, and lipophilic nutraceuticals, take part in the formulation of the organic phase (McClements and Rao, 2011). Mass physicochemical attributes (polarity, water dissolvability, interfacial strain, the refractive list, thickness, and synthetic stability) of the oil stage influence the arrangement, reliability, and property of the nanoemulsions.

10.2.2 Aqueous Phase

Primarily, water comprises this phase of nanoemulsions; anyway, the assortment of other chemical segments can be available in this phase like acids, bases, minerals, proteins, starches, and co-solvents (Acosta, 2009). Stability, formation, and physicochemical behaviour of nanoemulsions are specifically affected by the composition because the components affect the polarity, pH, density, interfacial tension, refractive index, rheology, and ionic force of the aqueous phase (Chime et al., 2014). Particle size is a standout amongst the most essential highlights of nanoemulsions. Nanoemulsions have a different droplet size, which ranges between 20 nm and 500 nm; however, to differentiate nanoemulsions from conventional emulsions, in the literature, there is no identified distinct size range (Solans and Sole, 2012). Stability, optical property, and rheology-based characteristics of the nanoemulsion are mainly affected by the extent of the droplet size. Achieving a nanometer range of particle size with high uniformity (monodispersed system) is the main goal of nanoemulsion studies. A measure for the size conveyance is known as the Polydispersity Index (PDI). When the PDI value is less than 0.200, the result shows the monodispersed amount of distribution is generally observed (Baboota et al., 2007). If multiple-sized particles are involved in an emulsion, emulsions are defined as polydispersed (multimodal system). Large particles and large size distributions indicate the instability of the formulation. After the formation of nanoemulsions, during storage some internal and external forces like gravitational, thermal, and interfacial stresses which cause the rapid amount of chemical and physical changes over time (Preetz et al., 2010).

10.2.3 Instability of Nanoemulsion

Instability can approach out of different types of decay mechanisms such as creaming, sedimentation, flocculation, Ostwald ripening, and coalescence (Figure 10.1; Klang et al., 2012). These instabilities lead to the conversion of monomodal size distribution to bimodal or multimodal type (Sole et al., 2012). In the coalescence process, small droplets in the emulsion formulation combine progressively to form larger droplets, and it results in complete phase separation (Tiwari and Tiwari, 2013). In flocculation, the larger droplets formed as a result of the coalescence dispersed phase separate the emulsion suspension into floccules. When the density of droplets is less than the continuous phase, droplets condensate at the nanoemulsion's upper surface. This situation is known as creaming (Tadros et al., 2004). Therefore, the phase of the constituent separates on top or bottom based on their differences in

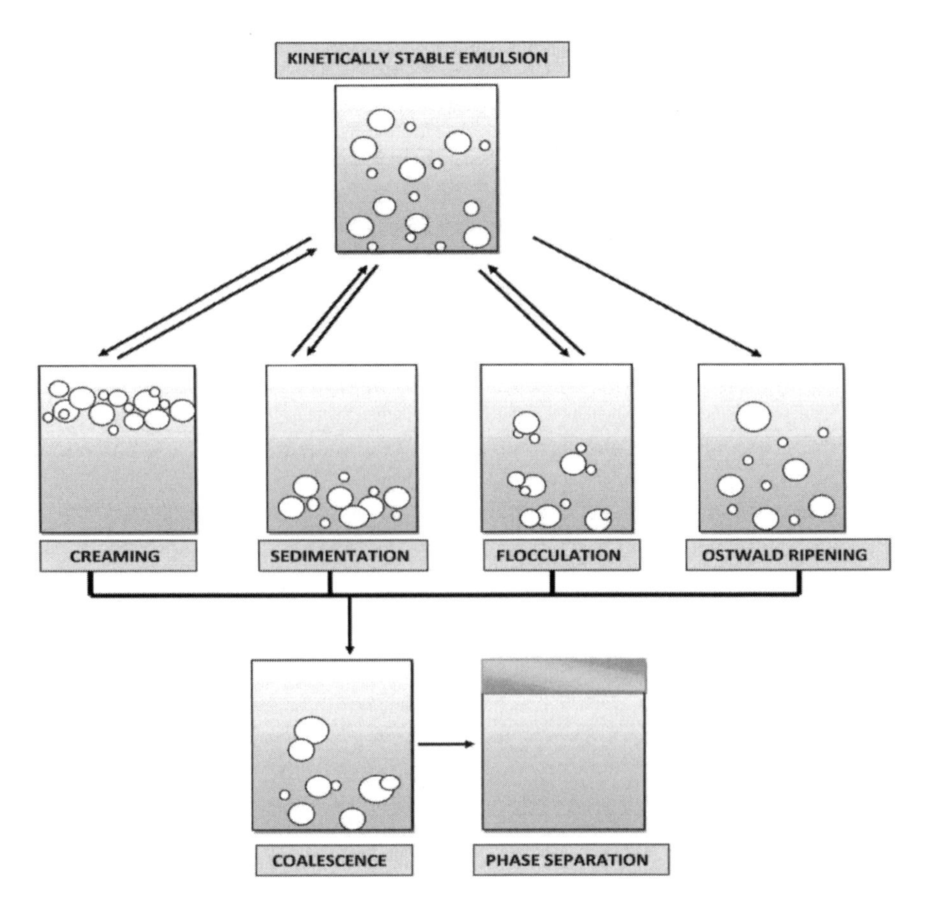

FIGURE 10.1 Mechanisms of emulsion destability.

densities. In Ostwald ripening, the larger droplets grow at the expense of smaller ones. In the continuous phase, the dispersed phase is partly soluble; then Ostwald ripening occurs because small droplets, which have high internal pressure, disperse into the continuous phase, and due to the mass transport of dispersed material, larger particles form (Delmas et al., 2011). Furthermore, McClements (2012) studied the rate of Ostwald ripening can be minimized by using an additional oil (long-chain triglycerides; e.g. corn oil) phase that has poor aqueous solubility or by adding weighting agents (polysaccharides; e.g. ester gum; Lim et al., 2011). Nanoemulsions are thermodynamically unstable systems and, therefore, less prone to decay mechanisms such as gravitational separation, flocculation, coalescence. But nanoemulsions are more prone to the Ostwald ripening process (Taylor, 1998).

10.3 ROLE OF SURFACTANTS

To avoid phase partition of an emulsion, surfactants are used. Surfactants upgrade the strength of emulsion because they frame a defensive layer to prohibit aggregation. Surfactants reduce interfacial tension and Laplace pressure that occurs among water and oil molecules (Grigoriev and Miller, 2009). Surfactant molecules incorporate at least one useful gathering that is not just firmly pulled into the mass medium but also have little attraction with the medium. Surfactants could be considered fewer than three major classifications. These are proteins, polysaccharides, and small surfactants (McClements, 2004).

10.3.1 Protein Emulsifiers

Some nourishment proteins, for example, soybean and whey protein segregate and β-lactoglobulin, are very utilized in the nourishment industry as surfactants because their nutritional values are high and generally recognized as safe (He et al., 2011). Proteins have amphiphilic characters, and they stabilize emulsions via electrostatic repulsion, steric hindrance, and generation of osmotic pressure at the external surface of protein molecules (Charoen et al., 2011).

10.3.2 Polysaccharide Emulsifiers

Gum arabic, some galactomannans, celluloses, modified starches, and few sorts of pectin are the most utilized polysaccharide emulsifiers in different foodstuffs. Depending on their functional groups, polysaccharides are named as non-ionic, anionic, and cationic (Dickinson, 2009). At the point when the monomer arrangement of polysaccharides is compared with proteins, polysaccharides are more uniform, and this provides specific functional and physicochemical properties like viscosity enhancement, binding properties, gelation, and solubility (Chanamai and McClements, 2001).

10.3.3 Small Surfactants

These types of surfactants having both hydrophilic (head group) and hydrophobic parts. These surfactants might be able to be examined under four subgroups, considering their hydrophilic functional groups in aqueous media. They are anionic, cationic, nonionic, and zwitterionic (Gupta and Kumar, 2012):

a. **Anionic surfactants:** These surfactants bear a negative charge. Anions of alkali metal salts of fatty acids, anions of long-chain sulfonates, sulfates, and phosphates are included in this group.
b. **Cationic surfactants:** These surfactants bear a positive charge. Generally, they consist of ammonium and pyridinium compounds (Chu et al., 2007).
c. **Nonionic surfactants:** These surfactants frequently refer to polyoxyethylene mixtures; however, sugar esters, amine oxides, and fatty alkanol amides are also included in this group (Komaiko and McClements, 2014).
d. **Zwitterionic surfac1tants:** These types of surfactants having both positive charges as well as negative charges in the hydrophilic component of the particle. Lecithin (long-chain phosphonyl cholines) belongs to this group (Azeem et al., 2009).

10.4 NANOEMULSION FORMATION

Arrangement of nanoemulsions requires energy that can be acquired from mechanical or stored chemical energy in the system (Gutierrez et al., 2008). Nanoemulsion preparation can be classified as either high-energy emulsification or low-energy emulsification (Figure 10.2; Maali and Mosavian, 2013).

10.4.1 High-Energy Emulsification

High-energy emulsification usually involves the utilization of mechanical devices that can generate intense disrupting force to reduce droplet size, such as higher-pressure homogenizers and microfluidizers (Swathy et al., 2018). Given its many advantages, for example easy scale-up, organic solvent–free, and high efficiency. High-pressure homogenizers are the most broadly used emulsifying device to fabricate stable nanoemulsions in the food-based industries (Reza, 2011). Ordinarily, coarse emulsions delivered by high-shear blenders are directed into a chamber in the homogenizer and afterwards constrained through a limited valve at a high weight (50–200 MPa; Gupta et al., 2010), which causes exceptionally troublesome powers, for example turbulence, hydraulic shear, and cavitation that are capable of breaking down large droplets into small ones (Lovelyn and Attama, 2011). Microfluidizers are another type of commonly used high-energy emulsifying device. Similar to the high-pressure homogenizer, microfluidizers also work at higher pressure (3–134 MPa; Thakur et al., 2012). Coarse emulsions are forced into an inlet channel and then separated into two different streams that interrupt on each other intensely in a cooperation chamber, where droplet disruption occurs under solid disorderly forces (Jaiswal et al., 2015). Microfluidizers can produce fine nanoemulsions with narrow droplet distributions (Dixit et al., 2008). Ultrasonic homogenizers deliver nanoemulsions by utilizing high-power ultrasonic waves, that is frequency greater than 20 kHz, that are

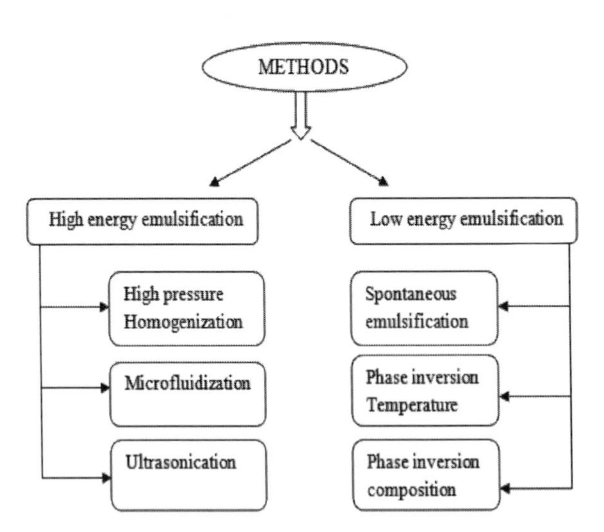

FIGURE 10.2 Formation methods of nanoemulsions.

extremely proficient in diminishing the droplet measure; however, they are more reasonable for little clumps and subsequently broadly utilized in research labs (Mahdi Jafari et al., 2006). An ultrasonic probe in the gadget changes over electrical waves into the extreme mass, which produces exceptional disrupting powers (Kentish et al., 2008). The emulsifying effectiveness altogether relies on the ultrasonication time at various amplitudes (Solans et al., 2005).

10.4.2 Low-Energy Emulsification

Low-energy emulsifications depend on the physicochemical qualities of surfactants and co-surfactants (Anton et al., 2008). Nanoemulsions can be able to impulsively form as the system compositions or ecological situation are adjusted. The two most frequently approaches are spontaneous emulsification and the phase inversion temperature (PIT; Setya et al., 2014). There is an increase in the low-energy approach; that is spontaneous emulsification is a very simple process (Bouchemal et al., 2004). Nanoemulsions are formed by mixing an organic phase that consists of oil, surfactant, and a water solvent and a pure aqueous phase at a particular temperature (Anton and Vandamme, 2009). This approach depends on the fast diffusion of water-miscible solvent like ethanol and acetone from the organic phase into the aqueous phase, which induces great unstable forces at the water or oil interface (Ichikawa et al., 2007). A disadvantage of this approach is that a high solvent/oil ratio is necessary to shape nano-droplets, which largely reduces the oil amount in the final last nanoemulsions (Fernandez et al., 2004). The PIT technique exploits changes in the affinities of nonionic surfactants for oil and water with respect to temperature (Devarajan and Ravichandran, 2011). At a low temperature, the surfactant is fully solubilized in water, which supports the development of oil-in-water (O/W) emulsions.

As the temperature is gradually increased above a particular temperature, the surfactant turns out to be more solvent in oil than in water; then O/W emulsions invert to water-in-oil (W/O) emulsions (Savardekar and Bajaj, 2016). Then an O/W nanoemulsion can be produced by rapidly cooling the system below the PIT. These procedures are reversible, as the temperature becomes raised the presence of clear nanoemulsion will end up turbid once more, which could be an issue in some nourishment and drink applications that need humid treatments (Anandharamakrishnan, 2014).

10.5 ANTIMICROBIAL ACTIVITY OF NANOEMULSIONS

Even though incredible assortments of preservation strategies are available, in the nourishment trade, food poisoning and spoilage by organisms are the main concerns. Foodborne sickness begins from an enormous population of microorganisms, for example *Escherichia coli*, *Bacillus cereus*, and *Listeria monocytogenes* (Skrinjar and Nemet, 2009). *E. coli* is a gram-negative, anaerobic, rod-shaped microscopic organism and belongs to the *Enterobacteriaceae* family that is, for the most part, found in soil and water, cause flavour and textural defects in nourishment. This organism also lives in the ordinary intestinal vegetation of people and other warm-blooded creatures, yet some cause sickness and many intestinal diseases in people (Donsì and Ferrari, 2016). In terms of their virulence properties, pathogenic mechanisms, clinical syndromes, and O: H serogroups, there are, most important, at least six classes of *E. coli* that cause foodborne gastrointestinal disease in humans (Sagis, 2015). These are Enteropathogenic *E. coli*, which is associated with infantile diarrheoa; Enterotoxigenic *E. coli* (ETEC), which is associated with traveller's diarrhoea and infant diarrhoea; Enteroinvasive *E. coli*, which causes dysentery like shigellosis; Enteroaggregative *E. coli*, which causes persistent diarrhoea; and Diffusely Adherent *E. coli* and Enterohemorrhagic *E. coli*, which are linked through bloody diarrhoea and hemolytic uremic disease (Donsì et al., 2012). To prevent foodborne microorganisms and to decrease the treatment of synthetic and antimicrobial compounds in foods through natural ways, essential oils are considered as alternative natural antimicrobial additives. However, they have low water solubility, high volatility, and strong odour. Essential oils are aromatic hydrophilic liquid products; therefore, O/W nanoemulsions have been thought to be efficient delivery systems (Liang et al., 2012).

10.5.1 Mechanism of Action Against Antimicrobials

Plant-based oils are a prospective source of antimicrobials such as 1, 8-cineole, perillaldehyde, cinnamaldehyde, and carvacrol (Wang et al., 2009). These different types of phytochemicals expand the timeframe of the realistic usability of prepared nourishment items by preventing it from antimicrobials and lipid oxidation (Singh et al., 2007). The antibacterial action of these mixes in the oil owes to their hydrophobic nature, and they are probably going to break down in the hydrophobic area of the cytoplasmic film of bacterial cells (Paranagama et al., 2001). Cinnamon oil has a few components, for example cinnamaldehyde, pinene, and myrcene. Antibacterial action of cinnamon oil is identified with cinnamaldehyde (Zhou et al., 2007). Possible mechanisms for antibacterial movement of cinnamaldehyde are impacts on film penetrability and control of usage of glucose (Gill and Holley, 2004). There is evidence showing that cinnamon oil can inhibit bacterial activity (i.e. *Porphyromonas gingivalis*, *L. monocytogenes*, *Salmonella enterica*, and *E. coli*; Wang et al., 2018). The method of action of nanoemulsions towards microorganisms is explained based on the interaction with the oil phase and cells (Figure 10.3). Lipid-containing cell membranes of microorganisms are damaged by nanoemulsion particles due to active ingredients because of their hydrophobic nature. The lipophilic character of the emulsions causes accumulation on the cell membrane (Ma et al., 2016).

10.6 CINNAMON OIL NANOEMULSIONS AND THEIR ANTIMICROBIAL ACTIVITY

Cinnamon, as a spice, has been used for medicinal purposes and flavouring from ancient times to the present. Its botanical name is *Cinnamomum*, and there are 250 species, such as *C. verum* and *C. cassia* (Chinese cinnamon; Qin et al., 2003). Cinnamon is started first from Sri Lanka and India. It has been utilized in the traditional medicines of China (Goñi et al., 2009). Cinnamon oil has proved to have various natural functions, for example antimicrobial (Tzortzakis, 2009), antidiabetic (Ping et al., 2010), and antioxidant effects (Özcan and Arslan, 2011); since ancient times, it has been used for the treatment of arthritis, blood circulation disturbance,

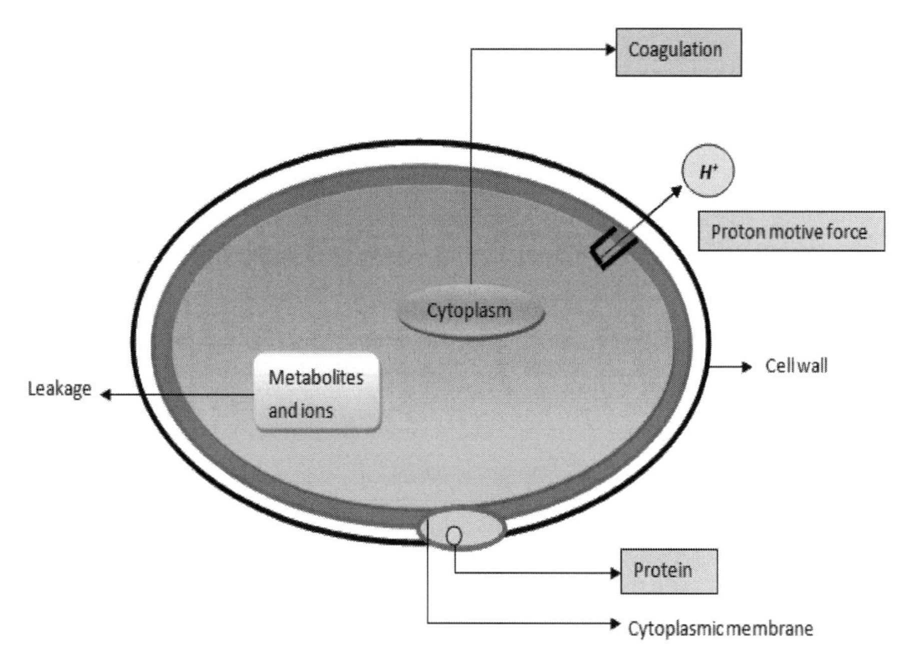

FIGURE 10.3 Mechanism of antimicrobial action of plant-based oils.

dyspepsia, gastritis, and various inflammatory disorders in many countries. It also has anxiolytic, antiulcerogenic, antipyretic, analgesic, antithrombotic, and anti-allergic activity (Tung et al., 2008). The formulation of nanoemulsion using plant oils is cheap, readily scalable, and environmentally-friendly. The plant oil nanoemulsion produced with a non-ionic surfactant is highly safe, stable, and biocompatible. The most widely used approaches for nanoemulsion formation with plant oils such are the ultrasonication and high-pressure homogenization methods (Figure 10.4). Cinnamon oil is obtained by the refining of bark and leaves of the cinnamon tree. These leaves contain a high amount constituents of eugenol (4-allyl-2-methoxyphenol), cinnamaldehyde (3-Phenyl-2-propenal), linalool, cinamyl acetate, geraniol, bornyl acetate, caryophyllene oxide, caryophyllene, and α-cubebene (Ghosh et al., 2013; Table 10.1).

These active constituents show advancement in the field of antimicrobial association against important foodborne pathogenic microbes such as *E. coli, B. cereus,* and *S. aureus*. Through the phospholipid layer of the bacterial cell membrane, the penetration of eugenol occurs, and it causes the alteration of membrane structures and accordingly permeability changes. For this reason, intracellular constituents of cell leak, and moreover, cell death occurs (Ghosh et al., 2014). Phenolic compounds furthermore interrupt the structure of the cell membrane while interchanging its hydroxyl group with other ions like potassium. Moreover, phenolic compounds ensue as a protonophore while carrying protons through lipid bilayers; therefore, they dissipate the proton motive force. Also, phenolic compounds affect the permeability of cell membranes and cause leakage of some substances such as amino acids, adenosine triphosphate (ATP), ions, and nucleic acids (Hamouda et al., 2001). In

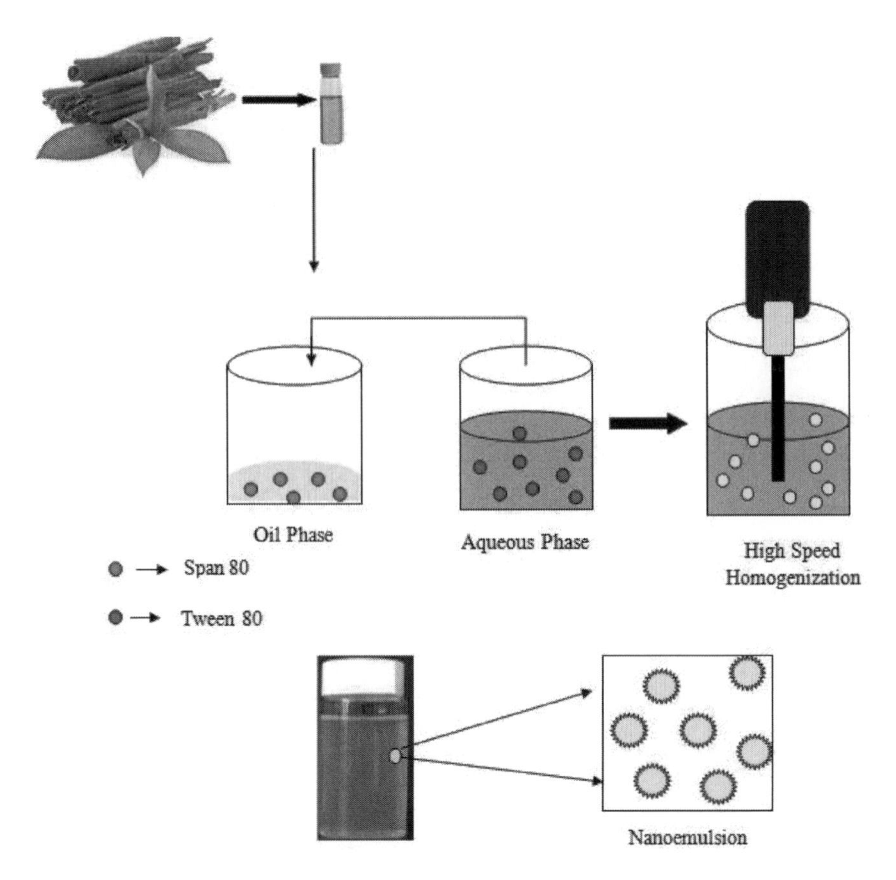

FIGURE 10.4 Formation of cinnamon oil–based nanoemulsion.

TABLE 10.1
Major Bioactive Constituents of Cinnamon Oil

Compound	Molecular Formula	Molecular Weight (g/mol)
Cinnamaldehyde	C_9H_8O	132.16
Eugenol	$C_{10}H_{12}O_2$	164.2
Linalool	$C_{10}H_{18}O$	154.25
Cinnamyl acetate	$C_{11}H_{12}O_2$	176.21
Caryophyllene	$C_{15}H_{24}$	204.35
Camphor	$C_{10}H_{16}O$	152.23
Caryophyllene oxide	$C_{15}H_{24}O$	220.35
Geraniol	$C_{10}H_{18}O$	154.25
Bornyl acetate	$C_{12}H_{20}O_2$	196.29
Guaiol	$C_{15}H_{26}O$	222.37
α-cubebene	$C_{15}H_{24}$	204.35

another study, membrane damage of *E. coli* O157: H7 was analysed by fatty acid profiling Gas chromatography (GC analysis of fatty acid methyl esters FAMEs) after exposure to eugenol (phenolic compound) and cinnamaldehyde (aliphatic aldehyde). Changes in rod morphology were practically observed by SEM. When the cells were treated with cinnamaldehyde, a reduction in unsaturated fatty acid (UFA) profile of *E. coli* was observed (Karam et al., 2013). Moreover, SEM analysis showed that cinnamaldehyde caused the structural alterations in the membrane. Big holes and white spots were observed on the cellular wall. Besides, it was investigated that eugenol caused cell wall deterioration and a high degree of cell lysis (Chang et al., 2013). However, with GC analysis, it was found that eugenol had a slight effect on the change of fatty acid profiling of *E. coli*, and the reason for this case was explained as eugenol having hydroxyl groups and these groups binding to proteins. There is a different method of action that affected the membrane protein (Di Pasqua et al., 2007). Because of customer's request, nourishment makers mean to process added substance-free, ordinary-tasting nourishment which has an additional drawn-out period of usability alongside microbiological well-being. In this way, there is increasing enthusiasm for the convention of antimicrobials of the characteristic source for the conservation of simply handled nourishments. In another investigation by Zhang et al. (2017), cinnamon oil nanoemulsions were readied utilizing ethanol and Tween 80 as a surfactant. The nanoemulsion demonstrated a stable state with a normal molecule size of 8.69 nm. The nanoemulsion was steady after centrifuging at 10,000 rpm for 20 min, put away at 60°C for almost one month, or even warmed at 80°C for 30 min, individually. Contrasted with the non-nanoemulsion parts, the individual nanoemulsion indicated higher antimicrobial movement against four microorganisms, *E. coli*, *B. subtilis*, *S. typhimurium*, and *S. aureus*, even at far lower fixations. In 2016, a study by Tian et al. likewise announced that normal cinnamaldehyde was enclosed in self-emulsifying emulsion frameworks. After the optimization process, the strength of the enclosed cinnamaldehyde was assessed by the dynamic light scattering. The phase separation was seen following after 12 days of ability under 37°C. The antimicrobial movement of cinnamaldehyde was researched by the least hindrance focus and time-kill measure. Consequently, the antimicrobial movement of cinnamaldehyde enclosed in nanoemulsion could give a long-lasting restraint on the bacterial development of *E. coli* in comparison with pure cinnamaldehyde.

In another study by Yildiri et al. (2017), the objective was to figure out the stable cinnamon oil nanoemulsions showing high antimicrobial movement by utilizing the low-energy method compared with two high-energy methods. To set up the nanoemulsions, the oil stage containing cinnamon oil and surfactant (Tween 80) at 10% (w/w) was titrated into a watery stage (refined water). For antimicrobial action, agar plate dispersion strategy with *E. coli* as the microorganism was utilized. Nanoemulsions figured by low vitality demonstrate higher antimicrobial movement against *E. coli*, while the antimicrobial action was not affected by cinnamon oil by high-energy technique. There is growing enthusiasm inside the nourishment industry for the usage of nanoemulsions as colloidal conveyance frameworks for the delivery of bioactive components, for example, antimicrobials, flavours, micronutrients, and nutraceuticals (Kong et al., 2011). The formulation of essential oils through the advancement of nanoemulsions can be useful in improving the reliability and capability of oils. Research has been done more recently indicating the upgraded stability and strength in vitro antimicrobial viability of essential oil nanoemulsion when contrasted with free essential oils against a scope of nourishment pathogens. Thus, a few features like the convergence of essential oils and emulsifier, alongside the sort of nanoemulsions utilized for arrangement, impact its antimicrobial viability and dependability (McClements, 2015). Be that as it may, a dominant part of antimicrobial activity investigations of essential oil nanoemulsion, have been assessed in invitro frameworks, and its real viability in the nutrition framework is yet incomplete. In this way, this showed the fabrication of cinnamon oil nanoemulsion may be able to possibly be created as a characteristic antimicrobial operator for food protection.

10.7 ADVANTAGES OF PLANT-BASED ANTIMICROBIALS

Antimicrobial aggravates that originate from plant sources display remarkable viability against most pathogenic microorganisms capable of foodborne sicknesses and nourishment decay (Tiwari et al., 2009). There is a solid consumer discernment that normal additives have fewer symptoms to well-being that their non-common associates, even though, in a few cases, the focus required to accomplish an antimicrobial impact is more prominent than that required with manufactured additives (Carocho et al., 2014). Antimicrobials got from plants are substances started from their supplementary incorporation, which assumes a defensive job against predators or focusing on conditions (Solórzano-Santos and Miranda-Novales, 2012).

The most critical gathering of plant-based antimicrobials is difficult blends of unstable mixes which must be present in the numerous herbs and flavours (Burt, 2004). The primary gatherings of mixes in charge of their antimicrobial properties incorporate different constituents like phenolic acids, quinones, saponins, and many others (BassoléI and Juliani, 2012). Despite the expanding eagerness for applying essential oil for nourishment protection, there are only some elements influencing their antimicrobial movement, for example, the poor water solvency, mass exchange, unpredictability, or reactivity can impact its adequacy in sustenance frameworks (Prakash et al., 2015). What's more, the utilization of oil fundamentally changes the organoleptic profile of nutrition or might be harmful at high focuses (Dima et al., 2015). The antimicrobial adequacy of essential oil might be likewise affected by the pH, fat substance, or water movement present in the nutrition complex (Lai and Roy, 2004). Plant-based antimicrobials may tie to lipids, proteins, and sugars in foodstuffs, requiring higher fixations than those utilized in vitro concentrates to accomplish a similar impact (Weiss et al., 2015).

10.8 CONCLUSION

This chapter features the capability of nanoemulsions to effectively encased and convey antimicrobial compounds. Taking everything into account, antimicrobial nanoemulsions are non-phospholipid-based, modest, stable, non-lethal, and non-particular antimicrobial agents that have clinical applications. Cinnamon oil contains exceptionally dynamic mixes and to use these materials, nanoemulsions can be formulated. Nanoemulsions may hypothetically be composed in a wide range of approaches to fill in as carriers of antimicrobials to enhance the safety and nature of foods. The utilization of fundamental oil stacked nanoemulsions in nourishment applications can be promising because they demonstrate basic impacts on the definite antimicrobial action of the items while taking care of customer requirements with natural items that are free of manufactured added substances. Besides, even though the fact that the entire hindrance isn't assured, these sorts of details can be utilized for improving the antimicrobial impact in medications with the anti-infection and essential oils blend. Hence, the exploitation of antibiotics can be decreased by acquiring a synergistic impact.

REFERENCES

Acosta, E. (2009). Bioavailability of nanoparticles in nutrient and nutraceutical delivery. *Current Opinion in Colloid & Interface Science*, 14(1), 3–15.

Anandharamakrishnan, C. (2014). Nanoencapsulation of food bioactive compounds. In *Techniques for Nanoencapsulation of Food Ingredients* (pp. 1–6). Springer, New York, NY.

Anton, N., Benoit, J. P., and Saulnier, P. (2008). Design and production of nanoparticles formulated from nano-emulsion templates—A review. *Journal of Controlled Release*, 128(3), 185–199.

Anton, N., and Vandamme, T. F. (2009). The universality of low-energy nano-emulsification. *International Journal of Pharmaceutics*, 377(1–2), 142–7.

Azeem, A., Rizwan, M., Ahmad, F. J., Iqbal, Z., Khar, R. K., Aqil, M., and Talegaonkar, S. (2009). Nanoemulsion components screening and selection: A technical note. *AAPS Pharmscitech*, 10(1), 69–76.

Baboota, S., Shakeel, F., Ahuja, A., Ali, J., and Shafiq, S. (2007). Design, development and evaluation of novel nanoemulsion formulations for transdermal potential of celecoxib. *Acta Pharmaceutica*, 57(3), 315–332.

Bassoléĺ, H., and Juliani, H. R. (2012). Essential oils in combination and their antimicrobial properties. *Molecules*, 17(4), 3989–4006.

Bouchemal, K., Briançon, S., Perrier, E., and Fessi, H. (2004). Nano-emulsion formulation using spontaneous emulsification: Solvent, oil and surfactant optimisation. *International Journal of Pharmaceutics*, 280(1–2), 241–251.

Burt, S. (2004). Essential oils: Their antibacterial properties and potential applications in foods—A review. *International Journal of Food Microbiology*, 94(3), 223–253.

Carocho, M., Barreiro, M. F., Morales, P., and Ferreira, I. C. (2014). Adding molecules to food, pros and cons: A review on synthetic and natural food additives. *Comprehensive Reviews in Food Science and Food Safety,* 13(4), 377–399.

Chanamai, R., and McClements, D. J. (2001). Depletion flocculation of beverage emulsions by gum arabic and modified starch. *Journal of Food Science*, 66(3), 457–463.

Chang, Y., McLandsborough, L., and McClements, D. J. (2013). Physicochemical properties and antimicrobial efficacy of carvacrol nanoemulsions formed by spontaneous emulsification. *Journal of Agricultural and Food Chemistry*, 61(37), 8906–8913.

Charoen, R., Jangchud, A., Jangchud, K., Harnsilawat, T., Naivikul, O., and McClements, D. J. (2011). Influence of biopolymer emulsifier type on formation and stability of rice bran oil-in-water emulsions: Whey protein, gum arabic, and modified starch. *Journal of Food Science*, 76(1), e165–e172.

Chawla, P., Kumar, N., Bains, A., Dhull, S. B., Kumar, M., Kaushik, R., and Punia, S. (2020). Gum arabic capped copper nanoparticles: Synthesis, characterization, and applications. *International Journal of Biological Macromolecules*, 146, 232–242.

Chawla, P., Kumar, N., Kaushik, R., and Dhull, S. B. (2019). Synthesis, characterization and cellular mineral absorption of nanoemulsions of *Rhododendron arboreum* flower extracts stabilized with gum arabic. *Journal of Food Science and Technology*, 56(12), 5194–5203.

Chime, S. A., Kenechukwu, F. C., and Attama, A. A. (2014). Nanoemulsions—advances in formulation, characterization and applications in drug delivery. In: Ali Demir Sezer (ed.) In *Application of nanotechnology in drug delivery*, (vol. 3) InTech, London, UK.

Chu, B. S., Ichikawa, S., Kanafusa, S., and Nakajima, M. (2007). Preparation of protein-stabilized β-carotene nanodispersions by emulsification–evaporation method. *Journal of the American Oil Chemists' Society*, 84(11), 1053–1062.

Delmas, T., Piraux, H., Couffin, A. C., Texier, I., Vinet, F., Poulin, P., Cates, M. E., and Bibette, J. (2011). How to prepare and stabilize very small nanoemulsions. *Langmuir*, 27(5), 1683–1692.

Devarajan, V., and Ravichandran, V. (2011). Nanoemulsions: As modified drug delivery tool. *International Journal of Comprehensive Pharmacy*, 4(1), 1–6.

Dhull, S. B., Anju, M., Punia, S., Kaushik, R., and Chawla, P (2019). Application of gum arabic in nanoemulsion for safe conveyance of bioactive components. In: Prasad, R., Kumar, V., Kumar, M., and Choudhary, D. (eds) *Nanoemulsions: A promising tool for dairy sector. In Nanobiotechnology in Bioformulations* (pp. 85–98). Springer, Cham.

Di Pasqua, R., Betts, G., Hoskins, N., Edwards, M., Ercolini, D., and Mauriello, G. (2007). Membrane toxicity of antimicrobial compounds from essential oils. *Journal of Agricultural and Food Chemistry*, 55(12), 4863–4870.

Dickinson, E. (2009). Hydrocolloids as emulsifiers and emulsion stabilizers. *Food Hydrocolloids*, 23(6), 1473–1482.

Dima, Ş., Dima, C., and Iordăchescu, G. (2015). Encapsulation of functional lipophilic food and drug biocomponents. *Food Engineering Reviews*, 7(4), 417–438.

Dixit, N., Kohli, K., and Baboota, S. (2008). Nanoemulsion system for the transdermal delivery of a poorly soluble cardiovascular drug. *PDA Journal of Pharmaceutical Science and Technology*, 62(1), 46–55.

Donsì, F., Annunziata, M., Vincensi, M., and Ferrari, G. (2012). Design of nanoemulsion-based delivery systems of natural antimicrobials: Effect of the emulsifier. *Journal of Biotechnology*, 159(4), 342–350.

Donsì, F., and Ferrari, G. (2016). Essential oil nanoemulsions as antimicrobial agents in food. *Journal of Biotechnology*, 233, 106–120.

Duncan, T. V. (2011). Applications of nanotechnology in food packaging and food safety: Barrier materials, antimicrobials and sensors. *Journal of Colloid and Interface Science*, 363(1), 1–24.

El Asbahani, A., Miladi, K., Badri, W., Sala, M., Addi, E. A., Casabianca, H., El Mousadik, A., Hartmann, D., Jilale, A., Renaud, F. N., and Elaissari, A. (2015). Essential oils: From extraction to encapsulation. *International Journal of Pharmaceutics*, 483(1–2), 220–243.

Fernandez, P., Andre, V., Rieger, J., and Kühnle, A. (2004). Nano-emulsion formation by emulsion phase inversion. *Colloids and Surfaces A: Physicochemical and Engineering Aspects*, 251(1–3), 53–58.

Ghosh, V., Mukherjee, A., and Chandrasekaran, N. (2014). Eugenol-loaded antimicrobial nanoemulsion preserves fruit juice against, microbial spoilage. *Colloids and Surfaces B: Biointerfaces*, 114, 392–397.

Ghosh, V., Saranya, S., Mukherjee, A., and Chandrasekaran, N. (2013). Cinnamon oil nanoemulsion formulation by ultrasonic emulsification: Investigation of its bactericidal activity. *Journal of Nanoscience and Nanotechnology*, 13(1), 114–122.

Gill, A. O., and Holley, R. A. (2004). Mechanisms of bactericidal action of cinnamaldehyde against *Listeria monocytogenes* and of eugenol against *L. monocytogenes* and *Lactobacillus sakei*. *Applied and Environmental Microbiology*, 70(10), 5750–5755.

Goñi, P., López, P., Sánchez, C., Gómez-Lus, R., Becerril, R., and Nerín, C. (2009). Antimicrobial activity in the vapour phase of a combination of cinnamon and clove essential oils. *Food Chemistry*, 116(4), 982–989.

Grigoriev, D. O., and Miller, R. (2009). Mono-and multilayer covered drops as carriers. *Current Opinion in Colloid & Interface Science*, 14(1), 48–59.

Gupta, P. K., Pandit, J. K., Kumar, A., Swaroop, P., and Gupta, S. (2010). Pharmaceutical nanotechnology novel nanoemulsion-high energy emulsification preparation, evaluation and application. *The Pharma Research*, 3(1), 117–138.

Gupta, S., and Kumar, P. (2012). Drug delivery using nanocarriers: Indian perspective. *Proceedings of the National Academy of Sciences, India Section B: Biological Sciences*, 82(1), 167–206.

Gutierrez, J., Barry-Ryan, C., and Bourke, P. (2008). The antimicrobial efficacy of plant essential oil combinations and interactions with food ingredients. *International Journal of Food Microbiology*, 124(1), 91–97.

Hamouda, T., Myc, A., Donovan, B., Shih, A. Y., Reuter, J. D., and Baker, J. R. (2001). A novel surfactant nanoemulsion with a unique non-irritant topical antimicrobial activity against bacteria, enveloped viruses and fungi. *Microbiological Research*, 156(1), 1–7.

He, W., Tan, Y., Tian, Z., Chen, L., Hu, F., and Wu, W. (2011). Food protein-stabilized nanoemulsions as potential delivery systems for poorly water-soluble drugs: Preparation, in vitro characterization, and pharmacokinetics in rats. *International Journal of Nanomedicine*, 6, 521.

Holley, R. A., and Patel, D. (2005). Improvement in shelf-life and safety of perishable foods by plant essential oils and smoke antimicrobials. *Food Microbiology*, 22(4), 273–292.

Huang, Q., Yu, H., and Ru, Q. (2010). Bioavailability and delivery of nutraceuticals using nanotechnology. *Journal of Food Science*, 75(1), R50–R57.

Ichikawa, H., Watanabe, T., Tokumitsu, H., and Fukumori, Y. (2007). Formulation considerations of gadolinium lipid nanoemulsion for intravenous delivery to tumors in neutron-capture therapy. *Current Drug Delivery*, 4(2), 131–140.

Jaiswal, M., Dudhe, R., and Sharma, P. K. (2015). Nanoemulsion: An advanced mode of drug delivery system. *3 Biotech*, 5(2), 123–127.

Kanatt, S. R., Rao, M. S., Chawla, S. P., and Sharma, A. (2012). Active chitosan–polyvinyl alcohol films with natural extracts. *Food Hydrocolloids*, 29(2), 290–297.

Karam, L., Jama, C., Dhulster, P., and Chihib, N. E. (2013). Study of surface interactions between peptides, materials and bacteria for setting up antimicrobial surfaces and active food packaging. *Journal of Materials and Environmental Science*, 4(5), 798–821.

Kentish, S., Wooster, T. J., Ashok Kumar, M., Balachandran, S., Mawson, R., and Simons, L. (2008). The use of ultrasonics for nanoemulsion preparation. *Innovative Food Science & Emerging Technologies*, 9(2), 170–175.

Klang, V., Matsko, N. B., Valenta, C., and Hofer, F. (2012). Electron microscopy of nanoemulsions: An essential tool for characterisation and stability assessment. *Micron*, 43(2–3), 85–103.

Komaiko, J., and McClements, D. J. (2014). Optimization of isothermal low-energy nanoemulsion formation: Hydrocarbon oil, non-ionic surfactant, and water systems. *Journal of Colloid and Interface Science*, 425, 59–66.

Kong, M., Chen, X. G., Kweon, D. K., and Park, H. J. (2011). Investigations on skin permeation of hyaluronic acid based nanoemulsion as transdermal carrier. *Carbohydrate Polymer*, 86(2), 837–843.

Kumar, N., and Mandal, A. (2018). Surfactant stabilized oil-in-water nanoemulsion: Stability, interfacial tension and rheology study for enhanced oil recovery application. *Energy & Fuels.*, 32(6), 6452–6466

Lai, P. K., and Roy, J. (2004). Antimicrobial and chemopreventive properties of herbs and spices. *Current Medicinal Chemistry*, 11(11), 1451–1460.

Liang, R., Xu, S., Shoemaker, C. F., Li, Y., Zhong, F., and Huang, Q. (2012). Physical and antimicrobial properties of peppermint oil nanoemulsions. *Journal of Agricultural and Food Chemistry*, 60(30), 7548–7555.

Lim, S. S., Baik, M. Y., Decker, E. A., Henson, L., Popplewell, L. M., McClements, D. J., and Choi, S. J. (2011). Stabilization of orange oil-in-water emulsions: A new role for ester gum as an Ostwald ripening inhibitor. *Food Chemistry*, 128(4), 1023–1028.

Lovelyn, C., and Attama, A. A. (2011). Current state of nanoemulsions in drug delivery. *Journal of Biomaterials and Nanobiotechnology*, 2(5), 626.

Luykx, D. M., Peters, R. J., van Ruth, S. M., and Bouwmeester, H. (2008). A review of analytical methods for the identification and characterization of nano delivery systems in food. *Journal of Agricultural and Food Chemistry*, 56(18), 8231–8247.

Ma, Q., Davidson, P. M., and Zhong, Q. (2016). Antimicrobial properties of microemulsions formulated with essential oils, soybean oil, and Tween 80. *International Journal of Food Microbiology*, 226, 20–25.

Maali, A., and Mosavian, M. H. (2013). Preparation and application of nanoemulsions in the last decade (2000–2010). *Journal of Dispersion Science and Technology*, 34(1), 92–105.

Mahdi Jafari, S., He, Y., and Bhandari, B. (2006). Nano-emulsion production by sonication and microfluidization—a comparison. *International Journal of Food Properties*, 9(3), 475–485.

McClements, D. J. (2004). Protein-stabilized emulsions. *Current Opinion in Colloid & Interface Science,* 9(5), 305–313.

McClements, D. J. (2011). Edible nanoemulsions: Fabrication, properties, and functional performance. *Soft Matter*, 7(6), 2297–2316.

McClements, D. J. (2012). Nanoemulsions versus microemulsions: Terminology, differences, and similarities. *Soft Matter*, 8(6), 1719–1729.

McClements, D. J. (2015). *Food emulsions: Principles, practices, and techniques*. CRC press in Contemporary Food Science, United States.

McClements, D. J., and Rao, J. (2011). Food-grade nanoemulsions: Formulation, fabrication, properties, performance, biological fate, and potential toxicity. *Critical Reviews in Food Science and Nutrition*, 51(4), 285–330.

Özcan, M. M., and Arslan, D. (2011). Antioxidant effect of essential oils of rosemary, clove and cinnamon on hazelnut and poppy oils. *Food Chemistry*, 129(1), 171–174.

Panghal, A., Chhikara, N., Anshid, V., Charan, M. V. S., Surendran, V., Malik, A., and Dhull, S. B. (2019). Nanoemulsions: A promising tool for dairy sector. In: Prasad, R., Kumar, V., Kumar, M., and Choudhary, D. (eds) *In Nanobiotechnology in Bioformulations* (pp. 99–117). Springer, Cham.

Paranagama, P. A., Wimalasena, S., Jayatilake, G. S., Jayawardena, A. L., Senanayake, U. M., and Mubarak, A. M. (2001). A comparison of essential oil constituents of bark, leaf, root and fruit of cinnamon (*Cinnamomum zeylanicum* Blum) grown in Sri Lanka. *Journal of the National Science Foundation of Sri Lanka*, 29(3–4), 147–153.

Ping, H., Zhang, G., and Ren, G. (2010). Antidiabetic effects of cinnamon oil in diabetic KK-Ay mice. *Food and Chemical Toxicology*, 48(8–9), 2344–2349.

Prakash, B., Kedia, A., Mishra, P. K., and Dubey, N. K. (2015). Plant essential oils as food preservatives to control moulds, mycotoxin contamination and oxidative deterioration of agri-food commodities–Potentials and challenges. *Food Control*, 47, 381–391.

Preetz, C., Hauser, A., Hause, G., Kramer, A., and Mader, K. (2010). Application of atomic force microscopy and ultrasonic resonator technology on nanoscale: Distinction of nanoemulsions from nanocapsules. *European Journal of Pharmaceutical Sciences*, 39(1–3), 141–151.

Punia, S., Sandhu, K. S., Siroha, A. K., and Dhull, S. B. (2019). Omega 3-metabolism, absorption, bioavailability and health benefits-a review. *Pharma Nutrition*, 10, 100162.

Qian, C., Decker, E. A., Xiao, H., and McClements, D. J. (2012). Physical and chemical stability of β-carotene-enriched nanoemulsions: Influence of pH, ionic strength, temperature, and emulsifier type. *Food Chemistry*, 132(3), 1221–1229.

Qin, B., Nagasaki, M., Ren, M., Bajotto, G., Oshida, Y., and Sato, Y. (2003). Cinnamon extract (traditional herb) potentiates in vivo insulin-regulated glucose utilization via enhancing insulin signaling in rats. *Diabetes Research and Clinical Practice*, 62(3), 139–148.

Reza, K. H. (2011). Nanoemulsion as a novel transdermal drug delivery system. *International Journal of Pharmaceutical Sciences and Research*, 2(8), 1938.

Sagis, L.M., editor (2015). *Microencapsulation and microspheres for food applications*. Academic Press, Cambridge, United States.

Savardekar, P., and Bajaj, A. (2016). Nanoemulsions a review. *IJRPC*, 6(2), 312–322.

Setya, S., Talegaonkar, S., and Razdan, B. K. (2014). Nanoemulsions: Formulation methods and stability aspects. *World Journal of Pharmacy and Pharmaceutical Sciences*, 3(2), 2214–2228.

Singh, G., Maurya, S., and Catalan, C. A. (2007). A comparison of chemical, antioxidant and antimicrobial studies of cinnamon leaf and bark volatile oils, oleoresins and their constituents. *Food and Chemical Toxicology*, 45(9), 1650–1661.

Singh, H. (2016). Nanotechnology applications in functional foods; opportunities and challenges. *Preventive Nutrition and Food Science*, 21(1), 1.

Skrinjar, M. M., and Nemet, N. T. (2009). Antimicrobial effects of spices and herbs essential oils. *Acta Periodica Technologica*, (40), 195–209.

Solans, C., Izquierdo, P., Nolla, J., Azemar, N., and Garcia-Celma, M. J. (2005). Nano-emulsions. *Current Opinion in Colloid & Interface Science*, 10(3–4), 102–110.

Solans, C., and Sole, I. (2012). Nano-emulsions: Formation by low-energy methods. *Current Opinion in Colloid & Interface Science*, 17(5), 246–254.

Sole, I., Solans, C., Maestro, A., Gonzalez, C., and Gutierrez, J. M. (2012). Study of nano-emulsion formation by dilution of microemulsions. *Journal of Colloid and Interface Science*, 376(1), 133–139.

Solórzano-Santos, F., and Miranda-Novales, M. G. (2012). Essential oils from aromatic herbs as antimicrobial agents. *Current Opinion in Biotechnology*, 23(2), 136–141.

Swathy, J. S., Mishra, P., Thomas, J., Mukherjee, A., and Chandrasekaran, N. (2018). Antimicrobial potency of high-energy emulsified black pepper oil nanoemulsion against aquaculture pathogen. *Aquaculture*, 491, 210–220.

Tadros, T., Izquierdo, P., Esquena, J., and Solans, C. (2004). Formation and stability of nano-emulsions. *Advances in Colloid and Interface Science*, 108, 303–318.

Taylor, P. (1998). Ostwald ripening in emulsions. *Advances in Colloid and Interface Science.*, 75(2), 107–163.

Thakur, N., Garg, G., Sharma, P. K., and Kumar, N. (2012). Nanoemulsions: A review on various pharmaceutical application. *Global Journal of Pharmacology*, 6(3), 222–225.

Tian, W. L., Lei, L. L., Zhang, Q., and Li, Y. (2016). Physical stability and antimicrobial activity of encapsulated cinnamaldehyde by self-emulsifying nanoemulsion. *Journal of Food Process Engineering*, 39(5), 462–471.

Tiwari, A., and Tiwari, A., editors (2013). *Nanomaterials in drug delivery, imaging, and tissue engineering*. John Wiley and Sons, United States.

Tiwari, B. K., Valdramidis, V. P., O'Donnell, C. P., Muthukumarappan, K., Bourke, P., and Cullen, P. J. (2009). Application of natural antimicrobials for food preservation. *Journal of Agricultural and Food Chemistry*, 57(14), 5987–6000.

Troncoso, E., Aguilera, J. M., and McClements, D. J. (2012). Fabrication, characterization and lipase digestibility of food-grade nanoemulsions. *Food Hydrocolloids*, 27(2), 355–363.

Tung, Y. T., Chua, M. T., Wang, S. Y., and Chang, S. T. (2008). Anti-inflammation activities of essential oil and its constituents from indigenous cinnamon (*Cinnamomum osmophloeum*) twigs. *Bioresource Technology*, 99(9), 3908–3913.

Tzortzakism, N. G. (2009). Impact of cinnamon oil-enrichment on microbial spoilage of fresh produce. *Innovative Food Science & Emerging Technologies*, 10(1), 97–102.

Wang, R., Wang, R., and Yang, B. (2009). Extraction of essential oils from five cinnamon leaves and identification of their volatile compound compositions. *Innovative Food Science & Emerging Technologies*, 10(2), 289–292.

Wang, Y, Zhang, Y., Shi, Y. Q., Pan, X. H., Lu, Y. H., and Cao, P. (2018). Antibacterial effects of cinnamon (*Cinnamomumzeylanicum*) bark essential oil on *Porphyromonasgingivalis*. *Microbial Pathogenesis,* 116, 26–32.

Weiss, J., Loeffler, M., and Terjung, N. (2015). The antimicrobial paradox: Why preservatives lose activity in foods. *Current Opinion in Food Science*, 4, 69–75.

Yildiri, S. T., Oztop, M. H., and Soyer, Y. (2017). Cinnamon oil nanoemulsions by spontaneous emulsification: Formulation, characterization and antimicrobial activity. *LWT-Food Science and Technology*, 84, 122–128.

Zhang, S., Zhang, M., Fang, Z., and Liu, Y. (2017). Preparation and characterization of blended cloves/cinnamon essential oil nanoemulsions. *LWT-Food Science and Technology*, 75, 316–322.

Zhou, F., Ji, B., Zhang, H., Jiang, H. U., Yang, Z., Li, J., Li, J., and Yan, W. (2007). The antibacterial effect of cinnamaldehyde, thymol, carvacrol and their combinations against the foodborne pathogen *Salmonella typhimurium*. *Journal of Food Safety*, 27(2), 124–133.

11 Applications, Formulations, Antimicrobial Efficacy, and Regulations of Essential Oils Nanoemulsions in Food

Anil Panghal, Nitin Kumar, Sunil Kumar, Arun Kumar Attkan, and Mukesh Kumar Garg
Chaudhary Charan Singh Haryana Agricultural University, India
Navnidhi Chhikara
Guru Jambheshwar University of Science and Technology, India

CONTENTS

11.1 INTRODUCTION

Natural compounds are in high demand in the growing food industry to develop novel food ingredients including the food preservatives. Among these preservatives, natural antimicrobial agents against spoilage and pathogenic microorganisms are in high demand. This increase in demand of natural antimicrobial agents is due to artificial chemical compounds used in food processing to control food spoilage, which is responsible for causing various health hazards such as respiratory disease, cancer, or other health risks. Essential oils from natural sources are complex mixtures of different volatile and nonvolatile organic molecules like alcohols, aldehydes, esters, ketones, phenols, and terpenes. These are generally recognized as safe (GRAS) and are potential antimicrobial agents. Researchers, scientists, and food industries have investigated these organic components extensively for their antimicrobial efficacy against spoilage and pathogenic microorganisms to improve food safety (Kehinde et al., 2020). Different essential oils exhibited effective antimicrobial action against foodborne pathogenic and spoilage microbes. In addition to antimicrobial actions, these organic compounds present in essential oil impart an additional function to food and these food products are called functional foods (Kaur et al., 2019). Essential oils are also good flavouring, antioxidant, insecticidal, anti-inflammatory, antiallergic, and anticancer agents. Essential oils are introduced to food processing as a component for disease prevention or health promotion values of food in addition to their nutritional value.

Essential oils have been used in different therapies for thousands of years due to their bioactive nature. The importance of essential oils from aromatic plants was realized in early ages. Arabs were extracting essential oils from aromatic plants in the 10th century and using them medicinally. During the spread of bubonic plague in England in the 14th century, the burning of plants containing essential oils in the streets at night to prevent the spread of microbes was ordered. The historical records exhibited the remarkable role of essential oils in traditional medicine. Food researchers are exploring essential oils applications as a food additive also due to recent growing interest in view of their strong antimicrobial properties. Foodborne pathogens are of major concern due to illnesses, hospitalizations, and deaths of human beings. The foodborne illness cost of world is more than 300 billion US dollars annually for immediate medical care and long-term health. Statistical data provided by Centers for Disease Control and Prevention of the United States indicate that 3000 deaths and 48 million illnesses in United States are due to bacterial contamination of food. The condition is worsening in the developing world. Improving the effectiveness of antimicrobial activity and reducing microbial load from food with natural agents like essential oils can help to prevent foodborne pathogens, improve food safety, and further reduce food waste.

During incorporation into food systems, the hydrophobic binding and low water solubility of essential oils with food components like proteins and lipids reduce their activity. The food matrix intrinsic factors (such as carbohydrates, fat, protein, pH, water activity, and salts) and extrinsic factors (like temperature, gaseous composition, and types of microorganisms) could reduce the antimicrobial action of essential oils. In food systems, essential oils may bind to carbohydrates, lipids and proteins and, therefore, require high concentrations for antimicrobial efficacy that may lead to changes in sensory properties (Weiss, Loeffler, and Terjung, 2015). The intense aroma and pungent flavour of majority essential oils, even at low doses, may negatively influence the organoleptic and sensorial properties (like colour, flavour, taste, and texture) of food items lead to rejection at consumer perspective (Noori, Zeynali, and Almasi, 2018). Therefore, the application of essential oils in form of nanoemulsions may overcome these disadvantages and can improve binding of essential oils in food matrix without compromising bioactive properties.

Emulsions are of different types are formed when one immiscible liquid is dispersed uniformly in the other liquid as tiny spherical droplets. Emulsions are defined by International Union of Pure and Applied Chemistry as "a fluid colloidal system in which liquid droplets and/or liquid crystals are dispersed in a liquid". The emulsion is oil-in-water and denoted by the symbol O/W if the continuous phase of the emulsion is an aqueous solution, whereas the emulsion is referred to as water-in-oil (W/O) if the continuous phase is oil. O/W emulsions of essential oils are used for the antimicrobial actions in the food matrix. Nevertheless, it is evident that the inefficient long-term stability and low bioavailability of essential oil emulsions may not sustain their benefits. Nanoemulsions are emulsions which have the dispersed phase droplets diameter are in the range of 1 to 100 nm. There is an increasing interest in the formulation, preparation, and utilization of essential oil nanoemulsions due to their novel physicochemical properties and higher bioavailability of the encapsulated essential oils. Nanoemulsions exhibited higher antimicrobial activity in comparison with non-nanoemulsion counterparts. The O/W nanoemulsions of essential oils are exploited to encapsulate, protect, and deliver hydrophobic components in a wide variety of industries, such as cosmetics, agrochemicals, food, and pharmaceuticals (Panghal et al., 2019a). Nanoemulsions with their unique subcellular size can effectively increase the distribution of antimicrobial agents in food matrices where microorganisms are preferably located, grow, and proliferate, and therefore, greater antimicrobial activity could be achieved (Zhang et al., 2017). Nanoemulsions can be dissolved and hold the essential oils well in the aqueous medium to exert long-time continuous functionalities like antimicrobial activity. Therefore, nanoemulsions are potentially studied and used as antimicrobial preservatives to enhance food safety in the food industry. The impact of nanotechnological advances utilization is also direct in several areas, like improved food packaging, efficient incorporation of food ingredients and additives, wastewater treatment, and nanosensors development for food safety and traceability (Chhikara et al., 2018). Moreover, nanoemulsions have various other advantages, like a low impact on the organoleptic properties of the food products along with improved bioactivity with better diffusion due to subcellular size (Sharma et al., 2019). Nanoemulsions help in the mass transfer of essential oil to the cell membranes of microbes, and the antimicrobial action supports the researchers to adopt the application of essential oils in food systems. Therefore, the delivery system for essential oils with micro- and nanoemulsions is studied extensively. These delivery systems distributed essential oil uniformly in complex food matrices and reduce the negative impacts on sensory, as well as other profiles, of food and became one of the most interesting delivery systems in food industry. Nanoemulsion-based delivery systems also improve the bioavailability of the encapsulated bioactive components and increase food stability. Due to the aforementioned advantages and characteristic properties, nanoemulsions of essential oils are widely utilized in food processing industries over the conventional emulsions. Thyme oil was used with sodium casinate and lecithin in nanoemulsion formation and shows antimicrobial activity when tested against bacterial pathogens in tryptic soy broth

(Xue et al., 2015). Oregano oil nanoemulsion was applied on fresh lettuce and found an effective antimicrobial agent against *Escherichia coli* O157:H7, *Listeria monocytogenes*, and *Salmonella typhimurium* (Bhargava et al., 2015). Carvacol nanoemulsions were studied against *Salmonella enterica* Enteritidis and *E. coli* in different fresh products like alfalfa seeds, broccoli, mung beans, and radish and found effective in all fresh products except broccoli (Landry et al., 2015). Chitosan-based nanoemulsions of mandarin oil were used in minimal processing for green beans and have shown promising results (Severino et al., 2014). Different methods for generation of nanoemulsions delivery systems of essential oils are reported membrane emulsification, spontaneous emulsification, phase inversion composition, ultrasonication, high-pressure homogenization, and high-energy emulsification technologies like microfluidization (Panghal et al., 2019b).

The preparation of nanoparticulate with different materials involving the shrinkage of large raw material objects to nanosized materials on exposure to diverse electrochemical, chemical, and mechanical processes to the molecular or atomic-scale dimensional agglomeration of materials to the nano-phase is most challenging and stability of these matrixes upon storage is still a grey area to explore.

11.2 METHODS FOR FORMULATION AND PREPARATION OF NANOEMULSIONS

11.2.1 Properties and Composition of Essential Oils

The stable nanoemulsions are usually of the O/W type, which consists of oil in the dispersed phase, with mean droplet diameter ranging from 30 to 250 mm, in a suitable dispersion medium, usually aqueous. This matrix is essentially stabilized by an emulsifying agent, which can be a food-grade surfactant like lecithins and polysorbates, or it can be a biopolymer like starch, gum, and protein, depending on the application of nanoemulsion. These surfactants not only reduce the surface tension of the solution but also provide them with desired attributes like specific interfacial behaviour and load-bearing capacity (Donsì et al., 2011; Malathi et al., 2017; Kumar et al., 2020). Essential oils usually consist of 20 to 40 different types of compounds in different concentration but are dominated by unsaturated hydrocarbons (>70%). The major antimicrobial agents present in these O/W nanoemulsions are identified by their similarity to alcohols, terpenes, phenols, and aldehyde. For instance, in clove oil, eugenol is the main constituent, which is a phenol. The menthol and terpineol are present in sage and peppermint oils. There is a huge variation in structure and composition of these compounds, thus their antimicrobial activity solely depends upon the configuration of their structure (Pathania et al., 2018; Mwaurah et al., 2020). Sometimes antimicrobial agents' activity is also affected by various unit operations performed especially in the food industry (Table 11.1).

Emulsions are generally thermodynamically unstable leading to their separation with time. Unlike them, the stability of an essential oil nanoemulsion (EON) is determined by the Brownian motion which dominates over gravitational force. Additionally, the EON is more stable when the net attractive forces between the droplets are more. As per a general trend, the strength of these forces reduces with

TABLE 11.1
Effect of Various Packaging Unit Operations on Antimicrobial Agents' Activity

Packaging Unit Operation	Antimicrobial Agent	Target Microorganism	Results	References
Thermo-compression Extrusion molding	Chitosan	*Escherichia coli* *Staphylococcus aureus* *Salmonella enteritidis* *Bacillus cereus*	o Processing improved mechanical properties of packaging o Antimicrobial activities of films were not affected by processing	Tomé et al. (2012)
Thermoforming Extrusion molding	Lysozymes	*Micrococcus lysodeikticus*	o High-temperature processing inactivated lysozymes activities o Increase in plasticizer favored lysozyme activity due to their increase in cross-linking	Mlalila et al. (2018); Jbilou et al. (2014)
Compression molding	Potassium sorbate (KS)	None (research was aimed to check the stability of KS under compression molding)	o Processing retained antimicrobial agent in films o Mechanical and barrier properties improved	Mlalila et al. (2018); Ortega-Toro et al. (2015); Kumar et al. (2020)
Extrusion molding	KS	*Zygosaccharomyces bailii*	o KS was not affected during processing	Flores et al. (2010)

the reduction in droplet size, thereby reducing the aggregation in an EON (Landry et al., 2014; Kumar et al., 2017). Furthermore, when the droplet diameter is made smaller than the wavelength of visible light (400–700 nm), the droplets are unable to scatter light and become transparent which makes them suitable for incorporating in shampoos and clear beverages. This phenomenon of optical transparency is achieved when the mean droplet diameter is less than 38 nm. Between 38 and 90 nm, an EON appears hazy and finally white when the mean droplet size crosses 90 nm due to induced reflection of light (Mason et al., 2007). Generally, the reactivity of an EON increases with the decrease in droplet size due to increased surface area–to–volume ratio which increases the activity of lipophilic compounds present in EONs (Donsì et al., 2012). The antimicrobial action and applications of essential oil nanoemulsions and their correlation with the formulation and mean droplet size are provided in Table 11.2.

11.2.2 Method of Fabrication of Essential Oil Nanoemulsions

The top-down and bottom-up approaches are used to fabricate the EON. In the former, the oil phase is disrupted into uniform, fine, and homogeneous droplets by mechanical means while the latter emphasizes gathering and assembling of molecules into a structure.

11.2.2.1 Top-Down Fabrication

This process employs a mechanical mean for reducing the size of droplet, thereby generating fluid mechanical stresses within EONs. The emulsification is achieved by reducing the size of droplets followed by absorption of emulsifying agent on these droplets to avoid their reaggregation and to provide stability to EONs. The amount of the emulsifying agent used, dispersed phase and dispersion medium concentration and processing conditions of temperature and pressure are the key factors in the fabrication of EONs (Silva and Cerqueira, 2015; Devgan et al., 2019). Some of the processes involved in top-down fabrication include ultrasonication, colloid milling, membrane emulsification, and high-pressure homogenization. The top-down fabrication processes are characterized by their high-energy input. Ultrasonication process is widely used in laboratories to fabricate EONs. It produces low-and high-pressure sound waves alternatively at high frequency (>19kHz) which produces bubbles in the EON matrix. The bubbles then implode inside the surrounding liquid, thus releasing energy in the form of shock waves which locally increase the pressure in the liquid to as high as 1.3 MPa (Donsì et al., 2011). Ultrasonication can produce uber-fine nanoemulsions, but local hotspots produced during this process might affect the antimicrobial activity and reactivity of EONs (Salvia-Trujillo et al., 2015a; Kumar, 2018). High-pressure homogenization comprises compressing the fluid to high pressures (60–400 MPa) and forcing it through micrometric homogenization chamber. This results in the formation of acute turbulence, cavitation, and shear stress which disrupt the droplets into a fine and uniform size. To further maintain the uniformity in the size of droplets, multiple passes of EON fluid have been recommended through micrometric homogenization chamber. The size of the droplets has been found directly proportional to the pressure applied during this process (Ghosh et al.,

TABLE 11.2
Antimicrobial Action, Applications of Essential Oil Nanoemulsions and their Correlation with Formulation and Mean Droplet Size

Type of Essential Oil	Purpose	Formulae	Reference/Tested Microorganism(s)	Surfactant	Mean Droplet Size (mm)	Nanoformulation Antimicrobial Effect	References
Eugenol/Clove oil	Food (Fruit juice), pharma (anti-inflammatory and antiseptic properties)	$C_{10}H_{12}O_2$	*Listeria monocytogenes, Escherichia coli*	Polysorbate 20,80, maltodextrin conjugates	130	Negative	Ghosh et al. (2014); Shah et al. (2012)
Mandarin oil	Pharma and cosmetic industries (aromatic and anti-depression properties)	$C_{10}H_{16}$	*L. innocua*	Polysorbate 80	90	Slightly positive	Severino et al. (2014); Donsì and Ferrari (2016)
Oregano oil (carvacrol +thymol+ cymene)	Pharma industries (antiviral, antioxidant, antidiabetic, anti-inflammatory and cancer suppressant)	$C_{10}H_{14}O$	*L. monocytogenes, Salmonella typhimurium, E. coli*	Polysorbate 20	105	Slightly positive	Bhargava et al. (2015); Pathania et al. (2018)
Pure carvacrol/ cymophenol	Pharma industries (antimicrobial, antioxidant and antidiabetic,)	$C_{10}H_{14}O$	*E. coli, S. enteric Enteritidis*	Polysorbate 80, Lecithin	172	Positive	Landry et al. (2015)
Lemongrass oil	Food, cosmetic and pharma industries (treatment of high blood pressure and digestive problems	$C_{51}H_{84}O_5$	*S. typhimurium, E. coli*	Polysorbate 80	15	Positive	Kim et al. (2013); Pathania et al. (2018)
Cinnamaldehyde	Food (flavouring agent)	C_9H_8O	*Lactobacillus delbrueckii, E. coli, S. cerevisiae*	Polysorbate 20	83	Slightly positive	Donsì and Ferrari (2016)

(Continued)

TABLE 11.2 *(Continued)*
Antimicrobial Action, Applications of Essential Oil Nanoemulsions and their Correlation with Formulation and Mean Droplet Size

Type of Essential Oil	Purpose	Formulae	Reference/Tested Microorganism(s)	Surfactant	Mean Droplet Size (mm)	Nanoformulation Antimicrobial Effect	References
Eucalyptus oil	Pharma industry (treatment of nasal and chest congestion, asthma, arthritis, and skin ulcers)	$C_{19}H_{18}O$	*Proteus mirabilis*, *E. coli*	Polysorbate 20	110	Positive	Saranya et al. (2012)
Basil oil	Food- and cosmetic-based applications	$C_{10}H_{160}$	*E. coli*	Polysorbate 80	140	Slightly positive	Ghosh et al. (2014); Pathania et al. (2018)
Palm oil	Food based applications (cooking, frying and as an emulsifying agent)	$C_{16}H_{32}O_2$ + $C_{18}H_{34}O_2$	*L. delbrueckii*, *E. coli*, *S. cerevisiae*	Polysorbate 80, glycerol and polysorbate 20	97	Slightly positive to neutral	Baldissera et al. (2013)
Peppermint oil	Food, Pharma and cosmetic industries	$C_{62}H_{108}O_7$	*L. monocytogenes*, *Staphylococcus aureus*	Starch and glycerol	149	Slightly negative	Donsì and Ferrari (2016)
Sage oil	Pharmaceuticals industry (therapeutic properties)	$C_{10}H_{16}O$ + $C_{10}H_{16}$	*L. delbrueckii*, *E. coli*, *S. cerevisiae*	Polysorbate 80, Lecithin	110–170	Positive	Flores et al. (2013)
Limonene/ Lemon oil	Food (as flavouring agent), pharmaceuticals (obesity, cancer, bronchitis)	$C_{10}H_{16}$	*Staphylococcus aureus*, *E. coli*, *L. delbrueckii*	Polysorbate 80, starch	90–185	Positive	Saranya et al. (2012); Ghosh et al. (2014)
Sunflower oil	Food and cosmetic (frying, cooking, emollient)	$C_{18}H_{32}O_2$	*L. delbrueckii*, *S. cerevisiae*	Polysorbate 20	78	Neutral to slightly positive	Donsì et al. (2011)

Type of Essential Oil	Purpose	Formulae	Reference/Tested Microorganism(s)	Surfactant	Mean Droplet Size (mm)	Nanoformulation Antimicrobial Effect	References
Thyme oil	Pharmaceuticals industry (antibiotic and antifungal properties)	$C_{50}H_{82}O_4$	*S. aureus*, *E. coli*, *L. monocytogenes*	Lecithin, sorbitan monolaurate, polysorbate 80	80–115	Slightly positive	Landry et al. (2015); Donsì et al. (2011)
Crabwood oil/ Andiroba oil	Pharmaceutical and chemical industries (treatment of arthritis, joint aches, anti-inflammatory properties, fuel, solvent for dissolving dyes, insect repellent)	–	*Trypanosoma evansi*	Polysorbate 80 and polysorbate 20	130	Slightly positive	Shah et al. (2012); Donsì and Ferrari (2016)
Canola oil	Food and industries (frying, baking, an ingredient in sauces and marinades, manufacturing of plastics, and adhesives and is used as a biofuel)	$C_{18}H_{34}O_2$ + $C_{18}H_{30}O_2$	*Staphylococcus aureus*, *E. coli*, *L. monocytogenes*	Starch	155	Neutral	Majeed et al. (2016); Pathania et al. (2018)
Neem oil/ Azadirachtin	Agricultural, cosmetics, food, and pharmaceuticals industries (manure, soil conditioner, fumigant, urea coating agent, food additive, emollient and wound healing properties)	$C_{35}H_{44}O_{16}$	*Vibrio vulnificus*, *E. coli*, *S. aureus*	Polysorbate 20	110–140	Positive	Ghotbi et al. (2014); Flores et al. (2013)

(Continued)

TABLE 11.2 (Continued)
Antimicrobial Action, Applications of Essential Oil Nanoemulsions and their Correlation with Formulation and Mean Droplet Size

Type of Essential Oil	Purpose	Formulae	Reference/Tested Microorganism(s)	Surfactant	Mean Droplet Size (mm)	Nanoformulation Antimicrobial Effect	References
Anethole/Anise oil	Food and pharmaceutical industries (flavouring agent in alcohols, meat and candies; manufacturing of soap, perfume, and sachets)	$C_{10}H_{12}O$	*Pseudomonas aeruginosa*, *L. monocytogenes*, *E. coli*	Triglycerides	120–145	Slightly positive	Orav et al. (2008); Donsì and Ferrari (2016)
Savory oil	Pharmaceutical and food industries (strong antiseptic, antidiuretic, treatment of diarrhoea, nausea, loss of appetite)	$C_{10}H_{14}O$ + $C_{10}H_{16}$	*Bacillus cereus*, *L. monocytogenes*, *S. aureus*	Polysorbate 20	110–170	Slightly positive	Tozlu et al. (2011); Pathania et al. (2018)

2014; Donsì and Ferrari, 2016; Donsì et al., 2012). Colloid milling essentially utilizes high-speed rotors to produce intense shear, vibration, and friction which collectively reduce the size of EONs. This type of system is scalable and cost-effective but unable to provide extremely fine droplet size, thereby limiting their applicability to the fabrication of only coarse nanoemulsions (Pan et al., 2014). Membrane emulsification comparatively requires less energy and thus produces low shear. It is utilized to fabricate O/W nanoemulsions in a narrow size range. The dispersion phase is forced to pass through a micrometric membrane into an aqueous phase containing a hydrophilic emulsifying agent by applying either positive pressure on one side or negative pressure on the other (Liu et al., 2011). Despite the simplicity and flexibility of this process, it is limited to the laboratories experimentation, and no industry application has been reported yet.

With respect to the industrial applicability, high-pressure homogenization offers major advantages in terms of simplicity of process, scalability, cost, and high throughput. Colloid milling has found its application in the food, pharmaceutical, and cosmetic industries, where the size of particles/droplets is kept comparatively large. The ultrasonication process confronts scalability issues when it comes to the industrial applicability, yet some of the paints, cosmetics, and food-processing industries have developed several prototypes as per their use (Salvia-Trujillo et al., 2015b; Donsì and Ferrari 2016; Mlalila et al., 2018).

11.2.2.2 Bottom-Up Fabrication

The bottom-up process for fabrication of essential oil–based nanoemulsions is a physicochemical process in which the structured oil molecules are encompassed by an emulsifying agent. This process is driven by the balanced repulsive and attractive forces making a thermodynamic equilibrium. The assembling of oil molecules in a structure depends on their intrinsic properties like geometry solubility and surface activity. Additionally, the environmental components like temperature and properties of the dispersed phase and dispersing medium such as ionic strength, pH, and concentration significantly affect the fabrication process (McClements and Rao, 2011; Sessa and Donsi, 2015; Kehinde et al., 2020). Some mechanical energy is also required for the purpose of agitation.

The bottom-up fabrication process is characterized by its low-energy requirement for producing extremely small droplets, utilizing the simple and scalable equipment. Nevertheless, a strict protocol needs to be followed in the selection of the emulsifying agent and oil type for formulating stable EONs. Additionally, the surfactant-to-oil ratio has to be kept lower than in the top-down fabrication process. Some of the processes involved in bottom-up fabrication include solvent demixing, phase inversion, and spontaneous emulsification. In the solvent demixing technique, the oil phase is dissolved in a preselected organic solvent followed by its segregation in nano form by the addition of an aqueous anti-solvent containing an emulsifying agent for stabilization (Baldissera et al., 2013). The organic solvent diffuses rapidly in the aqueous anti-solvent, leading to the formation of nanoemulsions with high degree of encapsulation. Precise intricacy is required in this process so some scientists have developed a comparatively more energy-intensive process like spray-drying of oil (thymol) in solvent (hexane) using maltodextrin, which results in a stable and transparent EON (Shah et al., 2012). Akin to solvent demixing, spontaneous emulsification includes mixing a preselected oil in a hydrophilic emulsifying agent to form stable O/W EONs. The mixing leads to interfacial turbulence which further gives rise to the formation of EONs (McClements and Rao, 2011). Phase inversion pertains to a process in which O/W emulsion phases back to W/O on agitation and vice versa depending on the environmental factors like temperature, pH, and concentration of surfactant. It produces extremely fine-sized droplets in a continuous phase. It has widespread application in pharmaceuticals, cosmetics, food processing, and detergent manufacturing (Liu et al., 2011; Donsì et al., 2011).

11.3 ANTIMICROBIAL EFFICACY OF ESSENTIAL OIL NANOEMULSIONS

Mostly additives used in the food industry are synthetic as these are cheaper and more stable to different processing and storage conditions. Increased awareness about the adverse side effects of synthetic additives has forced the industries to explore natural or near-to-nature additives (Panghal et al., 2018; Dhull et al., 2019). So EONs are gaining huge importance in preservation of food functioning as natural antimicrobial agents with better lethality against different microflora, along with various health benefits. EONs are considered GRAS and thus are a suitable natural alternative to chemical preservatives (Sharma et al., 2019). EOs are composite mixtures of different hydrophilic and hydrophobic components, and the antimicrobial performance is dependent on their composition, structural configuration, concentration, and the feasible sites of interaction. Different extrinsic and intrinsic parameters of food like pH, temperature, water activity, enzymes, relative humidity, and storage practices strongly affect the functionality of oil nanoemulsions (Panghal et al., 2019b). Furthermore, the negative side of some EOs is their impact on the sensory profile of food samples, especially colour and flavour.

11.4 MODES OF ACTIONS OF ESSENTIAL OILS

Different mechanisms have been suggested for the functioning of nanoemulsions. Essential oils intermingle with the fatty layer of cell and cell organelles, enhance the cell permeability, alter cell membrane ionic potential leading to ionic loss, and thus disturb the microbial metabolism, resulting in cell lysis and microbial death (Singh et al., 2020). Nanoemulsions of essential oil being very small in size have larger surface areas for interaction with the cell membrane and thus enhance the antimicrobial effect.

The antimicrobial effect can be enhanced by using a carrier possessing antimicrobial activity. These carriers generate the reactive oxygen component and influence the microbe's metabolism. These carriers also protect essential oils from the food matrix and help in delivering the EON to specific target sites. Different carriers have been used like sodium caseinate for thyme oil emulsification (Xue et al., 2015), or sodium alginate and Tween 80 mixture in lemongrass oil (Salvia-Trujillo et al., 2015a, b) and enhanced efficacy of EOs. However, reduced antimicrobial activity has been reported in peppermint oil–based nanoemulsions with starch

(Liang et al., 2012) or eugenol nanoemulsions with whey protein and maltodextrin mixture (Shah et al., 2013) as compared to non-encapsulated compounds. Different mechanisms have been proposed for the functionality of nanoemulsions in different food systems:

a. Hydrophilic nature of nanoemulsion: The passive transport through the external cell membrane and the higher surface area account for the very small size of nanoemulsion enables better interaction with the inner cytoplasmic membrane. This leads to the alteration in the integrity of different constituents (fatty acids, polysaccharides, and phospholipids) of cellular layers and interfere in the transport protein system. This disturbance in the cell results in the release of different cell organelles and thus death of the cell (Moghimi et al., 2016; Donsì et al., 2012). In gram-negative bacteria, the droplets of nanoemulsions with hydrophilic sites move through the cell membrane via the hydrophilic transmembrane channels of porin proteins (Nazzaro et al., 2013). In the case of gram-positive bacteria and molds, the nanoemulsion droplets account for better contact on surface and thus better accessibility to their action points of microbes (Majeed et al., 2016).
b. Emulsifiers: Emulsifiers decreases the interfacial tension and thus enhances the kinetic stability of nanoemulsions and prevents collision and coalescence between droplets (Mason et al., 2007). The nature of emulsifiers used might be anionic, cationic, zwitter ionic and non-ionic, for example surfactants, polyvinyl alcohol, phospholipids, proteins, and polysaccharides, depending on the food matrix in nature (Kralova and Sjöblom, 2009). The ionic emulsifier prevents the droplets' accumulation by electrostatic repulsion and the non-ionic surfactants decline the aggregation of droplets by stearic interference and hydration (Silva and Cerqueira, 2015). Emulsifiers of different natures can be used synergistically for nanoemulsion formation. The adhesion of the emulsifier droplets to the phospholipid bilayer assists in the controlled release of the essential oils at the specified sites. This targeted approach of using emulsifiers showed an elevation in the essential oil antimicrobial action (Salvia-Trujillo et al., 2015a,b). Different studies have been reported, for example Tween 80 or sodium dodecyl sulfate (SDS) emulsifier in eugenol nanoemulsion against *E. coli* (Li et al., 2015) or a sodium caseinate emulsifier in thyme oil nanoemulsion for *E. coli*; *Listeria monocytogenes* (Xue et al., 2015), modified starch, or Tween 80 as emulsifier in thyme oil nanoemulsion against *E. coli*, *S. aureus*; and *L. monocytogenes* (Majeed et al., 2016). However, some contradictory reports are also there. Reduced antimicrobial activity has been reported in peppermint oil nanoemulsions stabilized by modified starch (Liang et al., 2012) and eugenol nano-dispersed by whey protein-maltodextrin (Shah et al., 2013).
c. Higher activation efficiency: Nanoemulsions maintain equilibrium between hydrophobic and hydrophilic components. The regulated release of EOs from nanoemulsion system is monitored by the essential oil panel between oil globules and aqueous system and thus enhances the EO activity. Majeed et al. (2016) reported an initial lower inactivation rate of encapsulated essential oils than free essential oils, and the antimicrobial potential is dependent on the

amount of the oil phase in the nanoemulsion (Ziani et al., 2011; Chang et al., 2013; Pavoni et al., 2020). The emulsifier assists in better solubility of EOs through micellization mechanisms (Donsì et al., 2012).

d. Electrostatic interaction: Hydrophobic components of EO alter the ionic permeability for cations and thus changes in cations flow, pH, and overall cell chemical composition (Hyldgaard et al., 2012; da Cruz Cabral et al., 2013). The positively charged nanoemulsion droplets interact with negatively charged microbial cell walls, and thus, the amount of EOs is increased at the site of action (Chang et al., 2015). The permeability loss results in unbalanced intracellular osmotic pressure, followed by the disruption of cell compartments; the release of cellular contents, adenosine triphosphate (ATP) molecules; and, finally, cell death. The depolarization of the mitochondrial membrane occurs by essential oils with reduction in membrane potential, alteration in ionic channels, and a reduction in pH gradient (Bakkali et al., 2005). Some cationic surfactants, such as lauric arginate, being a strong antimicrobial, enhance the antimicrobial activity (Ziani et al., 2011; Chang et al., 2015; Xueet al., 2015). On other side, anionic surfactants neither reduce the clove oil nanoemulsion's antimicrobial potential (Majeed et al., 2016) nor elevate it as in case of SDS in eugenol nanoemulsion (Li et al., 2015). Kerekes et al. (2013) reported that the phenolic components of essential oils are highly active and results in disruption of bacterial cell membrane and thus death of the bacterial cell. Essential oils are more effective in controlling gram-positive organisms, such as *L. monocytogenes*, than gram-negative organisms.

11.5 ANTIMICROBIAL ACTIVITY OF NANOEMULSIONS AGAINST NUMEROUS MICROBES

Different work has been done in past few years by researchers to highlight the antimicrobial effectiveness of EONs. The effectiveness of essential oils is dependent predominantly on structural components of host cells as well as a minimum inhibitory concentration of microbiota.

11.5.1 Bacteria

The impact of essential oils on spoilage causing and pathogenic bacteria has been extensively studied in the past few years. Gram-positive bacteria showed comparatively higher sensitivity as compared to gram-negative bacteria (Seow et al., 2014), and this variation is due to structural differences in the cells wall of bacteria. Gram-positive bacteria contain a thick layer of peptidoglycans in the cell wall, whereas the cell walls of gram-negative bacteria contain a thin peptidoglycan layer but are structurally more complex due to additional phospholipid bilayer in the outer cell membrane. Essential oils can easily pass through gram-positive bacteria cell membranes, whereas the outer membrane of gram-negative bacteria is almost impermeable to them, limiting the access of essential oils to the cell membrane (Nazzaro et al., 2013).

Moghimi et al. (2016) studied the antibacterial activity of *Thymus daenensis* essential oil water-dispersible nanoemulsion produced with high-intensity ultrasound against *E. coli*. The leakage of potassium, proteins, and nucleic acids from the cell was studied to study the antibacterial strength of nanoemulsion. The study suggested that nanoemulsions amplified the antimicrobial activity of essential oils by elevating the essential oils' ability to rupture cell membrane and complete reduction in microbial load was observed within 5 min. Bhargava et al., 2015 used oregano oil nanoemulsions on fresh lettuce as preservative. The nanoemulsion ruptures the cell membranes and thus inhibited the growth of *Salmonella typhimurium*, *L. monocytogenes*, and *E. coli* O157:H7 on fresh lettuce.

11.5.2 Fungus

Nanoemulsions are high-energy droplets that fuse nonspecifically with lipids in the microbial outer membranes causing membrane damage and thus cell death. EONs contain highly volatile components, accounting for its suitability as a fumigant for fungus in food storage houses. Some researchers are of opinion that oil nanoemulsions may inactivate the essential enzymes in cells and a disturbance in the genetic material functionality occurs (Wu and Guy, 2009). Few are of opinion that oil nanoemulsions are involved in inhibition of the biosynthesis of ergosterol (de Oliveira Pereira et al., 2013). The fungal mycelium tips are swollen, damaged, and distorted and the inhibition of sporulation occurs on nanoemulsion application (Abd-Elsalam and Khokhlov, 2015). Essential oils cause different morphological changes like cytoplasmic coagulation, the formation of giant vacuoles, protoplast leakage, and shrivelling in mycelia, and thus, death of fungus occurs (Soylu et al., 2006). Sugumar et al. (2016) used orange oil nanaoemulsions in apple juice to inhibit the growth of *Saccharomyces cerevisiae*. Otoni et al. (2014) developed pectin based edible film containing cinnamaldehyde nanoemulsions and this inhibited the growth of *E. coli*, *S. enterica*, *L. monocytogenes*, and *S. aureus*. Different researchers have used nanoemulsions of different origin in different ways for controlling the growth of microorganisms (Table 11.2).

11.6 APPLICATION OF OIL NANOEMULSIONS IN THE FOOD INDUSTRY

Different types of microorganisms are responsible for food spoilage depending on the food's characteristics and storage practices. Within a food system, different microflora can coexist, so a wide-spectrum antimicrobial component is required for food preservation. Oil nanoemulsions have been extensively used to preserve and extend the storability of food samples (Table 11.3). Nanoemulsions are highly specific in their targets. Besides this, different components in the food matrix may interact, absorb, or bind the nanoemulsion and thus reducing their effectiveness. Soa comparatively higher concentration of nanoemulsions is required in a food sample in comparison to experimental studies (Seow et al., 2014). This higher concentration of essential oils, their volatility, and their reactivity with food components might interfere with the sensory characteristics of food (Kim et al., 2013; Gyawali and Ibrahim, 2014). So, in the application of nanoemulsions, better efficiency and minimal change in sensory attributes are the criteria for their food application (Gutiérrez et al., 2008;

TABLE 11.3
Application of Nanoemulsions in Different Food Samples

Food	Nanoemulsion Used	Microorganism	Impact	References
Fruits and Vegetables				
Cherry tomato	*Cinnamomum cassia* (500 ppm)	*Alternaria alternata*	34.2% reduction in microbial population	Feng and Zheng (2007)
Tomato Paste	*Thymus vulgaris* (350 ppm) *Satureja hortensis* (500 ppm)	*Aspergillus flavus*		Omidbeygi et al. (2007)
Strawberries	*Cinnamomum cassia* (800 ppm)	*Boytrtis cinerea*	40%	El-Mogy and Alsanius (2012)
Orange slices	Orange peel essential oil nanoemulsion based coating		Reduced microbial growth	Radi et al. (2018)
Fresh cut Fuji apples	Lemongrass essential oil nanoemulsion based edible coating	Natural microflora	Complete inhibition, shelf life extension up to 2 weeks	Salvia-Trujillo et al. (2015a,b)
Fruit juice	Eugenol loaded nanoemulsion containing sesame oil, non-ionic surfactant (Tween20/Tween80)	*Staphylococcus aureus*	Significant reduction in microbial load	Ghosh et al. (2014)
Fruit juice	Lemongrass oil	*Saccharomyces cerevisiae*		Tyagi et al. (2014)
Apple juice	Orange oil, Tween 80 (organic phase) and water (aqueous phase) by sonication technique, 2μL/mL	*S. cerevisiae*	Complete inhibition	Sugumar et al. (2016)
Salad dressing	Encapsulated oregano and clove essential oils, 1.95 mg/g	*Zygosaccharomyces bailii*	Complete inhibition	Ribes et al. (2019)
Grains				
Wheat and chickpea	*Cuminum cyminum* cumin seed EO	Fungal deterioration		Kedia et al. (2014)
Finger millet	0.5 μL/mL of *Cinnamomum zeylanicum* Blume EO	*Aspergillus flavus* and *A. niger*	60%–83% reduction in microbial growth	Kiran et al. (2016)

Food	Nanoemulsion Used	Microorganism	Impact	References
Maize seeds	0.01 and 0.05 μL/mL of *S. aromaticum* and *V. diospyroides* EO, respectively	*A. flavus* PSRDC-2	Complete inhibition	Boukaew et al. (2017)
Dairy Products				
Low fat cheese	Oregano essential oil nanoemulsion (carrier 2.0% (w/w) sodium alginate, 0.5% (w/w) mandarin fiber, 2.5% (w/w) Tween 80)	*Staphylococcus aureus*, psychrophilic bacteria, molds	Shelf life extended to 24 days	Artiga-Artigas et al. (2017)
Minas Padrão cheese	Nanoemulsions encapsulating essential oil of oregano (*Origanum vulgare*)	*Cladosporium* sp., *Fusarium* sp., and *Penicillium* sp.	Antifungal activity	Bedoya-Serna et al. (2018)
Milk	*Thymus capitatus* essential oil nanoemulsion	*S. aureus*	Shelf-life extension	Jemaa et al. (2017)
Milk	Organogel-nanoemulsion containing with D-limonene and nisin	*S. aureus*, *Bacillus subtilis*, *Escherichia coli*	Antimicrobial activity	Bei et al. (2015)
Milk	Eugenol emulsions	*Listeria monocytogenes*, *E. coli* O157:H7	Shelf-life extension	Gaysinsky et al. (2007)
Meat				
Chicken breast fillet	Nanoemulsion-based edible sodium caseinate coating containing ginger essential oil (GEO)	Total aerobic psychrophilic bacteria	Shelf life extension	Noori et al. (2018)
Fish	Thyme essential oil nanoemulsions	Spoilage bacteria of fish (*P. luteola*, *P. damselae*, *V. vulnificus*, *E. faecalis*, *S. liquefaciens*, and *P. mirabilis*)	Better storability	Ozogul et al. (2020)

Basak and Guha, 2018; Singh et al., 2020). Another concern associated with nanoemulsions is the non-uniform distribution in food matrices due to hydrophobic nature (Donsìet al., 2012; Hyldgaard et al., 2012; Prakash et al., 2018). On the contrary, the bioactive compounds and micronutrients present in food may play a protective role to stabilize and maintain microbial cells and thus nullify or reduce the effectiveness of oil nanoemulsions. Uhart et al. (2006) demonstrated that oil emulsions showed no or very low effectiveness against *L. monocytogenes* in the complex food system of ground beef.

Rapidly growing food industries and changing consumer demand have motivated researchers to develop new alternatives and approaches for health food formulation. This might be fortification, textural improvement, and stability of food product using nanoemulsions. Different industries have launched various products containing nanoemulsions; for example Unilever has formulated low-fat ice cream, Nestle patented a nanoemulsion for better thawing of frozen foods in the microwave, and nanotech-based ventures Aquanova and Zyme have developed omega-3 fatty acid nanocapsules by adopting nanoemulsion technology (Dasgupta et al., 2019; Möller et al., 2009). However, a lot of work has been done through researchers for shelf-life extension of food samples using nanoemulsions (Table 11.3), but no industry has come up with practical applications of these developed technologies.

11.7 REGULATIONS

Essential oils and oleoresins (solvent-free) are generally recognized as safe for their intended uses are mentioned in section 409 of Federal Food, Drug, and Cosmetic Act by the U.S. Food and Drug Administration (FDA). Various bioactive compounds extracted from essential oils and essential oils, as a whole, have been approved by the regulatory agencies such as European Commission and the FDA for their intended use as the ingredients or antimicrobial agents in the food products. However, food producers and food ingredient manufacturers are responsible for ensuring that food products are safe for human consumption and compliant with all applicable legal and regulatory requirements (FDA, 2014). As highlighted in this chapter, nanoemulsions of essential oils may behave differently from their emulsions. Although the potential health and functional benefits of the use of engineered nanoparticles (NPs) incorporated in the food matrix are arising, with a high prospect for the food industry, the risks associated with NPs are still unknown. Therefore, it is more important to make certain that the nanoemulsions unintentionally or intentionally present in commercial food products intended for oral consumption are safe (Wani et al., 2018). A huge number of research studies have been carried out to establish the potential toxicity of nanoemulsions, but far fewer studies have been carried out to evaluate the safety and toxicity of EONs. Therefore, much of our present understanding of the potential toxicity of nanoemulsions comes from the research on various other types of nanoscale materials. Hence, food regulation from global and local regulatory bodies is an important aspect that needs to be explored for the safety issues associated with the consumption of foods containing essential oils nanoemulsions. Particularly due to the variability in batch-wise composition of plant extracts, which reduces the possibility of their standardization, it is not possible to assign them acceptable daily intake (Donsì and Ferrari, 2016). Along with that, the potential associated risks with the use of highly reactive molecules, such as those constituting essential oils, should be monitored during the operations of food processing and storage. For instance, some essential oils used in tobacco products, including cinnamon bark oil, cinnamon leaf oil, and sage oil, have been shown in in vitro studies to be toxic or are believed to possess toxic properties based on the scientific literature. In consideration of usage potential of EONs in food preservation, it is desirable that specific International Organization for Standardization standards/global standards should be developed to assess the legal aspects to set out the definition, the general rules of usage, the labelling requirements, and the maximum levels authorized. In addition, the selection and concentration of ingredients in nanoemulsions and in food formulations have to meet the constraints coming from current regulations for food formulations, as well as economic viability for the food industry (FDA, 2015).

The potential toxicity of nanoemulsions may depend on their particle size distribution, particle shape, chemical nature (organic or inorganic), surface state (e.g. surface area, surface functionalization, and surface treatment), state of aggregation/agglomeration, and so on (Salvia-Trujillo et al., 2017). There is a factual need to study the biological fate and toxicology of essential oil NPs after digestion to determine their behaviour in the gastrointestinal tract and further possible bioaccumulation in human body tissues. The orientation of the scientific community is towards this purpose, and validation of the analytical techniques and experiments with different models like mice or humans are going on all around the world. The regulatory issues associated with their use in food systems are still under discussion and progress. The Codex Alimentatrius and different regulatory committees, like the Commission of European Parliament and the FDA, among others, have compiled a database for nanomaterial types and usage, including safety aspects. The available database on toxicological knowledge on EONs suggests that the toxicity relies on their nature, particle size, and reactivity with biological tissues.

The regulation of nanoemulsions is still in its infant stages and under progress, and there is no global regulation by any authority in this area. However, the European Union and Switzerland are the main regions where nano-regulations have been developed and enforced for agriculture and food (Kehinde et al., 2020). There is still a pressing need for regulatory authorities around the globe to develop robust global standard operating procedures for nanoemulsions and their products and establish their safety. The development of these regulations will ensure that the benefits of nanotechnology can be realized without adversely affecting the health of the consumer and the environment.

11.8 CONCLUSION

Nanoemulsions of essential oils have gained great attention and utilization in the last two decades due to their incomparable properties such as high surface area, smaller particle size, transparent appearance, robust stability, effectiveness in low dosage, and tunable rheology. Essential oils present in the dispersed phase are acting as natural preservative agents in food and are compounds of interest in the processing of food. This biologically functional material has received a substantial technological upgrade in recent years. These developments are particularly designed to check the inadequacies of materials, methodologies, products, and product functionalities in these subjects. EONs have various intrinsic and extrinsic critical challenges which impact their application as food preservatives. The scarcity of raw materials, chemotypic variation, inconsistent efficacy, lack of molecular mechanism of action, adverse impact on food matrix, low water solubility, high cost, and threat of biodiversity losses are some of the major challenges to EON-based preservatives.

The nanoemulsions of essential oils should be formulated and developed according to product properties to minimize the loss of organoleptic and sensory properties, along with their efficacy as antimicrobial agents. The targeted and controlled release of the essential oils from the nanoemulsions with a better understanding of the mechanism should be engineered for novel antimicrobial systems and strategies. Limited studies are performed on food models, and there exist several challenges in the application of this system in complex food matrices such as meat and meat products. The application of nanoemulsions in food models will also offer challenges to governments and industries. The food industry has to build consumer confidence for the acceptance of nano-food ingredients such as antimicrobial EONs. On the other side, regulatory agencies such as the FDA should ensure the safety of these antimicrobial delivery systems. The employment of polymeric nanocomposites in these regards has bolstered a significant revolution in industries related to food and bioprocessing. The development of natural antimicrobial nanoemulsions from essential oils has the essential impacts in food processing. The adoption of EONs as antimicrobial agents has a noteworthy impact on the bioprocessing specialization. Nonetheless, there is a crucial need for the integration of thorough scientific research and revolutionary industrialization for the benefits of EONs in these industries to be fully exploited.

REFERENCES

Abd-Elsalam, K. A., and Khokhlov, A. R. (2015). Eugenol oil nanoemulsion: Antifungal activity against *Fusarium oxysporum* f. sp. *vasinfectum* and phytotoxicity on cottonseeds. *Applied Nanoscience*, 5(2), 255–265.

Artiga-Artigas, M., Acevedo-Fani, A., and Martín-Belloso, O. (2017). Improving the shelf life of low-fat cut cheese using nanoemulsion-based edible coatings containing oregano essential oil and mandarin fiber. *Food Control*, 76, 1–12.

Bakkali, F., Averbeck, S., Averbeck, D., Zhiri, A., and Idaomar, M. (2005). Cytotoxicity and gene induction by some essential oils in the yeast *Saccharomyces cerevisiae*. *Mutation Research/Genetic Toxicology and Environmental Mutagenesis*, 585(1-2), 1–13.

Baldissera, M. D., Da Silva, A. S., Oliveira, C. B., Zimmermann, C. E., Vaucher, R. A., Santos, R. C., Rech, V. C., Tonin, A. A., Giongo, J. L., Mattos, C. B., and Koester, L. (2013). Trypanocidal activity of the essential oils in their conventional and nanoemulsion forms: In vitro tests. *Experimental Parasitology*, 134(3), 356–3561.

Basak, S., and Guha, P. (2018). A review on antifungal activity and mode of action of essential oils and their delivery as nano-sized oil droplets in food system. *Journal of Food Science and Technology*, 55(12), 4701–4710.

Bedoya-Serna, C. M., Dacanal, G. C., Fernandes, A. M., and Pinho, S. C. (2018). Antifungal activity of nanoemulsions encapsulating oregano (*Origanum vulgare*) essential oil: In vitro study and application in Minas Padrão cheese. *Brazilian Journal of Microbiology*, 49(4), 929–935.

Bei, W., Zhou, Y., Xing, X., Zahi, M. R., Li, Y., Yuan, Q., and Liang, H. (2015). Organogel nanoemulsion containing nisin and D-limonene and its antimicrobial activity. *Frontiers in Microbiology*, 6, 1010.

Bhargava, K., Conti, D. S., daRocha, S. R., and Zhang, Y. (2015). Application of an oregano oil nanoemulsion to the control of foodborne bacteria on fresh lettuce. *Food Microbiology*, 47, 69–73. doi: 10.1016/j.fm.2014.11.007

Boukaew, S., Prasertsan, P., and Sattayasamitsathit, S. (2017). Evaluation of antifungal activity of essential oils against aflatoxigenic *Aspergillus flavus* and their allelopathic activity from fumigation to protect maize seeds during storage. *Industrial Crops and Products*, 97, 558–566.

Chang, Y., McLandsborough, L., and McClements, D. J. (2013). Physicochemical properties and antimicrobial efficacy of carvacrol nanoemulsions formed by spontaneous emulsification. *Journal of Agricultural and Food Chemistry*, 61(37), 8906–8913.

Chang, Y., McLandsborough, L., and McClements, D. J. (2015). Fabrication, stability and efficacy of dual-component antimicrobial nanoemulsions: Essential oil (thyme oil) and cationic surfactant (lauric arginate). *Food Chemistry*, 172, 298–304.

Chhikara, N., Jaglan, S., Sindhu, N., Anshid, V., Charan, M. V. S., and Panghal, A. (2018). Importance of traceability in food supply chain for brand protection and food safety systems implementation. *Annals of Biology*, 34(2), 111–118.

daCruz Cabral, L., Pinto, V. F., and Patriarca, A. (2013). Application of plant derived compounds to control fungal spoilage and mycotoxin production in foods. *International Journal of Food Microbiology*, 166(1), 1–14.

Dasgupta, N., Ranjan, S., and Gandhi, M. (2019). Nanoemulsions in food: Market demand. *Environmental Chemistry Letters*, 17(2), 1003–1009.

deOliveira Pereira, F., Mendes, J. M., and deOliveira Lima, E. (2013). Investigation on mechanism of antifungal activity of eugenol against *Trichophyton rubrum*. *Medical Mycology*, 51(5), 507–513.

Devgan, K., Kaur, P., Kumar, N., and Kaur, A. (2019). Active modified atmosphere packaging of yellow bell pepper for retention of physico-chemical quality attributes. *Journal of Food Science and Technology*, 56(2), 878–888.

Dhull, S. B., Anju, M., Punia, S., Kaushik, R., and Chawla, P. (2019). Application of gum arabic in nanoemulsion for safe conveyance of bioactive components. In: Prasad, R., Kumar, V., Kumar, M., and Chaudhary, D. (Ed.), *Nanobiotechnology in Bioformulations* (pp. 85–98). Springer, Cham.

Donsì, F., and Ferrari, G. (2016). Essential oil nanoemulsions as antimicrobial agents in food. *Journal of Biotechnology*, 233, 106–120.

Donsì, F., Annunziata, M., Sessa, M., and Ferrari, G. (2011). Nanoencapsulation of essential oils to enhance their antimicrobial activity in foods. *LWT-Food Science and Technology*, 44(9), 1908–1914.

Donsì, F., Annunziata, M., Vincensi, M., and Ferrari, G. (2012). Design of nanoemulsion-based delivery systems of natural antimicrobials: Effect of the emulsifier. *Journal of Biotechnology*, 159(4), 342–350.

El-Mogy, M. M., and Alsanius, B. W. (2012). Cassia oil for controlling plant and human pathogens on fresh strawberries. *Food Control,* 28, 157–162.

FDA. (2014). *Guidance for Industry Considering whether an FDA–Regulated Product Involves the Application of Nanotechnology*. FDA. (accessed on March 10, 2020).

FDA. (2015). Guidance for Industry Use of Nanomaterials in Food for Animals. https://www.fda.gov/downloads/AnimalVeterinary/GuidanceComplianceEnforcement/GuidanceforIndustry/UCM401508.pdf. (accessed on March 10, 2020).

Feng, W., and Zheng, X. (2007). Essential oils to control *Alternaria alternata* in vitro and in vivo. *Food Control*, 18, 1126–1130.

Flores, F. C., De Lima, J. A., Ribeiro, R. F., Alves, S. H., Rolim, C. M., Beck, R. C., and Da Silva, C. B. (2013). Antifungal activity of nanocapsule suspensions containing tea tree oil on the growth of *Trichophytonrubrum. Mycopathologia,* 175(3-4), 281–286.

Flores, S. K., Costa, D., Yamashita, F., Gerschenson, L. N., and Grossmann, M. V. (2010). Mixture design for evaluation of potassium sorbate and xanthan gum effect on properties of tapioca starch films obtained by extrusion. *Materials Science and Engineering: C*, 30(1), 196–202.

Gaysinsky, S., Taylor, T. M., Davidson, P. M., Bruce, B. D., and Weiss, J. (2007). Antimicrobial efficacy of eugenol microemulsions in milk against *Listeria monocytogenes* and *Escherichia coli* O157: H7. *Journal of Food Protection*, 70(11), 2631–2637.

Ghosh, V., Mukherjee, A., and Chandrasekaran, N. (2014). Eugenol-loaded antimicrobial nanoemulsion preserves fruit juice against, microbial spoilage. *Colloids and Surfaces B: Biointerfaces*, 114, 392–397. doi: 10.1016/j.colsurfb.2013.10.034

Ghotbi, R. S., Khatibzadeh, M., and Kordbacheh, S. (2014). *Preparation of neem seed oil nanoemulsion*. In *Proceedings of the 5th International Conference on Nanotechnology: Fundamentals and Applications*, Prague, Czech Republic, August 11, 2014 (No. 150, pp. 11–13).

Gutiérrez, J. M., González, C., Maestro, A., Solè, I. M. P. C., Pey, C. M., and Nolla, J. (2008). Nano-emulsions: New applications and optimization of their preparation. *Current Opinion in Colloid & Interface Science*, 13(4), 245–251.

Gyawali, R., and Ibrahim, S. A. (2014). Natural products as antimicrobial agents. *Food Control*, 46, 412–429.

Hyldgaard, M., Mygind, T., and Meyer, R. L. (2012). Essential oils in food preservation: Mode of action, synergies, and interactions with food matrix components. *Frontiers in Microbiology*, 3, 12.

Jbilou, F., Galland, S., Telliez, C., Akkari, Z., Roux, R., Oulahal, N., Dole, P., Joly, C., and Degraeve, P. (2014). Influence of some formulation and process parameters on the stability of lysozyme incorporated in corn flour-or corn starch-based extruded materials prepared by melt blending processing. *Enzyme and Microbial Technology*, 67, 40–46.

Jemaa, M. B., Falleh, H., Neves, M. A., Isoda, H., Nakajima, M., and Ksouri, R. (2017). Quality preservation of deliberately contaminated milk using thyme free and nanoemulsified essential oils. *Food Chemistry*, 217, 726–734.

Kaur, S., Panghal, A., Garg, M. K., Mann, S., Khatkar, S. K., Sharma, P., and Chhikara, N., 2019. Functional and nutraceutical properties of pumpkin–a review. *Nutrition & Food Science*, 50(2), 384–401.

Kedia, A., Prakash, B., Mishra, P. K., and Dubey, N. K. (2014). Antifungal and antiaflatoxigenic properties of *Cuminum cyminum* (L.) seed essential oil and its efficacy as a preservative in stored commodities. *International Journal of Food Microbiology*, 168, 1–7.

Kehinde, B. A., Chhikara, N., Sharma, P., Garg, M. K., and Panghal, A. (2020, in press). Application of polymer nanocomposites in food and bioprocessing industries. In: Hussain, C. M. (Ed.), *Handbook of Polymer Nanocomposites for Industrial Applications*. Academic Press, United States of America.

Kerekes, E. B., Deák, É., Takó, M., Tserennadmid, R., Petkovits, T., Vágvölgyi, C., and Krisch, J. (2013). Anti-biofilm forming and anti-quorum sensing activity of selected essential oils and their main components on food-related micro-organisms. *Journal of Applied Microbiology*, 115, 933–942.

Kim, I. H., Lee, H., Kim, J. E., Song, K. B., Lee, Y. S., Chung, D. S., and Min, S. C. (2013). Plum coatings of lemongrass oil-incorporating carnauba wax-based nanoemulsion. *Journal of Food Science*, 78(10), E1551–E1559.

Kiran, S., Kujur, A., and Prakash, B. (2016). Assessment of preservative potential of *Cinnamomum zeylanicum* Blume essential oil against food borne molds, aflatoxin B1 synthesis, its functional properties and mode of action. *Innovative Food Science & Emerging Technologies*, 37, 184–191.

Kralova, I., and Sjöblom, J. (2009). Surfactants used in food industry: A review. *Journal of Dispersion Science and Technology*, 30(9), 1363–1383.

Kumar, N. (2018). Development of pectin-based bio nanocomposite films for food packaging (Doctoral dissertation, Punjab Agricultural University, Ludhiana). http://krishikosh.egranth.ac.in/handle/1/5810047779

Kumar, N., Kaur, P., and Bhatia, S. (2017). Advances in bio-nanocomposite materials for food packaging: A review. *Nutrition & Food Science*, 47(4), 591–606.

Kumar, N., Kaur, P., Devgan, K., and Attkan, A. K. (2020). Shelf life prolongation of cherry tomato using magnesium hydroxide reinforced bio-nanocomposite and conventional plastic films. *Journal of Food Processing and Preservation*, 44(4), e14379.

Landry, K. S., Chang, Y., McClements, D. J., and McLandsborough, L. (2014). Effectiveness of a novel spontaneous carvacrol nanoemulsion against *Salmonella enterica Enteritidis* and *Escherichia coli* O157: H7 on contaminated mung bean and alfalfa seeds. *International Journal of Food Microbiology*, 187, 15–21. doi: 10.1016/j.ijfoodmicro.2014.06.030

Landry, K. S., Micheli, S., McClements, D. J., and McLandsborough, L. (2015). Effectiveness of a spontaneous carvacrol nanoemulsion against *Salmonella enterica Enteritidis* and *Escherichia coli* O157: H7 on contaminated broccoli and radish seeds. *Food Microbiology*, 51, 10–17. doi: 10.1016/j.fm.2015.04.006

Li, W., Chen, H., He, Z., Han, C., Liu, S., and Li, Y. (2015). Influence of surfactant and oil composition on the stability and antibacterial activity of eugenol nanoemulsions. *LWT-Food Science and Technology*, 62(1), 39–47.

Liang, R., Xu, S., Shoemaker, C. F., Li, Y., Zhong, F., and Huang, Q. (2012). Physical and antimicrobial properties of peppermint oil nanoemulsions. *Journal of Agricultural and Food Chemistry*, 60(30), 7548–7555.

Liu, W., Yang, X. L., and Winston Ho, W. S., 2011. Preparation of uniform-sized multiple emulsions and micro/nano particulates for drug delivery by membrane emulsification. *Journal of Pharmaceutical Science*, 48(1), 43–60. doi: 10.1002/jps.22272

Majeed, H., Liu, F., Hategekimana, J., Sharif, H. R., Qi, J., Ali, B., Bian, Y. Y., Ma, J., Yokoyama, W., and Zhong, F. (2016). Bactericidal action mechanism of negatively charged food grade clove oil nanoemulsions. *Food Chemistry*, 197, 75–83. doi: 10.1016/j.foodchem.2015.10.015

Malathi, A. N., Kumar, N., Nidoni, U., and Hiregoudar, S. (2017). Development of soy protein isolate films reinforced with titanium dioxide nanoparticles. *International Journal of Agriculture, Environment and Biotechnology*, 10(1), 141–148.

Mason, T. G., Wilking, J. N., Meleson, K., Chang, C. B., and Graves, S. M. (2007). Nanoemulsions: Formation, structure, and physical properties. *Journal of Physics Condensed Matter*, 18(41), R635.

McClements, D. J., and Rao, J. (2011). Food-grade nanoemulsions: Formulation, fabrication, properties, performance, biological fate, and potential toxicity. *Critical Reviews in Food Science and Nutrition*, 51, 285–330. doi: 10.1080/10408398.2011.559558

Mlalila, N., Hilonga, A., Swai, H., Devlieghere, F., and Ragaert, P. (2018). Antimicrobial packaging based on starch, poly (3-hydroxybutyrate) and poly (lactic-*co*-glycolide) materials and application challenges. *Trends in Food Science & Technology*, 74, 1–11.

Moghimi, R., Ghaderi, L., Rafati, H., Aliahmadi, A., and McClements, D. J. (2016). Superior antibacterial activity of nanoemulsion of *Thymus daenensis* essential oil against *E. coli*. *Food Chemistry*, 194, 410–415.

Möller, M., Eberle, U., Hermann, A., Moch, K., and Stratmann, B. (2009). *Nanotechnology in the Food Sector*. TA-SWISS, Zürich.

Mwaurah, P. W., Kumar, S., Kumar, N., Attkan, A. K., Panghal, A., Singh, V. K., and Garg, M. K. (2020). Novel oil extraction technologies: Process conditions, quality parameters, and optimization. *Comprehensive Reviews in Food Science and Food Safety*, 19(1), 3–20.

Nazzaro, F., Fratianni, F., De Martino, L., Coppola, R., and De Feo, V. (2013). Effect of essential oils on pathogenic bacteria. *Pharmaceuticals*, 6(12), 1451–1474.

Noori, S., Zeynali, F., and Almasi, H. (2018). Antimicrobial and antioxidant efficiency of nano-emulsion-based edible coating containing ginger (*Zingiber officinale*) oil and its effect on safety and quality attributes of chicken breast fillets. *Food Control*, 84, 312–320.

Omidbeygi, M., Barzegar, M., Hamidi, Z., and Naghdibadi, H. (2007). Antifungal activity of thyme, summer savory and clove essential oils against *Aspergillus flavus* in liquid medium and tomato paste. *Food Control*, 18, 1518–1523.

Orav, A., Raal, A., and Arak, E. (2008). Essential oil composition of *Pimpinella anisum* L. fruits from various European countries. *Natural Product Research*, 22(3), 227–232.

Ortega-Toro, R., Morey, I., Talens, P., and Chiralt, A. (2015). Active bilayer films of thermo-plastic starch and polycaprolactone obtained by compression molding. *Carbohydrate Polymers*, 127, 282–290.

Otoni, C. G., de Moura, M. R., Aouada, F. A., Camilloto, G. P., Cruz, R. S., Lorevice, M. V., and Mattoso, L. H. (2014). Antimicrobial and physical-mechanical properties of pectin/papaya puree/cinnamaldehyde nanoemulsion edible composite films. *Food Hydrocolloids*, 41, 188–194.

Ozogul, Y., Boğa, E. K., Akyol, I., Durmus, M., Joe, Y. U., Regenstein, M., and Kosker, A. R. (2020). Antimicrobial activity of thyme essential oil nanoemulsions on spoilage bacteria of fish and food-borne pathogens. *Food Bioscience*, 36, 100635. doi: 10.1016/j.fbio.2020.100635

Pan, K., Chen, H., Davidson, P. M., and Zhong, Q. (2014). Thymol nanoencapsulated by sodium caseinate: Physical and antilisterial properties. *Journal of Agricultural and Food Chemistry*, 62, 1649–1657, doi: 10.1021/jf4055402

Panghal, A., Chhikara, N., Anshid, V., Charan, M. V. S., Surendran, V., Malik, A., and Dhull, S. B. (2019b). Nanoemulsions: A promising tool for dairy sector. In: Prasad, R., Kumar, V., Kumar, M., and Chaudhary, D. (Eds.), *Nanobiotechnology in Bioformulations* (pp. 99–117). Springer, Cham.

Panghal, A., Jaglan, S., Sindhu, N., Anshid, V., Charan, M. V. S., Surendran, V., and Chhikara, N. (2019a). Microencapsulation for delivery of probiotic bacteria. In: Prasad, R., Kumar, V., Kumar, M., and Chaudhary, D. (Eds.), *Nanobiotechnology in Bioformulations* (pp. 135–160). Springer, Cham.

Panghal, A., Yadav, D. N., Khatkar, B. S., Sharma, H., Kumar, V., and Chhikara, N. (2018). Post-harvest malpractices in fresh fruits and vegetables: Food safety and health issues in India. *Nutrition & Food Science*, 50, 2.

Pathania, R., Kaushik, R., and Khan, M. A. (2018). Essential oil nanoemulsions and their antimicrobial and food applications. *Current Research in Nutrition and Food Science Journal*, 6(3), 626–643.

Pavoni, L., Perinelli, D. R., Bonacucina, G., Cespi, M., and Palmieri, G. F. (2020). An over-view of micro-and nanoemulsions as vehicles for essential oils: Formulation, prepara-tion and stability. *Nanomaterials*, 10(1), 135.

Prakash, B., Kujur, A., Yadav, A., Kumar, A., Singh, P. P., and Dubey, N. K. (2018). Nanoencapsulation: An efficient technology to boost the antimicrobial potential of plant essential oils in food system. *Food Control*, 89, 1–11.

Radi, M., Akhavan-Darabi, S., Akhavan, H. and Amiri, S. (2018). The use of orange peel essential oil microemulsion and nanoemulsion in pectin-based coating to extend the shelf life of fresh-cut orange. *Journal of Food Processing and Preservation*, 42, e13441.

Ribes, S., Fuentes, A., and Barat, J. M. (2019). Effect of oregano (*Origanum vulgare* L. ssp. hirtum) and clove (*Eugenia* spp.) nanoemulsions on *Zygosaccharomyces bailii* survival in salad dressings. *Food Chemistry*, 295, 630–636.

Salvia-Trujillo, L., Rojas-Grau, M. A., Soliva-Fortuny, R., and Martin-Belloso, O. (2015a). Physicochemical characterization and antimicrobial activity of food-grade emulsions and nanoemulsions incorporating essential oils. *Food Hydrocolloids*, 43, 547–556.

Salvia-Trujillo, L., Rojas-Grau, M. A., Soliva-Fortuny, R., and Martin-Belloso, O. (2015b). Use of antimicrobial nanoemulsions as edible coatings: Impact on safety and quality attributes of fresh-cut fuji apples. *Postharvest Biology and Technology*, 105, 8–16.

Salvia-Trujillo, L., Soliva-Fortuny, R., Rojas-Graü, M. A., McClements, D. J., and Martin-Belloso, O. (2017). Edible nanoemulsions as carriers of active ingredients: A review. *Annual Review of Food Science and Technology*, 8, 439–466.

Saranya, S., Chandrasekaran, N., and Mukherjee, A. M. (2012). Antibacterial activity of eucalyptus oil nanoemulsion against *Proteus mirabilis*. *International Journal of Pharmacy and Pharmaceutical Sciences*, 4(3), 668–671.

Seow, Y. X., Yeo, C. R., Chung, H. L., and Yuk, H. G. (2014). Plant essential oils as active antimicrobial agents. *Critical Reviews in Food Science and Nutrition*, 54(5), 625–644.

Sessa, M., and Donsì, F. (2015). Nanoemulsion-based delivery systems. In: Sagis, L. M. C. (Ed.), *Microencapsulation and Microspheres for Food Applications* (pp. 79–94). Elsevier, London. doi: 10.1016/B978-0-12-800350-3.00007-8

Severino, R., Vu, K. D., Donsì, F., Salmieri, S., Ferrari, G., and Lacroix, M. (2014). Antibacterial and physical effects of modified chitosan based-coating containing nanoemulsion of mandarin essential oil and three non-thermal treatments against *Listeria innocua* in green beans. *International Journal of Food Microbiology*, 191, 82–88. doi: 10.1016/j.ijfoodmicro.2014.09.007

Shah, B., Davidson, P. M., and Zhong, Q. (2012). Nanocapsular dispersion of thymol for enhanced dispersibility and antimicrobial effectiveness against *Escherichia coli* O157: H7 and *Listeria monocytogenes* in model food systems. *Applied and Environmental Microbiology*, 78(23), 8448–8453.

Shah, B., Davidson, P. M., and Zhong, Q. (2013). Nanodispersed eugenol has improved antimicrobial activity against *Escherichia coli* O157: H7 and *Listeria monocytogenes* in bovine milk. *International Journal of Food Microbiology*, 161(1), 53–59.

Sharma, P., Panghal, A., Gaikwad, V., Jadhav, S., Bagal, A., Jadhav, A., and Chhikara, N. (2019). Nanotechnology: A boon for food safety and food defense. In: Prasad, R., Kumar, V., Kumar, M., and Chaudhary, D. (Eds), *Nanobiotechnology in Bioformulations* (pp. 225–242). Springer, Cham.

Silva, H. D., and Cerqueira, M. A. (2015). Influence of surfactant and processing conditions in the stability of oil-in-water nanoemulsions. *Journal of Food Engineering,* 167, 89–98. doi: 10.1016/j.jfoodeng.2015.07.037

Singh, V. K., Garg, M. K., Kalra, A., Bhardwaj, S., Kumar, R., Kumar, S., Kumar, N., Attkan, A. K., Panghal, A. and Kumar, D. (2020). Efficacy of microwave heating parameters on physical properties of extracted oil from turmeric (Curcuma longa L.). *Current Journal of Applied Science and Technology*, 39(25), 126–136.

Soylu, E. M., Soylu, S., and Kurt, Ş. (2006). Antimicrobial activities of the essential oils of various plants against tomato late blight disease agent *Phytophthora infestans*. *Mycopathologia*, 161, 119–128.

Sugumar, S., Singh, S., Mukherjee, A., and Chandrasekaran, N. (2016). Nanoemulsion of orange oil with nonionic surfactant produced emulsion using ultrasonication technique: Evaluating against food spoilage yeast. *Applied Nanoscience*, 6(1), 113–120.

Tomé, L. C., Fernandes, S. C., Sadocco, P., Causio, J., Silvestre, A. J., Neto, C. P., and Freire, C. S. (2012). Antibacterial thermoplastic starch-chitosan based materials prepared by melt-mixing. *BioResources*, 7(3), 3398–3409.

Tozlu, E., Cakir, A., Kordali, S., Tozlu, G., Ozer, H., and Akcin, T. A. (2011). Chemical compositions and insecticidal effects of essential oils isolated from Achilleagypsicola, Saturejahortensis, *Origanum acutidens* and *Hypericum scabrum* against broad bean weevil (*Bruchusdentipes*). *Scientia Horticulturae*, 130(1), 9–17.

Tyagi, A. K., Gottardi, D., Malik, A., and Guerzoni, M. E. (2014). Chemical composition, in vitro anti-yeast activity and fruit juice preservation potential of lemon grass oil. *LWT-Food Science and Technology*, 57(2), 731–737.

Uhart, M., Maks, N., and Ravishankar, S. (2006). Effect of spices on growth and survival of *almonella typhimurium* DT 104 in ground beef stored at 4 and 8C. *Journal of Food Safety*, 26(2), 115–125.

Wani, T. A., Masoodi, F. A., Jafari, S. M., and McClements, D. J. (2018). Safety of nanoemulsions and their regulatory status. In: Jafari, S. M., and McClements, D. J. (Eds.), *Nanoemulsions* (pp. 613–628). Academic Press, USA.

Weiss, J., Loeffler, M., and Terjung, N. (2015). The antimicrobial paradox: Why preservatives lose activity in foods. *Current Opinion in Food Science*, 4, 69–75.

Wu, X., and Guy, R. H. (2009). Applications of nanoparticles in topical drug delivery and in cosmetics. *Journal of Drug Delivery Science and Technology*, 19(6), 371–384.

Xue, J., Davidson, M., and Zhong, Q. (2015). Antimicrobial activity of thyme oil co-nano-emulsified with sodium caseinate and lecithin. *International Journal of Food Microbiology*, 210, 1–8.

Zhang, S., Zhang, M., Fang, Z., and Liu, Y. (2017). Preparation and characterization of blended cloves/cinnamon essential oil nanoemulsions. *LWT*, 75, 316–322.

Ziani, K., Chang, Y., McLandsborough, L., and McClements, D. J. (2011). Influence of surfactant charge on antimicrobial efficacy of surfactant-stabilized thyme oil nanoemulsions. *Journal of Agricultural and Food Chemistry*, 59(11), 6247–6255.

12 Antimicrobial Efficacy of Essential Oil Nanoemulsions

Anindita Behera
School of Pharmaceutical Sciences, Siksha 'O' Anusandhan Deemed to be University, India
Bharti Mittu
National Institute of Pharmaceutical Education and Research, India
Santwana Padhi
KIIT Technology Business Incubator, KIIT Deemed to be University, India
Ajay Singh
Mata Gujri College, India

CONTENTS

12.1 INTRODUCTION

Plants produce so many categories of constituents present in different parts of the same. "Oil" in the plants is a quite astonishing fact, but plants produce two types of oil, that is fixed oils (FOs) and essential oils (EOs). FOs are the ester derivatives of glycerol and fatty acids whereas the EOs are the combinations of volatile and semi-volatile organic compounds originated from a single source of the plant. The EOs provide the characteristic aroma, flavour, and fragrance to the plant or the plant part containing it (Sirousmehr et al., 2014). EOs of plant origin are used in various fields of health for therapy, cosmetics, and pharmaceuticals as adjuvants which enhance adaptability; agriculture as pesticides and as preservatives; food industry as flavouring agents. So many reports have been published regarding the extraction, isolation, characterization, and identification of many constituents of EOs. Nowadays, the EOs are more in discussion in the food industry due to their applicability as an antimicrobial agent which increases the life of the food products (Sanchez-Gonzalez et al., 2011). Plants contain two types of metabolites: primary metabolites and secondary metabolites. The EOs are produced as a defense mechanism for the plants to protect themselves from pathogens, ecological factors, and physiological stress. EOs not only work for defense but also enhance reproduction by attracting the pollinators (Hashemi and Sant'Ana, 2018). The number of EOs varies from species to species, and their classification can be based on the context of existence, origin, and chemical constituent. As per existence, the EOs are essences, balsams, or resins. As per the origin, they are natural, artificial, or synthetic and, as per the chemical constituents, derivatives of different functional groups or the content of heteroatoms (Thayumanavan and Sadasivam, 2003). EOs are the secondary metabolites of the plants having aromatic compounds with volatile properties. The percentage of EOs from a plant depends on so many factors, such as species of the plant, time of collection, nature of the soil, climatic condition and nutritional level of the plant. So the collection of EOs from the plant or the plant parts is done by two conventional techniques like distillation or extraction. The physical and chemical properties affect the extent of the collection. EOs are low molecular weight (usually less than 500 Da), having a high refractive index and less relative density with comparison to water, but some exceptions do exist. The solubility of EOs is more in fats, alcohol, and organic solvents (Li et al., 2014). Nowadays, extraction of EOs is done by so many modified techniques to overcome the problem of long duration in conventional methods. The new methods include steam distillation, hydro-distillation, organic solvent extraction, expression, enfleurage, microwave-assisted distillation, microwave hydro-diffusion and gravity, high-pressure solvent extraction, supercritical fluid extraction, ultrasonic extraction, and solvent-free microwave extraction process (Farhat et al., 2010; Okoh et al., 2010).

12.2 CHEMISTRY OF ESSENTIAL OILS

The fragrance of EOs are due to the presence of the volatile constituents belonging to anyone of the following classes:

1. Terpenes (hydrocarbons) and their oxygen derivatives (terpenoids)
2. Phenylpropanoids

3. Phenolic compounds and their derivatives
4. Other oxygenated molecules (alcohols, ethers, carboxylic acids)
5. Heterocyclic compounds containing nitrogen or sulfur atoms

Most of the EOs contain terpenoids and the second-largest chemical constituent is phenylpropanoids. Formation of terpenoids occurs by the head-to-tail condensation of the isoprene unit (2-methyl-1, 3-butadiene) or isoprenoids. Isoprenoids have the chemical formula C_5H_8, an unsaturated hydrocarbon. The number of isoprene units determines the property of the terpene or its oxygenated derivative, terpenoids (Berger, 2007). The different terpenes are hemiterpenes (single isoprene unit, C_5H_8), monoterpenes (two isoprene units, $C_{10}H_{16}$), sesquiterpenes (three isoprene units, $C_{15}H_{24}$), diterpenes (four isoprene units, $C_{20}H_{32}$), and so on (Baser and Buchbauer, 2010). Phenylpropanoids contain a benzene ring with one or more C_3–C_6 units. The ring contains a methyl ether group and a terminal propenyl group. Many of the EOs contain phenol or phenol ethers. EOs containing oxygenated compounds contain functional groups like acids, alcohols, aldehydes, esters, ethers, ketones, lactones, oxides, peroxides, and phenols (Punia et al., 2019). The molecular weight of these compounds is different from that of terpenoids. Nitrogen-containing EOs contain chemical nuclei like indole and methyl anthranilate. Sulfur-containing EOs contain sulfides, disulfides, sulfoxides, and isothiocyanates (Tisserand and Young, 2013).

12.3 LIMITATIONS IN THE USE OF ESSENTIAL OILS FOR BIOLOGICAL ACTIVITY

EOs are the best candidate as an antioxidant, preservative for food products, and food-packaging material (Tiwari et al., 2009; Kuorwel et al., 2011) and as a pesticide (Adorjan and Buchbauer, 2010). But traditional EOs have been used against protection from bacteria, viruses, fungi, parasites, and insects. Not only used for cultivation protection, but EOs are also used as an analgesic, anti-inflammatory, spasmolytic, and sedative (Bakkali et al., 2008; Adorjan and Buchbauer, 2010). But the most promising application in the current generation is in the field of health care and veterinary use (Franz, 2010). The common way of the use of EOs in humans is by oral or topical applications. The oral application is very limited in the form of gargle, mouthwashes, or gargles, although the EOs are safe to use.

- In topical applications, mostly EOs are used in aromatherapy, but some oils, specifically citrus, oil are ultraviolet-sensitive and may cause adverse effects by photosensitization (Bilia et al., 2014).
- In the case of EOs used as aerosols, eye irritation can occur, and their direct inhalation is also in some cases prohibited (Bilia et al., 2014).
- The EOs are readily metabolized in the body, and their relative distribution in the body is high. Most of the EOs are excreted through the kidney as polar compounds or by exhalation through lungs with carbon dioxide. For example menthol by oral application, 35% of it gets excreted through the kidney as menthol glucuronide, similarly thymol, carvacrol, eugenol, and limonene. Fast metabolism and short half-life cause no accumulation in the body (Kohlert

et al., 2002; Michiels et al., 2008). Severe adverse reactions like dermatitis and decreases the total lung vital capacity is caused due to higher doses of D-limonene. Pets like dogs, cats highly susceptible to allergy to the insecticides containing D-limonene, which causes ataxia, stiffness, central nervous system depression, tremor, and coma (Lis-Balchin, 2006).

- EOs are in low molecular weight, small and fat-soluble molecules that enhance their permeation through the skin and readily penetrate and drain into the systemic circulation and reach the target organ. So they are used for aromatherapy; their fragrance and the chemical constituents are the cure for many ailments. Serious side effects may result due to more absorption at a large exterior of the skin or wrecked skin if these EOs are used in excessive amount without dilution. Even reports of convulsions are reported as EOs enhance the permeation (Bilia et al., 2014).
- EOs are readily decomposed by heat, humidity, light, and oxygen. So when used, they must be available in an un-degraded form; otherwise, the change in chemical composition can alter their biological application (Turek and Stintzing, 2013). Some EOs contain furanocoumarin, a class of phytochemical that is responsible for the phyto photo-dermatitis or photo-toxicity (Averbeck et al., 1990).
- During the processing and storage of plant material, the EOs endure deterioration by enzyme-catalyzed or chemical processes like oxidation, isomerization, cyclization, or dehydrogenation. Due to this, change in organoleptic characters and viscosity causes the photosensitization, leading to allergic reactions or hypersensitivity reactions to the skin (Scott, 2005; Christensson et al., 2009; Divkovic et al., 2005)

To overcome the limitations of the EOs, EO-loaded nano-delivery systems are selected and developed for the following advantages:

- Local drug delivery system in a sustained and controlled release formulation
- As the active ingredients are present in nanometric size, enhances tissue penetration
- Increase in cellular uptake and subcellular trafficking
- Suppression of undesired effects of therapeutically active EOs

12.4 ENCAPSULATION OF ESSENTIAL OILS BY "NANOTECHNOLOGY" APPROACH

The encapsulation of bioactive compounds is a viable strategy for modulating drug release, increasing the physical stability of the entrapped active constituents, shielding them from the harsh biological environment, reducing their volatility, increasing their bioactivity, reducing associated toxicity, and improving patient compliance and comfort (Ravi Kumar and Kumar, 2001). The wide range of literature on the encapsulation of EOs deals with nanometric-sized carriers, intended for the prevention of interaction of active compounds with different external stress due to light, moisture, and oxygen. This decreases the volatility of oil and causes the transformation of the oil into powder form. Nano-systems using encapsulation can be an alternate suitable approach for overcoming these related issues. Nano-carriers increase the stability of EOs counter to degradation by enzymes, and the preferred therapeutic level is achieved in target tissues for the desired duration with a low dose and may guarantee an optimum pharmacokinetic profile to fulfil specific requirements. Effective nano-sized delivery across the cell membrane is influenced by particle size, shape, and surface properties of nanoparticles. Nano-systems with dimensions of 50 to 300nm, positive zeta-potential, and having hydrophobic surface were proved to possess superior activity as compared to their counterparts (Roger et al., 2010).

The class of nano-systems found apt for the successful delivery of EOs are discussed in the following sections.

12.4.1 Polymer-Based Delivery of EOs

The term *nanoparticle* (NP) is the combination of nanospheres and nanocapsules. Nanocapsules contain a liquid nucleus enclosed by a polymeric membrane and a core–shell-type structure within the membrane containing the active agent in liquid state forming the core enclosed by a membrane (the core–shell structure are often lipophilic or hydrophilic; Rodríguez et al., 2016). Whereas nanospheres are the colloidal fragments containing the bioactive agents that are chemically conjugated or physically adsorbed to the polymeric matrix by stepwise diffusion, trapping, and encapsulation, NPs are coated with non-ionic surfactants to reduce the immunologic and molecular interactions of the groups present on the particle surface by vander Walls, element bonding, and hydrophobic interactions (Chawla et al., 2019, 2020). The uptake of nanoparticles into the cells is acknowledged to be over that of alternative encapsulated systems. The oil reaches the target tissues by discharge and transport from these systems by diffusion or erosion mechanism. The major advantages conferred by polymer-based nano-systems include the controlled release of the entrapped active agent, enhanced apparent water solubility, and protection of EOs against deterioration by temperature or light which assures extending the final product shelf life (Hamad et al., 2018).

Numerous reports cite the successful entrapment of EOs in polymeric nano-systems. An eightfold higher retention of eugenol content, improvement in thermal stability, and radical scavenging activity up to 2.7- fold higher were obtained for eugenol encapsulated in chitosan nanoparticles than that of neat eugenol (Woranuch and Yoksan, 2013). The essential oils of turmeric and lemongrass have well-known pharmaceutical applications and their active constituents such as arturmerone and citral are widely used for antifungal, anti-mutagenic, and anti-carcinogenic activities. But their clinical use is limited due to problems with stability, volatility, and solubility in water. Therefore, chitosan-alginate nano-carrier was used for their encapsulation. Nanocapsules with EO were hemo-compatible and had appreciable anti-proliferative properties in A-549 cell lines as compared to their neat counterparts, suggesting their future use as medicine and pharmaceutical applications (Natrajan et al., 2015).

The antimicrobial activity of EOs is enhanced by nano-encapsulated polymeric particles. Iannitelli et al. (2011) reported the increased antimicrobial action of nanocapsules containing carvacrol conjugated by poly (lactic-*co*-glycolic acid; PLGA) and changing the rheological properties of bacterial biofilms. Recently, the use of EOs in nanoencapsulation form has explored markedly in the food industry as these natural preservatives are preferred to harmful synthetic food preservatives with regard to improvement in food safety and quality (Feyzioglu and Tornuk, 2016). Herculano et al. demonstrated the antimicrobial actions of eucalyptus oil encapsulated in nanoparticles on *Listeria monocytogenes* and *Salmonella enteritidis* bacteria. They concluded that the bactericidal activity of nano-encapsulated EOs was more effective against gram-positive than gram-negative bacteria, as enhanced activity was found against *S. enteritidis* (Herculano et al., 2015).

12.4.2 Lipid-Based Delivery of EOs

12.4.2.1 Solid-Lipid Nanoparticles

Solid-lipid nanoparticles (SLNs) are colloidal carrier systems consisted of a solid lipid in the sub-micron dimension, disseminated in water, or an aqueous surfactant solution. The lipid component consists of a broad range of lipid and lipid-like molecules such as tri-acyl glycerols or waxes (Mehnert, 2001). The benefit of SLNs is the delivery system for administering lipophilic compounds by immobilization of encapsulated drug molecules by the strong particle framework contributing to enhanced chemical safety, less leaching, and controlled release. This physical property enables greater management of both the physical (against recrystallization) and chemical (against degradation) stabilization of the entrapped moiety (Weiss et al., 2008).

The EOs acquired from frankincense and myrrh oil (FMO) exhibits antimicrobial, anti-inflammatory, and antitumor activities. The active constituent of FMO causes the stimulation of the gastrointestinal tract, so it is not suitable for oral administration leading to poor bioavailability. Nano-encapsulation of FMOs in SLNs resulted in better pharmacokinetics and anti-tumour efficacy in H22-bearing Kunming mice (Feng, 2012). In another study, lipid formulations containing eugenol and eugenyl acetate were spray-dried to produce solid dispersible powders. Then the formulation was subjected to study the influence of freeze- and spray-drying process. The freeze-dried SLN formulation portrayed significantly higher retention than the spray-dried eugenol and eugenyl acetate (Cortés-Rojas et al., 2014). Furthermore, AlHaj et al. developed a *Nigella sativa* EO-encapsulated SLN which had increased physical stability at different temperatures used for storage during the duration of 3 months. The observed result confirmed the suitability of SLNs to be an appropriate carrier for the food and pharmaceutical industry (AlHaj, 2010).

12.4.2.2 Liposomes

Liposomes are vesicular structures composed of an aqueous core and amphiphilic lipid bilayer (Walde and Ichikawa, 2001). The liposomes are classified based on their lamellarity and size such as multilamellar vesicles (MLV), small unilamellar vesicles (SUV), and large unilamellar vesicles (Walde and Ichikawa, 2001). Their size can range from a few nanometers to several micrometres depending on the formulation technique and components chosen, which is essential to define their physicochemical properties such as surface charge, size, and stability. Liposomes improve the pharmacokinetics and biodistribution of the entrapped drug with more specificity, leading to reduced toxicity, and are known to be biodegradable, biocompatible, non-toxic, and non-immunogenic (Sherry et al., 2013). The stability of liposomes encapsulating *Santolina insularis* was found to be stable for more than 1 year at a temperature of 4 to 5°C, where 96 ± 1.05% and 94 ± 0.93% of the EO was encapsulated in the liposomes for MLVs and SUVs, respectively. Also, the size of the liposomes remained approximately the same for 390 days for both MLVs and SUVs (Valenti et al., 2001).

Certain reports also suggested that liposomes enhance the antimicrobial efficacy of entrapped EOs. The antifungal activity of *Eucalyptus camaldulensis* leaf EOs was increased when loaded into liposomes (Moghimipour et al., 2012). In the same way, the antiherpetic activity of EO of *Artemisia arborescens* L. was also seen to be increased when incorporated into MLVs (Sinico et al., 2005).

12.4.3 Molecular Complex Delivery of EOs

A molecular complex is made up of a combination of two chemical entities by physical association. Usually, cyclodextrins (CDs) are employed for the encapsulation of EOs, forming a molecular inclusion complex. The CD molecules have a cavity which is hydrophobic with a hydrophilic surface that forms inclusion complex with a wide variety of active agents. The encapsulation of bioactive agents with CDs and their derivatives helps in the protection of the encapsulated agents against environmental stress conditions. It improves the functionalization of the product by increasing the aqueous solubility.

Inclusion complexes of β cyclodextrin (β-CD) with eugenol, trans-cinnamaldehyde, clove extract, and cinnamon bark extract were more effective against *S. enterica*, *S. typhimurium*, and *L. innocua* as compared to free EO at same concentrations. This confirmed that the inclusion complexes protect the lipophilic compounds in a better way (Hill et al., 2013).

The volatility and less aqueous solubility of β-Caryophyllene (BCP) restrict its pharmaceutical application. The higher C_{max} and increase in bioavailability up to 2.6-fold of inclusion complex of BCP with β-CD was comparable to free BCP. The results confirmed the superior advantages offered by the β-CD complex in improving the bioavailability by the oral route of the drug administration in rats when compared to free BCP (Liu et al., 2013).

12.4.4 Micro- and Nanoemulsion Delivery of EOs

Micro-emulsions are homogeneous, thermodynamically stable, translucent dispersions of two immiscible liquids stabilized by an emulsifier. Nanoemulsions are fine oil-in-water (O/W) –type dispersions with an intrinsic property to differentiate into the constituent phases (Dhull et al., 2019; Panghal et al., 2019). They demonstrate relatively high kinetic stability as they have a small size range and require a low concentration of surfactants for their formulation, which makes it a better option for use as a carrier system than micro-emulsion (Blanco-Padilla et al., 2014).

The advantages embodied by the nanoemulsions along with their potential application in antimicrobial therapy are discussed next.

Advantages of Nanoemulsion: The various advantages conferred by nanoemulsions include the following:

1. The solubility of drugs in the biological milieu plays a dominant role in imparting the desired therapeutic efficacy. Unfortunately, not all drugs are aqueous soluble. As per Biopharmaceutical Classification System, drugs categorized under Class II and Class IV have a limitation in aqueous solubility, and Class IV has an additional drawback of limited permeation across biological membranes (Savjani et al., 2012). Due to limited solubility and permeability, high doses of drugs are administered to achieve the targeted therapeutic benefit which often ends up in imparting undesirable side effects. In such a scenario, nanocarriers such as nanoemulsions can act as a suitable drug delivery carrier for lipophilic and amphiphilic drugs (Groo et al., 2017). The most probable mechanism increase in solubility is attained is by the substantial fraction of the oily phase in the structure in which the drug becomes soluble, as high surface area due to the nano-size of the system and the existence of a thin layer of surfactants at the interface of the two phases (internal and external phases).
2. The smaller size of droplets in nanoemulsions prevents the conjoining. In the conjoining or coalescence process, smaller droplets agglomerate to form a large droplet with augmented size, leading to variability in the instability of the emulsion. Nanoemulsions containing small droplet sizes prevent coagulation, leading to the formation of a more stable nano-system.
3. Nanoemulsion has very high dispersibility when compared to other variants of emulsion formulation due to small-sized droplets. It stops the coagulation of droplets making a dispersed system without any phase separation.
4. Nanoemulsions are an excellent carrier system for brain targeted drug delivery, especially to brain micro-vessel endothelial cells. They act as a useful tool particularly for bypassing the P-gp pump present in the blood-brain barrier.
5. Nanoemulsion encompasses a clear and liquid-like property that enhances the patient acquiescence and is found to be safer for administration due to the lack of any thickening agent and colloidal particles.
6. They are non-toxic and non-irritants.
7. Nanoemulsion can be administered through various routes such as oral, parenteral, nasal, ocular, and transdermal, among others.
8. They enhance the bioavailability of less water-soluble drugs (Fofaria et al., 2016).
9. Nanoemulsions can be formulated with a low concentration of surfactants as compared to microemulsions which reduce the surfactant-associated toxicity to a certain extent.
10. Due to low interfacial tension of the O/W droplets, they impart good wetting, spreading, and penetration ability which makes them an apt choice for transdermal drug delivery.
11. The physical property of the encapsulated agent is modified to allow easier handling.

12.5 MECHANISM OF ACTION OF NANOEMULSIONS ENCAPSULATING ESSENTIAL OILS

EOs exhibit a broad spectrum of potential antibacterial effects against both types of bacteria, that is gram-negative and gram-positive bacteria. The vulnerability of EOs as an antimicrobial agent is more for gram-positive bacteria than the gram-negative bacteria due to the thick outer peptidoglycan layer of gram-positive bacteria that permits the easy penetration of hydrophobic molecules into the cells (Nazzaro et al., 2013). Whereas gram-negative bacteria have hydrophilic lipopolysaccharides in their cell membrane which restricts the uptake of macromolecules and hydrophobic molecules resulting in a lower antimicrobial efficacy (Hyldgaard et al., 2012).

The mode of action of EOs as an antimicrobial agent is ascribed to a series of reactions within the whole cell which results in inhibition of the growth and toxic bacterial metabolites are produced. EOs perturb permeability of cell membrane as it causes depolarization of cell membrane which is concentration-dependent and leading to an increase in the outflow of cellular adenosine triphosphate and K^+. The bacterial structure is damaged by the hydrophobic nature of EOs leading to interruption of the cytoplasmic membrane and coagulation of cytoplasm and membrane proteins, thereby increasing permeability and causing cell lysis and death (Figure 12.1).

12.6 ANTIMICROBIAL ACTIVITY OF NANOEMULSIFIED ESSENTIAL OILS

The growing resistance of microorganisms to conventional drugs, antibiotics, and disinfectants has become a serious problem globally. Antimicrobial resistance in bacterial pathogens is a challenge that is associated with the incidence of high morbidity and mortality (Pearson, 2017). The observed resistance of microbes to a certain group of medicaments has eventually encouraged research for the identification of

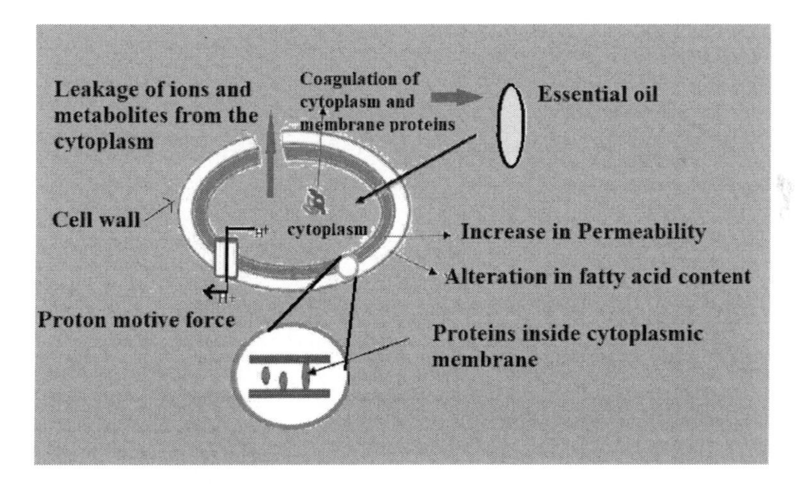

FIGURE 12.1 Sequence of processes exhibited by EOs in a bacterial cell leading to cell lysis and cell death. [*Source:* Ashakirin et al., 2017.]

new biocides with broad-spectrum activity. Reports suggest the potential application of several nanoemulsions encapsulating EOs which have shown appreciable antimicrobial properties (Kalemba and Kunicka, 2003).

The nanoemulsion attaches to the lipid membrane of the microorganism thermodynamically. This fusion is boosted by the electrostatic attraction of the cation charge of the cell membrane and the anion charge of the emulsion on the target pathogen. When an ample amount of nanoemulsion combines with the pathogens, the entrapped energy of the nanoemulsion gets released leading to lysis of the cell followed by cell death (Verma et al., 2017).

Due to the proven efficacy of EOs encapsulated nanoemulsion against a wide variety of pathogens, we have categorized their application according to the specificity of the microbial population. In the following section, we have broadly discussed the antibacterial, antifungal, antiviral, and antiprotozoal efficacy of nanoemulsion-encapsulating EOs.

12.6.1 Antibacterial Effects of Nanoemulsified Essential Oils

Several antibiotics are accessible to treat multiple bacterial pathogens. However, enhanced multidrug resistance has resulted in an increased incidence of diseases induced by bacterial pathogens. Additionally, poor immunity in host cells and the capacity of bacteria to build biofilm-associated drug resistance have further elevated the number of life-threatening human bacterial infections. This has impelled researchers to explore alternative molecules against bacterial strains. In this scenario, EOs and their major chemical constituents are prospective candidates to be utilized as suitable antibacterial agents.

Moghimi et al. reported the enhanced antibacterial activity of the nanoemulsion entrapping the essential oil of *Thymus daenensis* against *E. coli*. It was attributed due to the entry of the EOs to the bacterial cells at ease. Evaluation of the kinetics of bacterial deactivation demonstrated that the nanoemulsion killed *E. coli* in 5 min, but only a 1-log reduction was observed for its pure counterpart. The nanoemulsion augmented the antibacterial efficacy of *T. daenensis* against *E. coli* by enhancing the disruption of the cell membrane (Moghimi et al., 2016). Buranasuksombat et al. reported the antimicrobial activity which was influenced by the size of the droplet. O/W nanoemulsions were formulated from lemon myrtle oil (LMO) having antimicrobial property and soybean oil without antimicrobial properties. Results indicated the fact that all the encapsulated LMO nanoemulsions had an equal level of antimicrobial effects against the bacterial strains whereas the soybean oil–encapsulated nanoemulsions displayed no antimicrobial activity. Hence, the nanoemulsion containing the active agent (EOs) is responsible for the antimicrobial activity rather than the higher surface tensions and cell wall diffusivity of nano-sized emulsion droplets (Buranasuksombat et al., 2011).

Salvia-Trujillo et al. prepared nanoemulsions using a variety of EOs formulated by a high-shear homogenization technique. Here, Tween 80 and sodium alginate were used as the stabilizing agents. Nanoemulsions encapsulating EOs like clove, thyme, lemongrass, tea tree, geranium, sage, mint, marjoram, palmarosa, and rosewood were included for the study. The lemongrass or clove essential oil nanoemulsion exhibited a quicker and improved bactericidal activity as compared to the control ones (nanoemulsion without EOs). The observed results formed a proof of concept for the encouraging benefits of nanoemulsions in the food industry where it acts as carriers of flavourants and preservatives (Salvia-Trujillo et al., 2015).

In the food industry, different strains of bacteria like *Salmonella typhi*, *L. monocytogenes*, and *Staphylococcus aureus* were well managed with promising results by nanoemulsions entrapping the EO of sunflower (Joe et al., 2012).

12.6.2 Antifungal Effects of Nanoemulsified Essential Oils

Fungi attack susceptible plants by infecting the roots, so the vascular system is blocked by mycelium or spores which causes the discolouration of vascular and leaves and ultimately leading to the death of the plant (McGovern, 2015). Synthetic fungicides are the only choice of treatment modality for the said disease condition, but the associated toxicity and the observed microbial resistance limit its widespread usage (Gatto et al., 2011). In such a scenario, plant-derived EOs as crop protectants signify a developing bio-resource, and nanoemulsion as a vehicle option holds a promising solution.

A synergistic antifungal combination was found in the oil in water nanoemulsion of clove oil and lemongrass oil against *Fusarium oxysporum* f sp. *lycopersici* (FOL). The obtained nanoemulsion showed particle size of 76.73 nm and the minimum inhibitory concentration (MIC) of the optimized nanoemulsion was found to be 4,000 mg/L against FOL and the logistic kinetic model representing its fast fungicidal activity. The plant assay confirmed that the nano-formulation decreased the severity of FOL incidence up to 70.6% as compared to the control group which was untreated with nanoemulsion (Sharma et al., 2018). Flores et al. (2013) used the EO of *Melaleuca alternifolia* (tea tree oil) for formulating the nanocapsules and nanoemulsions. These formulations were further assessed for antifungal effectiveness in an onychomycosis model. Two *in vitro* models of dermatophytes nail infection were chosen for the antifungal activity against *Trichophyton rubrum*. The results were analysed by measuring the diameter of the fungal colony and found to be 2.88 ± 2.08 mm^2, 14.59 ± 2.01 mm^2, and 40.98 ± 2.76 mm^2 for the nanocapsules, nanoemulsion, and emulsion, respectively, containing tea tree oil and 38.72 ± 1.22 mm^2 for the control (untreated nail). The study signified that the nanoencapsulation of EO was able to decline the growth of *T. rubrum* in both the nail infection models (Cassella, 2002). The enhanced antifungal activity even at lower concentrations (IC_{90}), as observed in the present study, opened up wider options for the use of nanoemulsion of EOs as a commercially sustainable and eco-friendly bio-pesticide.

Aqueous emulsion of EOs from mustard, cassia, and clove was found to be effective and significant in decreasing the population thickness of *F. oxysporum* in soil up to 97.5 to 99.99% when compared to the untreated control group (Bowers and Locke, 2004).

12.6.3 Antiviral Effects of Nanoemulsified Essential Oils

Despite the technological advances of antiviral therapy for antimicrobial efficacy over the past decades, novel approaches towards improved management of infected cases are still required. Reports have suggested the wider application of nanoemulsion in carrying EOs for antiviral therapy.

The nanoemulsion of active constituents (poly-prenols) derived from leaves of ginkgo was formulated which was found to be very stable and had very potent antiviral activity against influenza AH_3N_2 and hepatitis B virus. The *in vitro* assay of the nanoemulsions by the MTT [(3-(4,5-dimethylthiazol-2-yl)-2,5-diphenlytetrezolium bromide)] method proved the potency (Wang et al., 2015).

The antiviral effectiveness of oregano oil along with its active component (carvacrol) was also evaluated against a human norovirus surrogate, the non-enveloped murine norovirus (MNV). The results demonstrated the fact that carvacrol was efficient in deactivating MNV within a period of 1h of exposure. It kills the virus by the direct attack to the capsid and then subsequently destroys the RNA. This study provided a strong supportive framework regarding the antiviral properties of these two active components (oregano oil and carvacrol) against MNV and illustrated the considerable potency of carvacrol to control human norovirus as a natural food and surface (fomite) sanitizer (Gilling et al., 2014).

12.6.4 Antiprotozoal Effects of Essential Oils

Antiparasitic activity of nanoemulsion of EOs of *Lavandula* species was reported by Shokri et al. The nanoemulsified EOs of *L. angustifolia* and *Rosmarinus officinalis* were found to be effective as antileishmanial against *Leishmania major. Leishmania* is the causative agent of cutaneous leishmaniasis in Iran. The major constituents responsible for the antileishmanial effect of *L. angustifolia* are 1, 8-cineol (22.29%) and linalool (11.22%). The nanoemulsion of EOs of *L. angustifolia* and *R. officinalis* was formulated and tested on promastigote. The IC_{50} value was found to be 0.11 μL/mL and 0.08 μL/mL, respectively, for *L. angustifolia* and *R. officinalis* as an antileishmanial agent whereas the nanoemulsified EOs of *R. officinalis* had $IC_{50} = 2.6$ μL/mL, which illustrated the increased strength of the nanoemulsion to have the mortality of the parasite (Shokri et al., 2017). Albuquerque and Echeverría reported the antiparasitic activity of EOs as controller of vectors and intermediate hosts responsible for infection of human beings and veterinary animals. Synthetic antiparasitic agents have the residual adverse effect which can be overcome by the use of bioactive compounds as EOs. They reported the dual mechanism of EOs in control of causative agents who are significant in transmission during the progression of disease spread and the pertinent mechanism of anti-pathogenic activity. The problems of EOs in solubility and stability can be modified to a better adaptable form that is nanoemulsion. Nanoemulsified EOs can be an alternative that shows multiple activities against larvae, insects, acaricides, and parasites and can be used as an effective agent against the causative agents or the vector transmitting the disease (Albuquerque and Echeverría, 2019).

12.7 FUTURE PERSPECTIVES

EOs have demonstrated to have antimicrobial effectiveness against a wide range of pathogens. However, the associated stability issues and their low molecular weight, along with high volatility, have restricted their use by researchers and clinicians. EOs' potential antimicrobial behavior motivated scientists to utilize nanotechnology, other plant essential oils, or antibiotics as future or next-generation medicaments. Although nano-entrapment has substantially enhanced the antimicrobial efficacy of EOs by controlled release mechanism and engaging them closely with the microbes, still are prevalent lacunae in our understanding regarding their mechanism of action in terms of synergism and antagonism. Hence, it becomes imperative on the part of researchers to discover the mechanism of action of each EO component along with the underlying synergism/antagonism between the components to have an in-depth understanding of the potential application of EOs in varied field. Therefore, novel strategies crafted for nano-encapsulation and focused research on understanding the underlying mode of action of EOs can endow on a remarkable platform in the said research domain.

12.8 CONCLUSION

EOs have very good antimicrobial activity against gram-positive and gram-negative bacteria. The extent of antimicrobial activity influenced by the type of bacterial strain and the chemical constituents present in the EOs. As the gram-positive bacteria have a thicker outer cell wall comprising several layers of peptidoglycan, they are highly vulnerable to antimicrobial agents compared to gram-negative bacterial strains. The approach of nanotechnology to get the EOs in nano-encapsulated form enhances not only their activity, but the limitations are also overcome with reduced dose and long-term safety.

REFERENCES

Adorjan, B., and Buchbauer, G. (2010). Biological properties of essential oils: An updated review. *Flavour and Fragrance Journal*, 25(6), 407–426. doi: 10.1002/ffj.2024.

Albuquerque, R.D.D.G., and Echeverría, J. (2019). Nanoemulsions of essential oils: New tool for control of vector-borne diseases and in vitro effects on some parasitic agents. *Medicines*, 6, 42–52. doi: 10.3390/medicines6020042.

AlHaj, O.A. (2010). Characterization of *Nigella sativa* L. essential oil-loaded solid lipid nanoparticles. *American Journal of Pharmacology and Toxicology*, 5(1), 52–57. doi: 10.3844/ajptsp.2010.52.57.

Ashakirin, S.N., Tripathy, M., Patil, U.K., and Majeed, A.B. (2017). Antimicrobial activity of essential oils: Exploration on mechanism of bioactivity. *International Journal of Pharmaceutical Science and Research*, 8(8), 3187–3193. doi:10.13040/IJPSR.0975-8232.

Averbeck, D., Averbeck, S., Dubertret, L., Young, A.R., and Morliere, P. (1990). Genotoxicity of bergapten and bergamot oil in *Saccharomyces cerevisiae*. *Journal of Photochemistry and Photobiology B*, 7(2–4), 209–229. doi:10.1016/1011-1344(90)85158-S.

Bakkali, F., Averbeck, S., Averbeck, D., and Idaomar, M. (2008). Biological effects of essential oils – A review. *Food and Chemical Toxicology*, 46(2), 446–475. doi: 10.1016/j.fct.2007.09.106.

Baser, K., and Buchbauer, G. (2010). *Handbook of Essential Oils: Science, Technology, and Applications*. Boca Raton, FL: CRC Press.

Berger, R.G. (2007). *Flavours and Fragrances: Chemistry, Bioprocessing and Sustainability*. Heidelberg: Springer Science and Business Media.

Bilia, A.R., Guccione, C., Isacchi, B., Righeschi, C., Firenzuoli, F., and Bergonzi, M.C. (2014). Essential oils loaded in nanosystems: A developing strategy for a successful therapeutic approach. *Evidence-Based Complementary and Alternative Medicine*, 2014, 1–14. doi: 10.1155/2014/651593

Blanco-Padilla, A., Soto, K., Hernández Iturriaga, M., and Mendoza, S. (2014). Food antimicrobials nanocarriers. *The Scientific World Journal*, 2014, 1–11. doi: 10.1155/2014/837215
Bowers, J., and Locke, J. (2004). Effect of formulated plant extracts and oils on population density of *Phytophthora nicotianae* in soil and control of *Phytophthora blight* in the greenhouse. *Plant Disease*, 88(1), 11–16. doi: 10.1094/pdis.2004.88.1.11.
Buranasuksombat, U., Kwon, Y., Turner, M., and Bhandari, B. (2011). Influence of emulsion droplet size on antimicrobial properties. *Food Science and Biotechnology*, 20(3), 793–800. doi: 10.1007/s10068-011-0110-x.
Cassella, S. (2002). Synergistic antifungal activity of tea tree (*Melaleuca alternifolia*) and lavender (*Lavandula angustifolia*) essential oils against dermatophyte infection. *International Journal of Aromatherapy*, 12(1), 2–15. doi: 10.1054/ijar.2001.0127.
Chawla, P., Kumar, N., Bains, A., Dhull, S.B., Kumar, M., Kaushik, R., and Punia, S. (2020). Gum arabic capped copper nanoparticles: Synthesis, characterization, and applications. *International Journal of Biological Macromolecules*, 146, 232–242.
Chawla, P., Kumar, N., Kaushik, R., and Dhull, S.B. (2019). Synthesis, characterization and cellular mineral absorption of nanoemulsions of *Rhododendron arboreum* flower extracts stabilized with gum arabic. *Journal of Food Science and Technology*, 56(12), 5194–5203.
Christensson, J.B., Forsström, P., Wennberg, A.-M., Karlberg, A.-T., and Matura, M. (2009). Air oxidation increases skin irritation from fragrance terpenes. *Contact Dermatitis*, 60(1), 32–40. doi:10.1111/j.1600-0536.2008.01471.x.
Cortés-Rojas, D., Souza, C., and Oliveira, W. (2014). Encapsulation of eugenol rich clove extract in solid lipid carriers. *Journal of Food Engineering*, 127, 34–42. doi: 10.1016/j.jfoodeng.2013.11.027.
Dhull, S.B., Anju, M., Punia, S., Kaushik, R., and Chawla, P. (2019). Application of gum arabic in nanoemulsion for safe conveyance of bioactive components. In: Prasad, R., Kumar, V., Kumar, M., and Choudhary, D. (Eds.), *Nanobiotechnology in Bioformulations* (pp. 85–98). Cham: Springer. doi:10.1007/978-3-030-17061-5_3.
Divkovic, M., Pease, C.K., Gerberick, G.F., and Basketter, D.A. (2005). Hapten-protein binding: From theory to practical application in the in vitro prediction of skin sensitization. *Contact Dermatitis*, 53(4), 189–200. doi: 10.1111/j.0105-1873.2005.00683.x.
Farhat, A., Fabiano-Tixier, A.S., Visinoni, F., Romdhane, M., and Chemat, F. (2010). A surprising method for green extraction of essential oil from dry spices: Microwave dry-diffusion and gravity. *Journal of Chromatography A*, 1217(47), 7345–7350. doi: 10.1016/j.chroma.2010.09.062.
Feng, N. (2012). Preparation and characterization of solid lipid nanoparticles loaded with frankincense and myrrh oil. *International Journal of Nanomedicine*, 7, 2033–2043. doi: 10.2147/ijn.s30085.
Feyzioglu, G., and Tornuk, F. (2016). Development of chitosan nanoparticles loaded with summer savory (Satureja hortensis L.) essential oil for antimicrobial and antioxidant delivery applications. *LWT - Food Science and Technology*, 70, 104–110. doi: 10.1016/j.lwt.2016.02.037.
Flores, F., de Lima, J., Ribeiro, R., Alves, S., Rolim, C., Beck, R., and da Silva, C. (2013). Antifungal Activity of Nanocapsule Suspensions Containing Tea Tree Oil on the Growth of Trichophyton rubrum. *Mycopathologia*, 175(3-4), 281–286. doi: 10.1007/s11046-013-9622-7
Fofaria, N., Qhattal, H., Liu, X., and Srivastava, S. (2016). Nanoemulsion formulations for anti-cancer agent piplartine — Characterization, toxicological, pharmacokinetics and efficacy studies. *International Journal of Pharmaceutics*, 498(1–2), 12–22. doi: 10.1016/j.ijpharm.2015.11.045.
Franz, C.M. (2010). Essential oil research: Past, present and future. *Flavour and Fragrance Journal*, 25(3), 112–113. doi: 10.1002/ffj.1983.

Gatto, M., Ippolito, A., Linsalata, V., Cascarano, N., Nigro, F., Vanadia, S., and DiVenere, D. (2011). Activity of extracts from wild edible herbs against postharvest fungal diseases of fruit and vegetables. *Postharvest Biology and Technology*, 61(1), 72–82. doi: 10.1016/j.postharvbio.2011.02.005.

Gilling, D., Kitajima, M., Torrey, J., and Bright, K. (2014). Antiviral efficacy and mechanisms of action of oregano essential oil and its primary component carvacrol against murine norovirus. *Journal of Applied Microbiology*, 116(5), 1149–1163. doi: 10.1111/jam.12453.

Groo, A., DePascale, M., Voisin-Chiret, A., Corvaisier, S., Since, M., and Malzert-Fréon, A. (2017). Comparison of 2 strategies to enhance pyridoclax solubility: Nanoemulsion delivery system versus salt synthesis. *European Journal of Pharmaceutical Sciences*, 97, 218–226. doi: 10.1016/j.ejps.2016.11.025.

Hamad, A., Han, J., Kim, B., and Rather, I. (2018). The intertwine of nanotechnology with the food industry. *Saudi Journal of Biological Sciences*, 25(1), 27–30. doi: 10.1016/j.sjbs.2017.09.004.

Hashemi, S.M.B., Khaneghah, A.M., and Sant'Ana, A.D. (2018). *Essential Oils in Food Processing: Chemistry, Safety and Applications*. Oxford, UK: John Wiley & Sons Ltd.

Herculano, E., dePaula, H., deFigueiredo, E., Dias, F., and Pereira, V. (2015). Physicochemical and antimicrobial properties of nanoencapsulated *Eucalyptus staigeriana* essential oil. *LWT - Food Science and Technology*, 61(2), 484–491. doi: 10.1016/j.lwt.2014.12.001.

Hill, L., Gomes, C., and Taylor, T. (2013). Characterization of beta-cyclodextrin inclusion complexes containing essential oils (trans-cinnamaldehyde, eugenol, cinnamon bark, and clove bud extracts) for antimicrobial delivery applications. *LWT - Food Science and Technology*, 51(1), 86–93. doi: 10.1016/j.lwt.2012.11.011.

Hyldgaard, M., Mygind, T., and Meyer, R. (2012). Essential oils in food preservation: Mode of action, synergies, and interactions with food matrix components. *Frontiers in Microbiology*, 3, 12.

Iannitelli, A., Grande, R., Stefano, A., Giulio, M., Sozio, P., Bessa, L., et al. (2011). Potential antibacterial activity of carvacrol-loaded poly (DL-lactide-*co*-glycolide) (PLGA) nanoparticles against microbial biofilm. *International Journal of Molecular Sciences*, 12(8), 5039–5051. doi: 10.3390/ijms12085039.

Joe, M., Bradeeba, K., Parthasarathi, R., Sivakumaar, P., Chauhan, P., Tipayno, S. et al. (2012). Development of surfactin based nanoemulsion formulation from selected cooking oils: Evaluation for antimicrobial activity against selected food associated microorganisms. *Journal of The Taiwan Institute of Chemical Engineers*, 43(2), 172–180. doi: 10.1016/j.jtice.2011.08.008.

Kalemba, D., and Kunicka, A. (2003). Antibacterial and antifungal properties of essential oils. *Current Medicinal Chemistry*, 10(10), 813–829. doi: 10.2174/0929867033457719.

Kohlert, C., Schindler, G., März, R.W., Abel, G., Brinkhaus, B., Derendorf, H., Gräfe, E.U., and Veit, M. (2002). Systemic availability and pharmacokinetics of thymol in humans. *Journal of Clinical Pharmacology*, 42(7), 731–737.

Kuorwel, K.K., Cran, M.J., Sonneveld, K., Miltz, J., and Bigger, S.W. (2011). Essential oils and their principal constituents as antimicrobial agents for synthetic packaging films. *Journal of Food Science*, 76(9), R164–R177. doi: 10.1111/j.1750-3841.2011.02384.x.

Li, Y., Fabiano-Tixier, AS., and Chemat, F. (2014) Essential Oils: From Conventional to Green Extraction. In: *Essential Oils as Reagents in Green Chemistry. Springer Briefs in Molecular Science*. Cham: Springer. doi: 10.1007/978-3-319-08449-7_2.

Lis-Balchin, M. (2006). *Aromatherapy Science: A Guide for Healthcare Professionals*. London: Pharmaceutical Press.

Liu, H., Yang, G., Tang, Y., Cao, D., Qi, T., Qi, Y., and Fan, G. (2013). Physicochemical characterization and pharmacokinetics evaluation of β-caryophyllene/β-cyclodextrin inclusion complex. *International Journal of Pharmaceutics*, 450(1-2), 304–310. doi: 10.1016/j.ijpharm.2013.04.013.

McGovern, R. (2015). Management of tomato diseases caused by *Fusarium oxysporum*. *Crop Protection*, 73, 78–92. doi: 10.1016/j.cropro.2015.02.021.

Mehnert, W. (2001). Solid lipid nanoparticles Production, characterization and applications. *Advanced Drug Delivery Reviews*, 47(2–3), 165–196. doi: 10.1016/s0169-409x(01)00105-3.

Michiels, J., Missotten, J., Dierick, N., Fremaut, D., Maene, P., and de Smet, S. (2008). *In vitro* degradation and *in vivo* passage kinetics of carvacrol, thymol, eugenol and *trans*-cinnamaldehyde along the gastrointestinal tract of piglets. *Journal of the Science of Food and Agriculture*, 88(13), 2371–2381. doi: 10.1002/jsfa.3358.

Moghimi, R., Ghaderi, L., Rafati, H., Aliahmadi, A., and McClements, D. (2016). Superior antibacterial activity of nanoemulsion of *Thymus daenensis* essential oil against *E. coli*. *Food Chemistry*, 194, 410–415. doi: 10.1016/j.foodchem.2015.07.139.

Moghimipour, E., Aghel, N., Mahmoudabadi, A., Ramezani, Z., and Handali, S. (2012). Preparation and characterization of liposomes containing essential oil of *Eucalyptus camaldulensis* leaf. *Jundishapur Journal of Natural Pharmaceutical Products*, 7(3), 117–122. doi: 10.5812/jjnpp.5261.

Natrajan, D., Srinivasan, S., Sundar, K., and Ravindran, A. (2015). Formulation of essential oil-loaded chitosan–alginate nanocapsules. *Journal of Food and Drug Analysis*, 23(3), 560–568. doi: 10.1016/j.jfda.2015.01.001.

Nazzaro, F., Fratianni, F., DeMartino, L., Coppola, R., and De Feo, V. (2013). Effect of essential oils on pathogenic bacteria. *Pharmaceuticals*, 6(12), 1451–1474. doi: 10.3390/ph6121451.

Okoh, O., Sadimenko, A., and Afolayan, A. (2010). Comparative evaluation of the antibacterial activities of the essential oils of *Rosmarinus officinalis* L. obtained by hydro-distillation and solvent free microwave extraction methods. *Food Chemistry*, 120(1), 308–312. doi:10.1016/j.foodchem.2009.09.084.

Panghal, A., Chhikara, N., Anshid, V., Charan, M.V.S., Surendran, V., Malik, A., and Dhull, S.B. (2019). Nanoemulsions: A promising tool for dairy sector. In: Prasad, R., Kumar, V., Kumar, M., and Choudhary, D. (Eds.), *Nanobiotechnology in Bioformulations* (pp. 99–117). Cham: Springer. doi: 10.1007/978-3-030-17061-5_4.

Pearson, M. (2017). The challenge of antimicrobial resistance: The hidden "filrouge" for healthcare policy. *OECD Observer*. doi: 10.1787/6ad65772-en10.3389/fmicb.2012.00012.

Punia, S., Sandhu, K.S., Siroha, A.K., and Dhull, S.B. (2019). Omega 3-metabolism, absorption, bioavailability and health benefits—A review. *Pharma Nutrition*, 10, 100162.

Ravi Kumar, M., and Kumar, N. (2001). Polymeric controlled drug-delivery systems: Perspective issues and opportunities. *Drug Development and Industrial Pharmacy*, 27(1), 1–30. doi: 10.1081/ddc-100000124.

Rodríguez, J., Martín, M., Ruiz, M., and Clares, B. (2016). Current encapsulation strategies for bioactive oils: From alimentary to pharmaceutical perspectives. *Food Research International*, 83, 41–59. doi: 10.1016/j.foodres.2016.01.032.

Roger, E., Lagarce, F., Garcion, E., and Benoit, J. (2010). Biopharmaceutical parameters to consider in order to alter the fate of nanocarriers after oral delivery. *Nanomedicine*, 5(2), 287–306. doi: 10.2217/nnm.09.110.

Salvia-Trujillo, L., Rojas-Graü, A., Soliva-Fortuny, R., and Martín-Belloso, O. (2015). Physicochemical characterization and antimicrobial activity of food-grade emulsions and nanoemulsions incorporating essential oils. *Food Hydrocolloids*, 43, 547–556. doi: 10.1016/j.foodhyd.2014.07.012.

Sanchez-Gonzalez, L., Vargas, M., Gonzalez-Martinez, C., Chiralt, A., and Chafer, M. (2011). Use of essential oils in bioactive edible coatings: A review. *Food Engineering Reviews*, 3(1), 1–16. doi: 10.1007/s12393-010-9031-3.

Savjani, K., Gajjar, A., and Savjani, J. (2012). Drug solubility: Importance and enhancement techniques. *ISRN Pharmaceutics*, 2012, 1–10. doi: 10.5402/2012/195727.

Scott, R.P.W. (2005). *Encyclopedia of Analytical Science*, London, UK: Elsevier.

Sharma, A., Sharma, N., Srivastava, A., Kataria, A., Dubey, S., Sharma, S., and Kundu, B. (2018). Clove and lemongrass oil based non-ionic nanoemulsion for suppressing the growth of plant pathogenic *Fusarium oxysporum* f.sp. *lycopersici*. *Industrial Crops and Products*, 123, 353–362. doi: 10.1016/j.indcrop.2018.06.077.

Sherry, M., Charcosset, C., Fessi, H., and Greige-Gerges, H. (2013). Essential oils encapsulated in liposomes: A review. *Journal of Liposome Research*, 23(4), 268–275. doi: 10.3109/08982104.2013.819888.

Shokri, A., Saeedi, M., Fakhar, M., Morteza-Semnani, K. Keighobadi, M. Teshnizi, S.H., Kelidari, H.R., and Sadjadi, S. (2017). Antileishmanial activity of *Lavandula angustifolia* and *Rosmarinus officinalis* EOs and nano-emulsions on *Leishmania major* (MRHO/IR/75/ER). *Iranian Journal of Parasitology*, 12(4), 622–631. PMID: 29317888, PMCID: PMC5756313.

Sinico, C., DeLogu, A., Lai, F., Valenti, D., Manconi, M., Loy, G. et al. (2005). Liposomal incorporation of *Artemisia arborescens* L. essential oil and in vitro antiviral activity. *European Journal of Pharmaceutics and Biopharmaceutics*, 59(1), 161–168. doi: 10.1016/j.ejpb.2004.06.005.

Sirousmehr, A., Arbabi, J., and Asgharipour, M.R. (2014). Effect of drought stress levels and organic manures on yield, essential oil content and some morphological characteristics of sweet basil (*Ocimum basilicum*). *Advances in Environmental Biology*, 8(4), 880–886.

Thayumanavan, B., and Sadasivam, S. (2003). *Molecular Host Plant Resistance to Pests*. New York: CRC Press.

Tisserand, R., and Young, R. (2013). *Essential Oil Safety: A Guide for Health Care Professionals*. Edinburgh: Elsevier Health Sciences.

Tiwari, B.K., Valdramidis, V.P., O'Donnell, C.P., Muthukumarappan, K., Bourke, P., and Cullen, P.J. (2009). Application of natural antimicrobials for food preservation. *Journal of Agricultural and Food Chemistry*, 57(14), 5987–6000. doi: 10.1021/jf900668n.

Turek, C., and Stintzing, F.C. (2013). Stability of essential oils: A review. *Comprehensive Reviews in Food Science and Food Safety*, 12(1), 40–53. doi: 10.1111/1541-4337.12006.

Valenti, D., DeLogu, A., Loy, G., Sinico, C., Bonsignore, L., Cottiglia, F., et al. (2001). Liposome-incorporated santolina insularis essential oil: Preparation, characterization and in vitro antiviral activity. *Journal of Liposome Research*, 11(1), 73–90. doi: 10.1081/lpr-100103171.

Verma, S., Vaishnav, Y., Verma, S., and Jha, A. (2017). Anhydrous nanoemulsion: An advanced drug selivery system for poorly aqueous soluble drugs. *Current Nanomedicine*, 7(1), 36–46. doi: 10.2174/2468187306666160926124713.

Walde, P., and Ichikawa, S. (2001). Enzymes inside lipid vesicles: Preparation, reactivity and applications. *Biomolecular Engineering*, 18(4), 143–177. doi: 10.1016/s1389-0344(01)00088-0.

Wang, C., Li, W., Tao, R., Ye, J., and Zhang, H. (2015). Antiviral activity of a nanoemulsion of poly-prenols from Ginkgo leaves against influenza A H3N2 and Hepatitis B virus in vitro. *Molecules*, 20(3), 5137–5151. doi: 10.3390/molecules20035137.

Weiss, J., Decker, E., McClements, D., Kristbergsson, K., Helgason, T., and Awad, T. (2008). Solid lipid nanoparticles as delivery systems for bioactive food components. *Food Biophysics*, 3(2), 146–154. doi: 10.1007/s11483-008-9065-8.

Woranuch, S., and Yoksan, R. (2013). Eugenol-loaded chitosan nanoparticles: II. Application in bio-based plastics for active packaging. *Carbohydrate Polymers*, 96(2), 586–592. doi: 10.1016/j.carbpol.2012.09.099.

13 Nanotechnologies in Food Microbiology

Overview, Recent Developments, and Challenges

Sook Chin Chew and Suk Kuan Teng
Xiamen University Malaysia Campus, Malaysia
Kar Lin Nyam
UCSI University, Malaysia

CONTENTS

13.1 INTRODUCTION

Foodborne illness is a major issue caused by the pathogens that results in moderate to severe illness and death via bacterial infections and intoxications in contaminated foods (Scallan et al., 2011). Moreover, spoilage microorganisms will bring a negative influence on food quality and shelf life that leads to food deterioration. Therefore, there is an increasing awareness of the protection of the food system against pathogenic and spoilage microorganisms to promote food quality, safety, and security. At present, many physical food preservation techniques are commonly being practised. These include thermal processing, food irradiation, pulsed electric field processing, high-pressure processing, plasma processing, and others (Fu et al., 2016). However, physical food preservation techniques may affect the qualitative aspects of the food system, like texture, flavour, and nutritional value. Recently, natural antimicrobial agents have gained public attention due to greater awareness of food quality and safety among consumers. Antimicrobials are food additives that inactivate or suppress the growth of pathogenic microorganisms in food systems (Tiwari et al., 2009). The addition of antimicrobial agents in the food system can overcome the negative impacts of physical processing techniques.

Therefore, there is an increasing interest in the search for new approaches to food preservation. For instance, essential oils can be used as safe preservatives to inhibit bacterial activity (Donsì et al., 2011). The addition of antimicrobial agents suppresses microbial growth through environmental control. However, antimicrobial agents are chemically reactive species, which can create potential problems when applied to the food system. For example, antimicrobial agents would cause negative impacts in the physical stability or degrade the bioactive compounds of the food system. Thus, considerations on the concentrations of antimicrobial agents to be applied in the food system are important to ensure the qualitative aspects of the food product are not affected (Donsì et al., 2011; Weiss et al., 2009).

Encapsulation is a process whereby the active agent is entrapped by wall material, yielding particles in nanometres (nano-encapsulation), micrometres (micro-encapsulation), or millimetres. Microencapsulated particles have sizes ranging from 1 to 1000 μm, while nano-encapsulated particles have sizes ranging from 1 to 100 nm. Currently, nanotechnology is actively applied in the food industry, with the promising benefits in processing, packaging, storage, transport, functionality, and food safety. Industry and academic researchers are giving attention to nanotechnology in countering matters corresponding to food and nutrition in the process of encapsulating, protecting, and releasing functional active agents (Ezhilarasi et al., 2013). There are many nano-encapsulation technologies with the challenges of selecting a suitable method with optimal parameters to obtain the nanostructures with recognizing the type of nanomaterial perfect for a desired bioactive compound.

Nanoencapsulation is receiving attention recently to encapsulate various antimicrobial agents such as phytochemicals, alkaloids, and essential oils. This technique helps to reduce the microbial contamination in the food system with enhanced food sensory characteristics (colour, flavour, taste, and texture). Nanotechnology may offer innovative and economic growth for food safety and security shortly. The selection of wall material is an important factor in the nanoencapsulation technique. The wall material of nanocapsule acts as a protective film to isolate the core materials against the exposure of an extreme environment. Both the physico-chemical properties of the active agent and the intended applications have to be considered in the selection of wall materials. There are a various natural and synthetic wall materials available such as polyethylene, carbohydrates (starch, pectin, cellulose, and chitosan), proteins (casein, whey protein, albumin, and gelatin), lipids (fatty acids, wax, phospholipids, and paraffin), and gums (alginate, carrageenan, and gum arabic) (Da Silva et al., 2014; Ribeiro-Santos et al., 2017). Nevertheless, the use of natural polymers is preferred as their non-toxicity and lower costs than synthetic polymers (Prajapati et al., 2013).

13.2 NANOENCAPSULATION TECHNOLOGIES

13.2.1 Nanoliposomes

Liposomes are microscopic colloidal vesicles constituted of phospholipid bilayers which surround either aqueous or hydrophobic material. Liposomes have a hydrophilic core with one or more phospholipid bilayers surrounding the centre (Sebaaly et al., 2016; Tamjidi et al., 2013). Liposomes with a single-layer membrane are known as single (<30 nm) or large unilamellar vesicles (30–100 nm), whereas those with multilayer membranes are known as multilamellar or multivesicular vesicles (Taylor et al., 2005).

The constitution of liposomes is based on the hydrophilic–hydrophobic interaction between phospholipids and water-soluble materials. Figure 13.1 shows the structure of the phospholipid. Phospholipids can form a membranous vesicle when dispersed in an aqueous solution, with their hydrophilic heads interacting with the aqueous environment and the long hydrophobic tails interacting with one another (Bozzuto and Molinari, 2015). Owing to this amphipathic nature of phospholipids, nanoliposomes are the liposomes in nano-size (<200 nm) which are appropriate for the encapsulation of hydrophilic, hydrophobic, and amphiphilic compounds (Akhavan et al., 2018; Ghorbanzade et al., 2017). The hydrophilic compounds can be entrapped into the core region, hydrophobic compounds can be entrapped within the hydrophobic bilayer, and amphiphilic compounds can be entrapped into the lipid/water interface, as shown in Figure 13.2.

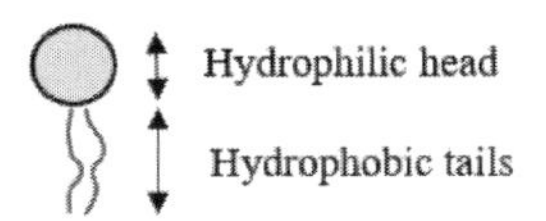

FIGURE 13.1 Structure of phospholipid.

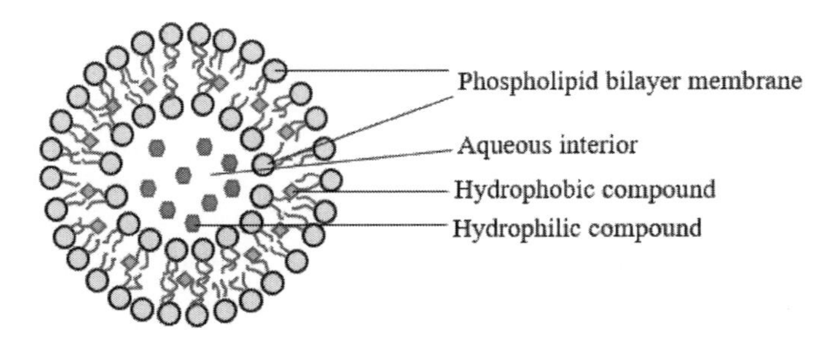

FIGURE 13.2 Basic structure of nanoliposome (Modified from Fakhravar et al., 2016).

Thin-film hydration, reverse phase evaporation, extrusion, supercritical fluid, and lyophilization methods can be utilized for the preparation of liposomes (Asbahani et al., 2015; Rezaei et al., 2019). The lipid is first dissolved in an organic solvent, and the solvent is then evaporated in the thin-film hydration method, which is also called the Bangham method. The lipid film is formed and then dispersed in an aqueous solution. In the reverse phase evaporation method, the lipid components are dissolved in an organic solvent and mixed with an aqueous solution, followed by sonication. After that, the solvent is evaporated under partial vacuum pressure. Finally, the lipid presents as a gel and is hydrated to produce vesicles. In the extrusion method, the lipid is dissolved in an organic solvent like chloroform. The mixture is then dried until a thin film is formed. The thin lipid film is added to a buffer solution containing the biocompound. Sonication, freeze-drying, and, last, extrusion are employed to obtain nano-sized liposomes (Pisoschi et al., 2018).

Kadimi et al. (2007) reported that the stability of liposomes synthesized by the supercritical fluid is better than the ultrasonication method. This is because the probe sonicator might contaminate the liposomes, which represents a limitation of the ultrasonication method. On the other hand, lyophilization can increase the shelf life of liposomes, especially for liposomes containing heat-sensitive compounds (Chen et al., 2010). The addition of cryoprotectants such as sugars into the liposomes can prevent phase transition and leakage of the encapsulated agents from the liposomes during the lyophilization process. Trehalose has been reported to exhibit the best shielding effect towards liposomes among the disaccharides (Heikal et al., 2009; Kawai and Suzuki, 2007).

Liposomes can protect the bioactive compounds against light, oxygen, temperature, and volatility and promote the solubility of poor-soluble compounds, as well as improve their bioavailability in the gastrointestinal tract (Detoni et al., 2012; Rezaei et al., 2019). The size, lamellarity, and encapsulation efficiency of liposomes are affected by the processing technique (Pattni et al., 2015). However, liposomes may pose some disadvantages like unstable in the aspects of physical and chemical factors that lead to degradation, oxidation, hydrolysis, and aggregation of phospholipids (Tantisripreecha et al., 2012). Moreover, liposomes have a fast release of core agents from the lipid bilayer in the gastrointestinal tract which restricts their use for encapsulation of poor-soluble compounds (Maestrelli et al., 2005). Besides that, liposomes are sensitive to mechanical and environmental factors like osmotic pressure alterations and are difficult to scale up in industrial production (Roos and Livney, 2017).

A previous study conducted on the antimicrobial film of eugenol and cinnamon leaf essential oil using lecithin liposomes has shown that the film extensibility had improved with enhanced water vapour barrier capacity compared to unencapsulated compounds (Valencia-Sullca et al., 2016). Distearoyl phosphatidylcholine and distearoyl phosphatidylethanolamine liposomes used to encapsulate *Artemisia afra*, *Eucalyptus* globules, and *Melaleuca alternifolia* essential oils to increase their shelf life as antimicrobial agents (Van Vuuren et al., 2010). Detoni et al. (2012) reported that the liposome successfully promoted the thermal stability and bioactivity of essential oils. The previous study showed that surface alteration of liposomes can facilitate their application, such as a nanocarrier coating with chitosan or thiol-derivatized chitosan, can facilitate the release property of encapsulated compounds in liposomes (Li et al., 2017; Wang et al., 2010). Chitosan coating can be applied by the drop-wise addition of a chitosan solution into the liposome dispersion (Zaru et al., 2009). Chitosan coating modified the ionic charge on the liposome surface and slightly increased its particle size. This surface modification extended the release behaviour under in vitro conditions and showed an improvement in the stability (Laye et al., 2008; Li et al., 2009). The optimal level of cholesterol in the liposome membrane helps to repair the membrane. Moreover, cholesterol dries the lipid–water interface, which improved the stability of the liposomes against oxidation and acyl ester hydrolysis (Parasassi et al., 1994; Samuni et al., 2000). Liposomes coated with polyethylene glycol have been reported to maintain its stability against the digestion by bile salts (Iwanaga et al., 1999; Li et al., 2003) and increase the bioavailability of the encapsulated agents (Fathi et al., 2012).

13.2.2 Nanoemulsions

The emulsion is two totally or partially immiscible liquids in which one is dispersed in droplets form in another liquid, such as oil-in-water (O/W) emulsion and water-in-oil (W/O) emulsion, as shown in Figure 13.3. There are multiple emulsions such as water-in-oil-in-water (W/O/W) and oil-in-water-in-oil (O/W/O), which resulted in the interaction of different layers of emulsifiers with opposite electrostatic charges (Chawla et al., 2019, 2020). These different layers of emulsifiers may give stability to the functional compounds that encapsulated in the emulsion system (McClements et al., 2009). Thus, emulsions are widely used to protect bioactive compounds against oxidation in food systems and to encapsulate essential oil components to enhance antimicrobial activity (Fu et al., 2016).

Microemulsions are colloidal dispersions of oil and water stabilized by an interfacial layer of surfactants, with particle sizes presented in micro-sizes (Rao and McClements, 2011; Slomkowski et al., 2011). Micro-emulsion systems have been used to encapsulate the essential oils like eugenol, monolaurin, and cinnamon against various pathogenic bacteria or spoilage moulds in food products like milk, noodles, and pears (Fu et al., 2008; Gaysinsky et al., 2007; Wang et al., 2014).

Nanoemulsion is colloidal dispersions comprising two immiscible liquids with droplet size ranging from 50 to 1000 nm, of which one is being dispersed in the other (Sanguansri and Augustin, 2006). Lipophilic bioactives such as carotenoids,

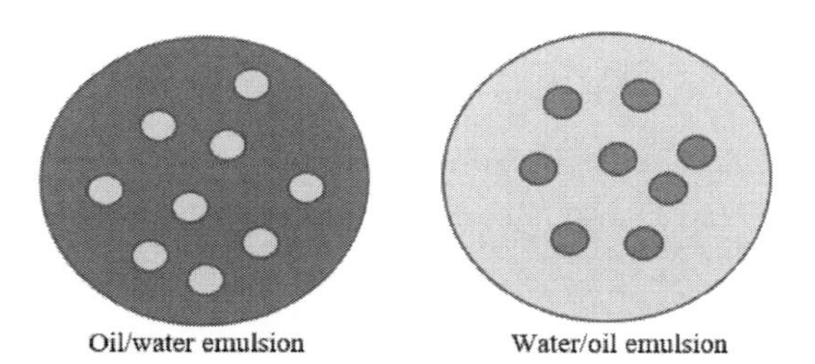

FIGURE 13.3 A simple scheme of oil/water emulsion and water/oil emulsion.

unsaturated fatty acids, tocopherols, and phytosterols, which have poor water solubility, can be encapsulated in this approach to offer lipophilic nutraceuticals or functional food supplements. Nanoemulsions can be used directly in its liquid phase or can be dried by additional drying methods after emulsification such as spray-drying or freeze-drying. Nanoemulsions can be formed through micro fluidization, high-pressure homogenization, and ultrasonication (Ezhilarasi et al., 2013).

In micro fluidization, very high pressure up to 20, 000 psi is applied in the process to push the liquid through an interaction chamber that consists of microchannels. The formation of the fine nanoscale emulsion droplets can be created when the emulsion feeds through the microchannels in a collision chamber (Fathi et al., 2012). High-pressure homogenization is a technique in which the emulsion is forced with high pressure with 100 to 2000 bar and high shear stress, resulting in the breakdown of the particles into nanoscale (Ezhilarasi et al., 2013). Due to their nano-size, the forces of attraction acting in between the droplets also decrease, resulted in nanoemulsions exhibited high kinetic stability against droplet flocculation and coalescence (Solans et al., 2005; Sonneville-Aubrun et al., 2004). On the other hand, ultrasonication is a technique using sound energy with more than 20 kHz frequency to break apart particle agglomerates. This is due to the disruptive forces created by the combination of cavitation, turbulence, and interfacial waves (Aswathanarayan and Vittal, 2019). Cavitation is the formation and collapse of empty spaces in a flowing liquid, caused by the decreased pressure because of the local velocity changes. Then the collapse of these cavities causes high pressure and turbulence waves to break the dispersed liquid (Jafari et al., 2006). The type of emulsifier and the nanoemulsion production technique will greatly affect the antimicrobial efficacy. The previous studies reported the different sizes for nanoemulsions, like less than 100 nm (Shakeel and Ramadan, 2010), 10 to 100 nm (Talegaonkar et al., 2010), 100 to 500 nm (Anton et al., 2008), and 100 to 600 nm (Sakulku et al., 2009). There are three components in the nanoemulsions, which are oil, water, and emulsifier. In the O/W nanoemulsions, oil droplets are dispersed in water. The oil and water can retain several other components by using one or multiple types of emulsifiers.

Nanoemulsion presents a better solubility than micellar dispersions and has higher kinetic stability, which is suitable to apply in food and pharmaceutical industries as an aqueous medium for targeted transportation of bioactive (Rostamabadi et al., 2019). Surfactants improved the permeability of the nanoemulsions and dispersion of the small droplets in the stomach. This enhances the solubility and bioavailability of insoluble bioactive compounds in the gastrointestinal tract (Talegaonkar et al., 2010). Nanoemulsions can be applied in food and beverage formulations to safeguard the hydrophobic bioactive compounds like essential oils and reduce the needed concentration of the compounds (Sugumar et al., 2016). Also, the nano-size of nanoemulsions leads to weak light scattering, making it suitable to be incorporated into colourless drinks and foods (Mehrnia et al., 2016). The oil phase is easily broken down into small droplets as its lower viscosity and interfacial tension (Qian and McClements, 2011). Thus, essential oils and flavour oils are inclined to form nanoemulsions as they have lower viscosity and interfacial tension properties (Rezaei et al., 2019). Nanoemulsions are widely used currently to encapsulate the food antimicrobial agents like basil oil, carvacrol, eugenol, and thymol to suppress the growth of pathogenic microorganisms and food spoilage (Bhargava et al., 2015; Donsì, et al., 2014; Ghosh et al., 2013). Pickering emulsions are emulsions that are stabilized by solid particles but not surfactants. Pickering emulsions preserve the nature of conventional emulsions but improve its stability. Pickering emulsions offer better merits compared to conventional emulsions as they are more stable against coalescence, flocculation, and Ostwald ripening, which normally occur in the emulsions. Moreover, Pickering emulsions can be applied in different fields as they are prepared without surfactants, which may limit its application (Aveyard et al., 2003; Chevalier and Bolzinger, 2013). Pickering emulsions are used to encapsulate the bioactives, including flavonoids (Luo et al., 2011), bioactive peptides (Shimoni et al., 2013), and antimicrobials, such as nisin and thyme oil (Bi et al., 2011; Wu et al., 2014). Overall, there is a developing interest in the application of Pickering emulsions for the delivery of functional components in food systems. Therefore, nanoemulsion gains vast attention in the food industries as a carrier agent for antimicrobial compounds.

13.2.3 Solid-Lipid Nanoparticles

Solid-lipid nanoparticles (SLNs) have attracted global interest recently in the pharmaceutical and food science industries. SLNs are nanoparticles (NPs) that prepared by the lipids that stay in the solid state at room temperature. SLNs are produced by homogenizing the oils and surfactants solution at a temperature higher than the oils' melting temperature to form an emulsion. After that, SLNs are formed when the emulsion is cooled below the temperature of crystallization. SLNs combine the benefits of nanocapsules, liposomes, and emulsion systems. A solid lipid is selected instead of a liquid lipid as it achieves the controlled release of active agents and improves the stability of active agents against oxidation (Fu et al., 2016). SLNs do not involve the use of organic solvents in the processing and able to achieve high encapsulation efficiency and feasibility to produce at an industrial scale (Fathi et al., 2012).

This system has been commonly used to entrap chemically unstable hydrophobic compounds. SLNs represented excellent physical stability with a lower rate of evaporation of essential oils (Lai et al., 2006) and maintained its structure for up to 3 months (Al-Haj et al., 2010). Also, SLNs have found applications to encapsulate bioactive compounds in food applications and synthesize edible film for fruit protection (Fu et al., 2016). This system has the powerful ability to be the carrier for antimicrobial agents, but low loading capacity and expulsion risk of the coated agents are its disadvantages.

Hot homogenization and cold homogenization are the two main methods to produce SLNs. In the hot homogenization method, the lipid is melted at 5 to 10°C above its melting point and the active agents are fixed into the liquid oil. After that, an aqueous surfactant solution is added and mixed at the same temperature, followed by the homogenization method using a high-pressure homogenizer at a controlled temperature. After that, SLNs are produced as the emulsion is cooled, and the lipids are recrystallized. Lyophilization can be used in recrystallization. However, the hot homogenization method cannot efficiently incorporate all the active agents into the solid matrix due to the leakage of active agents into the water phase. This method is not suitable for heat-sensitive components like enzymes (Fathi et al., 2012).

The cold homogenization method is comparable to the hot homogenization method, in which the active agent is stabilized into a lipid above its melting point. Then the oil is cooled and solidified, followed by grounded by a mortal mill. The produced lipid particles are scattered in a cold surfactant solution at room temperature and homogenized at room temperature or cooler than room temperature. The solid state of the matrix lessens the compartmentalization of the active agents to the liquid state. However, the expulsive release of active agents is the major limitation of SLNs. The use of a low processing temperature and low surfactant concentration would help to reduce the expulsive release of active agents from the model (Müller et al., 2002). Lipid formulation, surfactant type, surfactant concentration, droplet size, and cooling parameters are the main factors that influence the structure and characteristics of SLNs (Weiss et al., 2008).

On the other hand, SLNs can also be prepared by emulsification evaporation, together with the sonication method. In this method, liquid lipid, organic solvent, emulsifier, and bioactive compounds are mixed to produce a coarse emulsion. After that, this emulsion is subjected to sonication at a temperature higher than the melting point of lipid for a duration of time. Finally, the produced nanoemulsion is added and mixed into cold water containing surfactant, followed by the solvent evaporation method (Fathi et al., 2012).

There are three types of models for SLNs, which are homogenous matrix, bioactive-enriched shell, and bioactive-enriched core models, as shown in Figure 13.4. Formulation and processing parameters determine the type of obtained model. A homogenous matrix model is typically produced while using the cold homogenization method and while fixing the lipophilic compounds into SLNs using the hot homogenization method. The liberation of the active agents in this model depends on the dissolution mechanism. Phase separation takes place during the cooling process from the liquid oil droplets that might yield a bioactive-enriched shell model. This model presents the active agents in the shell, which might lead to an expulsive release behaviour. A bioactive-enriched core model can be formed while carrying out the opposite conditions of the bioactive-enriched shell model, which means the bioactive compounds start to precipitate first and produce a shell with fewer encapsulated compounds. This structure model leads to a membrane with controlled release property (Fathi et al., 2012; Müller et al., 2002).

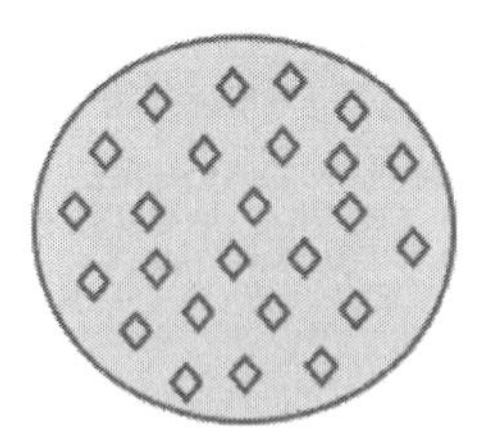

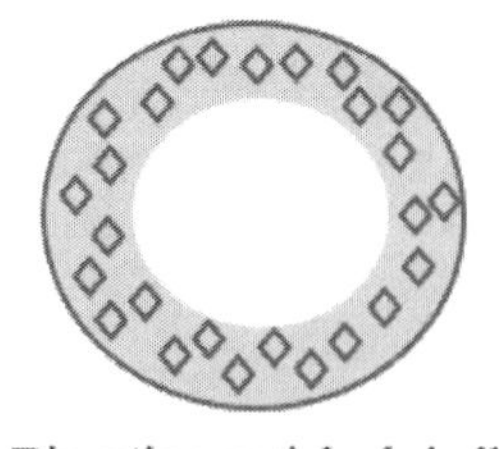

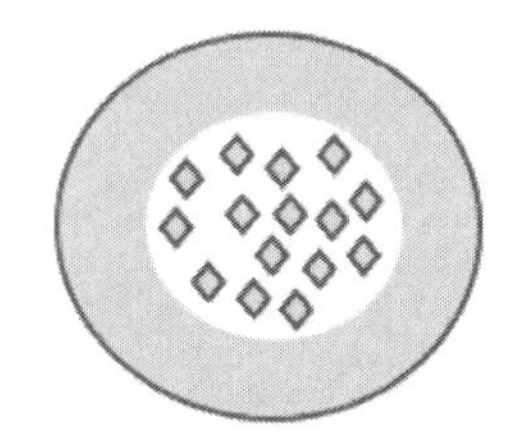

FIGURE13.4 Different types of solid-lipid nanoparticles. *Source*: Modified from Fathi et al. (2012).

13.2.4 Nanostructured Lipid Carriers

Nanostructured lipid carriers (NLCs) are believed as the new generation of SLNs. NLCs are produced from a solid matrix enclosing various liquid nanocomponents whereas SLNs are produced from pure solid lipids (Müller et al., 2002). NLCs possess good biocompatibility, controlled release, production on a large industrial scale compared to SLNs. The limitations of SLNs such as limited loading capacity, recrystallization potential, and expulsion risk during storage can be solved by the NLC technique. NLCs represents an alternative method for SLNs to transport lipophilic substances for food applications (Fu et al., 2016).

The NLCs technique is receiving interest as an effective encapsulation technology in the cosmetic, food, and pharmaceutical industries. The distorted matrix structure of NLCs produces more room to entrap active agents. Thus, NLCs can be applied to accommodate ultraviolet filters in sun care products, which makes it superior in cosmetic applications (Chu et al., 2019). NLCs prevent the expulsive release of active agents from the structure, compared to SLNs, as shown in Figure 13.5 (Rezaei et al., 2019).

13.2.5 Nanocapsules

Nanocapsules are the capsules presented in nano-size, ranging from 10 to 1000 nm, and consist of natural or synthetic wall materials. Nanocapsules are vesicular systems in which the active agents are entrapped in the centre by a nanocarrier membrane. Microencapsulation may assure the outstanding protection of essential oils toward degradation or evaporation and, at the same time, does not influence its antimicrobial activity. On the other hand, nanoencapsulation may promote passive cellular absorption mechanisms, thus lowering mass transfer resistances and increasing antimicrobial activity (Donsì et al., 2011). There are various techniques to produce nanocapsules, which include coacervation, ionic gelation, and spray-drying.

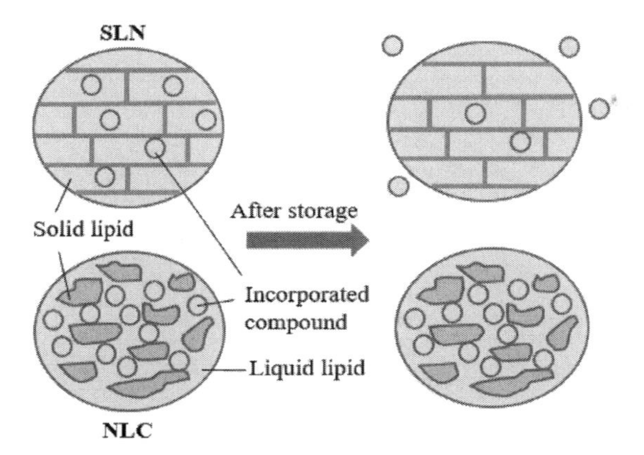

FIGURE13.5 Structures of solid-lipid nanoparticles and nanostructured lipid carriers after storage. *Source*: Modified from Rezaei et al. (2019).

13.2.5.1 Coacervation

Complex coacervation is a recognized technology based on the electrostatic attraction between two or more oppositely charged biopolymers. It is usually produced from the combination of protein and polyanions of polysaccharide to achieve high encapsulation yields, mainly related to the high loading capacity. Complex coacervation normally occurs via the reciprocal interaction of two oppositely charged wall materials, resulting in the phase separation and formation of a coacervate outer part around the core substances (Koupantsis et al., 2016). Hydrogen bonding and hydrophobic interactions are other weak interactions devoted to complex coacervation (Ach et al., 2015).

Gelatin is the most common protein used in the complex coacervation, and gum arabic, sodium dodecyl sulfate, pectin, or chitosan are used as the anionic polymers. However, more researches are needed to explore the alternative protein sources from soybean, peas, and other plants to replace the use of gelatin. This is due to the limitation of the use of animal protein in certain phenomena (Dias et al., 2017). Formaldehyde and glutaraldehyde are the common cross-linking agents used in the complex coacervation process. However, these chemicals are considered toxic and forbidden in food applications. Therefore, transglutaminase, tannic acid, and genipin were used as the cross-linking agents in the complex coacervation process due to their non-toxic property and environmental friendliness (Peng et al., 2014).

The ionic charge, pH, the ratio of protein to the polysaccharide, concentration and molecular weight of wall materials, charge density, flexibility and conformation of wall materials, stirring, pressure, and temperature are the factors that affect the attractive or repulsive interactions of the two biopolymers in an aqueous environment (Yang et al., 2012). Microcapsules or nanocapsules formed by complex coacervation have an outstanding controlled release behaviour, are heat resistant, and have promising encapsulation efficiency (Yang et al., 2014). A previous study reported that the nanoencapsulation of capsaicin by complex coacervation achieved 81.2% of encapsulation efficiency and exhibited good dispersion performance (Xing et al., 2004).

13.2.5.2 Ionic Gelation

Sodium alginate is the most common wall material used in ionic gelation because of its chemical stability, low toxicity, low immunogenicity and can form a rigid gel in the presence of Ca^{2+} (Chew and Nyam, 2016). Ionic gelation is carried out by incorporating the active agents into the alginate solution and then using the extrusion method by dropping droplets of the mixture solution into a hardening bath. A pipette, a syringe, an atomizing disk, a jet cutter, a vibrating nozzle, coaxial airflow, or an electric field can be used as the dripping tool (Nedovic et al., 2011). Hardening solution likes Ca^{2+} from $CaCl_2$ solution neutralizes the repulsive charges of carboxylate groups and results in the cross-linking of the alginate chains (Sun-Waterhouse et al., 2012). The low temperature used in this technique represents its advantage to prevent unavoidable reactions and product volatilization.

Besides that, co-extrusion that uses an encapsulator equipped with a concentric nozzle allows alginate solution and core solution pumped into the encapsulator simultaneously to produce uniform-size microcapsules. Vibrating nozzle technology is applied using an encapsulator, whereby a laminar liquid jet splits into droplets by a superimposed vibration on the nozzle and dripped into a $CaCl_2$ solution. This process can be performed under mild, non-toxic conditions and can easily be scaled up. The selection of wall material, concentration of wall material, nozzle size, flow rate, vibrational frequency, electrode tension, and drying methods are factors that affect the co-extrusion process. However, highly viscous fluids and the flow rate are limited by the nozzle, which presented as the limitations of this technique (Chew et al., 2015).

This technique avoids the use of organic solvent, high temperature, and extreme pH conditions. Besides, the selected wall materials can improve the antimicrobial activity of encapsulated essential oils via enhancing the cellular interactions exhibited with the pathogen. The previous study showed that the nanoencapsulation of essential oils by ionic gelation helped improve the stability of essential oils and enhanced antifungal activity (Mohammadi et al., 2015). Also, the thyme essential oil–in–water emulsion prepared by ionic gelation technique helped suppress the growth of *Enterobacteriaceae* and *Staphylococcus aureus* (Ghaderi-Ghahfarokhi et al., 2016).

13.2.5.3 Spray-Drying

Spray-drying is based on the principle of the atomization of emulsions into a high-temperature drying medium which leads to rapid evaporation of water and the formation of quick crust and quasi-instantaneous entrapment of the core material. Spray-drying is the most common technology used for the encapsulation process in the food industry due to its ability to cope with heat-sensitive materials, low cost, and available equipment. The spray-dried powders produced can have good reconstitution characteristics, have low water activity, protect the core material against undesirable reactions, and are suitable for transportation and storage (Phisut, 2012). Spray-drying consists of three steps, which are feed atomization, droplet drying, and powder recovery. The liquid feeds are pumped into the drying chamber via an atomizer which will atomize the feed into an enormous number of tiny droplets by the nozzle. This serves to establish a maximum heat transferring surface between the dry air and the liquid in a short time. As droplets of the atomized feed are released through the nozzle, they come in contact with the hot air, where evaporation of water begins. Thus, the liquid feeds are turned into dried powder form at this stage of droplet–air contact or droplet drying. Following the completion of drying, the dried powder can be separated from the drying air by cyclones, recovered, and collected at the collecting vessel (Patel et al., 2009).

Many factors affect the spray drying process, which includes atomization airflow, inlet air temperature, liquid flow rate, aspirator suction velocity, wall materials, and solid concentration (Roccia et al., 2014). The common wall materials used in the spray-drying process, which include polysaccharides (starches, maltodextrin, corn syrup, and gum arabic), lipids (stearic acid, mono- and diglycerides), and proteins (gelatin, casein, whey protein, soy, and wheat) (Saenz et al., 2009). Maltodextrin is the most commonly used wall material in the spray drying process. However, the high glycemic index (>130) of maltodextrin has discouraged its utilization as wall material for developing functional food products. Dietary fibre is encouraged for use as wall material in the encapsulation process due to its health advantage and protective advantage on the active agent (Chew et al., 2018).

13.3 BENEFITS AND APPLICATIONS

Nanoencapsulation is being applied in food, pharmaceutical, and cosmetic industries to protect the unstable active agents against extreme conditions during processing, storage, and transport to enhance their quality and bioactivity (Assadpour and Jafari, 2018). It increases the physical and chemical stabilities of bioactive compounds against oxygen, light, pH, heat, moisture, and gastric digestion. Some bioactive compounds like flavouring substances and essential oils are volatile and need protection. Moreover, some essential oils, fish oil, and nutraceuticals have an unpleasant flavour that restricts their consumer preference. Nanoencapsulation helps decrease the volatility and improve their sensory characteristics, as well as mask the unpleasant flavour in foods (Ghorbanzade et al., 2017; Wen et al., 2016). Thus, nanoencapsulation facilitates the invention of packaging materials with antimicrobial and antioxidant activities. All of these benefit scan be achieved by decreasing the size of NPs, thus increasing the surface area per unit volume (Bazana et al., 2019).

Bioactive compounds such as carotenoids, essential oils, unsaturated fatty acids, and fat-soluble vitamins are normally hydrophobic and poorly soluble in water. Thus, nanoencapsulation enhances the bioaccessibility and bioavailability of these compounds, as well as controls their liberation behaviour (Cheong et al., 2016). Nanoencapsulation improves the solubility of hydrophobic compounds in an aqueous solution and enables the extended absorption of nutrients. On the other hand, nanocarriers can extend the activity of therapeutic agents in blood and cellular uptake (Jafari and McClements, 2017; Kumariet al., 2010).

Essential oils with poor water solubility may affect their antimicrobial activity. Antimicrobial compounds such as thymol and carvacrol are easier to evaporate in nature. Rapid evaporation and chemical interactions like oxidation would make the antimicrobial compounds reduce their antimicrobial activity, which leads to the loss of functionality in food preservation (Kamimura et al., 2014). With the development of nanoencapsulation technologies, the application of antimicrobials would be more predominant.

Nanoencapsulation offers a larger surface area to enhance the solubility and bioavailability. Besides, it improves active agents' controlled release property, as well as the targeting of these agents, when compared to microencapsulation. There are two types of controlled release mechanisms for the release behaviour of active agents from nanocapsules, which are delayed release and sustained released. Delayed release is the release of active agents that is delayed from a bounded "lag time" up to a point when its release is desired and is no longer inhibited. This delivery mechanism is normally used in encapsulation of flavour in ready meals, colour in beverages, or targeted release of nutrients in the intestine. However, sustained release is the release of active agents at its target site at a constant concentration. Sustained release is suitable for extending the release of active agents (Donsì et al., 2011).

13.4 PERSPECTIVES AND CHALLENGES IN NANOTECHNOLOGY

Despite the promising benefits of nanotechnology, the application of nanotechnology may associate with potential toxicity. The toxicological effects of nanoencapsulation on consumers and the environment should be concerned. This negative impact is mostly due to the special characteristics of NPs, like spherical shape, nano-size, chemical composition, structure, and larger surface area (Rezaei et al., 2019; Walters et al., 2016). NPs can enter and assess the circulatory and lymphatic systems of the body tissues and organs. Some NPs can damage cells irreversibly by oxidation or lead to organelle injury (Buzea et al., 2007). Besides that, NPs that consumed by humans should be entirely digested and absorbed. This is because immunological reactions may occur if the NPs are partially digested and cross the biological barrier into the bloodstream (Rezaei et al., 2019). NPs maybe absorbed through inhalation or across the skin layer. This can result in oxidation and inflammation that lead to cardiovascular or extra pulmonary complications, which cause the accumulation of toxicity in the human body (Naseer et al., 2018; Sia, 2017).

The application of NPs in the food system is not fully authorized by the regulatory agencies. A lack of knowledge on the unexpected risks, consumers' safety, and ethical concerns are some of the main challenges in nanotechnology. This is due to the increasing public concerns about the safety issues in food nanotechnology. Therefore, it is necessary to understand the consequences of NPs to human health through in vitro and in vivo studies with the advancement of nanotechnology. The NPs must go through safety evaluations before the commercialization. A comprehensive evaluation of toxicity and potential health risks is necessary. However, there is no standard guideline to monitor the safety assessment of nanomaterials in the market. Thus, further studies are required to develop regulations to ensure food safety (Bazana et al., 2019).

Interaction of the antimicrobial agents with the nanocarrier and the food systems is also one of the challenges. Different food substances like enzymes, glutathione, sodium metabisulfite, and titanium dioxide may significantly deteriorate the antimicrobial agent (such as peptides) via specific and non-specific interactions (Bhatti et al., 2004). Plant-derived antimicrobial agents such as essential oils (thymol, eugenol, carvacrol, and *trans*-cinnamaldehyde) can deteriorate in food systems. These antimicrobial compounds may bind with the hydrophobic food compounds, such as lipids and proteins, which limit their effective amount against the target microorganism (Shah et al., 2012). Thus, it is necessary to understand the interaction between the NPs and the food system when they are incorporating in the food system. This can further explore the use of nanoencapsulation in improving the antimicrobial activity of antimicrobial agents.

13.5 CONCLUSION

Nanoencapsulation is a highly encouraged technique to encapsulate antimicrobial agents and to be applied in the food systems. It helps to protect the antimicrobial agent from deterioration, improve their water solubility, control their release behaviour, and enhance their antimicrobial activity. There are various nanoencapsulation technologies with their respective advantages and limitations. Appropriate antimicrobial encapsulating systems should be developed based on the fundamental concepts of the structures and functions of nanocarriers selected, as well as their interactions with the active agents. Future research of nanoencapsulation can be focused on the physicochemical properties of the nanocarriers, interaction between nanocarriers and active agents, interaction between NPs and food products, and release behaviour of encapsulated active agents in NPs. Also, the efforts to substantially reduce the cost of nanoencapsulation of antimicrobial agents in the food, pharmaceutical, and cosmetic industries are required.

REFERENCES

Ach, D., Briancon, S., Dugas, V., Pelletier, J., Broze, G., and Chevalier, V. (2015). Influence of main whey protein components on the mechanism of complex coavervation with Acacia gum. *Colloids and Surfaces A: Physicochemical and Engineering Aspects*, *481*, 367–374.

Akhavan, S., Assadpour, E., Katouzian, I., and Jafari, S. M. (2018). Lipid nano scale cargos for the protection and delivery of food bioactive ingredients and nutraceuticals. *Trends in Food Science & Technology*, *74*, 132–146.

Al-Haj, N. A., Shamsudin, M. N., Alipiah, N. M., et al. (2010). Characterization of *Nigella sativa* L. essential oil-loaded solid lipid nanoparticles. *American Journal of Pharmacology and Toxicology*, *5*, 52–57.

Anton, N., Benoit, J. P., and Saulnier, P. (2008). Design and production of nanoparticles formulated from nano-emulsion templates—A review. *Journal of Controlled Release*, *128*, 185–199.

Asbahani, A. E., Miladi, K., Badri, W., et al. (2015). Essential oils: From extraction to encapsulation. *International Journal of Pharmaceutics*, *483*, 220–243.

Assadpour, E., and Jafari, S. M. (2018). A systematic review on nanoencapsulation of food bioactive ingredients and nutraceuticals by various nanocarriers. *Critical Reviews in Food Science and Nutrition*, *59*, 3129–3151.

Aswathanarayan, J. B., and Vittal, R. R. (2019). Nanoemulsions and their potential applications in food industry. *Frontiers in Sustainable Food Systems*, *3*, 1–21. doi:10.3389/fsufs.2019.00095.

Aveyard, R., Binks, B. P., and Clint, J. H. (2003). Emulsions stabilised solely by colloidal particles. *Advances in Colloid and Interface Science*, *100–102*, 503–546.

Bazana, M. T., Codevilla, C. F., and de Menezes, C. R. (2019). Nanoencapsulation of bioactive compounds: Challenges and perspectives. *Current Opinion in Food Science*, *26*, 47–56.

Bhargava, K., Conti, D. S., daRocha, S. R., and Zhang, Y. (2015). Application of an oregano oil nanoemulsion to the control of foodborne bacteria on fresh lettuce. *Food Microbiology*, *47*, 69–73.

Bhatti, M., Veeramachaneni, A., and Shelef, L. A. (2004). Factors affecting the antilisterial effects of nisin in milk. *International Journal of Food Microbiology*, *97*, 215–219.

Bi, L., Yang, L., Bhunia, A. K., and Yao, Y. (2011). Carbohydrate nanoparticle-mediated colloidal assembly for prolonged efficacy of bacteriocin against food pathogen. *Biotechnology and Bioengineering*, *108*, 1529–1536.

Bozzuto, G., and Molinari, A. (2015). Liposomes as nanomedical devices. *International Journal of Nanomedicine*, *10*, 975–999.

Buzea, C., Blandino, I. I. P., and Robbie, K. (2007). Nanomaterials and nanoparticles: Sources and toxicity. *Biointerphase*, *2*, MR17–MR172.

Chawla, P., Kumar, N., Bains, A., Dhull, S. B., Kumar, M., Kaushik, R., and Punia, S. (2020). Gum arabic capped copper nanoparticles: Synthesis, characterization, and applications. *International Journal of Biological Macromolecules*, *146*, 232–242.

Chawla, P., Kumar, N., Kaushik, R., and Dhull, S. B. (2019). Synthesis, characterization and cellular mineral absorption of nanoemulsions of *Rhododendron arboreum* flower extracts stabilized with gum arabic. *Journal of Food Science and Technology*, *56*(12), 5194–5203.

Chen, C., Han, D., Cai, C., and Tang, X. (2010). An overview of liposome lyophilisation and its future potential. *Journal of Controlled Release*, *142*, 299–311.

Cheong, A. M., Tan, C. P., and Nyam, K. L. (2016). *In vitro* evaluation of the structural and bioaccessibility of kenaf seed oil nanoemulsions stabilised by binary emulsifiers and β-cyclodextrin complexes. *Journal of Food Engineering*, *189*, 90–98.

Chevalier, Y., and Bolzinger, M. A. (2013). Emulsions stabilized with solid nano-particles: Pickering emulsions. *Colloids and Surfaces A: Physicochemical and Engineering Aspects*, *439*, 23–34.

Chew, S. C., and Nyam, K. L. (2016). Microencapsulation of kenaf seed oil by co-extrusion technology. *Journal of Food Engineering*, *175*, 43–50.

Chew, S. C., Tan, C. P., and Nyam, K. L. (2018). *In-vitro* digestion of refined kenaf seed oil microencapsulated in β-cyclodextrin/gum Arabic/sodium caseinate by spray drying. *Journal of Food Engineering*, *225*, 34–41.

Chew, S. C., Tan, C. P., Long, K., and Nyam, K. L. (2015). *In-vitro* evaluation of kenaf seed oil in chitosan coated-high methoxyl pectin-alginate microcapsules. *Industrial Crops and Products*, 76, 230–236.

Chu, C. C., Tan, C. P., and Nyam, K. L. (2019). Development of nanostructured lipid carriers (NLCs) using pumpkin and kenaf seed oils with potential photoprotective and antioxidative properties. *European Journal of Lipid Science and Technology*, *121*, 1900082–1900091.

Da Silva, P. T., Fries, L. L. M., De Menezes, C. R., et al. (2014). Microencapsulation: Concepts, mechanisms, methods and some applications in food technology. *Ciência Rural*, *44*, 1304–1311.

Detoni, C. B., de Oliveira, D. M., Santo, I. E., et al. (2012). Evaluation of thermal-oxidative stability and antiglioma activity of *Zanthoxylumtingoassuiba* essential oil entrapped into multi- and unilamellar liposomes. *Journal of Liposome Research*, *22*, 1–7.

Dias, D. R., Botrel, D. A., Fernandes, R. V. D. B., and Borges, S. V. (2017). Encapsulation as a tool for bioprocessing of functional foods. *Current Opinion in Food Science*, *13*, 31–37.

Donsì, F., Annunziata, M., Sessa, M., and Ferrari, G. (2011). Nanoencapsulation of essential oils to enhance their antimicrobial activity in foods. *Food Science and Technology*, *44*, 1908–1914.

Donsì, F., Cuomo, A., Marchese, E., and Ferrari, G. (2014). Infusion of essential oils for food stabilization: Unraveling the role of nanoemulsion-based delivery systems on mass transfer and antimicrobial activity. *Innovative Food Science & Emerging Technologies*, *22*, 212–220.

Ezhilarasi, P. N., Karthik, P., Chhanwal, N., and Anandharamakrishnan, C. (2013). Nanoencapsulation techniques for food bioactive components: A review. *Food and Bioprocess Technology*, *6*, 628–647.

Fakhravar, Z., Ebrahimnejad, P., and Akbarzadeh, A. (2016). Nanoliposomes: Synthesis methods and applications in cosmetics. *Journal of Cosmetic and Laser Therapy*, *18*, 174–181.

Fathi, M., Mozafari, M. R., and Mohebbi, M. (2012). Nanoencapsulation of food ingredients using lipid based delivery system. *Trends in Food Science and Technology*, *23*, 13–27.

Fu, X. W., Huang, B., and Feng, F. Q. (2008). Shelf life of fresh noodles as affected by the food grade monolaurin microemulsion system. *Journal of Food Process Engineering*, *31*, 619–627.

Fu, Y., Sarkar, P., Bhunia, A. K., and Yao, Y. (2016). Delivery systems of antimicrobial compounds to food. *Trends in Food Science & Technology*, *57*, 165–177.

Gaysinsky, S., Taylor, T. M., Davidson, P. M., Bruce, B. D., and Weiss, J. (2007). Antimicrobial efficacy of eugenol microemulsions in milk against *Listeria monocytogenes* and *Escherichia coli* O157:H7. *Journal of Food Protection*, *70*, 2631–2637.

Ghaderi-Ghahfarokhi, M., Barzegar, M., Sahari, M. A., and Azizi, M. H. (2016). Nanoencapsulation approach to improve antimicrobial and antioxidant activity of thyme essential oil in beef burgers during refrigerated storage. *Food and Bioprocess Technology*, *9*, 1187–1201.

Ghorbanzade, T., Jafari, S. M., Akhavan, S., and Hadavi, R. (2017). Nano-encapsulation of fish oil in nano-liposomes and its application in fortification of yogurt. *Food Chemistry*, *216*, 146–152.

Ghosh, V., Mukherjee, A., and Chandrasekaran, N. (2013). Ultrasonic emulsification of food-grade nanoemulsion formulation and evaluation of its bactericidal activity. *Ultrasonics – Sonochemistry*, *20*, 338–344.

Heikal, A., Box, K., Rothnie, A., Storm, J., Callaghan, R., and Allen, M. (2009). The stabilisation of purified, reconstituted P-glycoprotein by freeze drying with disaccharides. *Cryobiology*, *58*, 37–44.

Iwanaga, K., Ono, S., Narioka, K., et al. (1999). Application of surface-coated liposomes for oral delivery of peptide: Effect of coating the liposome's surface on the GI transit of insulin. *Journal of Pharmaceutical Sciences*, *88*, 248–252.

Jafari, S. M., He, Y., and Bhandari, B. (2006). Nano-emulsion production by sonication and microfluidization—A comparison. *International Journal of Food Properties*, *9*, 475–485.

Jafari, S. M., and McClements, D. J. (2017). Nanotechnology approaches for increasing nutrient bioavailability. *Advances in Food and Nutrition Research*, *81*, 1–30.

Kadimi, U. S., Balasubramanian, D. R., Ganni, U. R., Balaraman, M., and Govindarajulu, V. (2007). In vitro studies on liposomal amphotericin B obtained by supercritical carbon dioxide-mediated process. *Nanomedicine*, *3*, 273–280.

Kamimura, J. A., Santos, E. H., Hill, L. E., and Gomes, C. L. (2014). Antimicrobial and antioxidant activities of carvacrol microencapsulated in hydroxypropyl-beta-cyclodextrin. *LWT-Food Science and Technology*, *57*, 701–709.

Kawai, K., and Suzuki, T. (2007). Stabilizing effect of four types of disaccharide on the enzymatic activity of freeze-dried lactate dehydrogenase: Step by step evaluation from freezing to storage. *Pharmaceutical Research*, *24*, 1883–1890.

Koupantsis, T., Pavlidou, E., and Paraskevopoulou, A. (2016). Glycerol and tannic acid as applied in the preparation of milk proteins—CMC complex coacervates for flavour encapsulation. *Food Hydrocolloids*, *57*, 62–71.

Kumari, A., Yadav, S. K., and Yadav, S. C. (2010). Biodegradable polymeric nanoparticles based drug delivery system. *Colloids and Surfaces B: Biointerfaces*, *75*, 1–18.

Lai, F., Wissing, S. A., Müller, R. H., and Fadda, A. M. (2006). *Artemisia arborescens* L. essential oil-loaded solid lipid nanoparticles for potential agricultural application: Preparation and characterization. *AAPS PharmSciTech*, *7*, 10–18.

Laye, C., McClements, D. J., and Weiss, J. (2008). Formation of biopolymer-coated liposomes by electrostatic deposition of chitosan. *Journal of Food Science*, *73*, 7–15.

Li, H., Song, J., Park, J., and Han, K. (2003). Polyethylene glycol-coated liposomes for oral delivery of recombinant human epidermal growth factor. *International Journal of Pharmaceutics*, *258*, 11–19.

Li, N., Zhuang, C., Wang, M., Sun, X., Nie, S., and Pan, W. (2009). Liposome coated with low molecular weight chitosan and its potential use in ocular drug delivery. *International Journal of Pharmaceutics*, *379*, 131–138.

Li, R., Deng, L., Cai, Z., et al. (2017). Liposomes coated with thiolated chitosan as drug carriers of curcumin. *Materials Science and Engineering C*, *80*, 156–164.

Luo, Z., Murray, B. S., Yusoff, A., Morgan, M. R., Povey, M. J., and Day, A. J. (2011). Particle-stabilizing effects of flavonoids at the oil-water interface. *Journal of Agricultural and Food Chemistry*, *59*, 2636–2645.

Maestrelli, F., González-Rodríguez, M. I., Rabasco, A. M., and Mura, P. (2005). Preparation and characterisation of liposomes encapsulating ketoprofen-cyclodextrin complexes for transdermal drug delivery. *International Journal of Pharmaceutics*, *198*, 55–67.

McClements, D. J., Decker, E. A., Park, Y., and Weiss, J. (2009). Structural design principles for delivery of bioactive components in nutraceuticals and functional foods. *Critical Reviews in Food Science and Nutrition*, *49*, 577–606.

Mehrnia, M. A., Jafari, S. M., Makhmal-Zadeh, B. S., and Maghsoudlou, Y. (2016). Crocin loaded nano-emulsions: Factors affecting emulsion properties in spontaneous emulsification. *International Journal of Biological Macromolecules*, *84*, 261–267.

Mohammadi, A., Hashemi, M., and Hosseini, S. M. (2015). Nanoencapsulation of *Zataria multiflora* essential oil preparation and characterization with enhanced antifungal activity for controlling *Botrytis cinerea*, the causal agent of gray mould disease. *Innovative Food Science and Emerging Technologies*, *28*, 73–80.

Müller, R. H., Radtke, M., and Wissing, S. A. (2002). Solid nanoparticle (SLN) and nanostructured lipid carrier (NLC) in cosmetic and dermatological preparations. *Advanced Drug Delivery Reviews*, *54*, 131–155.

Naseer, B., Srivastava, G., Qadri, O. S., Faridi, S. A., Islam, R. U., and Younis, K. (2018). Importance and health hazards of nanoparticles used in the food industry. *Nanotechnology Reviews*, *7*, 623–641.

Nedovic, V., Kalusevic, A., Manojlovic, V., Levic, S., and Bugarski, B. (2011). An overview of encapsulation technologies for food applications. *Procedia Food Science*, *1*, 1806–1815.

Parasassi, T., Di Stefano, M., Loiero, M., Ravagnan, G., and Gratton, E.(1994). Influence of cholesterol on phospholipid bilayers phase domains as detected by *Laurdan fluorescence*. *Biophysical Journal*, *66*, 120–132.

Patel, R. P., Patel, M. P., and Suthar, A. M. (2009). Spray drying technology: An overview. *Indian Journal of Science and Technology*, *2*, 44–47.

Pattni, B. S., Chupin, V. V., and Torchilin, V.P. (2015). New developments in liposomal drug delivery. *Chemical Reviews*, *115*, 10938–10966.

Peng, C., Zhao, S. Q., Zhang, J., Huang, G. Y., Chen, L. Y., and Zhao, F. Y. (2014). Chemical composition, antimicrobial property and microencapsulation of Mustard (*Sinapis alba*) seed essential oil by complex coacervation. *Food Chemistry*, *165*, 560–568.

Phisut, N. (2012). Spray drying technique of fruit juice powder: Some factors influencing the properties of product. *International Food Research Journal*, *19*, 1297–1306.

Pisoschi, A. M., Pop, A., Cimpeanu, C., Turcuş, V., Predoi, G., and Iordache, F. (2018). Nanoencapsulation techniques for compounds and products with antioxidant and antimicrobial activity—A critical view. *European Journal of Medicinal Chemistry*, *157*, 1326–1345.

Prajapati, V. D., Jani, G. K., Moradiya, N. G., and Randeria, N. P. (2013). Pharmaceutical applications of various natural gums, mucilages and their modified forms. *Carbohydrate Polymers*, *92*, 1685–1699.

Qian, C., and McClements, D. J. (2011). Formation of nanoemulsions stabilized by model food-grade emulsifiers using high-pressure homogenization: Factors affecting particle size. *Food Hydrocolloids*, *25*, 1000–1008.

Rao, J., and McClements, D. J. (2011). Food-grade microemulsions, nanoemulsions and emulsions: Fabrication from sucrose monopalmitate & lemon oil. *Food Hydrocolloids*, *25*, 1413–1423.

Rezaei, A., Fathi, M., and Jafari, S. M. (2019). Nanoencapsulation of hydrophobic and low-soluble food bioactive compounds within different nanocarriers. *Food Hydrocolloids*, *88*, 146–162.

Ribeiro-Santos, R., Andrade, M., and Sanches-Silva, A. (2017). Application of encapsulated essential oils as antimicrobial agents in food packaging. *Current Opinion in Food Science*, *14*, 78–84.

Roccia, P., Martínez, M. L., Liabot, J. M., and Ribotta, P. D. (2014). Influence of spray-drying operating conditions on sunflower oil powder qualities. *Powder Technology*, *254*, 307–313.

Roos, H. Y., and Livney, Y. D. (2017). *Engineering Foods for Bioactives Stability and Delivery*, New York: Springer-Verlag.

Rostamabadi, H., Falsafi, S. R., and Jafari, S. M. (2019). Nanoencapsulation of carotenoids within lipid-based nanocarriers. *Journal of Controlled Release*, *298*, 38–67.

Saenz, C., Tapia, S., Chavez, J., and Robert, P. (2009). Microencapsulation by spray drying of bioactive compounds from cactus pear (*Opuntiaficus-indica*). *Food Chemistry*, *14*, 616–622.

Sakulku, U., Nuchuchua, O., Uawongyart, N., Puttipipatkhachorn, S., Soottitantawat, A., and Ruktanonchai, U. (2009). Characterization and mosquito repellent activity of citronella oil nanoemulsion. *International Journal of Pharmaceutics*, *372*, 105–111.

Samuni, A., Lipman, A., and Barenholz, Y. (2000). Damage to liposomal lipids: Protection by antioxidants and cholesterol-mediated dehydration. *Chemistry and Physics of Lipids*, *105*, 121–134.

Sanguansri, P., and Augustin, M. A. (2006). Nanoscale materials development—A food industry perspective. *Trends in Food Science and Technology*, *17*, 547–556.

Scallan, E., Hoekstra, R. M., Angulo, F. J., et al. (2011). Foodborne illness acquired in the United States—Major pathogens. *Emerging Infectious Diseases*, *17*, 7–15.

Sebaaly, C., Charcosset, C., Stainmesse, S., Fessi, H., and Greige-Gerges, H. (2016). Clove essential oil-in-cyclodextrin-in-liposomes in the aqueous and lyophilized states: From laboratory to large scale using a membrane contactor. *Carbohydrate Polymers*, *138*, 75–85.

Shah, B., Davidson, P. M., and Zhong, Q. (2012). Nanocapsular dispersion of thymol for enhanced dispersibility and increased antimicrobial effectiveness against *Escherichia Coli* O157:H7 and *Listeria Monocytogenes* in model food systems. *Applied and Environmental Microbiology*, *78*, 8448–8453.

Shakeel, F., and Ramadan, W. (2010). Transdermal delivery of anticancer drug caffeine from water-in-oil nanoemulsions. *Colloids and Surfaces B: Biointerfaces*, *75*, 356–362.

Shimoni, G., Levi, C. S., Tal, S. L., and Lesmes, U. (2013). Emulsions stabilization by lactoferrin nano-particles under in vitro digestion conditions. *Food Hydrocolloids*, *33*, 264–272.

Sia, P. (2017). Nanotechnology among innovation, health and risks. *Procedia Social and Behavioral Sciences*, *237*, 1076–1080.

Slomkowski, S., Alemán, J. V., Gilbert, R. G., et al. (2011). Terminology of polymers and polymerization processes in dispersed systems (IUPAC Recommendations 2011). *Chemistry in Industry*, *83*, 2229–2259.

Solans, C., Izquierdo, P., Nolla, J., Azemar, N., and Garcia-Celma, M. J. (2005). Nanoemulsions. *Current Opinion in Colloid and Interface Science*, *10*, 102–110.

Sonneville-Aubrun, O., Simmonet, J. T., and Alloret, F. L. (2004). Nanoemulsions: A new vehicle for skin care products. *Advances in Colloid and Interface Science*, *108–109*, 145–149.

Sugumar, S., Ghosh, V., Mukherjee, A., and Chandrasekaran, N. (2016). Essential oil-based nanoemulsion formation by low- and high-energy methods and their application in food preservation against food spoilage microorganisms. In *Essential Oils in Food Preservation, Flavour and Safety*, ed. V. R. Preedy (pp. 93–100. London: Academic Press.

Sun-Waterhouse, D., Penin-Peyta, L., Wadhwa, S. S., and Waterhouse, G. I. N. (2012). Storage stability of phenolic-fortified avocado oil encapsulated using different polymer formulations and co-extrusion technology. *Food and Bioprocess Technology*, *5*, 3090–3102.

Talegaonkar, S., Mustafa, G., Akhter, S., and Iqbal, Z. I. (2010). Design and development of oral oil-in-water nanoemulsion formulation bearing atorvastatin: In vitro assessment. *Journal of Dispersion Science and Technology*, *31*, 690–701.

Tamjidi, F., Shahedi, M., Varshosaz, J., and Nasirpour, A. (2013). Nanostructured lipid carriers (nlc): A potential delivery system for bioactive food molecules. *Innovative Food Science and Emerging Technologies*, *19*, 29–43.

Tantisripreecha, C., Jaturanpinyo, M., Panyarachun, B., and Sarisuta, N. (2012). Development of delayed-release proliposomes tablets for oral protein drug delivery. *Drug Development and Industrial Pharmacy*, *38*, 718–727.

Taylor, T. M., Davidson, P. M., Bruce, B. D., and Weiss, J. (2005). Liposomal nanocapsules in food science and agriculture. *Critical Reviews in Food Science and Nutrition*, *45*, 587–605.

Tiwari, B. K., Valdramidis, V. P., O'Donnell, C. P., Muthukumarappan, K., Bourke, P., and Cullen, P. (2009). Application of natural antimicrobials for food preservation. *Journal of Agricultural and Food Chemistry*, *57*, 5987–6000.

Valencia-Sullca, C., Jiménez, M., Jiménez, A., Atarés, L., Vargas, M., and Chiralt, A. (2016). Influence of liposome encapsulated essential oils on properties of chitosan films. *Polymer International*, *65*, 979–987.

Van Vuuren, S. F., du Toit, L. C., Parry, A., et al. (2010). Encapsulation of essential oils within a polymeric liposomal formulation for enhancement of antimicrobial efficacy. *Natural Product Communications*, *5*, 1401–1408.

Walters, C., Pool, E., and Somerset, V., (2016). Nanotoxicity in aquatic invertebrates. In *Invertebrates Experimental Models in Toxicity Screening*, ed. M. L. Larramendy (pp. 13–34). Croatia: IntechOpen.

Wang, Y., Tu, S. L., Li, R. S., Yang, X. Y., Liu, L. R., and Zhang, Q. Q. (2010). Cholesterol succinyl chitosan anchored liposomes: Preparation, characterization, physical stability, and drug release behaviour. *Nanomedicine: Nanotechnology, Biology and Medicine*, *6*, 471–477.

Wang, Y., Zhao, R., Yu, L., Zhang, Y., He, Y., and Yao, J. (2014). Evaluation of cinnamon essential oil microemulsion and its vapour phase for controlling postharvest gray mold of pears (*Pyrus pyrifolia*). *Journal of the Science of Food and Agriculture*, *94*, 1000–1004.

Weiss, J., Decker, E. A., McClements, D. J., Kristbergsson, K., Helgason, T., and Awad, T. (2008). Solid lipid nanoparticles as delivery systems for bioactive food components. *Food Biophysics*, *3*, 146–154.

Weiss, J., Gaysinksy, S., Davidson, M., and McClements, J. (2009). Nanostructured encapsulation systems: Food antimicrobials. In *IUFoST World Congress Book Global Issues in Food Science and Technology*, ed. A. Mortimer, D. Lineback, W. Spiess, and K. Buckle (pp. 425–479). Amsterdam: Elsevier Inc.

Wen, P., Zhu, D. H., Wu, H., Zong, M. H., Jing, Y. R., and Han, S. Y. (2016). Encapsulation of cinnamon essential oil in electrospun nanofribrous film for active food packaging. *Food Control*, *59*, 366–376.

Wu, J. E., Lin, J., and Zhong, Q. (2014). Physical and antimicrobial characteristics of thyme oil emulsified with soluble soybean polysaccharide. *Food Hydrocolloids*, *39*, 44–150.

Xing, F., Cheng, G., Yi, K., and Ma, L. (2004). Nanoencapsulation of capsaicin by complex coacervation of gelatin, acacia, and tannins. *Journal of Applied Polymer Science*, *96*, 2225–2229.

Yang, Y., Anvari, M., Pan, C. H., and Chung, D. (2012). Characterisation of interactions between fish gelatine and gum arabic in aqueous solutions. *Food Chemistry*, *135*, 555–561.

Yang, Z., Peng, Z., Li, J., Li, S., Kong, L., Li, P., and Wang, Q. (2014). Development and evaluation of novel flavour microcapsules containing vanilla oil using complex coacervation approach. *Food Chemistry*, *145*, 272–277.

Zaru, M., Manca, M. L., Fadda, A. M., and Antimisiaris, S. G. (2009). Chitosan-coated liposomes for delivery to lungs by nebulisation. *Colloids and Surfaces B: Biointerfaces*, *71*, 88–95.

14 Nanocapsules as Potential Antimicrobial Agents in Food

Bababode Adesegun Kehinde
University of Kentucky, USA
Anil Panghal and Sunil Kumar
Chaudhary Charan Singh Haryana Agricultural University, India
Akinbode A. Adedeji
University of Kentucky, USA
Mukesh Kumar Garg
Chaudhary Charan Singh Haryana Agricultural University, India
Navnidhi Chhikara
Guru Jambheshwar University of Science and Technology, India

CONTENTS

14.1 INTRODUCTION

The International Union of Pure and Applied Chemistry describes a nanocapsule as a hollowed nanomaterial made of an outer solid framework that circumvents a core-forming region present to enmesh substances (Panghal et al., 2019a). They can also be defined as nano-sized materials, with surface regions or coatings having a distinct composition from their internal contents (Thies, 2012; Vert et al., 2012; Kehinde et al.,

2020a). Accordingly, they are outcomes of the finesse subject of nanotechnology, which is an important branch of contemporary technology, and have been employed in diverse fields of medicine and pharmaceutics, bioprocessing, food and health, cosmetics, and mechatronics, amongst several others. In the field of medical science, a layer-by-layer technique has been used to enhance the delivery of drugs using polyelectrolytes and non-cytotoxic core, and nanocapsules have been used as artificial tissues and as cellular coatings and generally for the targeted delivery of biomaterials to predefined organs and inhibition of undesirable immunological responses (Krol et al., 2004). Nanocapsules have been investigated as potential therapeutic agents for the management of metabolic syndrome disorders such as cancers and diabetes. In medical radiography, they have been used to aid ultrasonication diagnostics and for contrast imaging in magnetic resonance imaging (MRI); (Wang et al., 2012a). In the areas of food and bioprocessing, nanocapsule processing has been adopted in protein mineralization, improvement of the stability and morphology of bioactive components, preservation of fruits and vegetables, preparation of nanoemulsions, decontamination processes, development of biodegradable films, and delivery of functional components (Panghal et al., 2018; Sharma et al., 2019; Singh et al., 2020). For cosmetic purposes, nanocapsules, based on their advantage of permeability, have been used for dermatological needs, development of body steroids for muscular growth, and poly(ε-caprolacton) nanocapsules have been used to improve hair growth (Hwang and Kim, 2008). Researchers, scientists, and food industries have investigated these nanocapsule components extensively for their antimicrobial efficacy against spoilage and pathogenic microorganisms to improve food safety (Kehinde et al., 2020b; Chhikara et al., 2018).

14.2 FABRICATION OF NANOCAPSULES

The methodology adopted for preparing a nanocapsule depends on its target users. Considerations such as size, morphology, density, and stability of the nanocapsule are very important. Ezhilarasi et al. (2012) categorized the preparation of nanocapsules of food usage based on their preparation methods as top-down approach methods and bottom-up approach methods. The top-down techniques, which include emulsification and emulsification evaporation, were described as those involving a structure shaping and size reduction (from 1 mm to the desired nm) while the bottom-up approaches follow the opposite schematic of self-organization and self-construction of molecules (from 1 nm to the desired nm). Their preparation can also be classified based on the processing nature as polymerization preparation and ionic coacervation/pregelation preparation. The polymerization process generally entails incorporating bioactive materials into polymerization medium for encapsulation or simply by adsorption to nanocapsules after polymerization while the ionic class involves polyelectrolyte cross-linkage in the company of a counter ion (polyionic or cationic) to produce nanocapsules (Cerqueira et al., 2017).

One common method used is termed as the interfacial disposition. This procedure is used for nanocapsules with two different phases. An aqueous phase is mingled with an organic solvent (such as acetone) having the property of water miscibility and to this solution, a hydrophobic phase with substantial miscibility in the organic solvent is dispersed with a subsequent agitation for proper diffusion (Letchford and Burt,

2007). Other methods include the sol-gel method and polymerization techniques, such as in situ and mini-emulsion polymerization techniques (Liu et al., 2015). The mini-emulsion polymerization method, with the relative advantages of higher stability and lower energy requirements, involves the stable and tiny droplets of the nanoscale, containing initiators, emulsifiers, waters, and monomers, being subjected to extreme shear for the production of nanocapsules. The resulting characteristics and composition of the nanocapsule are remarkably related to the monomer used. The sol-gel method commences with a metal alkoxide which blends with the complexing agent, catalyst, and solvent homogenously; hydrolysis and condensation occur which give rise to the formation of a transparent and stable colloidal solution, and then a three-dimensional network construction of the nanocapsule occurs after the ageing of the sol (Liu et al., 2015). For in situ polymerization, as its name implies, the polymerization of the monomer which will form the polymeric wall material for the nanocapsule occurs outside the core material. For pharmaceutical and food applications, especially, the particle flocculation and chemical degradation concerns such as pH, heat, light, oxygen exposure, and heavy metals should be optimally considered (Thies, 2012).

Solvent evaporation has also proved to be a productive mechanism for the synthesis of nanocapsules. Briefly, double or single emulsions are prepared from solvents for the formulation of a nanoparticulate suspension. Subdivision operations using ultrasound or high-speed homogenization are then employed for the obtainment of particles of reduced sizes in the suspension (Nagavarma et al., 2012). Solvents are then ridded off by evaporation either by pressure reduction or by continuous stirring at room temperature. Precipitation of a polymer from an organic solution at the nano-level, with the organic solvent diffused thereafter for a spontaneous emulsification process, is termed as nanoprecipitation, and it is used for the nanoencapsulation of bioactive materials (Reis et al., 2006). The nanoencapsulation process conventionally requires some form of agitation or mixing or stirring. Processing techniques adopted to aid the formation of nanocapsules, especially in food and agriculture include ultrasonication (bath and probe), high-pressure homogenization, microfluidization, membrane extrusion, thin-film rehydration, freeze-drying, reverse phase evaporation, and detergent depletion (Taylor et al., 2005). For the inclusion complexation process, the bioactive material is a ligand that is supramolecular encapsulated into a space-bearing encapsulating material through hydrophobicity induced by entropy change, van der Waals force, or hydrogen bonding (Ezhilarasi et al., 2012).

The electrohydrodynamic spraying technique, which is also simply referred to as the electro-spraying technique, involves the application of a predetermined voltage to a liquid for its nanoscale reduction and the subsequent usage of that reduced liquid for nano-encapsulation. Important parameters to be considered in this process include the liquid flow rate, applied high voltage, the distance for the particle collection, and the density, viscosity, and concentration of the polymer used (Tapia-Hernández et al., 2017). The coacervation method involving phase separation of a mixture of a single polyelectrolyte from a solution is another prominent procedure for preparing nanocapsules. In this method, the freshly prepared coacervate phase surrounding the active ingredient can be crosslinked with a suitable enzyme such as (transglutaminase or glutaraldehyde) or chemical to enhance the coacervate robustness (Zuidam and Shimoni, 2010). Two or more polymers can be used, in which case, the process is referred to as a complex coacervation, but when a single polymer is used, it is called simple coacervation.

14.3 NANO-ENCAPSULATION IN THE FOOD INDUSTRY

Nanocapsules are more efficient for these uses in food processing relative to micro- or macroparticles based on their superior chemical and physical stability, larger surface area per unit volume, better compatibility with food matrices, subcellular sizes, more effective release systems, more suitable emulsifiers, and better nutraceutical delivery (Berekaa, 2015). In food processing, nanocapsules or nano-scale material carrier systems, usually in the form of proteins, micelles, or liposomes, and are useful as emulsifiers, bioavailability enhancers, nutrition supplements, dispersibility agents, antimicrobials, antioxidants, packaging material fabrication, and development of nutraceuticals (Table 14.1). Encapsulation of bioactive ingredients has been found to minimize undesirable intra-related ingredient interactions and offers better control over chemical reactions and mass transport phenomena, and for nutraceuticals, it improves bioactivity, for instance by enhancing the uptake and adsorption of components in the digestion process (Weiss et al., 2009; Figure 14.1).

14.3.1 Food Antimicrobial Nanocapsules

The U.S. Food and Drug Administration (FDA) has described antimicrobial agents as chemical constituents whose presence or addition to foods, their packages, contact surfaces, or processing environments retard the growth of, or inactivate spoilage or pathogenic microorganisms (Davidson et al., 2005).

14.3.1.1 Mechanism

The mechanism of action of food antimicrobial agents have been categorized into two fundamental groups: inhibition of dynamics for proton transfer and microbial membrane perturbation, although other pathways such as disruption of microbial enzymes may be followed (Weiss et al., 2009). The use of food-grade organic acids as antimicrobials usually performs based on inhibiting proton dynamics, and their

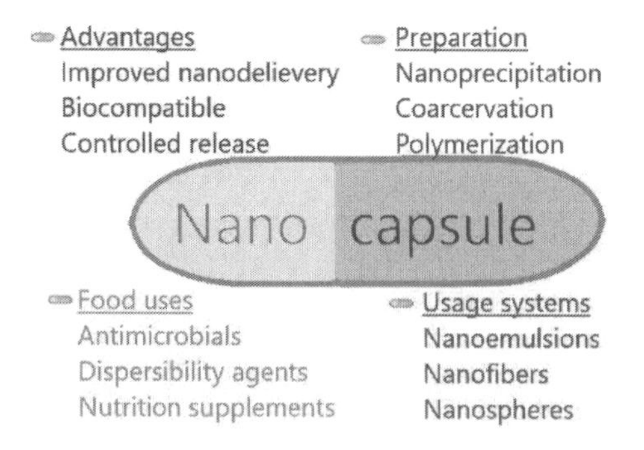

FIGURE 14.1 Brief details regarding nanocapsule usage.

TABLE 14.1
Applications of Nano-Encapsulation in Food Systems

Nanocapsule	Preparation Technique	Nanocapsule Properties	Encapsulation Effects	Reference
Curcumin encapsulated in casein nanocapsules	Spray drying	zeta-potential of −31.1 ± 0.9 mV at pH 6.8, and encapsulation efficiency of about 83.1%	Increased dispersibility, antioxidant activity and arrest of cell growth	Pan, Zhong, and Baek, 2013
Essential oil-loaded chitosane-alginate nanocapsules	Emulsification aided with ultrasonication	Zeta potential was greater than +35 mV and particle sizes of 256 and 226 nm for turmeric oil- and lemongrass oil-loaded nanocapsules	Significant decrease of cancerous activity using cultured A549 cells	Natrajan et al., 2015
Poly-caprolactone (PCL) nanocapsules containing -tocopherol	Nanoprecipitation	Mean particle size of our obtained PCL nanocapsules ranged from 184.4 to 219.1 nm and zeta-potential values ranging from −30.6 to −51.1 mV	High values for tocopherol recovery (in the range of 78.31%–90.34%) with encapsulation efficiency of 99.97%	Noronha et al., 2013, 2014
Chitosan hydrochloride (CHC)– and epigallocatechin gallate (EGCG)–loaded nanocapsules	Solution casting method		Enhancing the formation of rougher, more elastic, and harder film structures having advantageous optical properties	Liang et al., 2017
Gelatin/gum arabic nanocapsules	Complex coacervation	Size of 384.14 ±8.28 nm and zeta potential of 1.92 ±2.37 mV after 7 h with no hardening	Improved heat resistance	Lv et al., 2014
Poly(epsilon-caprolactone) nanocapsules stabilized by gelatin	Modified emulsion–diffusion method	Mean particle size of 197.5 nm, poly dispersibility index and zeta potential of0.12and 7.49mV	Aiding preparation of a freeze-dried food	Nakagawa et al., 2011

(Continued)

TABLE 14.1 (Continued)
Applications of Nano-Encapsulation in Food Systems

Nanocapsule	Preparation Technique	Nanocapsule Properties	Encapsulation Effects	Reference
Lycopene-loaded lipid-core nanocapsules with poly(ε-caprolactone) interfacial deposition	Interfacial deposition	z-average (193 ± 4.7 nm) and zeta potential (−11.5 ± 0.40 mV).	Satisfactory stability, presenting around 50% Lyc content after 14 days of storage at room temperature (25°C)	dos Santos et al., 2015
Chitosan/fucoidan nanocapsules encapsulating poly-L-lysine	Layer-by-layer deposition technique	Zeta-potential values generally higher than +30 mV or lower than −30 mV	Stable preparation with desirable release	Pinheiro et al., 2015
Nanocapsules obtained from a blend of β-carotene, α-carotene, and lutein	Interfacial deposition of preformed polymer	z-average and zeta-potential after 100 days of storage were, respectively, 166.53 ± 4.71 nm and−18.37 ± 2.06 mV	β-carotene content recovery of 67.62%±7.77%	da Silva et al., 2016
Lutein nanocapsules	Interfacial deposition of preformed polymer	Size distribution with a polydispersity index of 0.11 ± 0.02, z-average of 191.9 ± 3.24 nm, zeta potential of −5.14 ± 2.22 mV	Solubilization of lutein in the aqueous medium and increased the stability of lutein in different temperature	Brum et al., 2017

efficiency in this regard has been attributed to their pH and concentration. Food-grade acids would normally dissociate weakly into ions in foods, and this effect would weaken their antimicrobial tendency because it is only at their undissociated forms that they can penetrate microbial organelles. Dissociated cations would get repelled at microbial cell membranes and would exert no antimicrobial effects. At conditions of low pH and higher concentrations, however, the population of undissociated acids is higher and effective (Brul et al., 2002). On the other hand, disruption of the phospholipid bilayer of membranes will make the membranes fail in their purpose of serving as a cellular barrier protector against external factors (Weiss et al., 2009; Table 14.2). Juneja, Dwivedi, and Yan (2012) reviewed the concept of food antimicrobial agents of novelty and reported that natural antimicrobials have been isolated from plant (citric acid, essential oils, thymol, eugenol, benzoic acid, vanillin, carvacrol, saponins, and a host of flavonoids), animal (pleurocidin, ovotransferrin, lactotransferrin, chitosan, lysozyme, defensins, and protamine), and microbes (bacteriophages, natamycin, reuterin, nisin, glucose oxidase, and cystibiotics) and have been reported to act with diverse mechanisms such as the following:

1. Inhibition of DNA synthesis through ribonucleotide reductase activity inhibition, leading to membrane hyperpermeability and an eventual cellular lysis
2. Distortion of the cytoplasmic membrane, drainage of vital intracellular constituents, ensuing in cell death
3. Degradation of essential structural cell wall constituents of the microorganisms
4. Initiation of death of vegetative bacterial cells through the sequential order of coherence, enclosure, agglomeration, and pore formation
5. Dispossessing the target microbe of multivalent metals of paramount importance to its metabolism

14.3.1.2 Advantages

Nanoencapsulation of natural antimicrobials avails the advantages of protection against undesirable reactions with food ingredients and unwanted chemical changes, enhanced solubility, stability conservation during long periods of storage, regulation of delivery, and intensification of bioavailability and absorption (McClements et al., 2009). Although microencapsulation can achieve these goals to some extent, nanoencapsulation of food antimicrobials, due to their high surface area–to–volume ratio, are superior in terms of increasing the localization of antimicrobials to food locations densely populated with microbes, and the permeation of cellular organelles of food microorganisms.

Blanco-Padilla et al., 2014 also delineated some advantages of food nano-encapsulated antimicrobials, such as the following:

1. Biocompatibility with the human system: From ingestion to delivery, nanocapsules can be designed to meet body suitability. They can be rapidly absorbed, directed at specific targets, and designed for prolonged or controlled release.

TABLE 14.2
Application of nanocapsules as antimicrobial agents in food materials

Food Material	Nanocapsule Formulation	Nanocapsule Preparation Technique	Target Microorganisms	Study Period for Controlled Release	Inhibitory Period and Results against Microbes	Reference
Tomato juice	Nisin-loaded chitosan/ carageenan	Ionic complexation method	*Micrococcus luteus* MTCC (Microbial Type Culture Collection) 1809, *Pseudomonas aeruginosa* MTCC 424, *Salmonella enterica* MTCC 1253, and *Enterobacter aerogenes* MTCC2823	6 months	20 days	Chopra et al., 2014
Lime essential oil	Chitosan nanoparticles and chitosan nanocapsules incorporated with lime essential oil	Nanoprecipitation and nano-encapsulation (constant agitation)	*Staphylococcus aureus*, *Listeria monocytogenes*, *Shigella dysenteriae*, and *Escherichia coli*		After 24 h, highest antibacterial activity was observed against *Shigella dysenteriae*, attaining an inhibition halo (IH) value of 3.5 cm for 40 mL of minimum inhibitory volume (MIV)	Sotelo-Boyás et al., 2017a

Fresh tuna fish	Liquid smoke (LS) nanocapsules from coconut shell using chitosan and maltodextrin	Polyelectrolyte complexation	*Bacillus subtilis* FNCC 0059, *E. coli* FNCC 0091, *P. fluoroscens* FNCC 0070, and *S. aureus* FNCC 0047	0–48 h	The larger inhibition zones were observed for the gram-negative bacteria (*P. fluorescens* and *E. coli*) while the lower for gram positive (*B. subtilis* and *S. aureus*) after 48 h	Saloko et al., 2014
Milk whey	Zein/casein nanoparticles encapsulating eugenol and thymo	Spray drying	*E. coli* O157:H7 and *L. monocytogenes* Scott A		Bactericidal and bacteriostatic effects observed	Chen, Zhang, and Zhong, 2015
Bread	*Origanum vulgare* and *Thymus vulgaris* encapsulated in zein nanocapsules	Nanoprecipitation method	*L. monocytogenes* ATCC 7644, *S. aureus* ATCC 2593, *E. coli* ATCC 25922, *S. entericasero var typhimurium* ATCC 14028	90 days	Higher antimicrobial activity against gram-positive bacteria than against gram-negative ones considering inhibition halo	da Rosa et al., 2020

(*Continued*)

TABLE 14.2 (Continued)
Application of nanocapsules as antimicrobial agents in food materials

Curcumin	Curcumin/polyrhodanine nanocapsule		Gram-positive (*S. aureus*) and gram-negative (*E. coli*)	6 weeks	Minimum inhibitory concentration (MIC) value of 512 μg mL^{-1} against gram-negative bacteria with higher results for *E. coli*	Kook et al., 2016
Milk	Nisin and garlic extract co-encapsulated phosphatidylcholine nanoliposomes	Thin-film hydration method	*L. monocytogenes*, *S. enteritidis*, *E. coli* and *S. aureus*	0–25 days	Decreasing the bacterial counts for up to 6 log CFU/mL for *L. monocytogenes*, 5 log CFU/mL for *S. aureus*, and 3–4 log CFU/mL for *S. enteritidis* and *E. coli* at 10 h incubation.	Pinilla and Brandelli (2016)
Water melon juice	*trans*-cinnamaldehyde and Tween-20 (1:3)	High-energy emulsification at a high speed of 10,000 rpm and 20,000 psi high pressure	*E. coli* O157:H7 933, *S. typhimurium* KCCM 11862, and *S. aureus* ATCC 12692	2–48 h	Even at low concentration, nanocapsule was effective to inhibit microorganism growth and with a release rate having significant relation with growth rate of microorganism.	Jo et al., 2015

Thyme essential oil	Thyme essential oil--loaded chitosan nanoparticles	Solvent displacement technique	*S. aureus*, *L. monocytogenes*, *B. cereus*, *Salmonella typhi*, *Shigella dysenteriae* and *E. coli*	630 min	Highest inhibitory activity was observed against *S. aureus* with inhibition halo of 4.3 cm	Sotelo-Boyás et al., 2017b
Peppermint oil	Tragacanth gum encapsulating peppermint oil	Ultrasonication with magnetic stirring	*E. coli*, *S. aureus* and *C. albicans*	1–12 h	100% microbial reduction after 12 h stirring	Ghayempour et al., 2015
Composite essential oil	Cinnamon-thyme-ginger composite essential oil nanocapsules	Ionic gelification reaction	*E. coli* (ATCC 25922), *B. subtilis* (ATCC 6633), and *S. aureus* (ATCC 25923)	1–18 days	Composite essential oil nanoparticles possessed the long-term antimicrobial effect even after 7 days.	Hu et al., 2018

2. Conveyance of cumbersome bioactive materials, especially those of extreme hydrophobicity and heat sensitivity
3. Suitability for usage in aqueous foods such as beverages
4. Food-grade nanocapsules as antimicrobial agents can be manufactured from natural materials with minimal toxicity.
5. They are shelf-stable even when incorporated into food products of considerable moisture content.
6. They can be fabricated with natural materials such as proteins and polysaccharides with the advantages of biodegradability, bioavailability, and consumer acceptance.
7. They can be formulated and developed on a continuous, large scale.
8. They are employable for usage as various stages of food processing and can be patterned to meet diverse physical and barrier requirements such as high porosity, large surface area, and high gas permeability.

14.3.1.3 Shortcomings

The importance of antimicrobial agents in food products has normally necessitated food processors to use synthetic materials for that target. However, the negative consumer perception and reported after-effects of these additives have created the need for the adoption of natural materials as antimicrobials. Nonetheless, natural materials as food antimicrobials have been found to have some shortcomings such as gradual loss of antimicrobial potentials resulting from their gradual degradation, negative interaction with food components, and concerns of cost arising from their extraction (Blanco-Padilla et al., 2014). Gyawali and Ibrahim (2014) enumerated others as follow:

1. The insolubility of antimicrobials: Most antimicrobial materials are lipophilic, have poor water solubility, and cannot form hydrogen bonds in aqueous environments, thus limiting their diffusion into cells of microorganisms.
2. pH and temperature relationship with antimicrobial activity: Some antimicrobials require optimum and specific conditions of pH and temperature to exhibit their antimicrobial effects. Extreme conditions of these parameters would induce certain metabolic changes in them and retard their antimicrobial functionalities.
3. Poor organoleptic and sensory effects of antimicrobials on the addition to foods: Undesirable changes in flavour, taste, colour, and/or texture initiated in a food product by the addition of an antimicrobial substance would depreciate its quality and consumer acceptability.
4. The complexity of food matrices to accommodate antimicrobial agents: the network and layers of micro-components of some food materials make it difficult for proper incorporation and homogenization of antimicrobial compounds.

14.3.2 Nanocapsules and Foodborne Microorganisms

Nanocapsules used in food processing exist in models such as nanoemulsions, nanosphere, nanofibers, nanoliposomes, and solid-lipid nanoparticles (SLNs), amongst several others. Several research studies have examined these nanomaterials with respect to their activities against foodborne pathogens and spoilage microorganisms. Scientific investigations tend to discern the differences in antimicrobial effects of the prepared nanomaterials against gram-positive and gram-negative bacteria. Ingredients used in such formulations are usually those adopted for the fabrication of packages or sometimes directly in the processing of the food products. Emulsions are simply dispersions of two immiscible phases (usually oil and water) with one phase (the dispersed phase) being dispersed in the continuous phase. Several food products such as yoghurt, milk, mayonnaise, cream, and chocolate, amongst several others, contain immiscible phases as ingredients and are examples of emulsions (Mwaurah et al., 2020). Nanoemulsions differ fundamentally from micro-emulsion based on droplet size with nano-size ranging from 1 to 100 nm and micro-size ranging from 0.1 to 100 μm; nonetheless, an important distinction lies in their appearance because unlike microemulsions, nanoemulsions do not scatter light and are transparent due to their smaller sizes (Panghal et al., 2019b). Also, regarding emulsion stability and functionality, nanoemulsions are superior. Creaming, sedimentation, Ostwald ripening, flocculation, and coalescence, which are the prominent stability concerns of emulsions, are less for nanoemulsions. Furthermore, their reduced sizes imply better surface areas for them to impart their functionalities where required.

Foodborne oils and lipids have been used in several studies for the formulation of nanoemulsions of probable antimicrobial potentials. Few examples include carvacrol, peanut oil, d-limonene, sunflower oil, stearic acid, peppermint oil, medium-chain triglycerides, tea tree oil, eugenol, canola oil, lemongrass oil, eucalyptus oil, sesame oil, lemongrass oil, thyme oil, clove bud oil, and orange oil (Donsì and Ferrari, 2016). Li et al. (2020) prepared thymol nanoemulsions, incorporated them into gelatin films, and examined their antimicrobial efficiency against gram-negative and gram-positive bacteria (*Escherichia coli* and *Bacillus subtilis*, respectively). The nanoemulsions were prepared with varying concentrations of gelatin and soy lecithin and were then tested against the spoilage and pathogenic microorganisms. Results obtained from the study showed that thymol nanoemulsions incorporated into the films exhibited the best inhibition potentials after 48 h compared to treatments where they were used in isolation. Zhang and Zhong (2020) encapsulated thymine oil in sodium caseinate for the preparation of nanoemulsions and examined their antimicrobial effects against *Staphylococcus aureus* ATCC 25923 and *E. coli* O157:H7 ATCC 43895. The results were patterned based on minimum bactericidal concentrations (MBCs) and minimum inhibition concentrations (MICs) in milk and tryptic soy broth and were reported to decrease for the nanocapsulated treatment relative to the control, implying enhanced antimicrobial activity due to the nanoemulsification processing which improved the dispersibility and solubility of the treatments. Furthermore, the nanocapsules were found to be more effective against *E. coli* O157:H7 (gram-negative) in comparison with *S. aureus*, an outcome attributed possibly to disparities in the membrane structures of the bacteria, offering different permeabilities to the antimicrobial emulsions. Huang et al. (2019) prepared a nanoemulsion with high-pressure homogenization processing using chitosan and gelatin, blended it into a mixture of ε-poly-L-lysine (ε-PL) and rosemary extract, and applied them as an edible coating on ready-to-eat carbonado chicken. Results obtained showed a decrease in the Thiobarbituric acid reactive substances value, pH changes, mold, and yeast counts, and total viable bacterial counts compared to the control (uncoated) samples after a 16-day refrigerated storage (4°C). These outputs were attributed to the nano-processing, which aided the transportation of the antimicrobial agents to the microbes; the slow and sustained release pattern of the antimicrobials; the nanodroplets having increased surface area with better penetration into the microbes; and the positive charge of the ε-poly-L-lysine peptide with a remarkable antibacterial potential through DNA damage and oxidative stress induced on the microbes.

14.3.3 Nanofibers

These are nano-sized, ultra-thin fibre structures with average diameters less than 100 nm, usually produced by electrospinning and useful in food, pharmaceutics, and cosmetic manufacturing, amongst several others. The electrospinning process involves the application of a high-voltage electrical field on a polymer (usually biological polymers such as carbohydrates and proteins) in a molten or solution form with a forceful ejection and subsequent evaporation and deposition in the solid phase and at the nanoscale on a specific target (Weiss et al., 2009). Through these procedures, nanofibres can be fabricated as encapsulants and carriers of antimicrobials for food usage (Wang et al., 2015). Some unique attributes that make nanofibers as choice materials for food processing include high porosity, good mechanical strength, and large surface area–mass ratio all stemming from their enhanced polymeric orientation.

Liu et al. (2018) used electrospinning to fabricate polylactic acid/tea polyphenol composite nanofibers and examined their antibacterial activities against *S. aureus* and *E. coli*. The lactic acid and polyphenol were mixed at different formulation ratios of 2:1, 3:1, 4:1, and 5:1, respectively, and it was reported that though the tea polyphenol addition decreased the mechanical properties of the nanofiber especially the elongation at break and tensile strength, functional properties of the nanofibers such as antioxidant and antimicrobial activities were remarkably improved by it. Results obtained from the study showed that in solitary, polylactic acid showed no appreciable inhibition against *S. aureus* and *E. coli* with inhibition values of 3.11 ± 1.21% and 5.23 ± 1.85%, respectively. However, the 3:1 combination, which proved superlative of all treatments, had inhibition values of 94.58 ± 6.53% and 92.26 ± 5.93%, respectively, for both bacterial microorganisms. The results were ascribed possibly to the substantial solubility of the tea polyphenol which made it solubilize through the outer membranes of the microbes and impart its antimicrobial effects. In this context, Wang et al.(2015), through electrospinning encapsulated pleurocidin, an antimicrobial peptide with polyvinyl alcohol (PVA), to fabricate nanofibres and examined their in vitro antimicrobial effect against *E. coli* O157:H7. Two nanofibers of PVA different concentrations (0.25% and 0.03%) were prepared and used against the pathogen. At a detection limitation of 10 CFU/mL, a rapid inhibition for the 0.25-PVA nanofibre preparation was observed, and within a 28-day incubation period, no bacterial cell recovery was reportedly observed. For the 0.03-PVA nanofibre, the pathogen was reported as undetectable at 24 h incubation but reached an 8 log10–CFU/mL mark at the end of the 28-day incubation period. Nonetheless, PVA in solitude was found to have no antipathogenic effect, but its encapsulation in a nanofibre form was attributed to the likelihood of the protection against the environment and the immobilization of the antimicrobial peptide by PVA.

14.3.4 Solid-Lipid Nanoparticles

SLNs are nanoparticulate emulsions comprising a matrix solid-lipid shell and having the lipid phase partially or completely solidified by a composite or a singular solid lipid (Cerqueira et al., 2017). They are prepared from manufacturing nanoemulsions from melted lipids vulnerable to crystallization with a subsequent cooling under regulated conditions (Weiss et al., 2009). They can also be prepared using techniques such as homogenization (hot or cold), ultrasonication, and emulsification evaporation (Jafari, Fathi, and Mandala, 2015). Compared to other nano-sized systems, SLNs have the advantages of exhibiting relatively enhanced encapsulation efficiencies, not requiring organic solvents for their manufacture, slower release rates, and proficient stability (chemical and physical) for the encapsulation of hydrophobic bioactive components. In food processing, antimicrobials that are prone to degradation by food constituents are simply encapsulated into SLNs for their protection, targeted delivery, and controlled release. Prombutara et al. (2012) loaded SLNs with nisin and examined their inhibition against *Lactobacillus plantarum* TISTR 850 and *Listeria monocytogenes* DMST 2871. Fat compounds, including Imwitor 900, Cetylpalmitate, Softisan 154, and Witepsol E85, were used as the fats, and high-pressure homogenization was employed for the preparation of the SLNs with varying nisin concentrations. Results showed that for all nisin-loaded SLN preparations (with nisin concentrations of 0.5%, 1%, 2%, and 3%), *L. plantarum* TISTR 850 was inhibited for up to 15 days while *L. monocytogenes* DMST 2871 was inhibited for 20 days. The results obtained were attributed to the slow release effects of the nanocapsules prepared.

Furthermore, Wang et al. (2012b) encapsulated tilmicosin with hydrogenated castor oil using a hot homogenization and ultrasonication to prepare SLNs and examined their antipathogenic potentials against *S. aureus* in vitro using the colony-counting technique. An in vivo study was also conducted by examining its effect against mice infected by the pathogen by teat canal infusion. In vitro experiments showed that the MIC and MBC of the SLN preparation was higher than that of tilmicosin in solitary, and the colony count tests showed that the SLN preparation showed slow and controlled release with higher longevity of antimicrobial potential relative to the native tilmicosin. The inflammatory levels of the pathogen on the mice were used to evaluate the therapeutic potential of the SLN preparations. The investigation revealed that SLN-treated mice showed a reduction in inflammation, and 5 days after treatment, there was no significant appearance of the infection on the mammary glands of the mice. The results were all associated with the slow and controlled release mechanism of the tilmicosin-loaded hydrogenated castor oil–SLN preparations.

Also, Kalhapure et al. (2014) used hot homogenization and ultrasonication to encapsulate triethylamine neutralized vancomycin (an antibiotic) with linoleic acid using compritol 888 ATO as the fat. The SLN preparation was examined in vitro against methicillin-resistant *S. aureus* (MRSA) and *S. aureus*, and after 72 h, the preparation was found to still display inhibitory action against both strains relative to other preparations. The antimicrobial activity of the SLN was attributed to the likelihood of controlled release, the action of the linoleic acid in inhibiting protein reductase of the bacteria, and/or possibly due to the action of vancomycin in inhibiting the biosynthesis of peptidoglycan of the bacteria. The nanoparticulation of the antimicrobials into an integral SLN were opined to have improved the lipophilicity of the vancomycin–linoleic acid conjugate, the sustained release of vancomycin from the SLN due to ionic pair bonding with the anionic linoleic acid, and the nano-delivery of the antimicrobials.

Xie et al.(2017) encapsulated enrofloxacin with docosanoic acid using hot homogenization and ultrasonication to prepare SLNs and measured their antimicrobial activities against *Salmonella* intracellularly using RAW 264.7 cells. Results obtained from the study showed that the SLN preparations at the same extracellular concentrations were 27.06 to 37.71 times more efficient relative to free drugs. The nanoencapsulation was discussed to have aided in a controlled but effective intracellular release of enrofloxacin to impart its antimicrobial effects.

14.3.5 Nanospheres

They can be described as nanosized solid cores enclosed by dense polymers, having their sizes ranging from 10 to 1000 nm, and with the potential of entrapping, dispersing, adsorbing, chemically bonding, or encapsulating bioactive materials (Rangan, Manjula, and Satyanarayana, 2016; Kondiah et al., 2018; Wattanasatcha et al., 2012). Nanospheres have been discussed as the simplest forms of nanoparticles, having their radius as the only alterable geometric parameter (Ahmadivand, Karabiyik, and Pala, 2016). They are commonly used in the formulation of pharmaceutical materials due to their ease of oral or intravenous administration, long duration of circulation in the plasma, biocompatibility, and biodegradability. Synthetic polymers commonly used for the fabrication of nanospheres include polycaprolactone, poly (lactic acid), and poly (lactic-*co*-glycolic acid) which is the most applied due to its approval by the FDA.

Gomes, Moreira, and Castell-Perez (2011) used poly (DL-lactide-*co*-glycolide; PLGA) nanomaterial with entrapped eugenol and *trans*-Cinnamaldehyde to prepare nanospheres and examined their antimicrobial efficiency against *Listeria* spp. (gram-positive bacterium) and *Salmonella* spp. (gram-negative bacterium). The nanospheres were prepared in the sequence of emulsion evaporation, ultrafiltration, and freeze-drying with PVA used as the surfactant. Results indicated remarkable inhibition of both pathogens with MICs ranging from 10 to 20 mg/mL. The antimicrobial agents were found to still be released after 72 h. In addition to the slow release, the minimal sizes of the nanospheres were explained to have probably aided access to the cytomembranes of the pathogens.

Khatoon et al. (2017) synthesized silver nanospheres by reduction of silver nitrate ($AgNO_3$) using trisodium citrate ($C_6H_5Na_3O_7$) and examined their antibacterial and antifungal tendencies. The zones of inhibition were used to check their effects against fungal strains: *Candida albicans* and *Saccharomyces cerevisiae* and bacterial strains: *E. coli* (gram-negative) and *Bacillus subtilis* (gram-positive). The results derived from the study showed the strongest inhibition against *E. coli*, the gram-negative bacteria, followed by *B. subtilis*, and then the fungal strains. The results were explained to have resulted from the controlled release of silver ions from the nanospheres which act as reservoirs for the ions and as antimicrobial agents. Yin et al. (2016) prepared chitosan-1-hydroxymethyl-5,5-dimethylhydantoinnanospheres by their ionic cross-linking through gelation. Through the solvent evaporation technique, PVA hybrid films were manufactured with the nanospheres, chlorinated, and their antimicrobial properties against *E. coli* O157:H7 and *S. aureus* were evaluated. The study outcome indicated a 6-log inactivation of both bacteria within 5 min.

14.4 CONCLUSION AND FUTURE PROSPECTS

Research to date has demonstrated the excellent potential for nanocapsules to serve as a model system for food safety, antimicrobial properties, and cell membranes and to deliver a wide variety of quality-enhancing compounds to biological systems. The use of nanocapsules in the pharmaceutical industry as an antimicrobial agent has led to the successful treatment of a wide variety of diseases. These properties are yet to be explored more in food-processing industries for functional foods. In the area of food science and technology, the use of nanocapsules is just starting to be explored. The next leading edge for nanocapsules in food processing is the entrapment research of different food components and their target delivery will focus on the delivery of bioactive compounds in biological systems. The FDA and other global regulatory agencies have described antimicrobial agents as chemical constituents whose presence or addition to foods, their packages, contact surfaces, or processing environments retard the growth of, or inactivate spoilage, of pathogenic microorganisms; however, these constituents in form of nanocapsules are yet to be described. Food industries have to build consumer confidence in the acceptance of nanofood ingredients such as antimicrobial nanocapsules. On the other side, regulatory agencies such as the FDA should ensure the safety of these antimicrobial nanocapsule systems. The employment of these polymeric nanocomposites in these regards has bolstered significant revolution in industries related to food and bio-processing. The development of natural antimicrobial nanocapsules from food-grade materials has the essential impacts in food processing. The adoption of nanocapsules for antimicrobial agents is a noteworthy impact in the bioprocessing specialization. Nonetheless, there is a crucial need for the integration of thorough scientific research and revolutionary industrialization for the benefits of nanocapsules in these industries to be fully exploited. In the food industry, it is also important to consider the stability of encapsulated different bioactive and essential compounds under food-processing and food-matrix conditions. Another area to explore is the utilization of the encapsulation of antimicrobial agents in ready-to-eat foods after processing and during storage.

REFERENCES

Ahmadivand, A., Karabiyik, M., and Pala, N. (2016). Plasmonic photodetectors. In: Nabet B. (Ed.), *Photodetectors: Materials, Devices and Applications*, 157–193. Woodhead Publishing, Elsevier. doi: 10.1016/b978-1-78242-445-1.00006-3.

Berekaa, M. M. (2015). Nanotechnology in food industry; advances in food processing, packaging and food safety. *International Journal of Current Microbiology and Applied Sciences*, *4*(5), 345–357.

Blanco-Padilla, A., Soto, K. M., Hernández Iturriaga, M., and Mendoza, S. (2014). Food antimicrobials nanocarriers. *The Scientific World Journal*, *2014*, 1–11. doi: 10.1155/2014/837215.

Brul, S., Coote, P., Oomes, S., Mensonides, F., Hellingwerf, K., and Klis, F. (2002). Physiological actions of preservative agents: Prospective of use of modern microbiological techniques in assessing microbial behaviour in food preservation. *International Journal of Food Microbiology*, *79*(1–2), 55–64.

Brum, A. A. S., dos Santos, P. P., da Silva, M. M., Paese, K., Guterres, S. S., Costa, T. M. H., Pohlmann, A. R., Jablonski, A., Flôres, S. H., and de Oliveira Rios, A. (2017). Lutein-loaded lipid-core nanocapsules: Physicochemical characterization and stability evaluation. *Colloids and Surfaces A: Physicochemical and Engineering Aspects*, *522*, 477–484. doi: https://doi.org/10.1016/j.colsurfa.2017.03.041.

Cerqueira, M. Â., Pinheiro, A. C., Ramos, O. L., Silva, H., Bourbon, A. I., and Vicente, A. A. (2017). Advances in food nanotechnology. In: Busquets, R. (Ed.), *Emerging Nanotechnologies in Food Science*, 11–38. Elsevier, USA. doi: 10.1016/b978-0-323-42980-1.00002-9.

Chen, H., Zhang, Y., and Zhong, Q. (2015). Physical and antimicrobial properties of spray-dried zein–casein nanocapsules with co-encapsulated eugenol and thymol. *Journal of Food Engineering*, *144*, 93–102. doi: https://doi.org/10.1016/j.jfoodeng.2014.07.021.

Chhikara, N., Jaglan, S., Sindhu, N., Anshid, V., Charan, M. V. S., and Panghal, A. (2018). Importance of traceability in food supply chain for brand protection and food safety systems implementation. *Annals of Biology*, *34*(2), 111-118.

Chopra, M., Kaur, P., Bernela, M., and Thakur, R. (2014). Surfactant assisted nisin loaded chitosan-carageenan nanocapsule synthesis for controlling food pathogens. *Food Control*, *37*, 158–164. doi: https://doi.org/10.1016/j.foodcont.2013.09.024.

da Rosa, C. G., de Melo, A. P. Z., Sganzerla, W. G., Machado, M. H., Nunes, M. R., de Oliveira Brisola Maciel, M. V., Cleber Bertoldi, F., and Manique Barreto, P. L. (2020). Application in situ of zeinnanocapsules loaded with *Origanumvulgare* Linneus and *Thymus vulgaris* as a preservative in bread. *Food Hydrocolloids*, *99*, 105339, https://doi.org/10.116/j.foodhyd.2019.105339.

da Silva, M. M., Nora, L., Cantillano, R. F. F., Paese, K., Guterres, S. S., Pohlmann, A. R., Costa, T. M. H., and de Oliveira Rios, A. (2016). The production, characterization, and the stability of carotenoids loaded in lipid-core nanocapsules. *Food Bioprocessing Technology*, *9*, 1148–1158. doi: https://doi.org/10.1007/s11947-016-1704-3.

Davidson, P. M., Sofos, J. N., and Branen, A. L. (2005). *Antimicrobials in Food*, CRC Press, Boca Raton, FL.

Donsì, F., and Ferrari, G. (2016). Essential oil nanoemulsions as antimicrobial agents in food. *Journal of Biotechnology*, *233*, 106–120. doi: 10.1016/j.jbiotec.2016.07.005.

dos Santos, P. P., Paese, K., Guterres, S. S. Pohlmann, A. R., Costa, T. H., Jablonski, A., Flôres, S. H., and de Oliveira Rios, A. (2015). Development of lycopene-loaded lipid-core nanocapsules: Physicochemical characterization and stability study. *Journal of Nanoparticle Research*, *17*, 107. doi: https://doi.org/10.1007/s11051-015-2917-5.

Ezhilarasi, P. N., Karthik, P., Chhanwal, N., and Anandharamakrishnan, C. (2012). Nanoencapsulation techniques for food bioactive components: A review. *Food and Bioprocess Technology*, *6*(3), 628–647. doi: 10.1007/s11947-012-0944-0.

Ghayempour, S., Montazer, M., and Rad, M. M. (2015). Tragacanth gum as a natural polymeric wall for producing antimicrobial nanocapsules loaded with plant extract. *International Journal of Biological Macromolecules*, *81*, 514–520. doi: https://doi.org/10.1016/j.ijbiomac.2015.08.041.

Gomes, C., Moreira, R. G., and Castell-Perez, E. (2011). Poly (DL-lactide-*co*-glycolide) (PLGA) nanoparticles with entrapped *trans*-cinnamaldehyde and eugenol for antimicrobial delivery applications. *Journal of Food Science*, *76*(2), N16–N24. doi: 10.1111/j.1750-3841.2010.01985.x.

Gyawali, R., and Ibrahim, S. A. (2014). Natural products as antimicrobial agents. *Food Control*, *46*, 412–429. doi: 10.1016/j.foodcont.2014.05.047.

Hu, J., Zhang, Y., Xiao, Z., and Wang, X. (2018). Preparation and properties of cinnamon-thyme-ginger composite essential oil nanocapsules. *Industrial Crops and Products*, *122*, 85–92. doi: https://doi.org/10.1016/j.indcrop.2018.05.058.

Huang, M., Wang, H., Xu, X., Lu, X., Song, X., and Zhou, G. (2019). Effects of nanoemulsion-based edible coatings with composite mixture of rosemary extract and ε-poly-L-lysine on the shelf life of ready-to-eat carbonado chicken. *Food Hydrocolloids*, *102*, 105576. doi: 10.1016/j.foodhyd.2019.105576.

Hwang, S. L., and Kim, J. C. (2008). In vivo hair growth promotion effects of cosmetic preparations containing hinokitiol-loaded poly (ε-caprolacton) nanocapsules. *Journal of Microencapsulation*, *25*(5), 351–356.

Jafari, S. M., Fathi, M., and Mandala, I. (2015). Emerging product formation. In: Galanakis, C.M. (Ed.), *Food Waste Recovery: Processing Technologies and Industrial Techniques*, 293–317. Academic Press, Elsevier USA. doi: 10.1016/b978-0-12-800351-0.00013-4.

Jo, Y.-J., Chun, J.-Y., Kwon, Y.-J., Min, S.-G., Hong, G.-P., and Choi, M.-J. (2015). Physical and antimicrobial properties of *trans*-cinnamaldehyde nanoemulsions in water melon juice. *LWT – Food Science and Technology*, *60*(1), 444–451. doi: https://doi.org/10.1016/j.lwt.2014.09.041.

Juneja, V. K., Dwivedi, H. P., and Yan, X. (2012). Novel natural food antimicrobials. *Annual Review of Food Science and Technology*, *3*(1), 381–403. doi: 10.1146/annurev-food-022811-101241.

Kalhapure, R. S., Mocktar, C., Sikwal, D. R., Sonawane, S. J., Kathiravan, M. K., Skelton, A., and Govender, T. (2014). Ion pairing with linoleic acid simultaneously enhances encapsulation efficiency and antibacterial activity of vancomycin in solid lipid nanoparticles. *Colloids and Surfaces B: Biointerfaces*, *117*, 303–311. doi: 10.1016/j.colsurfb.2014.02.045.

Kehinde, B. A., Chhikara, N., Sharma, P., Garg, M. K., and Panghal, A. (2020a, in press). Application of polymer nanocomposites in food and bioprocessing industries. In: Hussain, C.M. (Ed.), *Handbook of Polymer Nanocomposites for Industrial Applications*, Academic Press, USA.

Kehinde, B. A., Majid, I., Hussain, S., and Nanda, V. (2020b). Innovations and future trends in product development and packaging technologies. In B. Prakash (Ed.), *Functional and Preservative Properties of Phytochemicals*, (pp. 377–409). Academic Press, USA. https://doi.org/10.1016/B978-0-12-818593-3.00013-0.

Khatoon, U. T., Nageswara Rao, G. V. S., Mohan, K. M., Ramanaviciene, A., and Ramanavicius, A. (2017). Antibacterial and antifungal activity of silver nanospheres synthesized by tri-sodium citrate assisted chemical approach. *Vacuum*, *146*, 259–265. doi: 10.1016/j.vacuum.2017.10.003.

Kondiah, P. P. D., Choonara, Y. E., Kondiah, P. J., Marimuthu, T., Kumar, P., du Toit, L. C., Modi, G., and Pillay, V. (2018). Nanocomposites for therapeutic application in multiple sclerosis. In: Inamuddin, Asiri A. M., and Mohammad, A. (Ed.), *Applications of Nanocomposite Materials in Drug Delivery*, (pp. 391–408). Woodhead Publishing, Elsevier. doi: 10.1016/b978-0-12-813741-3.00017-0.

Kook, J. -W., Kim, S., Lee, J. -Y., and Kim, J. -H. (2016). Synthesis of curcumin/polyrhodanine nanocapsules with antimicrobial properties by oxidative polymerization using the Fenton reaction. *Reactive and Functional Polymers*, *109*, 125–130. doi: https://doi.org/10.1016/j.reactfunctpolym.2016.10.016.

Krol, S., Diaspro, A., Cavalleri, O., Cavanna, D., Ballario, P., Grimaldi, B., Filetici, P., Ornaghi, P., and Gliozzi, A. (2004). Nanocapsules—Anovel tool for medicine and science. In: Buzaneva, E, and Scharff, P. (Ed.), *Frontiers of Multifunctional Integrated Nanosystems*, (pp. 439–446). Springer, Dordrecht.

Letchford, K., and Burt, H. (2007). A review of the formation and classification of amphiphilic block copolymer nanoparticulate structures: Micelles, nanospheres, nanocapsules and polymersomes. *European Journal of Pharmaceutics and Biopharmaceutics*, *65*(3), 259–269. doi: 10.1016/j.ejpb.2006.11.009.

Li, X., Yang, X., Deng, H., Guo, Y., and Xue, J. (2020). Gelatin films incorporated with thymol nanoemulsions: Physical properties and antimicrobial activities. *International Journal of Biological Macromolecules*, *150*, 161–168. doi: 10.1016/j.ijbiomac.2020.02.066.

Liang, J., Yan, H., Zhang, J., Dai, W., Gao, X., Zhou, Y., Wan, X., and Puligundla, P. (2017). Preparation and characterization of antioxidant edible chitosan films incorporated with epigallocatechin gallatenanocapsules. *Carbohydrate Polymers*, *171*, 300–306. doi: https://doi.org/10.1016/j.carbpol.2017.04.081.

Liu, C., Rao, Z., Zhao, J., Huo, Y., and Li, Y. (2015). Review on nanoencapsulated phase change materials: Preparation, characterization and heat transfer enhancement. *Nano Energy*, *13*, 814–826. doi: 10.1016/j.nanoen.2015.02.016.

Liu, Y., Liang, X., Wang, S., Qin, W., and Zhang, Q. (2018). Electrospun antimicrobial polylactic acid/tea polyphenol nanofibers for food-packaging applications. *Polymers*, *10*, 561.

Lv, Y., Yang, F., Li, X., Zhang, X., and Abbas, S. (2014). Formation of heat-resistant nanocapsules of jasmine essential oil via gelatin/gum arabic based complex coacervation. *Food Hydrocolloids*, *35*, 305–314. doi: https://doi.org/10.1016/j.foodhyd.2013.06.003.

McClements, D. J., Decker, E. A., Park, Y., and Weiss, J. (2009). Structural design principles for delivery of bioactive components in nutraceuticals and functional foods. *Critical Reviews in Food Science and Nutrition*, *49*(6), 577–606.

Mwaurah, P. W., Kumar, S., Kumar, N., Attkan, A. K., Panghal, A., Singh, V. K., and Garg, M. K. (2020). Novel oil extraction technologies: Process conditions, quality parameters, and optimization. *Comprehensive Reviews in Food Science and Food Safety*, *19*(1), 3–20.

Nagavarma, B. V. N., Yadav, H. K. S., Ayaz, A., Vasudha, L. S., and Shivakumar, H. G. (2012). Different techniques for preparation of polymeric nanoparticles—A review. *Asian Journal of Pharmaceutical and Clinical Research*, *5*(Suppl. 3), 16–23.

Nakagawa, K., Surassmo, S., Min, S. -G., and Choi, M. -J. (2011). Dispersibility of freeze-dried poly(epsilon-caprolactone) nanocapsules stabilized by gelatin and the effect of freezing. *Journal of Food Engineering*, *102*(2), 177–188. doi: https://doi.org/10.1016/j.jfoodeng.2010.08.017.

Natrajan, D., Srinivasan, S., Sundar, K., and Ravindran, A. (2015). Formulation of essential oil-loaded chitosan–alginate nanocapsules. *Journal of Food and Drug Analysis*, *23*(3), 560–568. doi: https://doi.org/10.1016/j.jfda.2015.01.001.

Noronha, C. M., Granada, A. F., de Carvalho, S. M., Lino, R.C., Matheus Vinicius de O.B. Maciel, and Barreto, P. L. M. (2013). Optimization of α-tocopherol loaded nanocapsules by the nanoprecipitation method. *Industrial Crops and Products*, *50*, 896–903. doi: https://doi.org/10.1016/j.indcrop.2013.08.015.

Noronha, C. M., de Carvalho, S. M., Lino, R.C., and Barreto, P. L. M. (2014). Characterization of antioxidant methylcellulose film incorporated with α-tocopherol nanocapsules. *Food Chemistry*, *159*, 529–535. doi: https://doi.org/10.1016/j.foodchem.2014.02.159.

Pan, K., Zhong, Q., and Baek, S. J. (2013). Enhanced dispersibility and bioactivity of curcumin by encapsulation in casein nanocapsules. *Agriculture and Food Chemistry*, *61*(25), 6036–6043. doi: https://doi.org/10.1021/jf400752a..

Panghal, A., Chhikara, N., Anshid, V., Charan, M. V. S., Surendran, V., Malik, A., and Dhull, S. B. (2019b). Nanoemulsions: A promising tool for dairy sector. In: Prasad, R., Kumar, V., Kumar, M., and Chaudhary, D. (Ed.), *Nanobiotechnology in Bioformulations*, (pp. 99–117). Springer, Cham.

Panghal, A., Jaglan, S., Sindhu, N., Anshid, V., Charan, M. V. S., Surendran, V., and Chhikara, N. (2019a). Microencapsulation for delivery of probiotic bacteria. In: Prasad, R., Kumar, V., Kumar, M., and Chaudhary, D. (Ed.), *Nanobiotechnology in Bioformulations*, (pp. 135–160). Springer, Cham.

Panghal, A., Yadav, D. N., Khatkar, B. S., Sharma, H., Kumar, V., and Chhikara, N. (2018). Post-harvest malpractices in fresh fruits and vegetables: Food safety and health issues in India. *Nutrition & Food Science*, *48*(4), 561–578.

Pinheiro, A. C., Bourbon, A. I., Cerqueira, M. A., Maricato, É., Nunes, C., Coimbra, M. A., and Vicente, A. A. (2015). Chitosan/fucoidan multilayer nanocapsules as a vehicle for controlled release of bioactive compounds. *Carbohydrate Polymers*, *115*, 1–9. doi: https://doi.org/10.1016/j.carbpol.2014.07.016.

Pinilla, C. M. B., and Brandelli, A. (2016). Antimicrobial activity of nanoliposomes co-encapsulating nisin and garlic extract against Gram-positive and Gram-negative bacteria in milk. *Innovative Food Science & Emerging Technologies*, *36*, 287–293. doi: https://doi.org/10.1016/j.ifset.2016.07.017.

Prombutara, P., Kulwatthanasal, Y., Supaka, N., Sramala, I., and Chareonpornwattana, S. (2012). Production of nisin-loaded solid lipid nanoparticles for sustained antimicrobial activity. *Food Control*, *24*(1-2), 184–190. doi: 10.1016/j.foodcont.2011.09.025.

Rangan, A., Manjula, M. V., and Satyanarayana, K. G. (2016). Trends and methods for nano-based delivery for nutraceuticals. In: Grumezescu, A. M. (Ed.), *Emulsions: Nanotechnology in Agri-Food Industry*, 573–609. Academic Press, USA. doi: 10.1016/b978-0-12-804306-6.00017-9.

Reis, C. P., Neufeld, R. J., Ribeiro, A. J., and Veiga, F. (2006). Nanoencapsulation I. Methods for preparation of drug-loaded polymeric nanoparticles. *Nanomedicine: Nanotechnology, Biology and Medicine*, *2*(1), 8–21.

Saloko, S., Darmadji, P., Setiaji, B., and Pranoto, Y. (2014). Antioxidative and antimicrobial activities of liquid smoke nanocapsules using chitosan and maltodextrin and its application on tuna fish preservation. *Food Bioscience*, *7*, 71–79. doi: https://doi.org/10.1016/j.fbio.2014.05.008.

Sharma, P., Panghal, A., Gaikwad, V., Jadhav, S., Bagal, A., Jadhav, A., and Chhikara, N. (2019). Nanotechnology: A boon for food safety and food defense. In: Prasad, R., Kumar, V., Kumar, M., and Chaudhary, D. (Ed.), *Nanobiotechnology in Bioformulations*, (pp. 225–242). Springer, Cham.

Singh, V. K., Garg, M. K., Kalra, A., Bhardwaj, S., Kumar, R., Kumar, S., Kumar, N., Attkan, A. K., Panghal, A., and Kumar, D. (2020). Efficacy of microwave heating parameters on physical properties of extracted oil from turmeric (Curcuma longa L.). *Current Journal of Applied Science and Technology*, *39*(25), 126–136.

Sotelo-Boyás, M., Correa-Pacheco, Z., Bautista-Baños, S., and Gómezy Gómez, Y. (2017b). Release study and inhibitory activity of thyme essential oil-loaded chitosan nanoparticles and nanocapsules against foodborne bacteria. *International Journal of Biological Macromolecules*, *103*, 409–414. doi: https://doi.org/10.1016/j.ijbiomac.2017.05.063.

Sotelo-Boyás, M. E., Correa-Pacheco, Z. N., Bautista-Baños, S., and Corona-Rangel, M. L. (2017a). Physicochemical characterization of chitosan nanoparticles and nanocapsules incorporated with lime essential oil and their antibacterial activity against food-borne pathogens. *LWT*, 77, 15–20. doi: https://doi.org/10.1016/j.lwt.2016.11.022.

Tapia-Hernández, J. A., Rodríguez-Félix, F., and Katouzian, I. (2017). Nanocapsule formation by electrospraying. In: Jafari, S.M., (Ed.), *Nanoencapsulation Technologies for the Food and Nutraceutical Industries*, 320–345. Academic Press, USA. doi: 10.1016/b978-0-12-809436-5.00009-4.

Taylor, T. M., Weiss, J., Davidson, P. M., and Bruce, B. D. (2005). Liposomal nanocapsules in food science and agriculture. *Critical Reviews in Food Science and Nutrition*, *45*(7–8), 587–605. doi: 10.1080/10408390591001135.

Thies, C. (2012). Nanocapsules as delivery systems in the food, beverage and nutraceutical industries. In: Huang, Q. (Ed.), *Nanotechnology in the Food, Beverage and Nutraceutical Industries*, 208–256. Woodhead Publishing, USA. doi: 10.1533/9780857095657.2.208

Vert, M., Doi, Y., Hellwich, K. H., Hess, M., Hodge, P., Kubisa, P., Rinaudo, M., and Schué, F. (2012). Terminology for biorelated polymers and applications (IUPAC Recommendations 2012). *Pure and Applied Chemistry*, *84*(2), 377–410.

Wang, X., Chen, H., Chen, Y., Ma, M., Zhang, K., Li, F., Zheng, Y., Zeng, D., Wang, Q., and Shi, J. (2012a), Perfluorohexane-encapsulated mesoporous silica nanocapsules as enhancement agents for highly efficient High Intensity Focused Ultrasound (HIFU). *Advanced Materials*, *24*, 785–791. doi: 10.1002/adma.201104033.

Wang, X., Yue, T., and Lee, T. (2015). Development of Pleurocidin-poly(vinyl alcohol) electrospun antimicrobial nanofibers to retain antimicrobial activity in food system application. *Food Control*, *5*, 150–157. doi: 10.1016/j.foodcont.2015.02.001.

Wang, X. F., Zhang, S. L., Zhu, L. Y., Xie, S. Y., Dong, Z., Wang, Y., and Zhou, W. Z. (2012b). Enhancement of antibacterial activity of tilmicosin against *Staphylococcus aureus* by solid lipid nanoparticles in vitro and in vivo. *The Veterinary Journal*, *191*(1), 115–120. doi: 10.1016/j.tvjl.2010.11.019.

Wattanasatcha, A., Rengpipat, S., and Wanichwecharungruang, S. (2012). Thymolnanospheres as an effective anti-bacterial agent. *International Journal of Pharmaceutics*, *434*(1–2), 360–365. doi: 10.1016/j.ijpharm.2012.06.017.

Weiss, J., Gaysinsky, S., Davidson, M., and McClements, J. (2009). Nanostructured encapsulation systems: Food antimicrobials. In: Barbosa-Cánovas, G., Mortimer, A., Lineback, D., Spiess, W., Buckle, K., and Colonna, P. (Eds.), *Global Issues in Food Science and Technology*, 425–479. Academic Press, USA. doi: 10.1016/b978-0-12-374124-0.00024-7.

Xie, S., Yang, F., Tao, Y., Chen, D., Qu, W., Huang, L., Liu, Z., Pan, Y., and Yuan, Z. (2017). Enhanced intracellular delivery and antibacterial efficacy of enrofloxacin-loaded docosanoic acid solid lipid nanoparticles against intracellular *Salmonella*. *Scientific Reports*, *7*(1), 41104. doi: 10.1038/srep41104.

Yin, M., Chen, X., Li, R., Huang, D., Fan, X., Ren, X., and Huang, T. -S. (2016). Preparation and characterization of antimicrobial PVA hybrid films with N-halamine modified chitosan nanospheres. *Journal of Applied Polymer Science*, *133*(46), 44204. doi: 10.1002/app.44204.

Zhang, Y., and Zhong, Q. (2020). Physical and antimicrobial properties of neutral nanoemulsions self-assembled from alkaline thyme oil and sodium caseinate mixtures. *International Journal of Biological Macromolecules*, *148*, 1046–1052. doi: 10.1016/j.ijbiomac.2020.01.233.

Zuidam, N. J., and Shimoni, E. (2010). Overview of microencapsulation use in food products or processes and methods to make them. In N. J. Zuidam and V. A. Nedovic (Eds.), *Encapsulation Technique for Active Food Ingredients and Food Processing,* (pp. 3–29). New York: Springer.

15 Nano-Starch Films as Effective Antimicrobial Packaging Material

Ritu Sindhu
Chaudhary Charan Singh Haryana Agricultural University, India
Shobhit Ambawat
Guru Jambheshwar University of Science and Technology, India

CONTENTS

15.1 INTRODUCTION

15.1.1 Biopolymers as Packaging Material

Any packaging, to be an efficient food packaging, must have the ability to carry out the functions of containing, protecting, communicating, and convenience for maintenance of safety and quality of packed products from packaging to the consumption of the product by its ultimate consumers. It is also expected to enhance the shelf life of a product by averting undesirable changes due to moisture and gases in environment, microbial or chemical contamination, light, and outside forces. To achieve these purposes, along with the suitable fundamental properties such as optical, mechanical, and thermal strength, packaging should be able to avoid microbial contamination, the permeation of gases, water vapours, and other volatile compounds. The incorporation of additional compounds in packaging, such as flavours, pigments, and active agents, like antimicrobial compounds to enhance the nutritional, sensory attributes, and microbial quality of packed products is the requirement of packaging to fulfil the demand of the present-day's consumer. Various types of materials like glass, paper, metals, and polymers or a combination of these materials are used for food packaging. Plastic polymers are the most commonly used as food-packaging material due to their low cost, versatile nature, good mechanical strength, and barrier properties.

However, environmental pollution is the major issue associated with the disposal of these polymers as these are non-biodegradable. So, biodegradable materials are preferred to non-biodegradable polymers for food-packaging purposes, such as edible films and biopolymers to meet the growing demand for environmental safety (Punia et al., 2019a, b). Polymers which are derived or prepared from biomass or naturally occurring polymers are known as biopolymers (Dhull et al., 2020). Biopolymers being biodegradable, easily available, biocompatible, and renewable can be used for packaging as a substitute for petroleum-derived non-biodegradable plastics (Rhim et al., 2013; Punia and Dhull, 2019). Molecules formed by covalent bonds between monomeric units in biopolymers can be degraded by naturally occurring organisms without producing any toxic compounds in the environment. Commercially used biopolymers are categorized as natural biopolymers derived from various natural sources and synthetic biopolymers produced or synthesized from biomass or petroleum products (Figure 15.1). Based on the sources from which these are derived natural biopolymers are further divided into two classes-: protein-based and carbohydrate-based biopolymers. Also, synthetic biopolymers can be further separated into two classes: biopolymers produced by means of microorganisms and biopolymers conventionally or chemically synthesized from biomass or petroleum products. The synthetic biopolymers being more durable, flexible, clear, and mechanically strong have the potential to form a sustainable industry. Early breakdown due to slight changes in surrounding environmental conditions is the major shortcoming of biopolymers. Untimely breakdown causes a failure in actions of biopolymers as packaging material. Also, biopolymers are susceptible to degradation at high temperature and therefore not applicable for products, processing, or storage at high temperature.

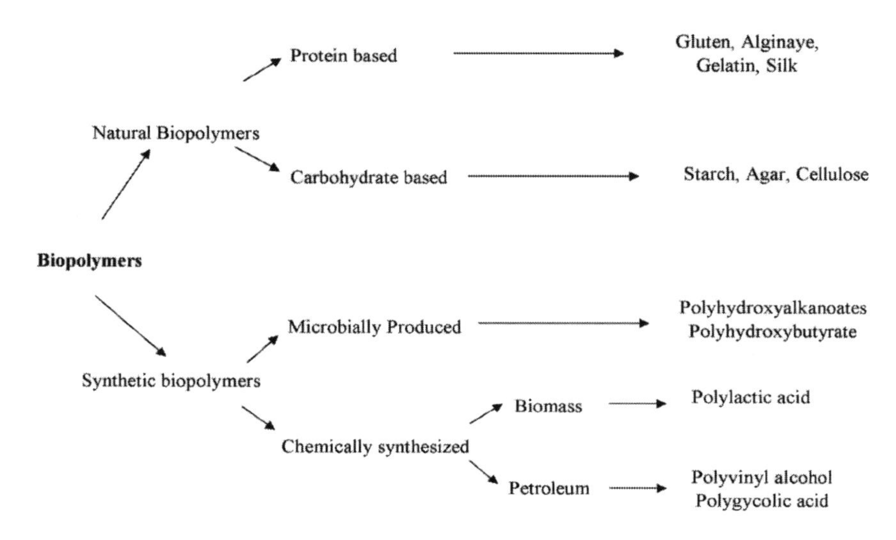

FIGURE 15.1 Classification of biopolymers.

15.1.2 Starch as Base for Antimicrobial Packaging

In recent years, growing wakefulness about environmental safety has impelled researchers to work on the development of eco-friendly packaging materials from biopolymers. Biopolymer-based biocompatible and biodegradable packaging materials are known as bioplastic or green packaging and have the potential to be substituted for synthetic polymers in the packaging industry. Among various natural biopolymers from different sources, starch is one which is widely used in different industries like pharmaceutical, cosmetics, and paper manufacturing. Starch is a naturally present polysaccharide and stands as a promising applicant for industrial use due to its low cost, availability, and thermoplastic characteristics. Sources of commercially used starch are cereals (maize, rice), tubers (potato), and roots (tapioca; Punia et al., 2019b, 2020a). Properties of starches and their products from different sources are distinctive depending on the size, shape, and arrangement of granules and ratio of amylose to amylopectin (Punia et al., 2020b). Starch and derivatives of starch are fit for human consumption which makes it safe packaging material for food products. Starch is not only completely degradable, but it also fuels the biodegradation of non-biodegradable materials when used in combination with them.

Fatefully, the utilization of starch for packaging purpose is restrained due to some limitations like weak thermal stability and low mechanical strength. Application of starch as biopolymer is significantly influenced by inherent hydrophilicity as well as non-resistance to microorganisms. The high hydrophilicity of starch is responsible for alteration in properties of starch-based packaging during processing or storage when it comes in contact with high moisture content. To conquer these limitations, a number of attempts have been made for improvement in the properties of starch biopolymers. Chemical, physical, and enzymatic modification of native starch to alter its properties; the incorporation of additives like plasticizers and antimicrobial agents; and the blending of other biopolymers with starch are some of the attempts taken in last few years for improvement of the starch matrix. The nano-size of starch particles has received much interest due to larger surface area and distinctive properties. The incorporation of nanoparticles (NPs) of starch alters the kinetics of crystallization, crystal shape, morphological features of crystalline, and crystallite dimension which ultimately results in enhanced mechanical strength and barrier properties of starch-based biopolymer. The integration of nanofillers having antimicrobial activity packaging films enhances food safety by preventing contamination and spoilage as well as and killing diseases causing microorganisms in the food. Usually, nanofillers having larger surface area allow for the attachment of a large number of microorganisms to nanofillers and hence boost the antimicrobial effectiveness of nanocomposite materials. Similarly, the larger surface area of NPs of starch as fillers provides more attachment sites for antimicrobial agents causing their uniformly dispersion in the nanocomposite matrix that results in improved antimicrobial effects. Therefore, starch NPs add to mechanical strength, barrier properties, and efficiency of antimicrobial efficiency in nanocomposite. The mechanism of action of antimicrobial compound integrated with NPs of starch in nanocomposite is depicted in Figure 15.2. Starch has been extensively studied and utilized as a raw material for film preparation due to rising cost and declining accessibility of conventional film-forming materials. Recently, investigations are going towards the development of starch NP (SNP)–filled packaging materials with antimicrobial properties suitable for food applications.

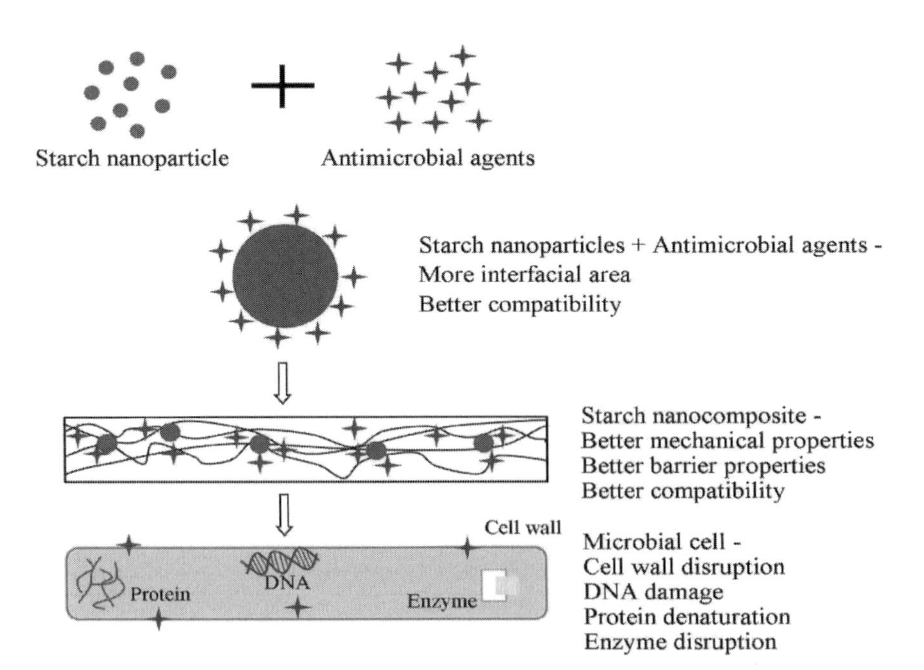

FIGURE 15.2 Mechanism of action of antimicrobial agents in starch nanocomposite.

15.2 CHARACTERIZATION OF STARCH NANOPARTICLES

NPs of starch formed using nanotechnology have achieved increasing interest in recent years due to abundantly presence of raw material in nature and non-toxicity to human health. SNPs with diverse properties from the native starch granules have been broadly utilized in various areas like in cosmetics, pharmaceuticals, medical and packaging industries. SNPs can be prepared by various methods including physical, chemical, and enzymatic or combinations of these. Chemical methods of SNP production include acid hydrolysis, nanoprecipitation, and emulsion cross-linking. High-pressure homogenization, ultrasonication, gamma irradiation, and reactive extrusion are some of the physical methods of SNP production. Enzymatic hydrolysis can be used alone or in combination with other techniques for the preparation of SNPs. Different processes of SNP production modify properties of starch granules in diverse ways which finally determine characteristics and applicability of NPs. Some of the important characteristics of SNPs produced by different methods are explained in the following.

15.2.1 Morphological Characteristics

These characteristics of SNPs depend mainly on the method of preparation and the botanical source. The botanical origin of starch influences the thickness of starch nanocrystals (SNC), crystalline organization, and, accordingly, the size of SNCs. LeCorre et al. (2012) reported that the morphology of SNPs relates to the crystalline structure of original starches. They reported that starches of A type (normal maize, waxy maize, wheat) and B type (high amylose maize, potato) produced square- and round-shaped NPs, respectively. The difference in shape is due to variation in the morphology of blocklets between starches of A and B types. Kim et al. (2012) reported that in spite of their starch origin, NPs had oval or round shapes. Moreover, NPs from starches of both B and C types (mungebean starch) were bigger in size than A-type starch.

Ultrasonication is one of the most utilized methods to produce SNPs with the size varying between 30 and 200 nm. Ahmad et al. (2020) revealed that such SNPs were found to be irregular in shape and size without an identical pattern, which may be attributed to the cavitation phenomenon giving high pressures and shear forces, leading to the production of small particles due to damage of the crystalline and amorphous layers on the exterior of starch granules. The high-pressure homogenization technique results in the formation of NPs 10 nm in size. In addition, these NPs have a high dispersibility, narrow size distribution, and spherical shape (Shi et al., 2011). Gamma radiation technique is used to develop SNPs with the capability to break large molecules into smaller ones (<100 nm diameter) as well as cleaving glycosidic linkages. The acid-modified starch and ultrasound-treated starch have a mean diameter of 21.8 and 454.3 nm with an average polydispersity of 0.202 and 0.380, respectively (Gonçalves et al., 2014). These polydispersity values showed a narrow size distribution as well as homogeneity of NPs. The higher size reduction under acid hydrolysis may be related to extended treatment time, as fragments of starch are sequentially released from the surface of the granule, resulting in small particle size.

15.2.2 Crystallinity

Crystallinity, an important characteristic of the starch nanostructure, generally depends on the preparation conditions of NPs. The relative crystallinity (RC) of NPs to the crystallinity of original starch granules is reported to be positively linked with the recovery yield of NPs. The crystallinity of NPs was higher in comparison to native starch due to the selective removal of amorphous regions by acid hydrolysis method (Kaur et al., 2018). Starch amylose content has been considered as the most important parameter governing the degree of crystallinity of SNPs. The RC of the SNPs' has been reported to increase with the reduction in amylose content. Treatment time and conditions also play an important role in changing the crystallinity of NPs of starch. In maize, the RC of the SNPs has been reported to increase substantially after 10 days of acid hydrolysis as compared to 4 to 6 days of hydrolysis. The increase in crystallinity index of SNCs produced by ultrasound-assisted acid hydrolysis was 21.6% in comparison to that synthesized by conventional acid hydrolysis (Hakke et al., 2020).

15.2.3 Rheological Properties

SNPs have different pasting behavior in comparison to native starch (Ma et al., 2008). The pasting viscosity of SNPs prepared by acid hydrolysis process is reported to be much lower as compared to that of native starch. On the other hand, with a rise in the SNPs' amount, the value of pasting viscosity also improved. Moreover, the viscosity of SNPs has been reported to increase with the increase in temperature up to 95°C. However, decreasing the temperature from 95°C to 50°C, pasting viscosity increased again. The suspensions containing vacuum-freeze-dried SNPs have higher shear thinning in comparison to spray-dried SNPs. Additionally, the suspension having vacuum-freeze-dried SNPs exhibited stronger elastic structure, while those containing spray-dried SNPs were found to be stiffer and showed a better tendency of recovery from deformation.

The SNPs produced by ultrasonication process have higher storage modulus ($G^{/}$) and loss modulus ($G^{//}$) as compared to native starch (Ahmad et al., 2020). It was reported that more elastic behavior of SNP suspensions could be due to higher variation between $G^{/}$ and $G^{//}$of SNPs as compared to native starch samples. The reduction of starch granule size to the nano level resulted in increased viscosity because rheological behaviour of starch is mainly dependent on the dimension, shape, and distribution of granules as well as on amylose content and granule- granule interaction.

15.2.4 Thermal Transition Properties

The melting temperatures, as well as enthalpy, depend on NP preparation parameters as well as the botanical source of starch. SNPs with different melting characteristics can be obtained by altering the conditions of preparation like temperature and time of hydrolysis (Kim et al., 2013). The SNPs' preparation by acid hydrolysis method exhibited endothermic melting transition peak in broadened ranges in comparison to native starch. The extensive melting range of SNPs was owing to their heterogeneity (Lamanna et al., 2013). Thermogravimetric analysis of the SNC formed by acid hydrolysis as well as ultrasonication revealed that SNC synthesized in the former method had a sharp drop owing to the existence of a higher amorphous fraction, while in a later one, the weight loss increases gradually around 300°C (Hakke et al., 2020). Water chestnut SNPs produced by ultrasonication method exhibited a sharp peak of heat flow at 135.8°C, which was equivalent to its melting point as well as crystal structure. Horse chestnut SNP curve exhibited extensive endotherm glass transition at 95°C, whereas, for lotus stem SNP, a sharp endothermic melting transition was observed at 160°C. Ultrasonication of starch granules disrupts the crystalline structure of clustered amylopectin and enhances the hydrocarbon chain which may be linked to increased van der Waal forces and hydrogen bonds having better stability, as well as high melting temperature, in fact resulting in NPs of low crystallinity or amorphous character (Ahmad et al., 2020).

15.3 PROPERTIES OF NANO-STARCH-BASED ANTIMICROBIAL BIOPOLYMERS

The fundamental functions for antimicrobial packaging should be the same as for the traditional food packaging, along with antimicrobial effect. The important parameters which determine the suitability of antimicrobial film to any particular application include mechanical properties (tensile strength and elongation at break), permeability (water vapour and gases permeability), and optical quality (colour and transparency). The quality of edible coating or film mainly depends on nature and concentration of raw material, ratio and type of additives, technique, and conditions used for film preparation (Hernandez et al., 2008). The addition of NPs of starch greatly affects the mechanical, barrier, appearance, and biodegradability of a starch-based nanocomposite. The nano-size of inserted starch particles increases the compactness in the matrix and decreases the rate of diffusion of vapours or gases, causing a reduction in the permeability of films. SNPs as fillers exhibit better compatibility in the starch matrix, due to similar inherent characters from starch granules, give a reinforcement effect in starch nanocomposite films. Production processes alter the crystallinity of SNPs and hence affect the light transmittance of films when incorporated in the film matrix. The incorporation of SNPs in films enhances the biodegradable due to the smaller size and more surface area of NPs than native granules of starch. The incorporation of antimicrobial substance in the biopolymer helps to prevent the growth of microorganisms while it may positively or negatively alter the mechanical strength, permeation capacity, and thermal stability of the ultimate packaging material. So these factors must be considered while making antimicrobial biopolymer-based packaging material. The data on mechanical and barrier properties of biodegradable antimicrobial films based on starch from different sources have been summarized in Table 15.1.

15.3.1 Mechanical Properties

The protection of food from physical harms like bruising, breaking, and denting is the primary function of packaging. The performance of this function of packaging film depends mainly on two mechanical parameters, namely tensile strength,

TABLE 15.1
Mechanical Properties and Water Vapour Permeability of Starch Based Antimicrobial Films

Base material	Antimicrobial Compounds	Mechanical Properties		Water Vapour Permeability ($\times10^{-10}$ g/m^2 s Pa)	References
		Tensile Strength (MPa)	Elongation at Break (%)		
Tapioca starch	Grape pomace extracts	2.58–3.55	22.5–45.8	13.42–20.22	Xu et al., 2018
Taro starch	saponin	24.10–28.18	29.77–92.74	–	Assefa and Admassu, 2013
Corn starch	Silver nanoparticles	–	–	1.30 ± 0.16	Abreu et al., 2015
Corn starch + Montmorillonite	Silver nanoparticles	–	–	1.01 ± 0.21	Abreu et al., 2015
High amylose starch film	Pomegranate peel	21.32–22.84	5.64–28.32	–	Ali et al., 2018
Starch	Potassium sorbate	20.92–36.25	38.74–43.67	–	Raigond et al., 2019

representing the highest strength that a film can bear up, and elongation at break, indicating the stretchability of film. The study of the mechanical characteristics of antimicrobial films gives the knowledge about the elasticity and the resisting capacity of film to the applied force that is helpful in predicting the efficacy of film throughout the handling and storage time. The smaller size and shape of incorporated NPs play an important role in the enhancement of the mechanical properties of starch films. Owing to their nanometric size, SNPs become able to occupy inter- and intramolecular sites, causing a densification of the matrix. Moreover, the higher surface area of NPs induces more interfacial interactions between additives or fillers and polymeric molecules, as well as within polymer complex, and results in improved mechanical strength of the matrix. Therefore, nanocomposite films filled with SNPs exhibit enhanced mechanical strength. The effects of adding SNPs to the mechanical properties of films are profoundly depend on the quantity or level of NPs filled in the matrix. SNCs incorporated at a high concentration in the matrix get aggregated and decrease the interactions between the molecules of filler and polymer, causing the weakening of the ultimate nanocomposite. Liu et al. (2016) reported improved tensile strength of corn starch films on the incorporation of SNPs 15% while the reverse effect on tensile strength was observed on the addition of a higher concentration (up to 25%). The incorporation of SNCs at the level of 5% improved tensile strength of a pea starch-based nanocomposite while the decrement in the tensile strength and increment in elastic modulus were observed on increasing the concentration of starch nanofillers (Li et al., 2015). Generally, the addition of NPs to films reduces the elongation at break value of nanocomposite films. Studies showed that the reinforcement of a rice starch-based film with the addition of SNCs up to 20% improved tensile strength and elongation at break (Piyada et al., 2013).

The mechanical properties of starch-based edible films may or may not be improved by incorporation of antimicrobial compounds, although the concentration of antimicrobial compounds strongly influences these properties. Poor elongation or low tensile strength of the film may cause the untimely breakdown or cracking of the film during preparation, application, or storage. Therefore, for production of antimicrobial films with sufficient mechanical strength, the quantity of antimicrobial agent must be taken into consideration prior to its preparation. The addition of nano-silicon (SiO_2) in starch-based nanocomposite films increased the mechanical strength and transparency while decreasing the water uptake capacity of biopolymers (Xiong et al., 2008). The reason behind it could be the decrement in intermolecular hydrogen bonding in the starch matrix.

15.3.2 Barrier Properties

The barrier properties of packaging films are important for the enhancement of storage shelf life of food products by prevention of deteriorative changes due to microorganisms, moisture, and oxidation. These properties of edible films or coatings indicate the extent of resistance for permeation of gases or water vapours through it. The most extensively studied barrier property of starch-based biodegradable antimicrobial films is water vapour permeability as moisture level is an important parameter for keeping the physiochemical quality of the product as well as retarding microbial growth on the surface. Generally, the low water vapour permeability of packaging film is required to avoid the transfer of moisture between the product and its neighbouring environment for increasing storage life and maintaining product quality. However, the specific requirement for the level of water vapour permeation of the film is based on various factors like the properties of food, expected storage life, and conditions of the surrounding environment.

The integration of NPs of starch gives a positive impact on barrier properties of starch films. Liu et al. (2016) reported the decrement in water vapour permeability of starch films on the incorporation of SNPs up to 15%. Cassava starch films reinforced with SNCs (2.5%) showed a 40% decrement in permeability for water vapours as compared with controls samples (Garcia et al., 2009). The improvement in barrier capacity nanocomposite for water vapours and oxygen by adding SNCs was reported by Angellier et al. (2005). Added nanocrystals in a film matrix cause a curved and complicated passageway for the vapours to diffuse through the films resulting in decreased penetration and permeability of water molecules through nanocomposite films. However, the concentration of nanofillers is important for the alteration of barrier properties of starch nanocomposite films. A higher concentration of SNCs in the film can cause the aggregation of starch particles and decrease the barrier capacity for water vapours as compared to films unfilled or filled with lower level of starch nanofillers. Pea starch nanocomposite films showed a rise in water vapour permeability on the addition of SNCs at a level exceeding 5% (Li et al., 2015).

Alteration of the barrier capacity of starch films on inclusion of antimicrobial compound depends on the starch's properties, type and the concentration of compound, and its interaction with other ingredients of the film matrix. The added antimicrobial agents may shrink the volume of free space in polymer, weaken the water binding capacity, or cause the blockage of diffusive pathway throughout the film, which ultimately results in decreased water vapour permeability value of the film (Rawdkuen et al., 2012; Hashemi et al., 2017; Kaewprachu et al., 2017; Xu et al., 2018). The enhancement in water vapour permeation of films occurred when added antimicrobial compounds act as a plasticizing agent, lead to increased intermolecular interactions, and raise the mobility of macromolecules in the film (Xu et al., 2018). Besides, the barrier properties of films also depend on the structural and chemical properties of macromolecules, other added compounds, and the hydrophilicity of the films. Mohan et al. (2017) prepared biofilm based on corn starch filled with nanoclay and compared its properties with unfilled cornstarch biopolymer. Results showed decreased water absorption capacity, lower moisture uptake, and less swelling while increasing the oxygen barrier for the biofilm containing a nanoclay filler as compared to the biofilm without nanofiller (Sadegh-Hassani and Nafchi et al., 2014). Increased mechanical strength and reduced water vapour permeability were noticed on the addition of nanoclay up to 5% in cassava starch-based organically modified nanocomposite (Muller et al., 2012). Slavutsky et al. (2012) reported that the incorporation of NPs of clay below the level of 10% improved the water barrier capacity of a corn starch-clay nanocomposite and found it suitable for the production of biopolymer films (Park et al., 2002). The incorporation of nanoclay and silver NPs in corn starch-based nanocomposites improved barrier capacity for water vapours and oxygen gas (Abreu et al., 2015).

15.3.3 Optical Properties

Optical properties of food packaging are very vital properties as these directly influence the appearance of products and also consumer's acceptance. The colour traits are generally expressed in terms of L* value (lightness to darkness), a* value (red to green), and b* value (yellow to blue). Starch films are appreciably transparent and types and concentrations of fillers or additives significantly affect its transparency. SNCs are composed of amorphous and crystalline zones. The crystalline regions decrease transmittance of light by causing reflection and refraction phenomenon. Therefore, the incorporation of SNPs changes the appearance of nanocomposite films by increasing opacity. The integration of SNPs up to the level of 25% reduced light transmittance by about 14% in the case of corn starch nanocomposite films (Liu et al., 2016).

The addition of antimicrobial compounds influences the appearance of packaging material by altering the transparency and colour values. The type and source of antimicrobial agent added into the biopolymer matrix and pigments inherent in it are mainly responsible for changes in the colour attributes of the final packaging material. Generally, the degree of change in appearance is directly related to the concentration of the antimicrobial agent added. The biopolymer-based films with less transparency are unsuitable as packaging material when transparent packaging and clear shine on product's surface are required. Although the colour attributes of starch-like biopolymer-based antimicrobial film may influence the consumers' opinion, there are also several advantages, such as protecting the packaged product to some extent from ultraviolet and visible light, causing discolouration, loss of nutrients, and having antimicrobial properties which are not present in the conventional packaging.

15.3.4 Antimicrobial Properties

The antimicrobial activity of starch films may be an inherent character of its base material or due to the addition of some antimicrobial compounds during preparation. The antimicrobial compound immobilized into the film performs its function either by contacting the food surface or by slowly releasing itself into the food matrix. Commonly, the agar disc diffusion method is followed to evaluate the antimicrobial effect of films. In this method, a portion of the film is positioned on the surface of agar already inoculated by the target microbes, subjected to the optimum environment for required periods, and then developed in an inhibition zone around the film's portion is observed. The activity of the antimicrobial biopolymer-based films observed against a broad range of microbes showed that the characteristics of the target microorganism, nature, and concentration of incorporated antimicrobial agents, as well as the chief active compounds present, are the major factors influencing it. Generally, efficacy is directly related to the concentration of active compound used. Different microorganisms show unique responses to any particular antimicrobial compound; for example, gram-positive bacteria are more sensitive to a majority of antimicrobial agents as compared to gram-negative bacteria. The reason behind it is the unique features of different microorganisms, like the strong protective covering of cell wall, allow the limited diffusion of an antimicrobial agent in case of gram-negative bacteria. In addition to the nature of the target microorganisms, there are some other factors such as film preparation method, storage conditions, composition, and the hydrophilic- hydrophobic balance of the antimicrobial agent, that influence the release or diffusion of antimicrobial material through the film. The phenomena of action of added active compound also depend on its interaction with the biopolymer molecules and food components. All these factors play important role and are able to be modified for production of an efficient antimicrobial packaging material. Various studies on starch- and clay-based nanocomposites incorporated with NPs of silver and copper reported significant antimicrobial activity, particularly antibacterial effects, of these biopolymers (Shankar et al., 2015, 2017; Carbone et al., 2016). However, copper is generally not added in polymers used for food packaging as it may cause the deterioration of food products by accelerating the reaction of chemical oxidation.

Abreu et al. (2015) prepared a nanostructured antimicrobial starch film using corn starch incorporated with NPs of montmorillonite and silver NPs. The film showed an antimicrobial effect for *Staphylococcus aureus*, *Escherichia coli*, and *Candida Albicans* with no major differences due to variation in concentration of silver NPs in the starch matrix. Negligible migration of components indicated that nanostructured material produced by incorporation of nano-silver and nanoclay particles in starch is suitable for packaging applications. Amjad et al. (2018) utilized high amylose starch (80% amylose) incorporated with pomegranate peel for film preparation and evaluated it for antibacterial effects. Results showed that the film inhibited the growth of *S. aureus* and *Salmonella*, indicating its potential to be used as a biodegradable antimicrobial biopolymer for packaging.

15.4 ANTIMICROBIAL AGENTS USED IN STARCH-BASED PACKAGING MATERIAL

Antimicrobial agents are the additives used for controlling the biological deterioration as well as to restrain the microbial growth, including pathogens. Many types of antimicrobial compounds, including natural extracts, chemical agents, enzymes, and probiotics, are potentially integrated into packaging materials. Majority of the antimicrobial compounds have been categorized as generally recognized as safe. Of late, the demand of consumers for the natural substances and chemical-free preservative food products is constantly increasing. The antimicrobial substances in large number may be integrated into the packaging materials to increase antimicrobial property to control the rate of growth of particular or a group of microbes in the headspace of a package, to ensure the safety of the food product, and to extend its shelf life. In culture media, strong antimicrobial activity against target microbes has been demonstrated by numerous antimicrobial agents. However, several antimicrobial agents used as food additives in packaging matrix have restricted effect on the microorganisms in foodstuff. The selection criteria of antimicrobial agents for food-packaging materials mainly depend on the characters of active agent and its mechanism of inhibition, organoleptic properties of active compounds, physico-chemical properties of food products, manufacturing process of packaging and its influence on the effectiveness of active agents, storage conditions, toxicity and regulatory issues, micro-flora of foods, physiology of target microbes, and release mechanism of the active substances into the foods (Coma, 2012). Thereafter, the classification of antimicrobial compounds can be done on the basis of their nature. Antimicrobial effects of various antimicrobial agents incorporated in starch-based biopolymers are presented in Table 15.2.

15.4.1 Chemical Agents

Generally, organic acids are the major sources of chemical antimicrobial compounds used in food products, considering their effectiveness and low price. Their production is done through the process of chemical synthesis or alteration of natural acids (Han, 2005). These are potentially integrated into the packaging matrix because of their antimicrobial effect for a wide range of microbes. The growth of microbes is inhibited on the addition of organic acids through a decrease in pH, acidification of cytoplasm, affecting proton gradient cross the membrane, and a hindrance to the transport of chemicals through cell membrane (Naidu, 2000). Studies have shown the development and use of organic acid- contained biopolymer films having antimicrobial activity.

Salleh and Muhamad (2010) added lauric acid in wheat starch film and observed antimicrobial activity in the liquid culture and the solid media. Results indicated that film had more antimicrobial efficacy for *Bacillus subtilis* as compared to *E. coli.* Shen et al. (2010) reported that the incorporation of potassium sorbate at the level of 15% (w/w) in sweet potato-starch film considerably reduced the count of *E. coli* on solid media as compared to control film.

15.4.2 Essential Oils

The aromatic and volatile oily extracts obtained from various plant parts are known as essential oils (Burt, 2004). However, due to a strong flavour, their direct inclusion into foods as preservatives is constantly limited. So, these oils may well be added to edible films or packaging films to evade the previously mentioned problem. In biobased films, normally used essential oils are cinnamon, clove, ginger, lemongrass, marjoram, thyme, oregano, sage, *Ziziphora clinopodioides*, and *Eucalyptus globules*. These proved quite effective against different microbes (Ahmad et al., 2012; Lee et al., 2015; Martucci et al., 2015; Hafsa et al., 2016; Ejaz et al., 2018). The antimicrobial activity of these essential oils may possibly be ascribed to their major phenolic or terpenic compounds, available in large quantity (up to 85%; Burt, 2004). Essential oils from different sources comprising different key compounds or a ratio of these compounds have diverse capacity to unite with membrane proteins of microbial cells as well as to alter membrane properties (Acevedo et al., 2015; Punia et al., 2019c). Pelissari et al. (2009) followed the agar disc diffusion method to explore the antimicrobial effect of oregano essential oil added into starch film and found significant inhibition in the growth of *E. coli*, *B. cereus*, and *S. enteritidis*. Thyme essential oil incorporated in biodegradable nanocomposite film formed of sweet potato starch and montmorillonite nanoclay was reported to be effective against *E. coli* and *S. typhi* (Issa et al., 2017).

TABLE 15.2
Antimicrobial Activity of Starch Based Antimicrobial Biopolymers

Coating Material	Antimicrobial Compounds	Microorganisms Used for Testing	Outcomes	References
Tapioca starch	Grape pomace extracts	*Staphylococcus aureus* and *Listeria monocytogenes*	Stronger inhibitory effect on *S. aureus* compared to *L. monocytogenes*	Xu et al., 2018
Corn starch-beeswax	Lauric arginate + natamycin	*Rhizopus stolonifer*, *Colletrotrichum gloeosporioides*, *Botrytis cinerea*, and *Salmonella Saintpau*	Completely inhibited all the tested microorganisms	Ochoa et al., 2017
Starch film	Chitosan	*S. enteritidis*	Inhibited tested microorganism	Durango et al., 2006
Starch film	Lauric acid	*Bacillus subtilis* and *E. coli*	Inhibited both the tested microorganisms	Salleh et al., 2007
Starch film	Lysozyme	*B. thermosphacta* B2	Inhibited tested microorganism	Nam et al., 2007
Starch film	Potassium sorbate	*E. coli* and *S. aureus*	Inhibited *E. coli* but not *S. aureus*	Shen et al., 2010
Starch film	Potassium sorbate	*S. typhimurium* and *E. coli*	Inhibited *S. typhimurium* and *E. coli* O157:H7 by 4 and 2 logs, respectively	Baron and Sumner, 1993
Starch-alginate	Lemongrass oil	*E. coli* O157:H7	Inhibited tested microorganism	Maizura et al., 2008
Starch-chitosan	Oregano Essential oils	*E. coli* O157:H7, *S. aureus*, *S. enteriditis*, and *B. cereus*	Inhibited all the tested microorganisms	Pelissari et al., 2009
Starch	Grape seed extract	*L. monocytogenes*, *E. coli*, *E. faecalis*, *E. faecium*, *S. typhimurium*, and *B. thermosphacta* B2	Reduced 1.3 log CFU mL^{-1} of *B. thermosphacta* B2 on pork loin; inhibited gram-positive bacteria on solid media but not gram-negative bacteria	Corrales et al., 2009
Starch	Pomegranate peel powder	*S. aureus* and Salmonella	Reduced the count of bacteria tested	Amjad et al., 2018

15.4.3 Plant and Spice Extracts

In recent times, an increasing interest has been shown in extracts of plant parts and/or spices due to their higher level of bioactive ingredients. The majority of these extracts have an appreciable antimicrobial effect against a broad range of microbes. The different parts of plants (roots, bark, leaves, buds, flowers, and seeds) and spices can be used to obtain extracts. The antimicrobial efficiency is normally ascribed to the kind and quantity of phenolic compound available in a particular extract. Commonly found phenolic compounds in various plant or spice extracts include tannin, catechin, gallic acid, ferulic acid, caffeic acid, and carvacrol. In addition to antimicrobial efficiency, additional benefits offered by extracts are their antioxidant capability and use as substitute medicines. Most of the plant and spice extracts are more efficient against gram-positive bacteria as compared to gram-negative bacteria owing to its more complex cell structure.

Allyl isothiocyanate (AIT), having strong inhibitory activity for several microbes, is the key flavouring constituent present in the various parts (roots, stem, leaves, and seeds) of plants included in the cruciferry family (mustard, wasabi and horseradish). AIT, being volatile in nature, is released steadily in surrounding space from the packaging materials, and thus inhibits the growth of detrimental microbes on the surface of foods. Grape fruit seed extract (GFSE), a naturally derived antimicrobial agent, contains plentiful of phenolic compounds like gallic acid, catechins, epicatechin, and procyanidins. GFSE is known to exhibit an inhibitory effect on the growth of gram-positive and gram-negative bacteria (Heggers et al., 2002; Cvetnic et al., 2004). GFSE was integrated with poly-ε-caprolactone, chitosan, and polyethylene for producing potential antibacterial packaging material suitable for food applications (Tan et al., 2015; Tong et al., 2018). GFSE could be well integrated into chitosan at various concentrations (0.5%, 1.0%, and 1.5% v/v) with no disturbance to film transparency. The incorporation of GFSE into a synthetic plastic film by the extrusion or solution-coating method, and its application to ground beef at refrigeration temperature showed an improved antimicrobial activity against numerous microbes.

15.4.4 Enzyme

Lysozyme, the commonly used antimicrobial enzyme, is a nutraceutical generally produced from egg white, milk, and blood. It was found more effective for gram-positive bacteria as compared to gram-negative bacteria by unraveling the bonds between N-acetylmuramic acid and N-acetylglucosamine of peptidoglycan in bacterial cell wall. Studies have shown that lysozyme restrained the growth of *Listeria innocua* and *Saccharomyces cerevisiae*; however, such antimicrobial effect on *E. coli* and *S. aureus* was not observed (Rawdkuen et al., 2012). It was further observed that the combination of lysozyme with other substances initiated the disruption of membrane. The combination of lysozyme with catechin in the ratio of 1:1 showed significant inhibition in the growth of a wider range of microbes as compared to lysozyme alone (Rawdkuen et al., 2012). The activity of lysozyme against gram-negative bacteria can be enhanced by use of a small quantity of ethylenediaminetetraacetic acid (EDTA; Branen and Davidson, 2004). As the substrates and the environmental parameters like temperature and pH highly influence the antimicrobial efficiency, therefore, these must be maintained and controlled very carefully while using an antimicrobial enzyme.

15.4.5 Bacteriocins

Bacteriocins are microbially produced peptides and have an inhibitory effect on growth of bacteria, particularly gram-positive bacteria, responsible for food spoilage. Naturally present bacteriocins, including nisin, lacticin, and pediocin, have the potential to be integrated in packaging materials to retard the growth of disease-causing and food spoilage-causing microbes. Nisin is well-known to show a stronger antimicrobial effect for gram-positive bacteria whereas it exhibited limited activity against gram-negative bacteria due to its inability to penetrate the cell (Naidu et al., 2000). Numerous studies showed that the antimicrobial activity of nisin against gram-negative bacteria improved when it was used in combination with chelating agent (EDTA) as it effectively destabilized the exterior membrane, disturbing the lipopolysaccharide layer which supports the admittance of nisin to the cytoplasmic membrane (Sivarooban et al., 2008; Prudêncio et a., 2016; Morsy et al., 2018). The antimicrobial effectiveness of bacteriocins is affected by their concentrations, number and types of microbes, inactivation or interaction by food components, and pH and temperature of the product (Naidu et al., 2000).

15.4.6 Probiotics

Probiotics are living microbes known for favorable health effects on consumption of adequate quantity (Dehghani et al., 2018). The majority of commonly consumed probiotic bacteria relate to the genera *Lactobacillus* and *Bifidobacterium*. The probiotics could effectively manage the competitive detrimental microbes and have ability to retard growth of other bacterial strains (Han et al., 2005). It has been observed that hydroxypropyl methylcellulose films containing microencapsulation of *Lactococcus lactis* subsp. Lactis was found effective in reducing the growth of *L. monocytogenes* by five-log cycle after 12 days of storage, as compared with control film (Bekhit et al., 2018). Moreover, probiotics, as a result of their safety and effectiveness, can be applied in edible biopolymers for food-packaging applications.

15.4.7 Ethylenediaminetetraacetic Acid

EDTA, is generally used as a chelator in the food and pharmaceutical industries. It also works as antimicrobial agent due to its limited cation availability and its capability to weaken cell membranes through the complexion of divalent cations that act as salt bridges linking the membrane and macro-molecules (Boziaris et al., 1999; Economou et al., 2009). EDTA enhances the antibacterial capacity of weak acids and nisin when used in combination. EDTA also helps in improving the sensitivity of gram-positive and gram-negative bacteria to lysozyme (Cha et al., 2004; Ntzimani et al., 2010). Studies revealed that the excellent antimicrobial effect was achieved with the use of combination of lysozyme and EDTA, which inhibited the growth of most of the tested spoilage causing bacteria in chilled buffalo meat (Cannarsi et al., 2008). Bhatia et al.

(2015) studied the antibacterial effects of starch films containing a combination of antimicrobial compounds, including nisin, EDTA, and lysozyme. They reported that nisin and EDTA collectively gave a synergistic effect on their final antimicrobial activity.

15.5 APPLICATIONS OF NANO-STARCH-BASED ANTIMICROBIAL PACKAGING MATERIAL

15.5.1 Food Industry

Nano-starch-based antimicrobial packaging material is mainly utilized in the form of packaging like nanocomposite and edible films in food industries. The packaging of food must serve various purposes such as safety and maintenance of freshness and sensory characteristics in addition to the containment and protection of foods. Ecological pollution caused due to disposal is the major disadvantage associated with traditional packaging materials, such as plastic polymers (Ozdemir and Floros, 2004). To take the advantages of conventional packaging materials, as well as overcome the problem of environmental pollution associated with it, nano-starch-based biodegradable packaging material can be a potential substitute for food packaging. Edible films based on starch are biodegradable and free from taste, colour, odour, and toxicity. They exhibit as good barriers against the transmission of oxygen when used under low relative humidity. Nanocomposites based on starch and clay are applicable as food packaging with enhanced barrier and mechanical properties along with a reduced migration of additives or polymer itself (Avella et al., 2005). Foamed material developed by a combination of starch and steam can be used to replace synthetic polymer packaging materials like polystyrene foam. This biodegradable starch foam can be moulded into trays or dishes and be easily disposed of in the environment without leaving any toxicity (Siracusa et al., 2008). On the other hand, directly adding antimicrobial agents to food can affect the activity or efficacy of the antimicrobial agents due to the interference by food components. Sometimes, the sensory characteristics of food products are altered by the addition of antimicrobial agents having strong colour or odour. Therefore, the starch films or composites containing NPs of starch as a base or fillers, along with added antimicrobial agents, seems to be a possible approach for supporting food safety by inhibiting microbial growth on food surface during extended storage as well as biodegradability of the biopolymer after use. Starch films containing potassium sorbate as antimicrobial compound were found to be effective in extension of shelf life of frozen cheese by inhibiting the growth of *S. aureus*, *Penicillium* spp., *Candida* spp., and *Salmonella*. However, the type of starch and different pH level showed an effect on kinetic release (Lopez et al., 2013). Applications of polysaccharide-based biopolymers containing antimicrobial agents on various food products are presented in Table 15.3.

15.5.2 Fruits and Vegetables

Fresh fruits and vegetables are very prone to biochemical, structural and textural changes during transportation and storage. Rapid loss of water and microbial action may accelerate these postharvest changes. The quality of fruits and vegetables is

TABLE 15.3
Applications of Polysaccharides Based Antimicrobial Biopolymers on Various Food Products

Material for Coating	Antimicrobial Compounds	Food as Substrate	Outcomes	References
Tapioca starch	Grape pomace extracts + cellulose nanocrystal	Chicken meat	*Listeria monocytogenes* growth reduced (1–2 log CFU/g) in storage period of 10 days	Kalaycıoğlu et al. (2017)
Cassava starch	Oregano oil + extract of pumpkin residue	Grounded beef	Coliform and *Salmonella* growth delayed	Caetano et al. (2017)
Chitosan	Olive oil residue extract	Apple	*Pemicillium expansum* and *Rhizopus stolonifer* growth delayed in cold storage period of 35 days	Khalifa et al. (2016)
-do-	-do-	Strawberry	*P. expansum* and *R. stolonifer* growth delayed in cold storage period of 16 days	Kraśniewska et al. (2014)
-do-	Ready to cook pork chops	Bamboo vinegar	Inhibited microbial growth against *Pseudomonas* spp, lactic acid bacteria and total viable count	Zhang et al. (2018)
-do-	Shrimp	Pomegranate peel extract	Inhibited natural microbiota of shrimp	Yuan et al. (2016)
-do-	Fish steak	Ginger oil	Inhibited *Brochothrix thermosphacta* and lactic acid bacteria	Remya et al. (2017)
Pullulan	*Satureja hortensis* extract	Apple	*S. aureus* growth reduced (1.4 log CFU/g at 16°C for 21 days)	Kraśniewska et al. (2014)
-do-	-do-	Pepper	*Aspergillus niger* growth reduced (1.33 log CFU/g at 16°C for 21 days)	Khalifa et al. (2016)
Alginate	Eugenol	Arbutus berries	Microbial spoilage reduced; storage of berries possible for 28 days at 0.5°C	Guerreiro et al. (2015)

drastically reduced by the outbreak of fungus (Vieira et al., 2016). The adverse changes in the quality of fruit and vegetable can be minimized by the use of starch-based edible coatings or films having antimicrobial compounds, resulting in enhanced shelf life, better quality, and attractive organoleptic properties of products. Garcia et al. (1998) reported an extension of shelf life of strawberries by 1 week after the application of a starch-based coating incorporated with glycerol. Mali and Grossmann (2003) utilized yam starch as a coating material on fresh fruits of strawberries and reported a positive effect on storage quality.

15.5.3 Seafood

Products of seafood are very perishable and have been classified into various categories with distinct spoilage patterns depending on inherent textural, composition and biochemical variations (Dehghani et al., 2018). The storage life of seafood can be enhanced by retarding the growth of spoilage, causing microorganisms which could affect the acceptability of the product by changing its organoleptic properties. In addition, seafood-associated foodborne pathogen outbreaks need attention. Foodborne diseases in several forms, such as infection, intoxication, or both, can be caused by consumption of seafood infected with disease causing microorganisms. Chief microbial contaminants of seafood are some pathogens such as *E. coli*, *Salmonella*, *Vibrio parahaemolyticus*, and *L. monocytogenes*. Thus, the application of biopolymer-based antimicrobial films can prove to be quite practical in controlling the growth of microorganisms, maintaining the quality, and extending the storage life of seafood. Echeverría et al. (2018) prepared a starch-based film incorporated with montmorillonite (MMT) and clove essential oil and studied its effect on storage quality of muscle fillets of bluefin tuna (*Thunnus thynnus*). It was observed that the shelf life of the product increased as the addition of the clay improved the antimicrobial activity, as well as enhanced the antioxidant effects of clove essential oil, whereas no migration of clay particles into the product was noticed (Echeverría et al., 2018).

15.5.4 Meat Products

The majority of the fresh and stored meat products are extremely prone to spoilage because of their biochemical and nutritional components. Normally, fresh meat has 12% to 20% of protein, 3% to 45% fat, and 0% to 6% carbohydrates. Actually, muscle tissue contains water in the range of 42% to 80% that is sufficient to support the microbial growth in meat. Edible films or packaging having antimicrobial action or the capacity to release antimicrobial compounds can be utilized for the purpose of inhibition of development and spread of slime-forming bacteria, pathogens, and quality-deteriorating microorganisms on the meat's surface. Starch-based nanocomposite films containing grape pomade extract showed antibacterial activity particularly against *L. monocytogenes* in the case of chicken during refrigeration storage (Xu et al., 2018). Considerable antimicrobial capacity has been shown by films based on chitosan/starch/cellulose nanofibrils during storage of fresh beef sirloin (Yu et al., 2017).

15.5.5 Medical

Good biocompatibility, biodegradability with toxic-free products, and sufficient mechanical strength are some of the properties which make nano-starch composites as effective packaging material in the medical field. Nano-starch-based composites have cementing properties that provide instant structural support and are widely utilized in bone tissue engineering. Starch-based packaging material utilized either as microsphere or hydrogels is pertinent for the proper delivery of medicine or drug (Balmayor et al., 2008; Reis et al., 2008). The surgical removal of the device needs not be done after drug depletion. The starch-based hydrogels have numerous properties, viz., hydrophilicity, permeability, biocompatibility, and, to some extent, similar to the flexible biological structure, making them suitable for diverse biomedical applications (Peppas et al., 2000). Moreover, its three-dimensional structure allows for sufficient absorption and holding up of water that help in maintaining the adequate mechanical property at the same time.

15.5.6 Agriculture

The starch-based packaging materials are mainly used in agriculture as greenhouse covering, mulching film, and coatings for fertilizers for the controlled release of nutrients (Dilara et al., 2000). In agriculture, the utilization of films is abundant, and they generally disposed of in landfills, or by recycling or burning. But these methods of disposal require more time, are less cost-effective, and cause environmental pollution (Bohlmann et al., 2004). On the other hand, the efficiency of used fertilizer is the key factor for the improvement of agricultural productions. However, the fertilizers escape into the environment in the forms of surface runoff, leaching and vapourization, leading their diseconomy and causing environmental problems (Dave et al., 1999; Guo et al., 2005). The eco-friendly polymers formed of starch can be exploited as the fertilizers controlled release matrices for slow release of nutrients to minimize the loss of fertilizers to reduce or avoid the environmental pollution (Kumbar et al., 2001; Chen et al., 2008). The starch-based films, after their use, can be ploughed into soil and disposed of directly. The degradation of starch by soil microorganisms without the production of harmful products makes it suitable for the separation of agricultural mulch films. Thus, the development of starch-based materials for agriculture applications is very useful and needs to be promoted on a large scale. To augment the mechanical properties and solvent or gas resistance, starch-based biodegradable materials are used in combination with some nano-grade additives such as titanium oxide, layered silicate, and MMT to form bio-nanocomposites (Wang et al., 2004; Scarascia-Mugnozza et al., 2006; Yew et al., 2006).

15.6 CONCLUSION

Increasing consumers' awareness about the safety of food, as well as the environmental concerns of industry, represents a critical challenge for the packaging industry to plan and provides innovative packaging keys for the maintenance of quality and safety of a product. Active packaging, along with the improved properties of biodegradable packaging materials, is a suitable strategy for fulfilling the current demand for environmental and food safety. Starch presents itself as a good candidate for the production as a packaging biopolymer with appreciable compatibility with active agents, such as antimicrobial compounds. However, native starch-based biopolymers have some limitations like low mechanical or barrier properties, which can be improved by using nanotechnology in the development of the biopolymer matrix. The exploitation of nanocomposites as food packaging has turned out to be the most advanced area in the food industry. Composites films incorporated with SNPs exhibit improved properties and have the potential to enhance the applicability of biodegradable films.

REFERENCES

Abreu, A. S., Oliveira, M., Rodrigues, R. M., Cerqueira, M. A., Vicente, A. A., and Machado, A. V. (2015). Antimicrobial nanostructured starch based films for packaging. *Carbohydrate Polymers*, *129*, 127–134. http://dx.doi.org/10.1016/j.carbpol.2015.04.021.

Acevedo-Fani, A., Salvia-Trujillo, L., Rojas-Graü, M.A., and Martín-Belloso, O. (2015). Edible films from essential-oil-loaded nanoemulsions: Physicochemical characterization and antimicrobial properties. *Food Hydrocolloids*, *47*, 168–177. 10.1016/j. foodhyd.2015.01.032.

Ahmad, M., Benjakul, S., Prodpran, T., and Agustini, T. (2012). Physico-mechanical and antimicrobial properties of gelatin film from the skin of unicorn leatherjacket incorporated with essential oils. *Food Hydrocolloids*, *28*, 189–199. 10.1016/j.foodhyd.2011.12.003.

Ahmad, M., Gani, A., Hassan, I., Huang, Q., and Shabbir, H. (2020). Production and characterization of starch nanoparticles by mild alkali hydrolysis and ultra-sonication process. *Scientific Reports*, *10*(1), 1–11.

Ali, A., Chen, Y., Liu, H., Yu, L., Baloch, Z., Khalid, S., Zhu, J., and Chen, L. (2018). Starch-based antimicrobial films functionalized by pomegranate peel. *International Journal of Biological Macromolecules*, *129*, 1120–1126. 10.1016/j.ijbiomac.2018.09.068.

Angellier, H., Molina-Boisseau, S., Lebrun, L., and Dufresne, A. (2005). Processing and structural properties of waxy maize starch nanocrystals reinforced natural rubber. *Macromolecules*, *38*(9), 3783–3792.

Assefa, Z., and Admassu, S. (2013). Development and Characterization of Antimicrobial Packaging Films. *Journal of Food Processing and Technology*, *4*(6), 235. doi: 10.4172/2157-7110.1000235.

Avella, M., de VliegerJ. J., Errico, M. E., Fischer, S., Vacca, P., and Volpe, M. G. (2005). Biodegradable starch/clay nanocomposite films for food packaging applications. *Food Chemistry*, *93*, 467–474. 10.1016/j.foodchem.2004.10.024.

Balmayor, E. R., Tuzlakoglu, K., Marques, A. P., Azevedo, H. S., and Reis, R. L. (2008). A novel enzymatically mediated drug delivery carrier for bone tissue engineering applications: Combining biodegradable starch based microparticles and differentiation agents. *Journal of Material Science: Materials in Medicine*, *19*, 1617–1623. 10.1007/s10856-008-3378-5.

Baron, J. K., and Sumner, S. S. (1993). Antimicrobial containing edible films as inhibitory system to control microbial growth on meat products. *Journal of Food Protection*, *56*, 916.

Bekhit, M., Arab-Tehrany, E., Kahn, C. J., Cleymand, F., Fleutot, S., and Desobry, S., et al. (2018). Bioactive films containing alginate-pectin composite microbeads with Lactococcus lactis subsp. lactis: Physicochemical characterization and antilisterial activity. *International Journal of Molecular Sciences*, *19*, 574. 10.3390/ijms 19020574.

Bhatia, S., and Bharti, A. (2015). Evaluating the antimicrobial activity of Nisin, Lysozyme and Ethylene diamine tetra acetate incorporated in starch based active food packaging film. *Journal of Food Science and Technology*, *52*, 3504–3512.

Bohlmann, G., and Toki, G. (2004). *Chemical Economics Handbook*, SRI International, Menlo Park.

Boziaris, I. S., and Adams, M. R. (1999). Effect of chelators and nisin produced in situ on inhibition and inactivation of Gram negatives. *International Journal of Food Microbiology*, *53*, 105–113.

Branen, J. K., and Davidson, P. M. (2004). Enhancement of nisin, lysozyme, and monolaurin antimicrobial activities by ethylene diamine tetra acetic acid and lacto ferrin. *International Journal of Food Microbiology*, *90*, 63–74. 10.1016/S0168-1605(03)00172-7.

Burt, S. (2004). Essential oils: Their antibacterial properties and potential applications in foods – A review. *International Journal of Food Microbiology*, *94*, 223–253. 10.1016/j.ijfoodmicro.2004.03.022.

Caetano, K. S, Hessel, C. T., Tondo, E. C., Flôres, S. H., and Cladera-Olivera, F. (2017). Application of active cassava starch films incorporated with oregano essential oil and pumpkin residue extract on ground beef. *Journal of Food Safety*, *37*, e12355. 10.1111/jfs.12355.

Cannarsi, M., Baiano, A., Sinigaglia, M., Ferrara, L., Baculo, R., and Del Nobile, M. A. (2008). Use of nisin, lysozyme and EDTA for inhibiting microbial growth in chilled buffalo meat. *International Journal of Food Science and Technology*, *43*, 573–578.

Carbone, M., Donia, D. T., Sabbatella, G., and Antiochia, R. (2016). Silver nanoparticles in polymeric matrices for fresh food packaging. *Journal of King Saud University-Science*, *28*, 273–279.

Cha, D. S., and Chinnan, M. S. (2004). Biopolymer-based antimicrobial packaging: A review. *Critical Review of Food Science and Nutrition*, *44*, 223–237.

Chen, L., Xie, Z. G., Zhuang, X. L., Chen, X. S., and Jing, X. B. (2008). Controlled release of urea encapsulated by starchg-poly (L-lactide). *Carbohydrate Polymers*, *72*, 342–348.

Coma, V. (2012). Antimicrobial and antioxidant active packaging for meat and poultry. In: KerryJP, editor. *Advances in Meat, Poultry and Seafood Packaging*, Cambridge: Woodhead Publishing, pp. 477–498.

Corrales, M., Han, H. J., and Tauscher, B. (2009). Antimicrobial properties of grape seed extracts and their effectiveness after incorporation into pea starch films. *International Journal of Food Science and Technology*, *44*(2), 425–433.

Cvetnić, Z., and Vladimir-Knežević, S. (2004). Antimicrobial activity of grapefruit seed and pulp ethanolic extract. *Acta pharmaceutica*, *54*(3), 243–250.

Dave, A. M., Mehta, M. H., Aminabhavi, T. M., Kulkarni, A. R., and Soppimath, K. S. (1999). A review on controlled release of nitrogen fertilizers through polymeric membrane devices. *Polymer-Plastics Technology and Engineering*, *38*(4), 675–711.

Dehghani, S., Hosseini, S. V., and Regenstein, J. M. (2018). Edible films and coatings in seafood preservation: A review. *Food Chemistry*, *240*, 505–513.

Dhull, S. B., Sandhu, K. S., Punia, S., Kaur, M., Chawla, P., and Malik, A. (2020). Functional, thermal and rheological behavior of fenugreek (*Trigonella foenum–graecum* L.) gums from different cultivars: A comparative study. *International Journal of Biological Macromolecules*, *159*, 406–414.

Dilara, P. A., and Briassoulis, D. (2000). Degradation and stabilization of low-density polyethylene films used as greenhouse covering materials. *Journal of Agricultural Engineering Research*, *76*(4), 309–321.

Durango, A. M., Soares, N. F. F., Benevides, S., Teixeira, J., Carvalho, M., Wobeto, C., and Andrade, N. J. (2006). Development and evaluation of an edible antimicrobial film based on yam starch and chitosan. *Packaging Technology and Science: An International Journal*, *19*(1), 55–59.

Echeverría, I., López-Caballero, M. E., Gómez-Guillén, M. C., Mauri, A. N., and Montero, M. P. (2018). Active nanocomposite films based on soy proteins-montmorillonite-clove essential oil for the preservation of refrigerated bluefin tuna (*Thunnus thynnus*) fillets. *International Journal of Food Microbiology*, *266*, 142–149.

Economou, T., Pournis, N., Ntzimani, A., and Savvaidis, I. N. (2009). Nisin–EDTA treatments and modified atmosphere packaging to increase fresh chicken meat shelf-life. *Food Chemistry*, *114*(4), 1470–1476.

Ejaz, M., Arfat, Y. A., Mulla, M., and Ahmed, J. (2018). Zinc oxide nanorods/clove essential oil incorporated Type B gelatin composite films and its applicability for shrimp packaging. *Food Packaging and Shelf Life*, *15*, 113–121.

García, M. A., Martino, M. N., and Zaritzky, N. E. (1998). Starch-based coatings: Effect on refrigerated strawberry (*Fragaria ananassa*) quality. *Journal of the Science of Food and Agriculture*, *76*(3), 411–420.

García, N. L., Ribba, L., Dufresne, A., Aranguren, M. I., and Goyanes, S. (2009). Physico-mechanical properties of biodegradable starch nanocomposites. *Macromolecular Materials and Engineering*, *294*(3), 169–177.

Gonçalves, P. M., Noreña, C. P. Z., da Silveira, N. P., and Brandelli, A. (2014). Characterization of starch nanoparticles obtained from Araucaria angustifolia seeds by acid hydrolysis and ultrasound. *LWT – Food Science and Technology*, *58*(1), 21–27.

Guerreiro, A. C., Gago, C. M., Faleiro, M. L., Miguel, M. G., and Antunes, M. D. (2015). The effect of alginate-based edible coatings enriched with essential oils constituents on *Arbutus unedo* L. fresh fruit storage. *Postharvest Biology and Technology*, *100*, 226–233.

Guo, M., Liu, M., Zhan, F., and Wu, L. (2005). Preparation and properties of a slow-release membrane-encapsulated urea fertilizer with superabsorbent and moisture preservation. *Industrial & Engineering Chemistry Research*, *44*(12), 4206–4211.

Hafsa, J., Ali Smach, M., Khedher, M. R. B., Charfeddine, B., Limem, K., Majdoub, H., and Rouatbi, S. (2016). Physical, antioxidant and antimicrobial properties of chitosan films containing *Eucalyptus globulus* essential oil. *LWT – Food Science and Technology*, *68*, 356–364.

Hakke, V. S., Bagale, U. D., Boufi, S., Babu, G., and Sonawane, S. H. (2020). Ultrasound Assisted Synthesis of Starch Nanocrystals and It's Applications with Polyurethane for Packaging Film. *Journal of Renewable Materials*, *8*(3), 239–250.

Han, J. H. (Ed.). (2005). *Innovations in Food Packaging*, Elsevier Academic Press, Oxford.

Hashemi, S. M. B., Khaneghah, A. M., Ghahfarrokhi, M. G., and Eş, I. (2017). Basil-seed gum containing *Origanum vulgare* subsp. viride essential oil as edible coating for fresh cut apricots. *Postharvest Biology and Technology*, *125*, 26–34.

Heggers, J. P., Cottingham, J., Gusman, J., Reagor, L., McCoy, L., Carino, E., ..., and Zhao, J. G. (2002). The effectiveness of processed grapefruit-seed extract as an antibacterial agent: II. Mechanism of action and in vitro toxicity. *The Journal of Alternative & Complementary Medicine*, *8*(3), 333–340.

Hernandez-Izquierdo, V. M., and Krochta, J. M. (2008). Thermoplastic processing of proteins for film formation – A review. *Journal of Food Science*, *73*(2), R30–R39.

Issa, A., Ibrahim, S. A., and Tahergorabi, R. (2017). Impact of sweet potato starch-based nanocomposite films activated with thyme essential oil on the shelf-life of baby spinach leaves. *Foods*, *6*(6), 43.

Kaewprachu, P., Rungraeng, N., Osako, K., and Rawdkuen, S. (2017). Properties of fish myofibrillar protein film incorporated with catechin-Kradon extract. *Food Packaging and Shelf Life*, *13*, 56–65.

Kalaycıoğlu, Z., Torlak, E., Akın-Evingür, G., Özen, İ., and Erim, F. B. (2017). Antimicrobial and physical properties of chitosan films incorporated with turmeric extract. *International Journal of Biological Macromolecules*, *101*, 882–888.

Kaur, J., Kaur, G., Sharma, S., and Jeet, K. (2018). Cereal starch nanoparticles—A prospective food additive: A review. *Critical Reviews in Food Science and Nutrition*, *58*(7), 1097–1107.

Khalifa, I., Barakat, H., El-Mansy, H. A., and Soliman, S. A. (2016). Improving the shelf-life stability of apple and strawberry fruits applying chitosan-incorporated olive oil processing residues coating. *Food Packaging and Shelf Life*, *9*, 10–19.

Kim, H. Y., Lee, J. H., Kim, J. Y., Lim, W. J., and Lim, S. T. (2012). Characterization of nanoparticles prepared by acid hydrolysis of various starches. *Starch-Stärke*, *64*(5), 367–373.

Kim, H. Y., Park, D. J., Kim, J. Y., and Lim, S. T. (2013). Preparation of crystalline starch nanoparticles using cold acid hydrolysis and ultrasonication. *Carbohydrate Polymers*, *98*(1), 295–301.

Kraśniewska, K., Gniewosz, M., Synowiec, A., Przybył, J. L., Bączek, K., and Węglarz, Z. (2014). The use of pullulan coating enriched with plant extracts from *Satureja hortensis* L. to maintain pepper and apple quality and safety. *Postharvest Biology and Technology*, *90*, 63–72.

Kumbar, S. G., Kulkarni, A. R., Dave, A. M., and Aminabhavi, T. M. (2001). Encapsulation efficiency and release kinetics of solid and liquid pesticides through urea formaldehyde crosslinked starch, guar gum, and starch+ guar gum matrices. *Journal of Applied Polymer Science*, *82*(11), 2863–2866.

Lamanna, M., Morales, N. J., García, N. L., and Goyanes, S. (2013). Development and characterization of starch nanoparticles by gamma radiation: Potential application as starch matrix filler. *Carbohydrate Polymers*, *97*(1), 90–97.

LeCorre, D., Vahanian, E., Dufresne, A., and Bras, J. (2012). Enzymatic pretreatment for preparing starch nanocrystals. *Biomacromolecules*, *13*(1), 132–137.

Lee, J. H., Lee, J., and Song, K. B. (2015). Development of a chicken feet protein film containing essential oils. *Food Hydrocolloids*, *46*, 208–215.

Li, X., Qiu, C., Ji, N., Sun, C., Xiong, L., and Sun, Q. (2015). Mechanical, barrier and morphological properties of starch nanocrystals-reinforced pea starch films. *Carbohydrate Polymers*, *121*, 155–162.

Liu, C., Jiang, S., Zhang, S., Xi, T., Sun, Q., and Xiong, L. (2016). Characterization of edible corn starch nanocomposite films: The effect of self-assembled starch nanoparticles. *Starch-Stärke*, *68*(3–4), 239–248.

López, O. V., Giannuzzi, L., Zaritzky, N. E., and García, M. A. (2013). Potassium sorbate controlled release from corn starch films. *Materials Science and Engineering: C*, *33*(3), 1583–1591.

Ma, X., Jian, R., Chang, P. R., and Yu, J. (2008). Fabrication and characterization of citric acid-modified starch nanoparticles/plasticized-starch composites. *Biomacromolecules*, *9*(11), 3314–3320.

Maizura, M., Fazilah, A., Norziah, M. H., and Karim, A. A. (2008). Film Incorporated with Lemongrass (*Cymbopogon citratus*) Oil. *International Food Research Journal*, *15*, 233–236.

Mali, S., and Grossmann, M. V. E. (2003). Effects of yam starch films on storability and quality of fresh strawberries (*Fragaria ananassa*). *Journal of Agricultural and Food Chemistry*, *51*(24), 7005–7011.

Martucci, J. F., Gende, L. B., Neira, L. M., and Ruseckaite, R. A. (2015). Oregano and lavender essential oils as antioxidant and antimicrobial additives of biogenic gelatin films. *Industrial Crops and Products*, *71*, 205–213.

Mohan, T. P., Devchand, K., and Kanny, K. (2017). Barrier and biodegradable properties of corn starch-derived biopolymer film filled with nanoclay fillers. *Journal of Plastic Film & Sheeting*, *33*(3), 309–336.

Morsy, M. K., Elsabagh, R., and Trinetta, V. (2018). Evaluation of novel synergistic antimicrobial activity of nisin, lysozyme, EDTA nanoparticles, and/or ZnO nanoparticles to control foodborne pathogens on minced beef. *Food Control*, *92*, 249–254.

Müller, C. M., Laurindo, J. B., and Yamashita, F. (2012). Composites of thermoplastic starch and nanoclays produced by extrusion and thermopressing. *Carbohydrate Polymers*, *89*(2), 504–510.

Naidu, A. S. (Ed.). (2000). *Natural Food Antimicrobial Systems*, CRC Press, pp. 1–16.

Nam, S., Scanlon, M. G., Han, J. H., and Izydorczyk, M. S. (2007). Extrusion of pea starch containing lysozyme and determination of antimicrobial activity. *Journal of Food Science*, *72*(9), E477–E484.

Ntzimani, A. G., Giatrakou, V. I., and Savvaidis, I. N. (2010). Combined natural antimicrobial treatments (EDTA, lysozyme, rosemary and oregano oil) on semi cooked coated chicken meat stored in vacuum packages at 4 C: Microbiological and sensory evaluation. *Innovative Food Science & Emerging Technologies*, *11*(1), 187–196.

Ochoa, T. A., Almendárez, B. E. G., Reyes, A. A., Pastrana, D. M. R., López, G. F. G., Belloso, O. M., and Regalado-González, C. (2017). Design and characterization of corn starch edible films including beeswax and natural antimicrobials. *Food and Bioprocess Technology*, *10*(1), 103–114.

Ozdemir, M., and Floros, J. D. (2004). Active food packaging technologies. *Critical Reviews in Food Science and Nutrition*, *44*(3), 185–193.

Park, H. M., Li, X., Jin, C. Z., Park, C. Y., Cho, W. J., and Ha, C. S. (2002). Preparation and properties of biodegradable thermoplastic starch/clay hybrids. *Macromolecular Materials and Engineering*, *287*(8), 553–558.

Pelissari, F. M., Grossmann, M. V., Yamashita, F., and Pineda, E. A. G. (2009). Antimicrobial, mechanical, and barrier properties of cassava starch–chitosan films incorporated with oregano essential oil. *Journal of Agricultural and Food Chemistry*, *57*(16), 7499–7504.

Peppas, N. A., Bures, P., Leobandung, W. S., and Ichikawa, H. (2000). Hydrogels in pharmaceutical formulations. *European Journal of Pharmaceutics and Biopharmaceutics*, *50*(1), 27–46.

Piyada, K., Waranyou, S., and Thawien, W. (2013). Mechanical, thermal and structural properties of rice starch films reinforced with rice starch nanocrystals. *International Food Research Journal*, *20*(1), 439–449.

Prudêncio, C. V., Mantovani, H. C., Cecon, P. R., Prieto, M., and Vanetti, M. C. D. (2016). Temperature and pH influence the susceptibility of Salmonella Typhimurium to nisin combined with EDTA. *Food Control*, *61*, 248–253.

Punia, S., and Dhull, S. B. (2019). Chia seed (*Salvia hispanica* L.) mucilage (a heteropolysaccharide): Functional, thermal, rheological behaviour and its utilization. *International Journal of Biological Macromolecules*, *140*, 1084–1090.

Punia, S., Dhull, S. B., Kunner, P., and Rohilla, S. (2020a). Effect of γ-radiation on physico-chemical, morphological and thermal characteristics of lotus seed (*Nelumbo nucifera*) starch. *International Journal of Biological Macromolecules*, *157*, 584–590.

Punia, S., Dhull, S. B., Sandhu, K. S., and Kaur, M. (2019a). Faba bean (*Vicia faba*) starch: Structure, properties, and in vitro digestibility—A review. *Legume Science*, *1*(1), e18.

Punia, S., Sandhu, K. S., Dhull, S. B., and Kaur, M. (2019b). Dynamic, shear and pasting behaviour of native and octenyl succinic anhydride (OSA) modified wheat starch and their utilization in preparation of edible films. *International Journal of Biological Macromolecules*, *133*, 110–116.

Punia, S., Sandhu, K. S., Dhull, S. G., Siroha, A. K., Purewal, S. S., Kaur, M., and Kidwai, M. K. (2020b). Oat starch: Physico-chemical, morphological, rheological characteristics and its application – A review. *International Journal of Biological Macromolecules*, *154*, 493–498.

Punia, S., Sandhu, K. S., Siroha, A. K., and Dhull, S. B. (2019c). Omega 3-Metabolism, Absorption, Bioavailability and health benefits – A review. *PharmaNutrition*, *10*, 100162.

Raigond, P., Sood, A., Kalia, A., Joshi, A., Kaundal, B., Raigond, B., ..., and Chakrabarti, S. K. (2019). Antimicrobial activity of potato starch-based active biodegradable nanocomposite films. *Potato Research*, *62*(1), 69–83.

Rawdkuen, S., Suthiluk, P., Kamhangwong, D. and Benjakul, S. (2012). Mechanical, physico-chemical, and antimicrobial properties of gelatin-based film incorporated with catechin-lysozyme. *Chemistry Central Journal*, *6*, 131. https://doi.org/10.1186/1752-153X-6-131.

Reis, A. V., Guilherme, M. R., Moia, T. A., Mattoso, L. H., Muniz, E. C., and Tambourgi, E. B. (2008). Synthesis and characterization of a starch-modified hydrogel as potential carrier for drug delivery system. *Journal of Polymer Science Part A: Polymer Chemistry*, *46*(7), 2567–2574.

Remya, S., Mohan, C. O., Venkateshwarlu, G., Sivaraman, G. K., and Ravishankar, C. N. (2017). Combined effect of O_2 scavenger and antimicrobial film on shelf life of fresh cobia (*Rachycentron canadum*) fish steaks stored at 2 C. *Food Control*, *71*, 71–78.

Rhim, J. W., Park, H. M., and Ha, C. S. (2013). Bio-nanocomposites for food packaging applications. *Progress in Polymer Science*, *38*(10–11), 1629–1652.

Sadegh-Hassani, F., and Nafchi, A. M. (2014). Preparation and characterization of bionanocomposite films based on potato starch/halloysite nanoclay. *International Journal of Biological Macromolecules*, *67*, 458–462.

Salleh, E., and Muhamad, I. I. (2010, March). *Starch-based Antimicrobial Films Incorporated with Lauric Acid and Chitosan. AIP Conference Proceedings* (Vol. *1217*, No. 1, pp. 432–436). American Institute of Physics.

Salleh, E., Muhamad, I. I., and Khairuddinr, N.2007. *Inhibition of Bacillus subtilis and Escherichia coli by antimicrobial starch-based film incorporated with lauric acid and Chitosan. Proceedings of the 3rd CIGR Section Vl International Symposium on Food and Agricultural Products: Processing and Innovation.* Naples, Italy.

Scarascia-Mugnozza, G., Schettini, E., Vox, G., Malinconico, M., Immirzi, B., and Pagliara, S. (2006). Mechanical properties decay and morphological behaviour of biodegradable films for agricultural mulching in real scale experiment. *Polymer Degradation and Stability*, *91*(11), 2801–2808.

Shankar, S., Teng, X., Li, G., and Rhim, J. W. (2015). Preparation, characterization, and antimicrobial activity of gelatin/ZnO nanocomposite films. *Food Hydrocolloids*, *45*, 264–271.

Shankar, S., Wang, L. F., and Rhim, J. W. (2017). Preparation and properties of carbohydrate-based composite films incorporated with CuO nanoparticles. *Carbohydrate Polymers*, *169*, 264–271.

Shen, X. L., Wu, J. M., Chen, Y., and Zhao, G. (2010). Antimicrobial and physical properties of sweet potato starch films incorporated with potassium sorbate or chitosan. *Food Hydrocolloids*, *24*(4), 285–290.

Shi, A. M., Li, D., Wang, L. J., Li, B. Z., and Adhikari, B. (2011). Preparation of starch-based nanoparticles through high-pressure homogenization and miniemulsion cross-linking: Influence of various process parameters on particle size and stability. *Carbohydrate Polymers*, *83*(4), 1604–1610.

Siracusa, V., Rocculi, P., Romani, S., and Dalla Rosa, M. (2008). Biodegradable polymers for food packaging: A review. *Trends in Food Science & Technology*, *19*(12), 634–643.

Sivarooban, T., Hettiarachchy, N. S., and Johnson, M. G. (2008). Physical and antimicrobial properties of grape seed extract, nisin, and EDTA incorporated soy protein edible films. *Food Research International*, *41*(8), 781–785.

Slavutsky, A. M., Bertuzzi, M. A., and Armada, M. (2012). Water barrier properties of starch-clay nanocomposite films. *Brazilian Journal of Food Technology*, *15*(3), 208–218.

Tan, Y. M., Lim, S. H., Tay, B. Y., Lee, M. W., and Thian, E. S. (2015). Functional chitosan-based grapefruit seed extract composite films for applications in food packaging technology. *Materials Research Bulletin*, *69*, 142–146.

Tong, S. Y., Lim, P. N., Wang, K., and Thian, E. S. (2018). Development of a functional biodegradable composite with antibacterial properties. *Materials Technology*, *33*(11), 754–759.

Vieira, J. M., Flores-López, M. L., de Rodríguez, D. J., Sousa, M. C., Vicente, A. A., and Martins, J. T. (2016). Effect of chitosan–Aloe vera coating on postharvest quality of blueberry (*Vaccinium corymbosum*) fruit. *Postharvest Biology and Technology*, *116*, 88–97.

Wang, Y. Z., Yang, K. K., Wang, X. L., Zhou, Q., Zheng, C. Y., and Chen, Z. F. (2004). Agricultural application and environmental degradation of photo-biodegradable polyethylene mulching films. *Journal of Polymers and the Environment*, *12*(1), 7–10.

Xiong, H., Tang, S., Tang, H., and Zou, P. (2008). The structure and properties of a starch-based biodegradable film. *Carbohydrate Polymers*, *71*(2), 263–268.

Xu, Y., Rehmani, N., Alsubaie, L., Kim, C., Sismour, E., and Scales, A. (2018). Tapioca starch active nanocomposite films and their antimicrobial effectiveness on ready-to-eat chicken meat. *Food Packaging and Shelf Life*, *16*, 86–91.

Yew, S. P., Tang, H. Y., and Sudesh, K. (2006). Photocatalytic activity and biodegradation of polyhydroxybutyrate films containing titanium dioxide. *Polymer Degradation and Stability*, *91*(8), 1800–1807.

Yu, Z., Alsammarraie, F. K., Nayigiziki, F. X., Wang, W., Vardhanabhuti, B., Mustapha, A., and Lin, M. (2017). Effect and mechanism of cellulose nanofibrils on the active functions of biopolymer-based nanocomposite films. *Food Research International*, *99*, 166–172.

Yuan, G., Lv, H., Tang, W., Zhang, X., and Sun, H. (2016). Effect of chitosan coating combined with pomegranate peel extract on the quality of Pacific white shrimp during iced storage. *Food Control*, *59*, 818–823.

Zhang, H., He, P., Kang, H., and Li, X. (2018). Antioxidant and antimicrobial effects of edible coating based on chitosan and bamboo vinegar in ready to cook pork chops. *LWT*, *93*, 470–476.

16 Starch Bio-Nanocomposite Films as Effective Antimicrobial Packaging Material

Nitin Kumar, Anil Panghal, Sunil Kumar, Arun Kumar Attkan, and Mukesh Kumar Garg
Chaudhary Charan Singh Haryana Agricultural University, India
Navnidhi Chhikara
Guru Jambheshwar University of Science and Technology, India

CONTENTS

16.1 INTRODUCTION

Food packaging is a technique which provides protection to the food material from influences like dust, temperature, light, microorganisms, odours, physical damage, shocks, and other environmental contaminations; also, it assures the safety and quality of the food and extends the shelf life, minimizing the wastage and loss of food. Packaging protects from chemical, biological, and physical deterioration. In recent years, biodegradable starch-based films prepared with nanocomposites have been very popular in the food industry, and research related to nanocomposite food packaging is carried out extensively due to their usage for improvement of the shelf life of food products and their compatibility with the environment (Kehinde et al., 2020). Nowadays, food manufacturers, researchers, and consumers are demanding more from packaging related to the freshness, performance, and well-being of foods. Starch-based packaging materials are derived from renewable agricultural sources (Khatkar et al., 2009). In the current scenario, consumers prefer the green composites with bio-based material as compared to petroleum-based composites due to the advantages of being natural, ecological, and renewable (Chhikara et al., 2019). The main benefits of using the bio-based packaging materials are that they allow an improved ecosystem and create rural employment along with the economic development of the agricultural community. The poor mechanical and barrier properties of these bio-based packaging materials are the limitations and also the thermal properties are also not comparable to petroleum-based products. The maximum utilization of nanotechnology in the food industry is in the area of food packaging and will increase exponentially in the future. It has been announced that between 400 and 500 nano-packaging items are being utilized in the food industry, in which nanotechnology is utilized as a part of the production of 25% of all food packing in the coming decade (Panghal et al., 2019a, 2019b). The main function of nano packaging is to improve the functional properties of packaging material including the extension of the shelf life of food products by enhancing the barrier nature of food packets and by reducing moisture, gas exchange and the exposure to ultraviolet (UV) light. The use of starch as the polymeric base material for the development of edible and bio-nano-composite packaging films has been widely studied due to its wide availability, low cost, non-toxicity, renewability, biodigestibility, and capacity to be processed by the types of equipment currently available in polymer and food industries (Kaur et al., 2020). Starch nanoparticles (NPs) are defined as particles that have at least one dimension smaller than 1000 nm but are larger than a single molecule. The uses of these NPs in food businesses with the help of nanocomposite bottles, nanolaminates, and containers having other NPs are extensively studied.

An increase in end-user demands for ready-to-eat food such as fresh-cut fruits, mixed salads, and sandwiches, motivating food manufacturers and researchers to find novel food preservation technology that can control the microbial growth and improve the shelf life of food products (Shapi'i et al., 2020). Antimicrobial packaging is a promising preservation technique that can suppress microbial growth in the food product and improve the shelf life of food. Nanocomposites have been integrated into the aspect of food packaging in intelligent or smart packaging; for the design of biosensors or nano-coatings which provides signals on the current state of enclosed food products, in bioactive or active packaging; for the impartment of biocatalytic, antioxidative, and/or antimicrobial effects and an improved packaging; and for the enhancement of barrier and mechanical properties such as barrier against gases and moisture and improvement of flexibility and stability against thermal stress (Vasile, 2018). On another note, concerns regarding the environmental impacts of plastics conventionally used for food packaging have spurred the interests of food technologists for the adoption of biodegradable poly-nanocomposites (PNCs) for food packaging. This trend, referred to as green or eco-friendly packaging, connotes the microbial degradability of such packages into byproducts of organic characteristics and/or the syntheses of such packages from bio-renewable and natural resources (Vilarinho et al., 2018). Furthermore, PNCs with adsorptive properties have been included in food-packaging materials to adsorb any releases from the food product in the course of storage that can enhance the shelf-life of such a product (Panghal et al., 2019a, 2019b). The wide applications of nanocomposite are limited by the lack of water resistance and poor mechanical properties. These bio-based films are significantly improved for mechanical properties, water resistance, transmittance, and biodegradability with the addition of different NPs. The antimicrobial properties of these bio-nanocomposites are also improved significantly. The incorporation of nanofiller into a biopolymer may produce bio-nanocomposites which exhibit good tensile strength, thermal stability, and barrier properties (Shapi'i et al., 2020). Nanoparticles such as zinc oxide and titanium dioxide can act as antimicrobial agents that can produce antimicrobial food-packaging material when used as filler. Antimicrobial food packaging incorporated with antimicrobial agents will gradually release the antimicrobial agents to the food surface to inhibit the microbial growth and extend the shelf life of food. However, there are many speculations on the toxicity of NPs due to its small dimension which has a high potential to penetrate human tissue and thus is harmful to human health. Alternatively, NPs that are synthesized from natural biopolymers, such as nano-starch, can be used as nanofillers. The starch is edible and not harmful to human health. It exhibits good antimicrobial properties and has been used in food packaging, food coatings, textile industries, and others.

16.2 BIODEGRADATION PROCESS

The biodegradation process is the chemical dissolution of organic material and inorganic nutrients by bacteria and other biological means. The microorganisms present in the soil and the environment consume the material and return it to nature by degradation. In the first step, the long-chain polymer molecules are triggered by the mechanical stress, UV light, and heat which breaks them into shorter lengths, and oxygen adds itself to the polymers, which makes the molecules hydrophilic. With time, the molecules become so small that the microorganisms can consume them and the biodegradation process starts. In the second step, the material is broken down into the residual product of the biodegradation process in the presence of moisture and microorganisms, found in the environment. The third step is the final step of biodegradation in which the material is consumed by the microorganisms and the carbon dioxide, biomass, water produced is returned to nature.

16.2.1 Biopolymers Extracted Directly From Agro-Industrial Wastes

These types of biopolymers are generally available in the market. These polymers are obtained from the plant wastes, marine, and animal wastes. For example, polysaccharides such as cellulose and starch from the plant wastes, chitin, and chitosan from marine wastes; whey protein, and casein from the dairy wastes; and animal wastes, which can be used to develop the biopolymers for packaging. Cellulose-based paper is the most predominant category that is used for food packaging. After cellulose, hemicelluloses are the second-most prevalent polymers which can be obtained from plant, and biomaterial research is in its early stages for the development of packaging.

16.2.2 Starch as a Biopolymer

Starch, a natural polymer, mainly composed of amylose and amylopectin, is chiefly derived from plants, such as potatoes, corn, rice, and wheat, used for the development of biopolymers. Starch is the main constituent in most of the commercially available biopolymers, potentially the most suitable biodegradable polymer for packaging material because of its low cost, availability, and its production from biobased sources (Kaur et al., 2019; Singh et al., 2020). Among the agricultural wastes, potato peels, cassava wheat, and rice wastes are the most abundant sources of the starch. The determination of extraction methods, yield, and purity of starch from different fruits' waste are discussed in Table 16.1.

16.3 BIO-NANOCOMPOSITE FILMS CONCEPT

A bio-nanocomposite film is essentially a multiple-phase hybrid solid film containing at least one phase as nanofillers (<5% by weight) with one dimension smaller than 100nm in a polymer matrix (Kumar et al., 2017). Due to the dispersed nanofillers in the polymer matrix, nanocomposite films have considerably improved thermal (melting point, glass transition temperature), mechanical (Young's modulus, tensile strength, elongation at break, loading capacity), barrier (gas and water vapour transmission rate, UV light), and physicochemical (transparency, lightweight, dimensional stability, chemical resistance) properties as compared to natural polymers and other composite films with micro-fillers. The concept of utilization of bio-nanocomposite films in food packaging is two decades old but subsequent improvements in the structure and properties have been reported by many studies over the time (Sinha et al., 2003; Liao and Wu, 2009; Sandri et al., 2015; Kirandeep et al., 2018; Devgan et al., 2019; Kumar et al., 2020). Bio-nanocomposite films have significant potential in the development of innovative food-packaging materials for preservation and extension of the shelf life of packaged products (Shankar and Rhim, 2016). A bio-nanocomposite film is composed of a polymer matrix (protein, carbohydrate, or other biodegradable polymers in 2% to 6% concentration dissolved in an aqueous solvent), nanofiller material (to enhance the attributes of polymer), a plasticizer (to impart stretchability, reduce brittleness, and increase thermoplasticity of polymer), and compatibilizers (to improve interfacial adhesion).

TABLE 16.1
Determination of Extraction Methods, Yield, and Purity of Starch from Different Fruits' Waste

Source	Method of Extraction	Yield (%)	Purity (%)
Kiwifruit	Extracted by milling (wet) with 0.3% $Na_2S_2O_5$ (sodium pyrosulfite or metabisulfite) w/v	Not determined	Not determined
Pineapple (stems)	Extracted by milling (wet) by mixing the sample with water (1:2) for 24 h	23.4	95.18
Mango (kernels)	Degreasing with Soxhet apparatus with hexane (16h; 75°C) followed by milling (wet) in 0.16% $NaHSO_3$ (sodium bisulfite) w/v	45	98.9
Litchi (seeds)	Treatment with $C_6H_8O_7$ (citric acid) with a concentration of 0.3% w/w Treatment with NaOH (sodium hydroxide) 0.5%, w/w	10.9 12.5	Not determined
Tamarind (seeds)	Extracted by milling (wet) with 0.16% $Na_2S_2O_5$ (sodium pyrosulfite) w/v	Not determined	Not determined
Longan (LG) and Loquat(LQ) (seeds)	Extracted by milling (wet) with 0.5% $Na_2S_2O_5$ (sodium pyrosulfite) w/v and keeping overnight at 4 °C	Not determined	85.0 (LG and LQ)
Annatto (seeds)	Depigmentation and defatting of seeds followed by milling (wet) in 500 ppm $Na_2S_2O_5$ (sodium pyrosulfite) w/v	66.0	Not determined
Jackfruit (seeds)	Treatment with 0.1–0.5% w/v NaOH (sodium hydroxide) Treatment with 0.1–0.5 g α-amylase	84.48 57.34 52.74	94.21 96.05 94.02
Avocado (seeds)	Treatment with 2 ml Tris, 7.5 ml 0.1 M Sodium chloride (NaCl) and 8 ml sodium bisulfite ($NaHSO_3$) Treatment with 150 ppm $NaHSO_3$ (sodium bisulfite) w/v	19.55 19.98	Not determined
Apple (pulp)	Extracted by milling (wet) in 0.3% $Na_2S_2O_5$ (sodium pyrosulfite) w/v	Not determined	Not determined
Banana (peel)	Extracted by milling (wet) in 1%–5% ascorbic acid w/w	28.9	69.9

Source: Adapted from Kringel et al. (2020).

16.3.1 Bio-Nanocomposite Materials for Food Packaging

The bio-nanocomposite materials are being used in the food-packaging industry for several years, but still, it is a relatively new field with a high potential for future modifications. The polymers used in the preparation of nanocomposite films for food packaging can be categorized as follows:

- **Naturally occurring polymers:** These include proteins like gluten, whey, casein, zein, and gelatin
- **Microbially derived polymers:** These include PLAs (polylactic or polygalactic acid), PCL (polycaprolactone), PGA (polyglycolide), PVOH (polyvinyl alcohol), and PBS (polybutylene succinate)
- **Chemically derived polymers:** These comprise LDPE (low-density polyethylene), PP (polypropylene), nylon, PET (polyethylene terephthalate), polyamides

A detailed classification of biopolymers for the preparation of bio-nanocomposite films is discussed in Figure 16.1.

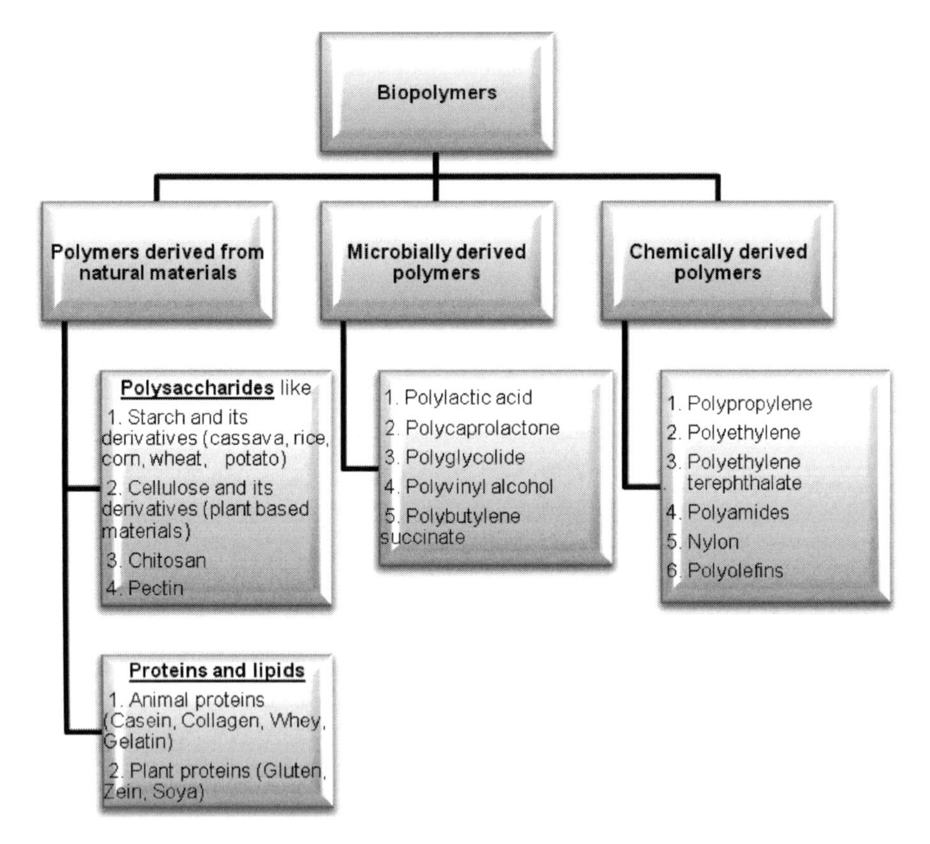

FIGURE 16.1 Classification of biopolymers for preparation of bionanocomposite films. (Adapted from Kumar et al. 2017.).

The typical **plasticizers** used in the preparation of bio-nanocomposite films are citrates, polyesters, benzoate ester, and phthalate esters. Nowadays, even bio-based plasticizers like castor, palm, soya bean, and linseed oil are predominantly used in some industries (Mwaurah et al., 2020). Compatibilization is a phenomenon under which blended properties and adhesion are increased between the different phases of bio-nanocomposite films, imparting strength, stabilizing morphology, and lessening the interfacial tension. The typical examples of **compatibilizers** are acrylic functions grafted on PP, LDPE, polyolefins, and PET.

16.3.2 Types of Nanofillers

The nanophase is divided into three categories based on the number of nano-sized dimensions. The first one includes nanospheres, nanocrystals, or NPs having three nanoscale dimensions. The second category includes nanowhiskers, nanofibrils, or nanotubes, having two nanoscale dimensions. The third category comprises nanosheets or nanoplatelets with a single nanoscale dimension. The categorization of nanofillers is dependent on several factors like the type of raw material, its source, and dimension. A comprehensive classification of different nanofillers for the preparation of bio-nanocomposite films has been provided in Figure 16.2. Nanofillers are usually classified as organic and inorganic materials which are discussed next.

16.3.2.1 Organic Nanofillers

The organic nanofillers mainly comprise naturally occurring cellulose and chitin in the form of nanocrystals, nanofibrils, nanowhiskers, or NPs. The attributes of nanofillers largely depend on the source and method of extraction.

16.3.2.1.1 Cellulose-Based Nanofillers

It is a well-known fact that cellulose is the most abundant natural polymer available on this planet. This is the reason why cellulose nanofillers are widely used in the preparation of bio-nanocomposite films. It is essentially a polysaccharide consist of a linear chain of numerous β 1, 4-glucopyranose units. The hydrogen bonding in linear chains makes several filaments which further consist of ordered crystalline and scattered amorphous parts. The former can be successfully separated into precisely ordered rod-like nanocrystals in nanoscale after removing the amorphous regions through processes like acid hydrolysis. These nanocrystals are called cellulose nanowhisker, nanofiber, or simply nanocellulose (de Souza and Borsali, 2004). The application of nanocellulose in the development of bio-nanocomposite films is continuously increasing in popularity, owing to its low cost and ability to improve the mechanical, physical, thermal, and barrier properties of the films (Sandri et al., 2015). Moreover, nanocellulose has a low molecular weight combined with biodegradability, compatibility, and renewability, making it a perfect fit for the food-packaging industry. Previously, the application of micro-cellulose has also been explored by conducting some studies, but nanocellulose has been proved far superior owing to its dispersing properties which covers a larger area and forms strong hydrogen bonds within the polymer matrix.

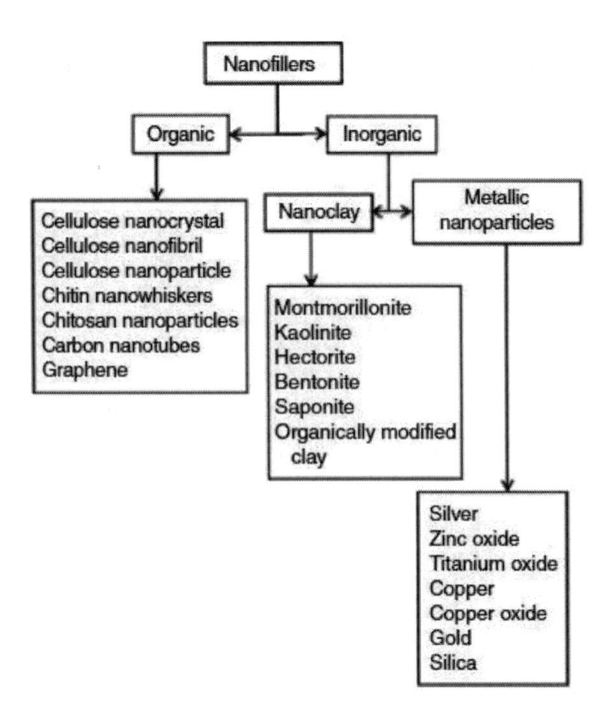

FIGURE 16.2 Classification of different nanofillers for preparation of bionanocomposite films. (Adapted from Shankar and Rhim 2016).

16.3.2.1.2 Chitin-Based Nanofillers

Chitin is another popular biopolymer, occupying second place in terms of abundance and availability. Similar to cellulose, chitin is also a long-chain polysaccharide consisting of several N-acetylglucosamine molecules which are derivatives of glucose. It is mainly obtained from animal sources like octopi, squid, insects, crabs, prawns, lobsters, and even from fungi. Just like cellulose, chitin can also form nanowhiskers and nanofibrils by acid hydrolysis, having numerous applications in the preparation of bio-nanocomposite films (Kumar et al., 2017). In a recent study, Shankar et al. (2015) has reported the use of chitin nanofibrils in improving the properties of carrageenan films.

16.3.2.2 Inorganic Nanofillers

The organic nanofillers mainly comprise metal oxide (silver, magnesium, titanium, gold, copper) and clay NPs (saponite, sauconite, montmorillonite). The attributes of NPs largely depend on the method of synthesis. These are further grouped as described in the following sections.

16.3.2.2.1 Metal Oxide–Based Nanofillers

Owing to the antimicrobial properties of metals and their oxides, these nanofillers offer a promising future in the field of antimicrobial food packaging systems. These properties have been sought by several scientists not only in the field of food packaging but also in biotechnology and biomedical engineering (Kumar et al., 2018). The inorganic nanofillers moreover have a high surface-to-volume ratio and specificity, making them a perfect fit for packaging applications. These can be synthesized either by physical (ultrasonication, radiolysis, spray pyrolysis, vacuum sputtering, thermal decomposition, vapor deposition, lithography, electrospinning, laser ablation), chemical (galvanic replacement, electrochemical, micelle-based, sol-gel, solvothermal), or biological (through use of bacteria, algae, fungi, amino acid, plants, peptides, protein, DNA, viruses) techniques (Shankar and Rhim, 2016).

16.3.2.2.2 Clay-Based Nanofillers

Clay nanofillers especially montmorillonite, saponite, sauconite, and hectorite are widely explored by the scientific fraternity as potential nanofillers in the preparation of bio-nanocomposite films for food packaging owing to their ease in preparation and processing, availability, low cost, and ability to impart superior thermal, mechanical, and barrier properties to the polymer matrix. After incorporation, these essentially form a clay layer (~1 nm thick) which imparts all these properties to the polymers. Before incorporation into the polymer matrix, the compatibility of clay nanofillers needs to be explored. Generally, the naturally occurring layered silicates have sodium and potassium cation, making the surface hydrophilic, in turn making them unsuitable for hydrophobic polymers. So an exchange reaction is followed to make the polymer surface hydrophobic by using organic cations. The organic cations further increase the interlayer spacing and interaction between layered silicate and polymer matrix, resulting in improved intercalated nanocomposite formation.

16.4 TECHNIQUES USED TO MANUFACTURE NANO-STARCH FILMS

The intercalation and uniform dispersion of nanofillers into the polymer matrix is the first step in the preparation of bio-nanocomposite films. The nanofiller materials tend to agglomerate owing to the presence of hydroxyl groups on their surface. This can result in the formation of poor nanophase and uneven dispersion of nanofillers within the polymer matrix, affecting the physical, chemical, mechanical, thermal, and barrier properties of the polymer. Several methods have been sought and successfully applied to avoid the self-aggregation and preparation of uniform nanocomposites which have been discussed as follows:

1. **Thermo-compression:** It is relatively simple and a mechanical means of disrupting the self-aggregation of the nanofillers. The polymer matrix is heated between two stainless steel plates lined with aluminum foils (non-stick surface) and subjected to a pressure of about 70 MPa for 4 to 5 min, depending on the type of polymer and nanofiller, followed by their peeling and cooling to the room temperature (~25°C) (Rhimet al., 2006; Ortega-Toro et al., 2015; Kumar et al., 2017).
2. **Melt intercalation:** In this process, the polymer matrix (polymer and nanofillers) is subjected to shear at a temperature above the melting point of the polymer, resulting in the intercalation of nanofillers (usually clay), thereby ensuring

their uniform dispersion. It is quite similar to the process of extrusion typically used in the processing of food (Alam et al., 2018). This method is popular owing to its versatility of using different polymers, flexibility, compatibility, simpler processing operation, and improved handling due to absence of solvents (Rhim et al., 2006; Tomé et al., 2012; Jbilou et al., 2014; Kumar et al., 2017).

3. **Insitu polymerization:** This is a simple method involving the mixing of nanofiller material (clay) with the monomer physically followed by their polymerization by employing heat, catalyst, or radiation. This results in the formation of the linear or cross-linked polymer matrix which can be used to prepare bio-nanocomposite films by adding a suitable plasticizer.
4. **Solution exfoliation:** This method essentially involves the dissolution of polymer in a solvent and pre-swelling of clay nanofillers in the same solvent independently. This is followed by the mixing of both solutions resulting in intercalation and arrangement of low-polarity polymers in a uniform layered lattice. It is a widely accepted technique for the production of thin bio-nanocomposite films with soluble natural polymers (Kumar et al., 2017).
5. **Solvent casting:** It is a fairly simple technique involving the dissolution of polymer in organic solvent or water (depending upon the polymer) either at room temperature (~25°C) or at an elevated temperature (up to 50°C for reducing processing times). It is followed by the addition of nanofiller material to the polymer matrix. After the dissolution of nanofillers, the solution is cast either on glass plates or any other non-stick surface, like one coated with Teflon, in the desired thickness and shape. After evaporation of solvent overnight, it leaves behind thin bio-nanocomposite films incorporated with nanofillers. The uniformity of incorporation depends on the type of nanofiller, polymer, conditions used, and precision of the process (Malathi et al., 2017; Kumar et al., 2018; Kumar et al., 2020).

16.5 EFFECTS OF NANOMATERIALS ON THE FUNCTIONAL PROPERTIES OF STARCH FILMS

Starch is a natural polymer and attracts researchers as a green alternative for food packaging applications. Naturally, the properties, that is physical, mechanical, barrier, and so on, of the starch polymer do not match the requirements of the products to be packed (Da Silva et al., 2018). Nanomaterial has an extraordinary potential in the improvement of properties of natural polymers, although there are concerns about its impact on the environment and human health. Researchers are continuing their investigation in unveiling its potential risks, and government agencies are in pursuit of forming the safety guidelines and legislation about the use of nanomaterials in human foods (Bumbudsanpharoke et al., 2015). Different materials, for example plasticizers, compatibilizers, and others, are being added in the starch for improvement in the functional properties of the starch films. It increases the cost of production of starch-based packing films and is currently unable to provide desired functional properties (Avella et al., 2005). However, with the advancement in technology and processes, this drawback will be overcome soon. Nanotechnology reduces the size of the active particles below 100 nm and increases their reactivity, controls their release as per requirement and make starch films stable, and prevents its failure in real situations.

16.5.1 Effects of Nanofillers on the Physical Properties

It has been reported that the incorporation of the chitin nanowhiskers in the starch films does not increase the thickness of the developed bio-nanocomposite films. Good interfacial interaction between the chitin nanowhiskers and the starch matrix has been observed (Qin et al., 2016). While the addition of the chitin nanowhiskers in the same starch matrix between 0 to 2% concentration decreased the water vapour permeability of film. This improvement of the film is attributed to the tortuous path for water molecule movement inside the film matrix. The swelling percentage represents the water intake by the starch films. The addition of chitin nanowhiskers into the starch matrix reduces its swelling degree from 103% to 82%, and it is attributed to low water uptake by the chitin nanowhiskers and strong hydrogen bonding (Qin et al., 2016). In other research, increasing the size of nano–silicon oxide (SiO_2) decreased the water vapor permeability and water uptake due to hydrogen bonds and good dispersion in the starch matrix. The incorporation of nano-SiO_2 in the starch matrix up to 100 nm did not affect the thickness of the starch film (Zhang, et al., 2018) (Table 16.1).

16.5.2 Effects of Nanofillers on the Antimicrobial Activity

Antimicrobial activity is the incorporation active ingredients that prevent the microbial growth in the packaging system without compromising the safety, shelf life, and quality of the packed products (Chhikara et al., 2018). Modern consumers are attracted to the natural antimicrobial agents embedded in the packaging material of natural sources instead of chemically synthesized or of animal sources (Sharma et al., 2019). Starch is a natural polymer and is compatible with a wide range of antimicrobial agents. The incorporation of the chitosan and potassium sorbate in starch-based films inhibits the growth of *Escherichia coli* and *Staphylococcus aureus*. Besides, these antimicrobial agents improve the barrier properties of the starch film (Mlalila et al., 2018). Bajpai et al. (2020) studied the effect of the zinc oxide NP on the starch film and noted the antibacterial action on the *E. coli* through zone inhibition method along with the improvement in water vapour permeation properties. Chitin nanowhiskers are known for their antibacterial activity and are the most cost-effective alternatives for food-packaging applications. It has strong antimicrobial activity against *Listeria monocytogenes* when used in maize starch films (Qin et al., 2016). Chitin nanowhiskers also improved the mechanical and barrier properties of the starch film. A list of recent publications in preparation and characterization of nano-starch films as an effective antimicrobial packaging material for food packaging has been present in Table 16.2.

TABLE 16.2
Recent Publications in Preparation and Characterization of Nano Starch Films as Effective Antimicrobial Packaging Material

Antimicrobial Agent and Type of Food	Target Organisms	Results	Research Articles
Silver nanospheres (AgNPs) in media	*E. coli* *Bacillus cereus* *Salmonella* sp. *Penicillium* sp. *Mycobacterium megmatis* *Staphylococcus aureus*	• Migration of individual components was well within permissible limit • Silver nanospheres modified montmorillonite (MMT) dispersion in the polymer matrix • Silver nanoparticles had pronounced effect on reducing the growth of gram (– and +) bacteria	Abreu et al. (2015); Mlalila et al. (2018); Mohanty et al. (2012); Yoksan and Chirachanchai (2010)
Potassium sorbate (KS) in dairy foods/media	*Listeria innocua* *Staphylococcus aureus* *Salmonella* sp. *Candida* sp. *Penicillium* sp. *E. coli*	• KS ameliorated the mechanical and barrier attributes of starch-based films such as tensile strength, Young's modulus, and elongation at break • The functionality of KS was found independent of food substrates and pH • Modification in the release of KS due to MMT nanowhiskers	Mlalila et al. (2018); Barzegar et al. (2014); Basch et al. (2013); López et al. (2013); Shen et al. (2010)
Chitosan/starch in media	*E. coli* *Staphylococcus aureus* *Bacillus* sp. *Aspergillus* sp. *Penicillium* sp.	• Films displayed antimicrobial activities against gram-negative bacteria • Films formed hydrogen bonding which greatly improved physico-mechanical attributes of the film	Mlalila et al. (2018); Bie et al. (2013); Shariatinia and Fazli (2015); Shen et al. (2010); Tomé et al. (2012)
Essential oils (thymol, cinnamaldehyde, eugenol, cinnamon clove, and carvone) in various cheese and bread/media	*E. coli* *S. aureus* *Penicillium commune* *Eurotiumamstelodami*	• Changes in color of films were recorded, but they retained transparency • Films exhibited increased elongation at break and reduced water vapour transmission rate (WVTR) • Essential oils affected microorganisms' population by encapsulating cheese and bread	Ghasemlou et al. (2013); Mlalila et al. (2018); Kechichian et al. (2010)

Extracts (plant) in media	*E. coli* *S. aureus*	• Extracts acted as good plasticizers to impart stretchability to films • WVTR and Young's modulus were reduced	Mlalila et al. (2018)
Zinc oxide nanoparticles in media	*S. aureus*	• ZnO Nanoparticles improved theviscosity of the polymer matrix and reduced WVTR • Reduced solubility	Nafchi et al. (2012); Mlalila et al. (2018)
Blend films in media	*E. coli* *S. aureus* *Aspergillus* sp.	• Enhanced antimicrobial properties of the films were recorded due to citric acid	Priya et al. (2014); Mlalila et al. (2018)
Nisin in cheese/media	*L. innocua* *S. cerevisiae* *Z. bailii*	• Enhanced antimicrobial effect of nisin and lysozymes • Only superficial antimicrobial effects were recorded	Basch et al. (2013); Mlalila et al. (2018); OlléResa et al. (2014)
Lysozymes in media	*L. monocytogenes*	• Lysozymes' release was dependent on temperature and it was most effective at 10 °C	Bhatia and Bharti, 2015; Mlalila et al. (2018); Fabra et al. (2014)
Lauric acid in media	*B. subtilis* *E. coli*	• Lauric acid release followed Fickian diffusion • Lauric acid and chitosan were found effective against microorganisms	Salleh et al. (2014); Mlalila et al. (2018)
Natamycin in cheese/media	*S. cerevisiae* *L. innocua*	• Natamycin proved to be effective against microorganisms	Mlalila et al. (2018); OlléResa et al. (2014)
Gold nanoparticles in media	*E. coli* *S. aureus*	• The prepared films exhibited inhibition in the growth of *E. coli* by 99% and *S. aureus* by 98%	Pagno et al. (2015); Mlalila et al. (2018)
Catechins in beef/media	*S. cerevisiae* *E. coli*	• Preservation of beef was found effective at temperatures <6°C • Increased amount of catechins in starch-based films decremented the elongation at break and tensile strength	Wu et al. (2010)

Source: Adapted from Mlalila et al. (2018).

16.5.3 Effects of Nanofillers on the Mechanical Properties

Several scientists have investigated the effect of nanofillers on the mechanical properties of the starch films and found that its incorporation improves the tensile strength and elongation at break and reduce the water vapour transmission rate (WVTR) without increasing thickness of the starch films (Mohanty et al., 2012; Barzegar et al., 2014; Qin et al., 2016; Kumar et al., 2018). It can be noted that above 1% incorporation of the nanofillers starts decreasing the tensile strength. As per another study, a synthetic water-soluble polymer named caprolactone with starch nanocrystals and Laponite RD were added to the starch films to improve its mechanical properties. It increased the Youngs' modulus and tensile strength of the developed film, although it decreased the elongation at break. On comparing the starch nanocrystals and Laponite RD individually, the starch nanocrystals performed better over Laponite RD in the improvement of the mechanical properties. The improvement of the mechanical properties can be attributed to the hydrogen bonding interaction, smaller size, good dispersion of nanofillers, restriction of the mobility of polymer chains, and larger surface area (Naderizadeh et al., 2019). Another study indicated that the nano-SiO_2 incorporation improves the tensile strength of the starch film, and it increased the size of the nano-SiO_2 without increasing the thickness of the starch film (Zhang et al., 2018). A detailed overview of the method of fabrication, barrier, and mechanical properties of biopolymers and bio-nanocomposite films used in food packaging has been provided in Table 16.3.

16.6 NANOPARTICLE MIGRATION

The packaging properties of the starch-based materials may be improved with the incorporation of NPs, but the incorporation of compounds at the nanoscale level into these packaging materials may lead to potential migration to the food matrix. The migration of particles from food packaging to the food matrix not only depends on the size of the NP but can also be affected by the loading. These migrations of NPs have raised concerns among consumers about the effects derived from the ingestion of these nano-compounds. Another concern noticed while reviewing the different studies was a deposition, toxicity, and action dynamics of these NPs inside the human body, along with their metabolism and elimination mechanisms, besides the definition of regulatory issues. Some researches tried to address this issue but studies are still demanded on this issue. Fick's second law, about the mass transfer by a diffusion process, can explain the migration of low molecular NPs, initially present in the packaging, into the food matrix. Specific migration tests and overall migration tests for the determination of release of substances from the material or article either into food or food simulant are described by the European Commission and other global regulatory agencies. The food simulant (ethanol 10, 20, 50% v/v; acetic acid 3%; vegetable oil and poly (2,6-diphenyl-p-phenylene) oxide) is used to study migrated substances into simulant media after a contact period through quantification or identification by analytical techniques or instruments such as spectroscopy and chromatography. The packaging materials used in a food-contact article (primary packaging material) or food-processing equipment that may migrate certain its compositional

TABLE 16.3
Method of Fabrication, Barrier, and Mechanical Properties of Biopolymers and Bio-Nanocomposite Films

		Barrier Properties		Mechanical Properties	
Polymer Type	**Method of Fabrication**	**Water Vapour Transmission Rate (WVTR)***	**Gas Barrier(O_2 and Co_2)****	**Tensile Strength (MPa)*****	**Elongation at Break(%)*****
Starch (plant-based)	Solvent casting (aqueous)/Solution exfoliation	Bad	Average	Average	Bad
Chitosan (animal-based)	Solvent casting (aqueous)	Bad	Good	Average	Average
Gluten (wheat)	Solvent casting (ethanol)	Average	Good	Average	Average
Soy protein (soybean)	Solution exfoliation/solvent casting (aqueous)	Bad	Good	Bad	Average
Zein (corn)	Solvent casting (ethanol)	Average	Average	Average	Average
Whey protein (milk)	Solvent casting (aqueous)	Bad	Good	Bad	Bad
Polylactic acid	Melt intercalation/ thermocompression	Average	Average	Good	Average
Starch/polycaprolactone	Melt intercalation	Average	Average	Average	Good
Starch bio-nanocomposite	Solvent casting (aqueous)/melt intercalation	Good	Good	Average	Bad
Soy protein bio-nanocomposite	Melt intercalation	Good	Good	Average	Average
Polylactic acid bio-nanocomposite	Solvent casting (aqueous)	Excellent	Good	Excellent	Good

Source: Adapted from Kumar et al. (2017)

Note: The results shown are valid for a thickness of polymer thickness ranging from 50 to 70 μm.

* Test Conditions: 39°C, 90/0%RH, Bad = 10–100 g*mm/m²*d*kPa, Average = 0.1–10 g* mm/m²*d*kPa, Good = 0.01–0.1 g*mm/m²*d*kPa; Excellent = <0.01 g*mm/m²*d*kPa.

** Test conditions: 25°C, 0%–50%RH, Bad = 100–1000 cm³*μm/m²*d*kPa, Average = 10–100 cm³*μm/m²*d*kPa, Good = 1–10 cm³*μm/m²*d*kPa.

*** Test conditions: 25°C, 50%RH, Bad = 10–100 MPa, Average = 10%–50%, (LDPE: Tensile strength = 13 Mpa, Elongation at break = 500%).

components or those constituents may be expected to migrate into food from these primary packaging material or food-processing types of equipment will be exempted from regulation as a food additive because it becomes a component of food at levels below the threshold of regulation if the substance satisfies certain criteria relating to (a) its potential carcinogenicity, (b) the estimated dietary exposure to the substance (c) its lack of technical effect in or on the food to which it migrates, and (d) its lack of significant adverse impact on the environment (USFDA, 2014).

For the migration of NPs, there is not much research in literature and no regulations are established by any regulatory authority to date. Several studies have been published on the migration of NPs out of polymers. Most of them investigated silver NPs (AgNPs), which is used as an antimicrobial agent. AgNPs have been used by researchers and food industries in the manufacturing process of food packaging (Ramos et al., 2016) due to their antimicrobial properties against a broad range of microorganisms such as bacteria, fungi, and viruses. With the help of toxicological studies, it was established that AgNPs are entered into the human body and cause various health-related problems (Cruz et al., 2017). Experiments conducted on rats and mice have shown that exposure to AgNP can cause toxicity in several organs, including the liver and brain, and changes in the activity of the immune system. The addition of montmorillonite during the extrusion process of the starch-based packaging material can prevent plasticizer migration and evaporation (McGlashan and Halley, 2003) and thus leads to stability of packaging material during the long storage period. Other benefits of montmorillonite in packaging material are (a) improve the compatibility between starch and polycaprolactone (Ikeo et al., 2006), (b) prevent starch recrystallization in starch–polyvinyl alcohol (PVA) blends (Dean et al., 2011), and (c) reduce the rate of embrittlement during storage. Bott et al. (2014) studied the migration of AgNPs containing polyolefins and found that AgNPs migrate in the ionic silver form (Ag^+) in contact with acidic and aqueous food simulants but does not lead to any measurable migration of NPs in any of the official food simulants. With different studies available, to date, it was not possible to assure that the antimicrobial agent present in food packages tends to migrate to the food matrix because of methodological inconsistences, little knowledge of the authors about the different types of polymers, a lack of proper analytical techniques, a lack of validation of the analytical methods used, the physicochemical properties of NPs, and deficiencies in the characterization of different antimicrobial NPs.

16.7 CONCLUSION

Nanotechnology in starch-based packaging materials provides prospects for novel research developments that can advantage researchers, industry, and consumers. The application of NPs in biodegradable films exhibits substantial advantages in improving the properties of packaging materials, even in the primary stages, and requires continuous focus and investments for research and development for the proper understanding of the disadvantages and advantages of nanocomposites use in starch-based packaging materials. The use of nanotechnology to fabricate food packaging with the antimicrobial agent can give numerous benefits in the range of advanced functional properties. The methodology adopted for preparing nano-starch films depends on their target usage. The NPs can bring to starch-based packaging materials with improved processing, shelf life, transportability, health and packaging functionalities, and reduced costs, as well as environmentally friendly. The U.S. Food and Drug Administration (FDA) and other global regulatory agencies have described antimicrobial agents as chemical constituents whose presence or addition to foods, their packages, contact surfaces, or processing environments retard the growth of, or inactivate, spoilage or pathogenic microorganisms; however, these constituents in form of NPs are yet to describe. Food industries have to build consumer confidence in the acceptance of nano-food ingredients, such as antimicrobial nanotechnological films. On the other side, regulatory agencies such as the FDA should ensure the safety of these antimicrobial systems in the starch-based packaging material. The employment of these polymeric nanocomposites in these regards has bolstered significant revolution in industries related to food and bioprocessing.

REFERENCES

Abreu, A. S., Oliveira, M., Sá, A., Rodrigues, R. M., Cerqueira, M. A., Vicente, A. A., and Machado, A. V. (2015). Antimicrobial nanostructured starch based films for packaging. *Carbohydrate Polymers*, *129*, 127–134.

Alam, M. S., Kumar, N., and Singh, B. (2018). Development of Sweet Lime (*Citrus limetta Risso*) pomace integrated ricebased extruded product: Process optimization. *Journal of Agricultural Engineering*, *55*(1), 47–53.

Avella, M., De Vlieger, J. J., Errico, M. E., Fischer, S., Vacca, P., and Volpe, M. G. (2005). Biodegradable starch/clay nanocomposite films for food packaging applications. *Food Chemistry*, *93*(3), 467–474.

Bajpai, S., Chand, N., and Lodhi, R. (2020). Water sorption properties and antimicrobial action of zinc oxide nano particles loaded sago starch film. *Journal of Microbiology, Biotechnology and Food Sciences*, *9*(4), 2368–2387.

Barzegar, H., Azizi, M. H., Barzegar, M., and Hamidi-Esfahani, Z. (2014). Effect of potassium sorbate on antimicrobial and physical properties of starch–clay nanocomposite films. *Carbohydrate Polymers*, *110*, 26–31.

Basch, C., Jagus, R., and Flores, S. (2013). Physical and antimicrobial properties of tapioca starch-HPMC edible films incorporated with nisin and/or potassium sorbate. *Food and Bioprocess Technology*, *6*, 2419–2428.

Bhatia, S. and Bharti, A. (2015). Evaluating the antimicrobial activity of Nisin, Lysozyme and Ethylenediaminetetraacetate incorporated in starch based active food packaging film. *Journal of Food Science and Technology*, *52*(6), 3504–3512.

Bie, P., Liu, P., Yu, L., Li, X., Chen, L., and Xie, F. (2013). The properties of antimicrobial films derived from poly (lactic acid)/starch/chitosan blended matrix. *Carbohydrate polymers*, *98*(1), 959–966.

Bott, J., Störmer, A. and Franz, R., (2014). A comprehensive study into the migration potential of nano silver particles from food contact polyolefins. In: Benvenuto, M.A., Ahuja, S., Duncan, T.V., Noonan, G.O., and Roberts-Kirchhoff, E.S. (Eds.) *Chemistry of Food, Food Supplements, and Food Contact Materials: from Production to Plate* (pp. 51–70). American Chemical Society, USA.

Bumbudsanpharoke, N., Choi, J., and Ko, S. (2015). Applications of nanomaterials in food packaging. *Journal of Nanoscience and Nanotechnology*, *15*(9), 6357–6372.

Chhikara, N., Jaglan, S., Sindhu, N., Anshid, V., Charan, M. V. S., and Panghal, A (2018). Importance of traceability in food supply chain for brand protection and food safety systems implementation. *Annals of Biology*, *34*(2), 111–118.

Chhikara, N., Kushwaha, K., Jaglan, S., Sharma, P., and Panghal, A. (2019). Nutritional, physicochemical, and functional quality of beetroot (*Beta vulgaris* L.) incorporated Asian noodles. *Cereal Chemistry*, *96*(1), 154–161.

Cruz, G. G. D. L., Rodrıguez-Fragoso, P., Reyes-Esparza, J., Rodrıguez-Lopez, A., Gomez-Cansino, R., and Rodriguez-Fragoso, L. (2017). Interaction of nanoparticles with blood components and associated pathophysiological effects. In: Gomes, AFC, and Sarria, M.P. (Eds.), *Unraveling the Safety Profile of Nanoscale Particles and Materials—From Biomedical to Environmental Applications*. (pp. 37–59), IntechOpen, United Kingdom.

Da Silva, N. M. C., de Lima, F. F., Fialho, R. L. L., Albuquerque, E. C. D. M. C., Velasco, J. I., and Fakhouri, F. M. (2018). Production and characterization of starch nanoparticles. In: Huicochea, E.F., and Rendon, R. (Eds.), *Applications of Modified Starches*, (pp. 41–48), IntechOpen, United Kingdom.

De Souza Lima, M. M., and Borsali, R. (2004). Rodlike cellulose microcrystals: Structure, properties, and applications. *Macromolecular Rapid Communications*, *25*(7), 771–787.

Dean, K.M., Petinakis, E., Goodall, L., Miller, T., Yu, L., and Wright, N. (2011). Nanostabilization of thermally processed high amylose hydroxylpropylated starch films. *Carbohydrate Polymers*, *86*, 652–658.

Devgan, K., Kaur, P., Kumar, N., and Kaur, A. (2019). Active modified atmosphere packaging of yellow bell pepper for retention of physico-chemical quality attributes. *Journal of Food Science and Technology*, *56*(2), 878–888.

Fabra, M. J., Sánchez-González, L., and Chiralt, A. (2014). Lysozyme release from isolate pea protein and starch based films and their antimicrobial properties. *LWT-Food Science and Technology*, *55*(1), 22–26.

Ghasemlou, M., Aliheidari, N., Fahmi, R., Shojaee-Aliabadi, S., Keshavarz, B., Cran, M. J., et al. (2013). Physical, mechanical and barrier properties of corn starch films incorporated with plant essential oils. *Carbohydrate Polymers*, *98*, 1117–1126.

Ikeo, Y., Aoki, K., Kishi, H., Matsuda, S., and Murakami, A. (2006). Nano clay reinforced biodegradable plastics of PCL starch blends. *Polymers for Advanced Technologies*, *17*, 940–944.

Jbilou, F., Galland, S., Telliez, C., Akkari, Z., Roux, R., Oulahal, N., ..., and Degraeve, P. (2014). Influence of some formulation and process parameters on the stability of lysozyme incorporated in corn flour-or corn starch-based extruded materials prepared by melt blending processing. *Enzyme and Microbial Technology*, *67*, 40–46.

Kaur, K., Chhikara, N., Sharma, P., Garg, M. K., and Panghal, A. (2019). Coconut meal: Nutraceutical importance and food industry application. *Foods and Raw materials*, *7*(2) 419–427.

Kaur, S., Panghal, A., Garg, M. K., Mann, S., Khatkar, S. K., Sharma, P., and Chhikara, N. (2020). Functional and nutraceutical properties of pumpkin–a review. *Nutrition & Food Science*. *50*(2), 384–401.

Kechichian, V., Ditchfield, C., Veiga-Santos, P., and Tadini, C. C. (2010). Natural antimicrobial ingredients incorporated in biodegradable films based on cassava starch. *Lebensmittel-Wissenschaft und – Technologie – Food Science and Technology*, *43*, 1088–1094.

Kehinde, B.A., Chhikara, N., Sharma, P., Garg, M.K., and Panghal, A. (2020). Application of polymer nanocomposites in food and bioprocessing industries. In: Hussain, C.M. (Ed.), *Handbook of Polymer Nanocomposites for Industrial Applications*. Academic Press, United Sates of America (In Press).

Khatkar, B. S., Panghal, A., and Singh, U. (2009). Applications of cereal starches in food processing. *Indian Food Industry*, *28*(2), 37–44.

Kirandeep, D., Kaur, P., Singh, B., and Kumar, N. (2018). Effect of temperature and headspace O_2 and CO_2 concentration on the respiratory behaviour of fresh yellow bell-pepper (Oribelli). *International Journal of Chemical Studies*, *6*(2), 3214–3420.

Kringel, D. H., Dias, A. R. G., Zavareze, E. D. R., and Gandra, E. A. (2020). Fruit wastes as promising sources of starch: Extraction, properties, and applications. *Starch-Stärke*, *72*(3–4), 1900200.

Kumar, N. (2018). *Development of pectin-based bio nanocomposite films for food packaging* (Doctoral dissertation, Punjab Agricultural University, Ludhiana). http://krishikosh.egranth.ac.in/handle/1/5810047779

Kumar, N., Kaur, P., and Bhatia, S. (2017). Advances in bio-nanocomposite materials for food packaging: A review. *Nutrition & Food Science*, *47*(4), 591–606.

Kumar, N., Kaur, P., Devgan, K., and Attkan, A. K. (2020). Shelf life prolongation of cherry tomato using magnesium hydroxide reinforced bio-nanocomposite and conventional plastic films. *Journal of Food Processing and Preservation*, *44*(4), e14379.

Liao, H. T., and Wu, C. S. (2009). Preparation and characterization of ternary blends composed of polylactide, poly (ε-caprolactone) and starch. *Materials Science and Engineering: A*, *515*(1–2), 207–214.

López, O. V., Giannuzzi, L., Zaritzky, N. E., and García, M. A. (2013). Potassium sorbate controlled release from corn starch films. *Materials Science and Engineering: C*, *33*, 1583–1591.

Malathi, A. N., Kumar, N., Nidoni, U., and Hiregoudar, S. (2017). Development of soy protein isolate films reinforced with titanium dioxide nanoparticles. *International Journal of Agriculture, Environment and Biotechnology*, *10*(1), 141–148.

McGlashan, S.A. and Halley, P.J., 2003. Preparation and characterisation of biodegradable starch-based nanocomposite materials. *Polymer International*, *52*(11), 1767–1773.

Mlalila, N., Hilonga, A., Swai, H., Devlieghere, F., and Ragaert, P. (2018). Antimicrobial packaging based on starch, poly (3-hydroxybutyrate) and poly (lactic-co-glycolide) materials and application challenges. *Trends in Food Science & Technology*, *74*, 1–11

Mohanty, S., Mishra, S., Jena, P., Jacob, B., Sarkar, B., and Sonawane, A. (2012). An investigation on the antibacterial, cytotoxic, and antibiofilm efficacy of starch-stabilized silver nanoparticles. *Nanomedicine: Nanotechnology, Biology and Medicine*, *8*, 916–924.

Mwaurah, P. W., Kumar, S., Kumar, N., Attkan, A. K., Panghal, A., Singh, V. K., and Garg, M. K. (2020). Novel oil extraction technologies: Process conditions, quality parameters, and optimization. *Comprehensive Reviews in Food Science and Food Safety*, *19*(1), 3–20.

Naderizadeh, S., Shakeri, A., Mahdavi, H., Nikfarjam, N., and Taheri Qazvini, N. (2019). Hybrid nanocomposite films of starch, poly (vinyl alcohol)(PVA), starch nanocrystals (SNCs), and montmorillonite (Na-MMT): Structure–properties relationship. *Starch-Stärke*, *71*(1–2), 1800027.

Nafchi, A. M., Alias, A. K., Mahmud, S., and Robal, M. (2012). Antimicrobial, rheological, and physicochemical properties of sago starch films filled with nanorod-rich zinc oxide. *Journal of Food Engineering*, *113*, 511–519.

Ollé Resa, C. P., Gerschenson, L. N., and Jagus, R. J. (2014). Natamycin and nisin supported on starch edible films for controlling mixed culture growth on model systems and Port Salut cheese. *Food Control*, *44*, 146–151.

Ortega-Toro, R., Morey, I., Talens, P., and Chiralt, A. (2015). Active bilayer films of thermoplastic starch and polycaprolactone obtained by compression molding. *Carbohydrate Polymers*, *127*, 282–290.

Pagno, C. H., Costa, T. M. H., De Menezes, E. W., Benvenutti, E. V., Hertz, P. F., Matte, C., Tosati, J. V., Monteiro, A. R., Rios, A. O. and Flores, S. H. (2015). Development of active biofilms of quinoa (*Chenopodium quinoa* W.) starch containing gold nanoparticles and evaluation of antimicrobial activity. *Food Chemistry*, *173*, 755–762.

Panghal, A., Chhikara, N., Anshid, V., Charan, M. V. S., Surendran, V., Malik, A. and Dhull, S. B. (2019a). Nanoemulsions: A promising tool for dairy sector. In: Prasad, R., Kumar, V., Kumar, M., and Chaudhary, D. (Eds.), *Nanobiotechnology in Bioformulations* (pp. 99–117). Cham: Springer.

Panghal, A., Jaglan, S., Sindhu, N., Anshid, V., Charan, M. V. S., Surendran, V., and Chhikara, N. (2019b). Microencapsulation for delivery of probiotic bacteria. In: Prasad, R., Kumar, V., Kumar, M., and Chaudhary, D. (Eds.), *Nanobiotechnology in Bioformulations* (pp. 135–160). Cham: Springer.

Priya, B., Gupta, V. K., Pathania, D., and Singha, A. S. (2014). Synthesis, characterization and antibacterial activity of biodegradable starch/PVA composite films reinforced with cellulosic fibre. *Carbohydrate Polymers*, *109*, 171–179.

Qin, Y., Zhang, S., Yu, J., Yang, J., Xiong, L., and Sun, Q. (2016). Effects of chitin nanowhiskers on the antibacterial and physicochemical properties of maize starch films. *Carbohydrate Polymers*, *147*, 372–378.

Ramos, K., Gomez-Gomez, M. M., Camara, C., and Ramos, L. (2016). Silver speciation and characterization of nanoparticles released from plastic food containers by single particle ICPMS. *Talanta*, *151*, 83–90.

Rhim, J. W., Mohanty, A. K., Singh, S. P., and Ng, P. K. (2006). Effect of the processing methods on the performance of polylactide films: Thermocompression versus solvent casting. *Journal of Applied Polymer Science*, *101*(6), 3736–3742.

Salleh, E., Muhammad, I. I., and Pahlawi, Q. A. (2014). Spectrum activity and lauric acid release behaviour of antimicrobial starch-based film. *Procedia Chemistry*, *9*, 11–22.

Sandri, D., Rinaldi, M. M., Ishizawa, T. A., Cunha, A. H., Pacco, H. C., and Ferreira, R. B. (2015). 'Sweet grape'tomato post harvest packaging. *Engenharia Agrícola*, *35*(6), 1093–1104.

Shankar, S., Reddy, J. P., Rhim, J. W., and Kim, H. Y. (2015). Preparation, characterization, and antimicrobial activity of chitin nanofibrils reinforced carrageenan nanocomposite films. *Carbohydrate Polymers*, *117*, 468–475.

Shankar, S., and Rhim, J. W. (2016). Polymer nanocomposites for food packaging applications. In: Desari, A. and Njuguna, J. (Eds.), *Functional and Physical Properties of Polymer Nanocomposites*, (pp. 29–55), Wiley & Sons Ltd., USA.

Shapi'i, R. A., Othman, S. H., Nordin, N., Basha, R. K., and Naim, M. N., 2020. Antimicrobial properties of starch films incorporated with chitosan nanoparticles: In vitro and in vivo evaluation. *Carbohydrate Polymers*, *230*, 115602.

Shariatinia, Z., and Fazli, M. (2015). Mechanical properties and antibacterial activities of novel nanobiocomposite films of chitosan and starch. *Food Hydrocolloids*, *46*, 112–124.

Sharma, P., Panghal, A., Gaikwad, V., Jadhav, S., Bagal, A., Jadhav, A., and Chhikara, N., 2019. Nanotechnology: A boon for food safety and food defense. In: Prasad, R., Kumar, V., Kumar, M., and Chaudhary, D. (Eds.), *Nanobiotechnology in Bioformulations* (pp. 225–242). Cham: Springer.

Shen, X. L., Wu, J. M., Chen, Y., and Zhao, G. (2010). Antimicrobial and physical properties of sweet potato starch films incorporated with potassium sorbate or chitosan. *Food Hydrocolloids*, *24*, 285–290.

Singh, V. K., Garg, M. K., Kalra, A., Bhardwaj, S., Kumar, R., Kumar, S., Kumar, N., Attkan, A. K., Panghal, A., and Kumar, D. (2020). Efficacy of microwave heating parameters on physical properties of extracted oil from turmeric (*Curcuma longa* L.). *Current Journal of Applied Science and Technology*, *39*(25), 126–136.

Sinha Ray, S., Yamada, K., Okamoto, M., and Ueda, K. (2003). Biodegradable polylactide/montmorillonite nanocomposites. *Journal of Nanoscience and Nanotechnology*, *3*(6), 503–510.

Tomé, L. C., Fernandes, S. C., Sadocco, P., Causio, J., Silvestre, A. J., Neto, C. P., and Freire, C. S. (2012). Antibacterial thermoplastic starch-chitosan based materials prepared by melt-mixing. *BioResources*, *7*(3), 3398–3409.

USFDA, 2014. United States food and drug administration. *CFR-Code of federal regulations title 21*. Maryland. Available from: http://www.fda.gov/MedicalDevices/DeviceRegulationandGuidance/Databases/ucm135680.htm. Accessed 2020 May 25.

Vasile, C., 2018. Polymeric nanocomposites and nanocoatings for food packaging: A review. *Materials*, *11*(10), 1834.

Vilarinho, F., Sanches Silva, A., Vaz, M. F., and Farinha, J. P., (2018). Nanocellulose in green food packaging. *Critical Reviews in Food Science and Nutrition*, *58*(9), 1526–1537.

Wu, J. G., Wang, P. J., and Chen, S. C. (2010). Antioxidant and antimicrobial effectiveness of catechin-impregnated PVA–starch film on red meat. *Journal of Food Quality*, *33*, 780–801.

Yoksan, R., and Chirachanchai, S. (2010). Silver nanoparticle-loaded chitosan–starch based films: Fabrication and evaluation of tensile, barrier and antimicrobial properties. *Materials Science and Engineering: C*, *30*, 891–897.

Zhang, R., Wang, X., and Cheng, M. (2018). Preparation and characterization of potato starch film with various size of nano-SiO_2. *Polymers*, *10*(10), 1172.

17 Biogenic Metal Nanoparticles and Their Antimicrobial Properties

Subramani Srinivasan
Annamalai University, India
Government Arts College for Women, India
Vinayagam Ramachandran
Thiruvalluvar University, India
Raju Murali
Annamalai University, India
Government Arts College for Women, India
Veerasamy Vinothkumar
Annamalai University, India
Devarajan Raajasubramanian
Annamalai University, India
Thiru. A. Govindasamy Government Arts College, India
Ambothi Kanagalakshimi
Annamalai University, India
Government Arts College for Women, India

CONTENTS

17.1 INTRODUCTION

In the 21st century, nanotechnology has become a more popularized technology that provides a range of nanoscale creatures with ultra-level optical, electronic, and catalytic properties. Strange physicochemical properties of nano-level materials have been introduced in many modern manufacturing industries which upgrades the fields of medicine, pharmacy, food, and biotechnology, as well. In reality, nanotechnology has provided a solution for many incurable problems with respect to preventing, diagnosing, and treating the diseases which may bring out impressive benefits to the healthcare staff, patients suffering from diseases, and the general society (Leso et al., 2019). Nanomaterials (NMs) have gained great importance in technological improvements as they possess effective physiochemical and biological properties with upregulated activity compared to their bulk analogs (Chawla et al., 2019, 2020). NMs are classified based on their size, content, morphology, and source. The capability to recognize the individual characteristics of NMs improves all its classification. As the increased level of manufacturing of NMs and its industrial usage, toxicity issues are irresistible. Nanoparticles (NPs) and nano-level structured materials indicate an active field of research and techno-economical division with the whole elaboration in several application sectors (Dhull et al., 2019; Panghal et al., 2019). NPs and nano-level structured materials have achieved some eminent technological developments due to the properties such as electrical, thermal, optical, conductivity, absorption, and scattering which offer better activity over its bulk analogs (Jeevanandam et al., 2018).

Recent developments in nanotechnology have got into the advancements of many types of nano-formulation for biomedical usages of diagnosis and therapeutics. Nowadays, many types of metallic NMs are in use such as zinc, magnesium, titanium, gold, and silver. NPs are used for a wide contemplates from medical treatments to various divisions of manufacturing industries like energy-conserving solar and oxide fuel batteries to a diverse integration of a range of materials of day to day usages (Singh et al., 2019). NPs used in diagnostics focus on viewing the pathological diversity and building a grasp of the extension of the fundamentals of the pathophysiology of many diseases and their treatments. In the clinical aspects, nano-diagnostics is in use for few cases as it requires the pharmacokinetics and ejection. Accordingly, the nanoformulations are now utilized in medical practices are tested for therapeutic schemes. Therapeutic NPs focus on the development of releasing aggregation of the pharmacologically active factors at the defect area, which improves therapy and decreases the occurrence and depth of side effects, thereby decreasing the unspecific action in normal tissues (Abdifetah and Na-Bangchang, 2019).

The effect of NPs triggers the performance in vivo. In particular, the morphological factors like shape and size help the NPs circulate within the body to get to their targets. These effects are accounted for the divergence in the destruction rate of NPs and the release of drug kinetics. The morphology of the NP also accounts for the targeted cell signals. The surface ligands of NPs also have apart in interacting with systemic biology. NPs are synthesized for many uses, including the delivery of drugs, bio-components, genes for biomedical applications like vaccination and cancer therapy, and so on (Chaudhary et al., 2019). NPs can be administered to patients through various routes, including oral, parenteral, nasal, and intraocular, which offers an effective delivery system in the research and development of biomedicines.

17.2 METAL NANOPARTICLES

Since ancient times, metals have had an important part in the lives of humans soon after the stone age began. Some metals, like copper, gold, zinc, and silver, are known to be efficient in fighting many infectious diseases. The ability to interact with metal NPs within the human physiology has a strong impact. The capacity to incorporate metal NPs into natural frameworks has had a tremendous effect on science and prescription. Some respectable metal NPs have been drawing in immense enthusiasm from mainstream researchers attributable to their marvellous properties and assorted variety of utilizations (Amini and Akbari, 2019). The customary utilization of metals uses, for the most part, the mass metallic properties, for example, quality, hardness, flexibility, high dissolving point, and warm and electrical conductivities. It has been conceived that metals and metal NMs will experience significant changes in numerous circles of our lives, science, innovation, and industry. This is because metal NPs display numerous properties that are valuable or that can be controlled for various new logical investigations and a large number of innovative applications.

17.3 METALLIC NANOPARTICLES

Metallic NPs are significant because of their potential applications in the fields of, for example, catalysis plasmonics, tranquilize conveyance, attractive recollections, biomedical imaging, and DNA discovery because of their low thickness and a powerful contact zone (Hakimian and Ghourchian, 2019). Metallic NPs have a groundbreaking ingestion range in unmistakable range, which is because of reasonable electron motions outside of particles, called surface plasmon reverberation (SPR). The SPR range of metallic NPs has various applications in the field of biotechnology and has drawn tremendous fascination as of late. Metallic NPs have captivated researchers for over a century and are presently vigorously used in biomedical sciences and designing. They are a focal point of intrigue on account of their colossal potential in nanotechnology. Today, these materials can be orchestrated and altered with different substance utilitarian gatherings which enable them to be conjugated with antibodies, ligands, and medications of intrigue and, in this way, open a wide scope of potential applications in biotechnology, attractive partition, and pre-grouping of target analytes; directed medication conveyance; vehicles for quality and medication conveyance; and, all the more significant, indicative imaging.

With the improvement of nanotechnology as of late, NMs have drawn a lot of consideration and have been the hotly debated issue because these materials have numerous wonderful properties contrasted and mass materials. Critically, an enormous number of specialists are progressively investigating the metal NPs, inferable from their great electrical properties, optical properties, and attraction, among others; metal NPs have been broadly applied in numerous regions of condition, material, science, and, particularly, medication. Also, metal NPs have been the subject of concentrated research over the world because of the distinctions in the optical properties they display in contrast with mass metals, which defeat the weaknesses of mass metals that do not have extraordinary versatility, homogeneity, and organic similarity (Zahran and Marei, 2019). Among every one of the NPs, metal NPs attracted extraordinary enthusiasm for biomedical applications. Particularly, copper, gold, silver, zinc NPs are of particular interest.

17.4 BIOGENIC METAL NANOPARTICLES

NPs synthesized using biological systems are referred to as biogenic NPs, and they are beneficial in the fact that they are water-soluble and biocompatible (Ingale and Chaudhari, 2013). The process of synthesis of the nanostructures by organisms is also known as biomineralization. Moreover, biogenic NPs are advantageous in having better stability owing to stabilization, biocompatible, and water-solubility. Furthermore, organically combined NPs have been demonstrated to be more proficient than the generally blended NPs. Regardless of the way that biogenic metallic NPs are biocompatible with high solidness and amiable for biomedical applications, a harmony between value, procedure, and adaptability is yet another impressive test. Even though diverse biogenic sources have been recognized for a combination of NPs, the system of amalgamation has yet to be unmistakably deciphered.

17.5 ANTIMICROBIAL EFFECTS OF BIOGENIC METAL NANOPARTICLES

Metal and metallic particles have been notable for the antimicrobial system for quite a long time, since 4000 BCE. Metallic NPs, when contrasted with their salts, can battle bacterial diseases. For the most part, the size of NPs has an impact on the antibacterial system (Nisar et al., 2019). Metals, for example copper, gold, silver, and zinc, and their different aspects have been utilized to treat and avoid a wide scope of irresistible infection conditions for quite a while. This chapter pointedly focuses on the recent scientific research pursuits in biogenic metal NPs, especially on the vast domain of antimicrobial efficacy and its vision for the future. The illnesses brought about by pathogenic bacterial contaminations have been influencing human well-being around the world, prompting higher dreariness and mortality. Ailments from bacterial diseases are serious medical issues, causing several passings every year. The advancement of anti-toxin obstruction has likewise become another test for the treatment of irresistible maladies. Research in the territory of antimicrobial movement has been recharged with gigantic enthusiasm because of the development of numerous medication-safe microscopic organisms. Scientists are attempting to create proficient antimicrobial medications that are proficient and can be utilized against numerous microorganisms (Shaaban and El-Mahdy, 2018). Critical biogenic metal NPs offer another helpful technique to treat pathogenic microbes. This chapter pointedly focuses on the recent scientific research pursuits in biogenic metal NPs, especially in the vast domain of antimicrobial efficacy and its vision for the future.

The metallic NMs as bacteriostatic agents have recently become a focus of intense research due to their steady and efficient antibacterial activity. Moreover, it is very important to study the antibacterial mechanisms of metallic NMs. A variety of mechanisms for the antibacterial activity of metallic NPs have been proposed, including the reactive oxygen species (ROS) production, mechanical damage to the cell membrane, entrapment, metal ion release, and so on (Hemeg, 2017). Conversely, the accurate mechanisms of these NPs as bactericide are not so far completely understood which could limit the

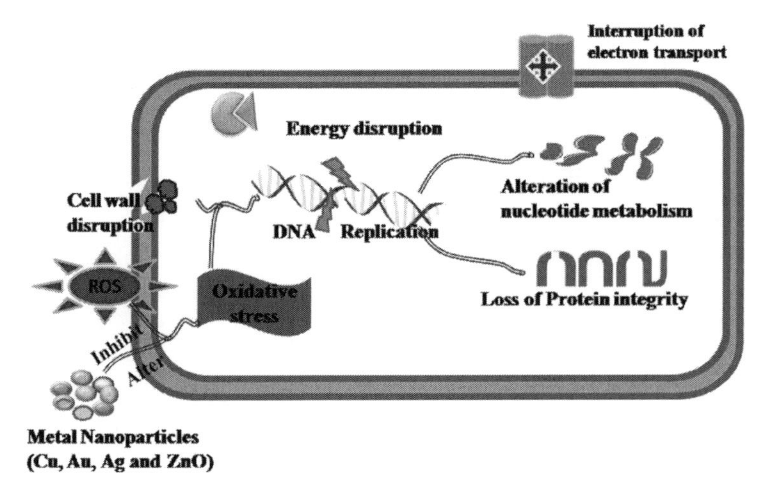

FIGURE 17.1 Mechanism of antimicrobial efficacy of nanoparticles.

application of NPs in the field of the antimicrobial efficacy. Therefore, the generation of metal NP with more efficient antimicrobial potentials is still expected (Figure 17.1).

17.6 COPPER NANOPARTICLES

There is substantial evidence from the ancient Egyptians in approximately 2500 BC that corroborates the relevance of copper in treating infections (Lotha et al., 2019). The antimicrobial properties of copper have been long documented and have profited human civilization for centuries. Copper nanoparticles (CuNPs) can dissolve quicker in a particular solution compared with huge particles, thereby releasing a well-built quantity of metal ions and presenting stronger antimicrobial effects, and CuNPs have been hypothetically and practically established for their antimicrobial activities (Toodehzaeim et al., 2018). CuNPs can stick to the bacterial cell wall, copper ions from these NPs obliterate the bacterial cell wall, and the cytoplasm is then corrupted and vanishes, resulting in cell death (Amiri et al., 2017). Furthermore, Mallick et al. (2012), hypothesized that CuNPs accumulate in the bacterial cell membrane, changing its permeability, and augmented permeability leads to the progressive release of intracellular materials and eventually to cell death. Gutierrez et al. (2018) put forward a comparable proposal, attributing CuNP antimicrobial activity to the improved release of Cu^{2+} ions in an aqueous medium. Tambosi et al. (2019) indicated that the antimicrobial properties of CuNPs may not be simply caused by augmented liberated copper ions, but they may also be a result of a straight contact of microorganisms with CuNPs. Recently, Deng et al. (2017) proved higher antibacterial activity against *Escherichia coli*, *Bacillus subtilis*, and *Staphylococcus aureus*. Furthermore, Rajeshkumar et al. (2019) reported that CuNPs have higher antimicrobial activity against human pathogens. Besides, a variety of biomedical applications of biosynthesized CuNPs against infectious microorganisms and phytopathogenic fungi were also examined (Ingle and Rai, 2017) that symbolize the promising uses of CuNPs being applied as a novel approach to administering these health-threatening problems.

17.7 GOLD NANOPARTICLES

The use of gold nanoparticles (AuNPs) possibly dates back to the 4th or 5th century BCE in Egypt and China for decoration and medical treatment purposes. In many fields, AuNPs have been utilized, namely as catalysts and antimicrobials to kill pathogenic microbes. It is considered that the AuNPs have exhibited effective near-infrared (NIR) adsorption and elevated photothermal transition, permitting the upregulated photo killing properties inhibits the growth of gram-positive bacteria. A recent report stated that AuNPs, other than *E. coli*, AgNPs showed a broad spectrum of activity for most of the pathogenic bacteria (Arya et al., 2019). It was also noted that the antimicrobial properties of AuNPs mediated with ethanolic extracts of *Ricinus communis* leaves (RcExt) showed a stronger resistance with a higher diameter ranging from 9.33 to 16.33 mm. It is also a fact that in much of the research where the extracts of *Euphorbia peplus* leaves were mediated, AuNPs exhibited as prominent antimicrobial agents (Ghramh et al., 2019). It has been proved that synthesized AuNPs using *Tragopogondubuis*, tested in an animal model affected with a burn wound and infected with *S. aureus*, proved to be effective wound-healing and antibacterial agents equivalent to tetracycline (Layeghi-Ghalehsoukhteh et al., 2018). Moreover, lipoic acid–mediated AuNPs exhibited a considerably good antibacterial activity (Guler et al., 2014). These findings also showcased chitosan-coated AuNps as a potentially effective antibacterial against *Pseudomonas aeruginosa*, *E. coli, S. aureus*, and *Bacillus sp.* (Sharma et al., 2019). Surprisingly, *Chenopodium formosanum* shell-AuNPs elicits a growth inhibitory effect on *E. coli* and *S. aureus* (Chen et al., 2019).

Jabir et al. (2018) revealed that linalool alone has shown low antibacterial efficacy in both gram-positive and -negative bacteria whereas AuNPs synthesized using linalool have shown to be effective against gram-positive bacteria. It could be of the interaction between the bacterial cell wall and led to the decreased integrity and increased permeability of the membranes, thereby inducing the release of ROS, causing DNA damage (Jabir et al., 2018). Finally, the analysis of antibacterial efficacy of AuNPs was studied by using the disc diffusion method; the minimum bactericidal concentration of gram-positive bacteria *Bacillus* and the minimum inhibitory concentration of gram-negative *E. coli* were taken (Lee and Lee, 2019).

17.8 SILVER NANOPARTICLES

Silver is the renowned antimicrobial agent since ancient times and it has been used as an efficient health additive in Indian Ayurveda medicine (Galib et al., 2011). Noteworthy is that, in India, silver vessels and formulations have also been assumed for centuries to display antimicrobial and therapeutic properties. These properties make silver a worthy choice for synthesizing silver NPs (AgNPs) for biomedical applications. Currently, AgNPs have proved proficient in antibacterial action due to its physicochemical property consequence into better antimicrobial efficacy. The bactericidal activity of AgNPs has been accredited to the collective action of numerous mechanisms, such as the production of ROS, induced gene modification, cell wall penetration, and damage (Durán et al., 2016). The exposition of bacterial cells to silver ions encourages alterations in the cell membrane structural constituents, leading to augmentations in its permeability and damage (Franci et al., 2015). It can influence the electrolytes and other metabolite transports, leading to the modification of cell basic functions and cell death. Haroon and his collaborators (2019) suggested that AgNPs have a multiplicity of applications in the diverse field and would be used as a prospective microbial agent to destroy the bacterial activity. Song and a research group (2018) found that polymeric silver nanocomposites capped with natural polymers have superior antimicrobial activities against yeast, bacteria, and fungi. Intriguingly, the antibacterial effect of starch-based AgNPs against human pathogens has been studied, and AgNPs were effectual against both types of bacteria (Lomeli-Marroquín et al., 2019). Fascinatingly, Noh et al. (2013) presented that chlorogenic acid–coated AgNPs are more toxic for gram-negative bacteria rather than gram-positive bacteria. Dehkordi et al. (2019) indicated that eugenol-coated AgNPs enhanced the antibacterial efficacy of both gram-negative and gram-positive bacteria in food ingredients. Hence, Song et al. (2019) exhibited curcumin-modified AgNPs presents compelling antibacterial activity against both gram-negative and gram-positive bacteria. Moreover, Liao and his research group (2019) found that AgNPs prepared by using *Pseudomonas aeruginosa* extract have extremely powerful antibacterial activity on human pathogens.

17.9 ZINC NANOPARTICLES

Zinc is indispensable micronutrients necessary for almost all living organisms, including humans. Sufficient substantiation is obtainable in the literature to specify the inhibitory efficacy of zinc against bacteria (Almoudi et al., 2018). There are several types of zinc NPs, such as zinc oxide (ZnO) and zinc sulfide (ZnS). Plausibly, the most extensive type of zinc NPs is ZnO. A substantial body of literature has specified that zinc oxide nanoparticles (ZnONPs) exhibit significant antimicrobial efficacy over a broad spectrum of bacterial species (Yusof et al., 2019). The antimicrobial effect of ZnONPs may be the promising interaction between NPs and bacteria. ZnONPs can agitate bacterial development either by interacting with the bacterial surface or penetrating in the bacterial cells. The anticipated antimicrobial mechanisms of ZnONPs could be due to the leaching of Zn^{2+} into the growth media and the intervention of Zn^{2+} ions with the enzyme organizations of the bacteria by dislocating magnesium ions which are critical for bacterial enzymatic actions (Auger et al., 2019). Zanet and co-workers (2019) hypothesized that the antibacterial efficiency of ZnONPs is mainly because of the release of Zn^{2+} ions. Padalia and Chanda (2017) also emphasized the antimicrobial activity of ZnONPs and proved that the release of Zn^{2+} ions was the cause of ZnO toxicity against yeast. Noteworthy is that researchers found that another mechanism of the antimicrobial action of ZnONPs is based on the formation of ROS which disrupts the integrity of the bacterial membrane (Wang et al., 2017). We can find in the investigation by Kadiyala et al. (2018) that the antimicrobial efficacy of ZnONPs against *S. aureus* bacteria is through the synthesis of ROS. Similarly, Alekish et al. (2018) also illustrated the antibacterial efficacy of

ZnONPs against *S. aureus* and the synthesis of intracellular ROS were the key rationale for bacterial cell death. In this milieu, scientists engrossed in nanomedicine projects based on zinc NPs should focus more on placing their materials in clinical trials so that novel zinc nanomedicines can evolve quickly.

17.10 CONCLUSION

Novel improvements in material science and nanotechnology have authorized NMs for the betterment of human health. In this chapter, we have discussed metal NPs and extensively reported on the antimicrobial activity mechanism. In the not-too-long research narration, particularly in the current decade, we have detected explosive improvement of this field. We covered the most extensively used stratagems and mechanistic particulars of copper, gold, silver, and zinc NPs and their antibacterial effect. As a consequence, this chapter encourages research in the captivating and very constructive area of metal NPs and their antimicrobial efficacy. We consider that this streamlined explanation of research will gain much importance in nanotechnology and material science.

REFERENCES

Abdifetah, O., and Na-Bangchang, K. (2019). Pharmacokinetic studies of nanoparticles as a delivery system for conventional drugs and herb-derived compounds for cancer therapy: A systematic review. *International Journal of Nanomedicine*, *14*, 5659–5677.

Alekish, M., Ismail, Z. B., Albiss, B., and Nawasrah, S. (2018). In vitro antibacterial effects of zinc oxide nanoparticles on multiple drug-resistant strains of Staphylococcus aureus and Escherichia coli: An alternative approach for antibacterial therapy of mastitis in sheep. *Veterinary World*, *11*, 1428–1432.

Almoudi, M. M., Hussein, A. S., Abu Hassan, M. I., and Mohamad Zain, N. (2018). A systematic review on antibacterial activity of zinc against *Streptococcus mutans*. *Saudi Dental Journal*, *30*, 283–291.

Amini, S. M., and Akbari, A. (2019). Metal nanoparticles synthesis through natural phenolic acids. *IET Nanobiotechnology*, *13*, 771–777.

Amiri, M., Etemadifar, Z., Daneshkazemi, A., and Nateghi, M. (2017). Antimicrobial effect of copper oxide nanoparticles on some oral bacteria and candida species. *Journal of Dental Biomaterials*, *4*, 347–352.

Arya, S. S., Sharma, M. M., Das, R. K., Rookes, J., Cahill, D., and Lenka, S. K. (2019). Vanillin mediated green synthesis and application of gold nanoparticles for reversal of antimicrobial resistance in *Pseudomonas aeruginosa* clinical isolates. *Heliyon*, *5*, 02021.

Auger, S., Henry, C., Péchaux, C., Lejal, N., Zanet, V., Nikolic, M. V., Manzano, M., and Vidic, J. (2019). Exploring the impact of Mg-doped ZnO nanoparticles on a model soil microorganism Bacillus subtilis. *Ecotoxicology and Environmental Safety*, *182*, 109421.

Chaudhary, Z., Subramaniam, S., Khan, G. M., et al. (2019). Encapsulation and controlled release of resveratrol within functionalized mesoporous silica nanoparticles for prostate cancer therapy. *Frontiers in Bioengineering and Biotechnology*, *7*, 225.

Chawla, P., Kumar, N., Bains, A., Dhull, S. B., Kumar, M., Kaushik, R., and Punia, S. (2020). Gum arabic capped copper nanoparticles: Synthesis, characterization, and applications. *International Journal of Biological Macromolecules*, *146*, 232–242.

Chawla, P., Kumar, N., Kaushik, R., and Dhull, S. B. (2019). Synthesis, characterization and cellular mineral absorption of nanoemulsions of Rhododendron arboreum flower extracts stabilized with gum arabic. *Journal of Food Science and Technology*, *56*(12), 5194–5203.

Chen, M. N., Chan, C. F., Huang, S. L., and Lin, Y. S. (2019). Green biosynthesis of gold nanoparticles using *Chenopodium formosanum* shell extract and analysis of the particles antibacterial properties. *Journal of the Science of Food and Agriculture*, *99*, 3693–3702.

Dehkordi, N. H., Tajik, H., Moradi, M., Kousheh, S. A., and Molaei, R. (2019). Antibacterial interactions of colloid nanosilver with eugenol and food ingredients. *Journal of Food Protection*, *82*, 1783–1792.

Deng, C. H., Gong, J. L., Zeng, G. M., et al. (2017). Graphene sponge decorated with copper nanoparticles as a novel bactericidal filter for inactivation of Escherichia coli. *Chemosphere*, *184*, 347–357.

Dhull, S. B., Anju, M., Punia, S., Kaushik, R., and Chawla, P. (2019). Application of gum arabic in nanoemulsion for safe conveyance of bioactive components. In *Nanobiotechnology in Bioformulations*, (pp. 85–98). Cham, Springer.

Durán, N., Durán, M., de Jesus, M. B., Seabra, A. B, Fávaro, W. J., and Nakazato, G. (2016). Silver nanoparticles: A new view on mechanistic aspects on antimicrobial activity. *Nanomedicine*, *12*, 789–799.

Franci, G., Falanga, A., Galdiero, S., Palomba, L., Rai, M., Morelli, G., and Galdiero, M. (2015). Silver nanoparticles as potential antibacterial agents. *Molecules*, *20*, 8856–8874.

Galib Barve, M., Mashru, M., Jagtap, C., Patgiri, B. J., and Prajapati, P. K. (2011). Therapeutic potentials of metals in ancient India: A review through Charaka Samhita. *Journal of Ayurveda and Integrative Medicine*, *2*, 55–63.

Ghramh, H. A., Khan, K. A., and Ibrahim, E. H. (2019). Biological activities of *Euphorbia peplus* leaves ethanolic extract and the extract fabricated gold nanoparticles (AuNPs). *Molecules*, *24*, 1431.

Guler, E., Barlas, F. B., Yavuz, M., et al. (2014). Bio-active nanoemulsions enriched with gold nanoparticle, marigold extracts and lipoic acid: In vitro investigations. *Colloids and Surfaces B: Biointerfaces*, *121*, 299–306.

Gutierrez, M. F., Malaquias, P., Matos, T. P., et al. (2018). Mechanical and microbiological properties and drug release modeling of an etch-and-rinse adhesive containing copper nanoparticles. *Journal of Dentistry*, *19*, 1–5.

Hakimian, F., and Ghourchian, H. (2019). Simple and rapid method for synthesis of porous gold nanoparticles and its application in improving DNA loading capacity. *Materials Science and Engineering: C Materials for Biological Applications*, *103*, 109795.

Haroon, M., Zaidi, A., Ahmed, B., Rizvi, A., Khan, M. S., and Musarrat, J., (2019). Effective inhibition of phytopathogenic microbes by eco-friendly leaf extract mediated silver nanoparticles (AgNPs). *Indian Journal of Microbiology*, *59*, 273–287.

Hemeg, H. A. (2017). Nanomaterials for alternative antibacterial therapy. *International Journal of Nanomedicine*, *12*, 8211–8225.

Ingale, A. G., and Chaudhari, A. N. (2013). Biogenic synthesis of nanoparticles and potential applications: An eco-friendly approach. *Journal of Nanomedicine and Nanotechnology*, *4*, 1–7.

Jabir, M. S., Taha, A. A., and Sahib, U. I. (2018). Linalool loaded on glutathione-modified gold nanoparticles: A drug delivery system for a successful antimicrobial therapy. *Artificial Cells, Nanomedicine, and Biotechnology*, *46*, 345–355.

Jeevanandam, J., Barhoum, A., Chan, Y. S., Dufresne, A., and Danquah, M. K. (2018). Review on nanoparticles and nanostructured materials: History, sources, toxicity and regulations. *Beilstein Journal of Nanotechnology*, *9*, 1050–1074.

Kadiyala, U., Turali-Emre, E.S., Bahng, J. H., Kotov, N. A., and VanEpps, J. S. (2018). Unexpected insights into antibacterial activity of zinc oxide nanoparticles against methicillin resistant Staphylococcus aureus (MRSA). *Nanoscale*, *10*, 4927–4939.

Layeghi-Ghalehsoukhteh, S., Jalaei, J., Fazeli, M., Memarian, P., and Shekarforoush, S. S. (2018). Evaluation of 'green' synthesis and biological activity of gold nanoparticles using *Tragopogon dubius* leaf extract as an antibacterial agent. *IET Nanobiotechnology*, *12*, 1118–1124.

Lee, B., and Lee, D. G. (2019). Synergistic antibacterial activity of gold nanoparticles caused by apoptosis-like death. *Journal of Applied Microbiology*, *127*, 701–712.

Leso, V., Fontana, L., and Iavicoli, I. (2019). Biomedical nanotechnology: Occupational views. *Nano Today*, *24*, 10–14.

Liao, S., Zhang, Y., Pan, X., Zhu, F., Jiang, C., Liu, Q., Cheng, Z., Dai, G., Wu, G., Wang, L., and Chen, L. (2019). Antibacterial activity and mechanism of silver nanoparticles against multidrug-resistant *Pseudomonas aeruginosa*. *International Journal of Nanomedicine*, *14*, 1469–1487.

Lomelí-Marroquín, D., Medina Cruz, D., Nieto-Argüello, A., Vernet Crua, A., Chen, J., Torres-Castro, A., Webster, T. J., and Cholula-Díaz, J. L. (2019). Starch-mediated synthesis of mono- and bimetallic silver/gold nanoparticles as antimicrobial and anticancer agents. *International Journal of Nanomedicine*, *14*, 2171–2190.

Lotha, R., Shamprasad, B. R., Sundaramoorthy, N. S., Nagarajan, S., and Sivasubramanian, A. (2019). Biogenic phytochemicals (cassinopin and isoquercetin) capped copper nanoparticles (ISQ/CAS@CuNPs) inhibits MRSA biofilms. *Microbial Pathogenesis*, *132*, 178–187.

Mallick, S., Sharma, S., Banerjee, M., Ghosh, S. S., Chattopadhyay, A., and Paul, A. (2012). Iodine-stabilized Cu nanoparticle chitosan composite for antibacterial applications. *ACS Applied Materials & Interfaces*, *4*, 1313–1323.

Nisar, P., Ali, N., Rahman, L., Ali, M., and Shinwari, Z. K. (2019). Antimicrobial activities of biologically synthesized metal nanoparticles: An insight into the mechanism of action. *JBIC Journal of Biological Inorganic Chemistry*, *24*, 929–941.

Noh, H. J., Kim, H. S., Jun, S. H., Kang, Y. H., Cho, S., and Park, Y. (2013). Biogenic silver nanoparticles with chlorogenic acid as a bioreducing agent. *Journal of Nanoscience and Nanotechnology*, *13*, 5787–5793.

Padalia, H., and Chanda, S. (2017). Characterization, antifungal and cytotoxic evaluation of green synthesized zinc oxide nanoparticles using *Ziziphus nummularia* leaf extract. *Artificial Cells, Nanomedicine, and Biotechnology*, *45*, 1751–1761.

Panghal, A., Chhikara, N., Anshid, V., Charan, M. V. S., Surendran, V., Malik, A., and Dhull, S. B. (2019). Nanoemulsions: A promising tool for dairy sector. In *Nanobiotechnology in Bioformulations*, (pp. 99–117). Cham: Springer.

Rajeshkumar, S., Menon, S., Venkat Kumar, S., et al. (2019). Antibacterial and antioxidant potential of biosynthesized copper nanoparticles mediated through *Cissus arnotiana* plant extract. *Journal of Photochemistry and Photobiology B*, *197*, 111531.

Shaaban, M., and El-Mahdy, A. M. (2018). Biosynthesis of Ag, Se, and ZnO nanoparticles with antimicrobial activities against resistant pathogens using waste isolate Streptomyces enissocaesilis. *IET Nanobiotechnology*, *12*, 741–747.

Sharma, R., Raghav, R., Priyanka, K., et al. (2019). Exploiting chitosan and gold nanoparticles for antimycobacterial activity of in silico identified antimicrobial motif of human neutrophil peptide-1. *Scientific Reports*, *9*, 7866.

Singh, A. P., Biswas, A., Shukla, A., and Maiti, P. (2019). Targeted therapy in chronic diseases using nanomaterial-based drug delivery vehicles. *Signal Transduction and Targeted Therapy*, *4*, 33.

Song, Y., Jiang, H., Wang, B., Kong, Y., and Chen, J. (2018). Silver-incorporated mussel-inspired polydopamine coatings on *Mesoporous Silica* as an efficient nanocatalyst and antimicrobial agent. *ACS Applied Materials & Interfaces*, *10*, 1792–1801.

Song, Z., Wu, Y., Wang, H., and Han, H. (2019). Synergistic antibacterial effects of curcumin modified silver nanoparticles through ROS-mediated pathways. *Materials Science and Engineering: C*, *99*, 255–263.

Tambosi, R., Liotenberg, S., Bourbon, M. L, et al. (2019). Silver and copper acute effects on membrane proteins and impact on photosynthetic and respiratory complexes in bacteria. *MBio*, *9*, 1535–1538.

Toodehzaeim, M. H., Zandi, H., Meshkani, H., and Hosseinzadeh Firouzabadi, A. (2018). The effect of CuO nanoparticles on antimicrobial effects and shear bond strength of orthodontic adhesives. *Journal of Dentistry*, *19*, 1–5.

Wang, D., Zhao, L., Ma, H., Zhang, H., and Guo, L. H. (2017). Quantitative analysis of reactive oxygen species photogenerated on metal oxide nanoparticles and their bacteria toxicity: The role of superoxide radicals. *Environmental Science & Technology*, *51*, 10137–10145.

Yusof, N. A. A., Zain, N. M., and Pauzi, N. (2019). Synthesis of ZnO nanoparticles with chitosan as stabilizing agent and their antibacterial properties against Gram-positive and Gram-negative bacteria. *International Journal of Biological Macromolecules*, *124*, 1132–1136.

Zahran, M., and Marei, A. H. (2019). Innovative natural polymer metal nanocomposites and their antimicrobial activity. *International Journal of Biological Macromolecules*, *136*, 586–596.

Zanet, V., Vidic, J., Auger, S., Vizzini, P., Lippe, G., Iacumin, L., Comi, G., and Manzano, M. (2019). Activity evaluation of pure and doped zinc oxide nanoparticles against bacterial pathogens and *Saccharomyces cerevisiae*. *Journal of Applied Microbiology*, *127*, 1391–1402.

18 Enhanced Antimicrobial Efficacy of Essential Oils–Based Nanoemulsions

Mukul Kumar, Samriti Guleria, and Ashwani Kumar
Lovely Professional University, India

CONTENTS

18.1 NANOEMULSIONS

A nanoemulsion is two immiscible liquids (water and oil) that are mixed to form a single-phase with low viscosity, optically transparent, and with high kinetic stability (Karthik et al., 2017; Dhull et al., 2019). According to the International Union of Pure and Applied Chemistry, an emulsion is the fluid colloidal system in which liquid droplets are dispersed in a liquid (Gutiérrez et al., 2008). An oil-in-water (O/W) emulsion has water as a continuous phase whereas in a water-in-oil (W/O) emulsion water is a dispersed phase (Figure 18.1) Many different methods like

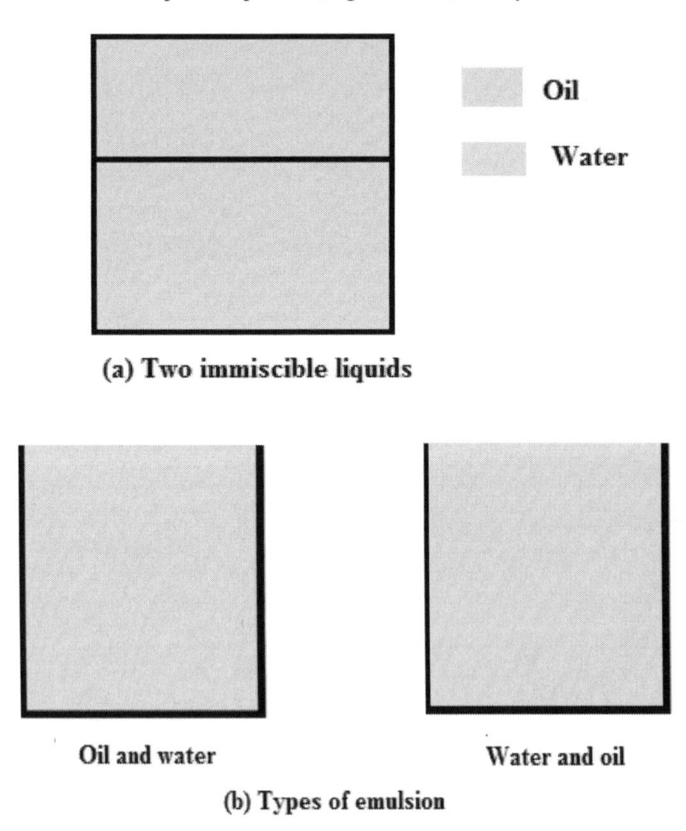

FIGURE 18.1 Schematic diagram of nanoemulsion preparation.

high-energy stirring, high-pressure homogenization, and ultrasound treatment can be used to create nanoemulsions (Singh et al., 2017a; Che et al., 2019). The particle size in nanoemulsions is very small and ranges between 20 and 200 nm (Silva et al., 2012). The small size of the particles gives them unique features like increased surface area and less sensitivity to physical and chemical changes (McClements, 2011; Panghal et al., 2019). Nowadays, it is one of the emerging fields in the pharmaceutical and food industries for enhancing the stability of food and improving the bioavailability of the encapsulated bioactive component (Silva et al., 2012; Chawla et al., 2019).

Nanoemulsion shave shown several advantages, such as solubilizing the lipophilic drugs; taste masking; thermodynamic stability; use in oral, topical, and intravenous delivery methods; increases in bioavailability; and rate of absorption (Basha et al., 2013; Dhull et al., 2019; Panghal et al., 2019). They have also shown several disadvantages, such as minimum solubility with high melting substances, the use of more co-surfactant, and surfactant in nanodroplets stability, and physical parameters affecting the stability of nanoemulsion (pH and temperature; Chawla et al., 2019, 2020; Wik et al., 2019).

18.2 DEVELOPMENT OF NANOEMULSIONS

A water phase, an oil phase, emulsifiers, a surfactant, a co-surfactant, and additives are used for the development of nanoemulsion (Gupta et al., 2016).

18.2.1 Water Phase

In the water phase, the nature of the size of the droplets and nanoemulsion stability was affected due to the presence of electrolytes in the system and pH medium (Karthik et al., 2017).

18.2.2 Oil Phase

Oil plays an important part in nanoemulsion preparation in the solubilized lipophilic substances (McClements et al., 2011). Several free fatty acids, triacylglycerols, diacylglycerols, monoacylglycerols (Gupta et al., 2016), and mineral oils are used to prepare the nanoemulsion which influences its property due to its chemical and physical characteristics (refractive index, water solubility, density, interfacial tension, chemical stability, and viscosity; Ávalos et al., 2017).

18.2.3 Surfactants

These are the non-ionic, cationic, anionic, and zwitterionic surfactants which are used for the formation of nanoemulsion, and they can reduce the interfacial tension between two immiscible liquids (Setya et al., 2014). Hydrophile–lipophile balance (HLB) plays an important role in the formation of W/O and O/W nanoemulsions. W/O and O/W nanoemulsions are prepared with a surfactant which has a low- and high-HLB (Aswathanarayan and Vittal, 2019). Surfactants also help reduce Laplace pressure and the stress through which the size of the droplet is reduced to a nanometer (Komaiko and McClements, 2016).

18.2.4 Co-Surfactants

These are amphiphilic molecules that can reduce the interfacial tension between the two liquid layers and decrease the viscosity of the interface (Borthakur et al., 2016). Several co-surfactants are used, such as transcutol p, glycerin, ethylene glycol, ethanol, propylene glycol, and propanol (Basha et al., 2013).

18.3 METHODS

Different methods are used for the preparation of nanoemulsion:

a. Low-energy methods
b. High-energy methods

18.3.1 Low-Energy Methods

In this system, internal chemical energy and the stirring process are used for the preparation of nanoemulsion (Solans and Solé, 2012). In this method, a huge amount of surfactant is used which has shown an adverse effect on the food quality (Komaiko and McClements, 2016). Several methods are used for the preparation of emulsions such as transitional phase inversion and catastrophic phase inversion (CPI; Luo et al., 2017).

18.3.2 High-Energy Methods

Higher energy is also used for the development of nanoemulsions. In this method, the high force breaks down the larger droplets into the nano-sized particles with the application of high kinetic energy. Several higher energy methods like high-pressure homogenization, microfluidization, and ultrasonication (Gonçalves et al., 2018) are used to prepare nanoemulsions. High-energy methods have more applications in the food industries to reduce the risk of spoilage and maintain stability, colour, rheological properties, and nutritional and sensory attributes (Gharibzahedi et al., 2019).

18.4 ESSENTIAL OILS

Essential oils are complex of several compounds (5000–7000 chemical constituents), such as volatile essential oils and aromatic organic compounds, obtained from a plant source, such as leaves, seed, flower, peel, berries, rhizome, root, wood, bark, petals, and resin, among others (Butnariu and Sarac, 2018; Punia et al., 2019). It contains oxides, alcohols, aldehydes, ketones, acids, esters, ethers, and etheric oils (Hassan and Soleimani, 2016; Saidi et al., 2017). Essential oil is used to protect from the pathogens, insects, light, and so on due to the synthesis of secondary metabolites (Asbahani et al., 2015). The essential oil contains several properties as shown in Figure 18.2.

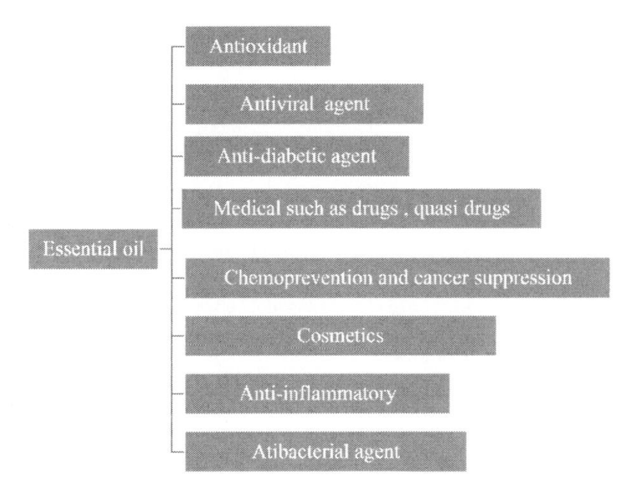

FIGURE 18.2 Several properties of essential oil. Source*:* Dima and Dima (2015); Donsì et al. (2011).

Essential oil plays an important role in food preservation, aroma, flavours, and essences due to the presence of effective compounds such as terpenes, monoterpenoids, and aromatic volatile compounds (Zerkaoui et al., 2018). It not only helps in food but also shows effective results against plant physiology and ecology, such as plant signalling, ozone quenching, protection against pathogens, reproduction of plants, and thermotolerance (Rehman et al., 2016; Naeem et al., 2018). Monoterpenoids, sesquiterpenoids, and phenylpropanoids are the different classes that showed highly functional properties in the essential oil (Sangwan et al., 2001). Essential oils are insoluble in water whereas, in alcohol, fixed oil, and ether, it showed a soluble nature (Zerkaoui et al., 2018). These oils are liquids that are colourless at room temperature (Hariri et al., 2017).

They work as natural aromas and essences due to their compounds which are obtained from the different extraction techniques such as effleurage, cold pressing, carbon dioxide extraction, solvent extraction (Zerkaou et al., 2018), steam distillation, hydro-distillation, and microwave-assisted process. Methods used for the extraction of essential oil are shown in Figure 18.3 (Hamid et al., 2011).

18.5 CHARACTERIZATION OF NANOEMULSIONS

Several different parameters are considered in the characterization of nanoemulsions as shown in Figure 18.4.

18.5.1 Droplet Size, Z-potential

Droplet size can also be determined by a disc centrifuge analyser with a range of 5 and 75 μm (Dasgupta et al., 2019). The analysis of particle size is done by dynamic light scattering, integrated with a Z-potential analyser to collect the information of surface charge of a particle (Chircov and Grumezescu, 2019). The dispersion

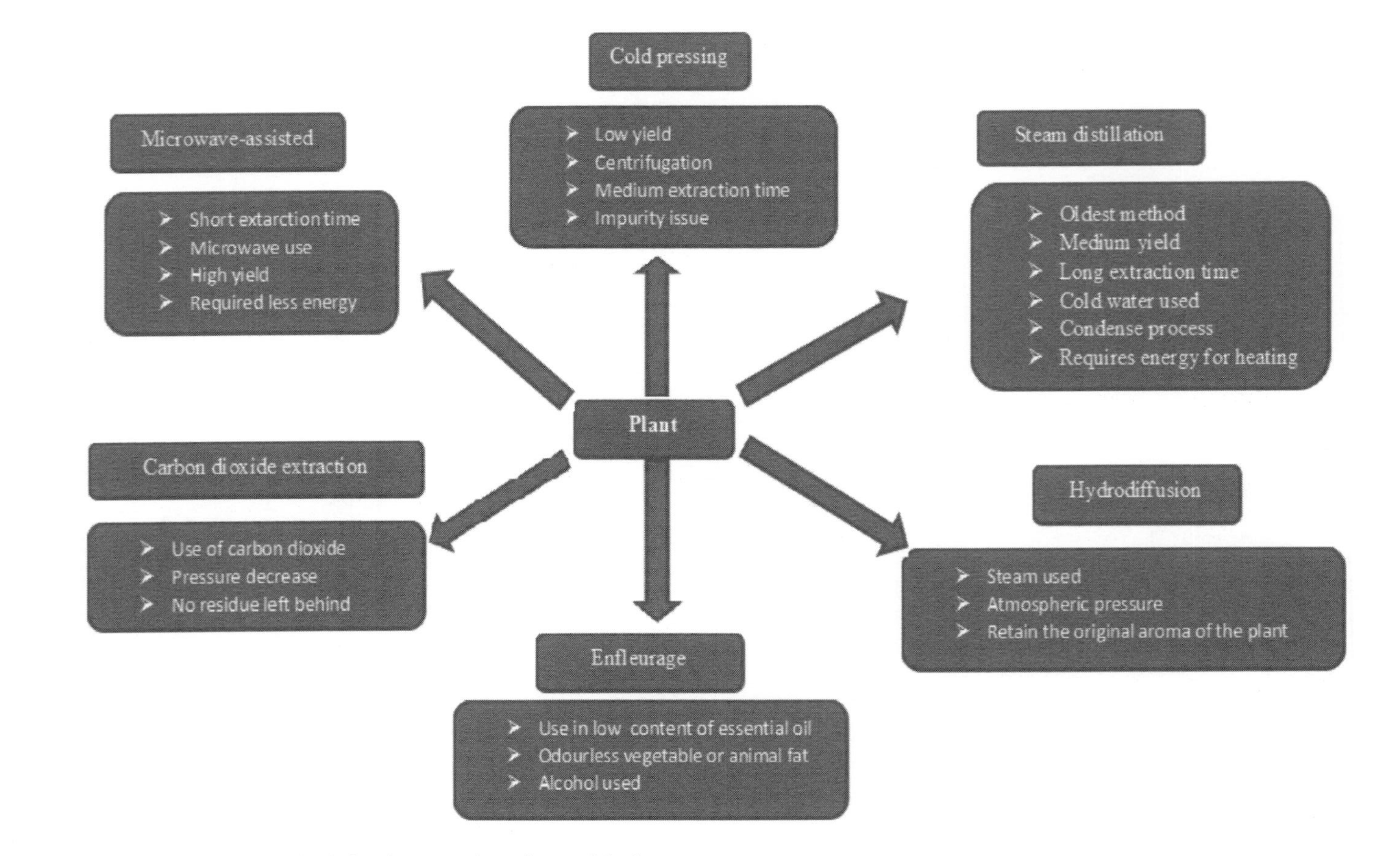

FIGURE 18.3 Different methods for the extraction of essential oils.

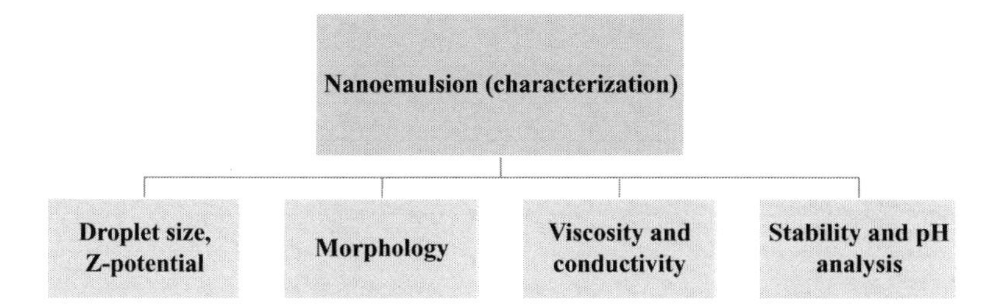

FIGURE 18.4 Characterization of nanoemulsion.

stability is predicted by the Z-potential, and the physicochemical properties, their adsorption, polymers, and electrolytes affect the value (Gurpreet and Singh, 2018).

18.5.2 Stability and pH Analysis

Nanoemulsion stability can be detected by looking at the destabilization mechanism such as flocculation, creaming, coalescence, and sedimentation (Chircov and Grumezescu, 2019). The size of the droplets was also indicated the stability of the emulsion. If the droplet size increases rapidly, then the system stability is low (Marzuki et al., 2019). pH plays an important role in emulsion stability due to the change in a chemical reaction (Dasgupta et al., 2019). The hydrolysis was responsible for the destabilization of nanoemulsion, which does not affect the quality of the product (Bernardi et al., 2011).

18.5.3 Viscosity and Conductivity

Viscosity and conductivity are important parameters of a nanoemulsion which provide information about the macroscopic level (Chircov and Grumezescu, 2019), micelles, and phase inversion mechanism (Chime et al., 2014). Several viscometers are used for the determination of the viscosity, such as Ostwald viscometer, Stormer viscometer, Brookfield viscometer (Chen et al., 2016), and Hoeppler falling ball viscometer (Malik et al., 2014).

18.5.4 Morphology

The morphology of a nanoemulsion is information about the size, distribution, and shape of the nanoparticles (Sugumar et al., 2016). Electron microscopy (transmission and scanning) and atomic force microscopy (AFM) were used to get information about the morphology of nanoemulsions (Chime et al., 2014; Dasgupta et al., 2019). Transmission electron microscopy helps with seeing the image of the nanoemulsion structure, such as coated droplets, spherical structure, aggregates, vesicles, and microstructure (Singh et al., 2017b). Scanning electron microscopy at the low magnification and high resolution shows the structure of oil droplets of nanoemulsion (Dasgupta et al., 2019). AFM is used to measure the surface morphology of a nanoemulsion by the change in the immobilized sample and shape probe (Chime et al., 2014).

18.6 APPLICATION OF NANOEMULSIONS AS ANTIMICROBIAL AGENTS

Nanoemulsions have shown antimicrobial properties due to the release of stored energy from an oil and detergent emulsion which was fused with the lipid bilayer of the cell membrane (Moghimi et al., 2016). It also destabilizes the lipid membrane of the bacteria such as *Escherichia coli* and *Staphylococcus aureus.* It shows potential activity against the different microorganisms which further classified as antibacterial, antifungal, sporicidal, antiviral, and antiprotozoal nanoemulsion (Krishnamoorthy et al., 2018).

18.6.1 Effect of Nanoemulsions Against Different Bacteria and Fungi

Nanoemulsions have shown effective results against different bacteria and fungi such as *Listeria monocytogenes*, *Proteus mirabilis*, *Streptococcus mutans*, *Leuconostoc mesenteroides*, and *Cryptococcus neoformans*, as shown in Table18.1.

18.6.2 Mechanism of Antimicrobial Efficacy of Nanoemulsions

Nanoemulsions can be prepared with oil, water, and antimicrobial agents which are stabilized by surfactants and alcohol (Bhargava et al., 2015). Several nanoemulsion bacteria activity studies were reported against the clinical and foodborne pathogen such as *Vibrio cholera*, *Streptococcus pneumonia*, *Neisseria gonorrhoeae*, *S. aureus*, *Pseudomonas aeruginosa*, *Haemophilus influenza*, *Bacillus cereus*, and *L. monocytogen* (Vijayalakshmi et al., 2014; Dasgupta et al., 2016). Several emulsifiers (lecithin, glycerol monooleate, pea proteins, and sugar ester) are used with the combination of Tween 20, which helps stabilize the nanoemulsion droplet size (100–200 nm) and shows a more effective result against bactericidal activity (Sugumar et al., 2015a; Sham and Sahari, 2016).

18.6.3 Fusing Methods (Bacterial Activity)

Several natural and chemical antimicrobial agents were emulsified and evaluated. It was observed for the reproduction and growth of the bacteria, water plays an important role (Dasgupta et al., 2016). An emulsion acts as an antimicrobial agent that affects the bacteria which cannot survive in pure oil or fat (Ma et al., 2016). The mechanism is the nanoemulsion droplets functioning with the fusion methods (with lipid bilayers of the cell membrane), as shown in Figure 18.5. The electrostatic attraction between the anionic (pathogen) and cationic charge (emulsion) has increased the fusion in a nanoemulsion droplet. Cell lysis and death occur due to the destabilizing of the lipid membrane of a pathogen (Pannu et al., 2009).

Nanoemulsion showed an effective result in the antifungal and sporicidal activity. Bacillus spores are killed when these spores were coated with the surfactant-based sunflower nanoemulsion which is responsible for the disruption of core (Landry et al., 2014). Fungi were effect by using a cinnamon nanoemulsion coating which disturbs the fungal hyphae and spore (Shivendu et al., 2016). Several studies are reported nanoemulsion showed effective results against the foodborne fungal pathogens and filaments such as *C. albicans*, *Fusarium oxysporum*, *Trichophyton rubrum*, *Microsporum gypseum*, *T. mentagrophytes*, and *Aspergillus fumigatus* (Joe et al., 2012).

TABLE 18.1
Effective Results of Nanoemulsion Against Different Bacteria and Fungi

Microorganisms	Nanoemulsion used	Result	References
Bacteria			
Listeria monocytogenes	BCTP nanoemulsion	Higher antibacterial activity (showed effective result against gram-positive bacteria)	Teixeira et al. (2007)
Proteus mirabilis	Eucalyptus oil	Enhance the activity of microorganism	Saranya et al. (2012)
Native cultivable bacteria	Eugenol	Increased antimicrobial activity (membrane permeability was altered)	Gyawali and Ibrahim (2014)
Listeria monocytogenes	Eugenol, thyme oil	Higher antibacterial activity (dispersion in the food matrix above the solubility limit is improved)	Xue et al. (2015)
Salmonella typhimurium	Tea tree oil and cinnamaldehyde	Enhanced antimicrobial activity	Jo et al. (2015)
Staphylococcus aureus	Eucalyptus globulus	Higher antibacterial activity	Sugumar et al. (2015b)
Streptococcus mutans	soybean oil	Showed effective results against fungal and bacterial activity.	Li et al. (2020)
Fungi			
Cryptococcus neoformans	Corn oil	Enhanced the antimicrobial activity (by the addition of LAE)	Chang et al. (2013)
Candida albicans	Amphotericin B and triglyceride		Donsì and Dima (2011)
Listeria monocytogenes	Anise oil	Higher antimicrobial activity	Topuz et al. (2016)
Candida albicans	Soybean oil	Increasing antifungal activity	Krishnamoorthy et al. (2018)
Leuconostoc mesenteroides	*Achillea* essential oils	Increased antimicrobial activity	Almadiy et al. (2016)

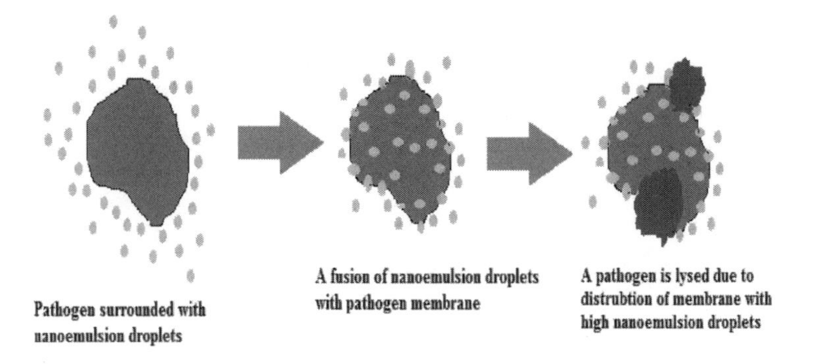

FIGURE 18.5 Killing of microbes with nanoemulsion droplets (fusing method).

18.7 CELLULAR ABSORPTION MECHANISMS

Bioactive compounds in nanoemulsion are used as natural antimicrobial agents. Several mechanisms, such as the ability of cellular absorption and the reduction the mass transfer resistance, effected antimicrobial activity (Salvia-Trujillo et al., 2013). The droplet size of a nanoemulsion is also responsible for the enhancing of antimicrobial activity by disturbing the surface of the lipid shroud (Dasgupta et al., 2016), as the antimicrobial activity of *Lactobacillus delbrueckii*, *E. coli*, and *Saccharomyces cerevisiae* has been reported (Vijayalakshmi et al., 2014).

18.7.1 Lipid Phase

In this phase, nanoemulsion droplets were loaded with the lipophilic antimicrobial by mass transport which prevents spoilage and maintains the quality of the food products as shown in Figure 18.6 (Weiss et al., 2009). In the aqueous phase, the lipid antimicrobial was soluble due to several characteristics, such as the chemical structure of antimicrobial, temperature, droplet size, and composition of the aqueous phase. Laplace effect is also responsible for the difference in antimicrobial concentration around the bulk phase and oil droplet (McClements, 2015).

18.7.2 Surface-Active Antimicrobial Activity

The surface-active antimicrobial is responsible for the stability of inert lipid and a nanoemulsion by the reduction in coalescence or flocculation mechanism as shown in Figure 18.7. The addition of emulsifiers improves the negatively charged

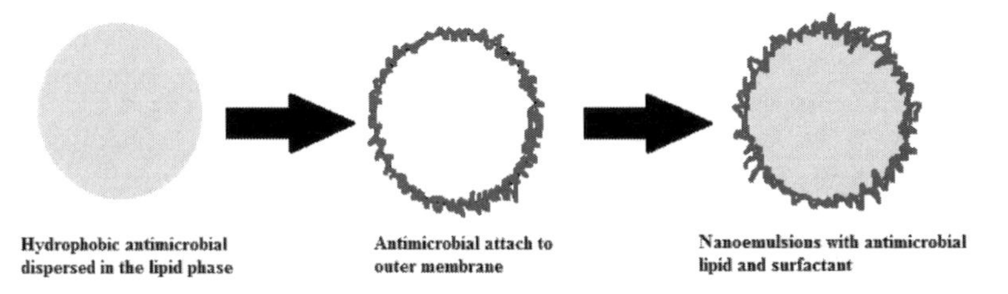

FIGURE 18.6 Nanoemulsion with antimicrobial lipid and surfactant.

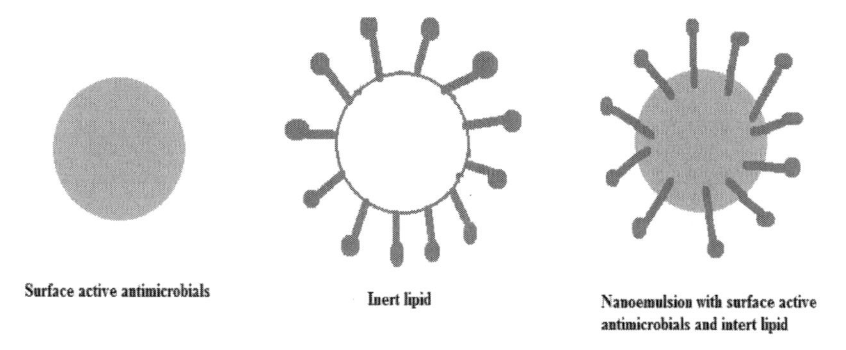

FIGURE 18.7 Function of nanoemulsion with surface-active antimicrobials and inert lipids.

microbial surfaces and electrostatic attraction between droplets (Sugumar et al., 2016). The size of the droplet is also responsible for increasing the collision between microorganisms (McClements, 2015).

18.7.3 Effect of Metal Oxide

Metalloproteins and metal transport systems are used to identify the bacterial cells (prokaryotic) from mammalian cells (eukaryotic; Jiménez et al., 2019). There are several mechanisms of nanomaterial which interact with the bacterial cell physically to kill the resistant bacteria and are shown in Figure 18.8.

18.7.4 Reactive Oxygen Species Generation

Reactive oxygen species (ROS) area group of reactive molecules by which microbes are oxidized by the superoxide anions (O_2). These are the highly reactive oxidizing radicals that react with macromolecule such as DNA, lipids, and protein (enzyme; Figure 18.9). ROS produced with the help of metal oxide nanoparticles undergo different processes such as oxidation and reduction (Guo et al., 2014). Oxidative stress produced in the bacterial cell due to the elimination of ROS and cause cell lysis in the bacterial

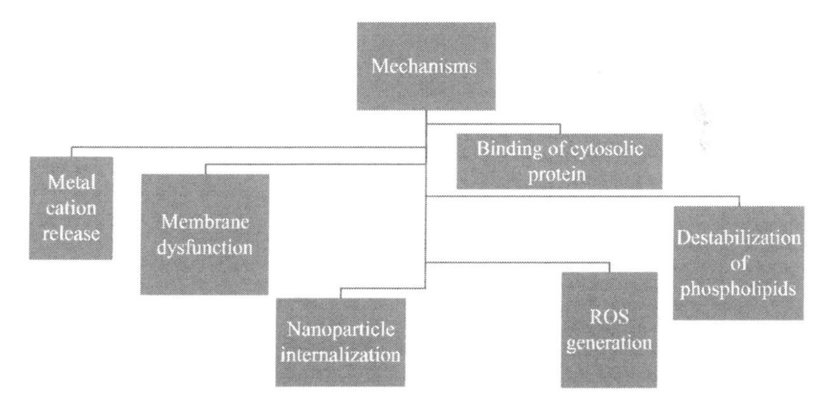

FIGURE 18.8 Several mechanisms of nanomaterial which interact with the bacterial cell. Source: Vega-Jiménez et al. (2019).

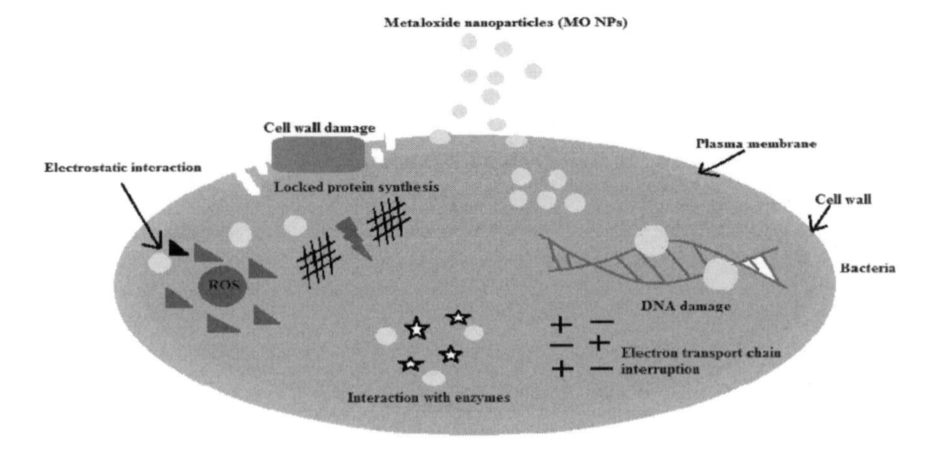

FIGURE 18.9 Mechanism of metal oxide nanoparticles in bacteria.

membrane. The endogenous antioxidants (superoxide and catalase enzymes) were neutralized by the O_2 and H_2O_2 which help in reducing the acute stress reaction (Vega-Jiménez et al., 2019). H_2O and O_2 also lead to the death of acute microbial electrons, reducing by oxidizing the dioxygen radical (Karton-Lifshin et al., 2011). Electrons were ejected from the valence band and the conduction band due to the photo-induced charge. ROS are also responsible for the damage of DNA due to the oxidation of amino acids and polyunsaturated fatty acids (Subhan, 2020). ROS formation increased with several metal oxides, such as magnesium oxide, zinc oxide, and gold, which were responsible for producing H_2O_2 from glucose oxidase. Several bacteria (*Klebsiella pneumoniae*, *Enterococcus faecalis* [Vega-Jiménez et al., 2019], *Streptococcus pyogenes*, *E. coli*, and *Pseudomonas aeruginosa*) growth were inhibited the antibiotic-resistant through the nitric oxide nanoparticles (McClements and Julian, 2015).

18.7.5 Metal Cation Release

A cationic charge is generated in metal oxide nanoparticle which showed positively charged and permeability towards the cell membrane. These metals showed effective properties against microbial growth which helps in metabolic functions (Vega-Jiménez et al., 2019). The structure of the molecules and functionality of the metabolic process with protease enzymes are affected because of the sulfhydryl, amino, and hydroxyl groups (Shim et al., 2016). These metal oxide NPs increase the permeability of the outer membrane and neutralize the charge. The hydroxyl group also helps in decaying the bacterial cell by the diffusion of metal ions. Adsorption, dissolution, and hydrolysis technique are used to release the metal ions, and they show a toxic nature towards bacteria and microbes (Awad et al., 2015).

18.7.6 Membrane Dysfunction

Metallic materials attract towards the cell wall due to the electrostatic attraction and positive charge (Sirelkhatim et al., 2015). These metals have blocked the pores with a different mechanism, such as the absorption process and transportation. It is also responsible for the deformation of cells and the killing of the microorganisms (Applerot et al., 2009).

18.7.7 Nanoparticle Internalization

Nanoparticles consist of a large surface area and high permeability which affect the cell wall of the microbes. These nanoparticles have shown a highly active nature due to their size (Shim et al., 2016). The transportation and malfunctioning were affected by the high adsorption on the cell membrane due to the high surface area. It showed a toxic nature against the microbes due to the exchange of energy and matter (Brayner et al., 2006).

18.7.8 Binding of Cytosolic Protein

Metaloxide nanoparticles are bind with the protein which shows antibacterial activity. The carbonyls are bound to protein when the metal ions are catalysing and help oxidize the side chain of amino acids (Brayner et al., 2006). The chemical reaction showed the degradation of protein when the catalytic activity of the enzyme was reduced. The enzyme and DNA are attached to the metal oxide NPs which induce the antimicrobial activity. The production of adenosine triphosphate (ATP) and metabolic pathways and respiratory inhibition are effected with the interaction of metal and protein. Replication and division of DNA were done due to the binding of enzymes with a silver (Sirelkhatim et al., 2015).

18.7.9 Destabilization of Phospholipids

The electrostatic attraction is responsible for the disruption of the cell membrane when it binds with metal oxide NPs. The interaction between the positive (metal oxide NPs) and negative charge (electrostatic) has shown distribution in the membrane, and bacterial protein is damaged due to the high oxidative stress (Subhan, 2020). The water is released from the cytosol due to the breakdown of the cell barrier, which helps the transport of electrons. Several metals are used for the destabilization of phospholipids such as zinc oxide, gold, magnesium oxide, silver, and titanium oxide (Nguyen et al., 2019).

18.8 SEVERAL STUDIES OF NANOEMULSIONS WITH DIFFERENT MICROORGANISMS

Several microorganisms such as *Sachromyces cerevisiae*, *E. coli*, *Lactobacillus delbrueckii*, *Zygosaccharomyces bailii*, *Brettanomyces bruxellensis*, *B. naardenensis*, *Pseudomonas aeruginosa*, and *Enterococcus faecalis*, among others, are used to see the effect of essential oil–based nanoemulsions in antimicrobial activity with different biopolymers and emulsifying agents, such as ethyl oleate; Tween 80, 60, 40, 20; glycerol; Q-natural; and starch, among others, as shown in Table 18.2.

TABLE 18.2
Use of Different Biopolymer, Emulsifying Agent and Techniques which Showed Effective Results Against Different Microorganism

Biopolymer/Emulsifying Agent	Essential Oil	Technique	Microorganism	Reference
• Tween 80 • Ethyl oleate • n-pentanol	Soya bean oil	Transition phase inversion	• *Salmonella typhimurium* • *Listeriamono cytogenes* • *Pseudomonas aeruginosa* • *Escherichia coli*	Teixeira et al. (2007)
• Soy lecithin • Glycerol • Monoolerate • Tween 20	Sunflower oil	High Pressure Homogenization	• *Saccharomyces cerevisiae* • *E. coli* • *Lactobacillus delbrueckii*	Donsi et al. (2011)
• Polyoxyethylenesorbitan • Monooleate	Lemon myrtle oil and soybean oil	Microfluidization	• *Pseudomonas aeruginosa* • *E. coli* • *Staphylococcus aureus*	Buranasuksombat et al. (2011)
• Modified starch	Pepper mint oil and MCT (medium-chain triacylglycerol)	High Pressure Homogenization	• *S. aureus* • *L. monocytogenes*	Liang et al. (2012)
	Sesame oil	Ultrasonication	• *E. coli* • *Bacillus strain* • *S. typhimurium*	Joe et al. (2012)
• Tween 60 • Tween 80	Soy bean oil	Microfluidizer	• *Baumanii strains* • *Acinetobacter baumanii*	Hwang et al. (2013)
• Polyoxyethylene sorbitan • Tween 80 • Tween 40	MCT oil	Spontaneous emulsification	• *Zygosaccharomyces bailii* • *Brettanomyces bruxellensis* • *Brettanomyces naardenensis* • *S. cerevisiae*	Chang et al. (2013)

• Tween 20	CIN,3-Phenyl-2 propenal	High-pressure homogenization	• *E. coli* • *S. typhimurium* • *S. aureus*	Jo et al. (2015)
• Polyoxyethylene sorbitan	*Thymus daenensis* oil	High-intensity sonication	• *E. coli*	Moghimi et al. (2016)
• Tween 80	Viscosa seed oil	Ultrasonication	• *E. coli* • *Pseudomonas aeruginosa* • *Klebsiella pneumoniae* • *Streptococcus pyogenes*	Krishnamoorthy et al. (2018)
• Ethylene glycol	Citral	Ultrasonication	• *Staphylococcus aureus* • *Escherichia coli* • *Pseudomonas aeruginosa* • *Enterococcus faecalis* • *S. typhimurium* • *L. monocytogenes*	Lu et al. (2018)
• Q-Naturale • Polyoxyethylene sorbitan	Carvacrol and MCT	Microfluidizer	• *Salmonella enterica*	Ryu et al. (2018)
• Tween 80	Canola oil and palm oil	Magnetic stir	• *S. enterica*	Ryu et al. (2018)
• Polypropylene glycol • Ethyl acetate		High-pressure homogenization	• *B. subtilis* • *E. coli*	Li et al. (2020)

18.9 CONCLUSION

A nanoemulsion is the combination of a water phase, an oil phase, emulsifiers, a surfactant, and a co-surfactant to form a single phase which helps in the protection of the food product by inhibiting the food pathogens. It is prepared through the low- and high-energy methods. Essential oil is used in nanoemulsions due to the presence of several compounds such as volatile, ethereal oils and organic compounds which protect food product from the pathogens and lights. The characterization of the nanoemulsion is measured through zeta-potential, stability, pH analysis, and morphology. It shows higher antimicrobial and antibacterial properties due to the embedded compounds which disturb the surface of the lipid membrane of bacteria and fungi such as *L. monocytogenes*, *Proteus mirabilis*, *S. mutans*, *Candida albicans*, and *Leuconostoc mesenteroides*. In future, the demand for nanoemulsions will increase due to their unique properties.

REFERENCES

Almadiy, Abdulrhman A., Gomah E. Nenaah, Basma A. Al Assiuty, Eman A. Moussa, and Nabila M. Mira (2016). "Chemical composition and antibacterial activity of essential oils and major fractions of four *Achillea* species and their nanoemulsions against foodborne bacteria." *LWT-Food Science and Technology,* 69, 529–537.

Applerot, Guy, Anat, Lipovsky, Rachel, Dror, Nina, Perkas, Yeshayahu, Nitzan, Rachel, Lubart, and Aharon, Gedanken (2009). "Enhanced antibacterial activity of nanocrystalline ZnO due to increased ROS-mediated cell injury." *Advanced Functional Materials*, 19(6), 842–852.

Aswathanarayan, Jamuna Bai, and Ravishankar Rai, Vittal (2019). "Nanoemulsions and their potential applications in food industry." *Frontiers in Sustainable Food Systems,* 3, 95.

Awad, A., Ahmed I. Abou-Kandil, I. Elsabbagh, M. Elfass, M. Gaafar, and E. Mwafy (2015). "Polymer nanocomposites part 1: Structural characterization of zinc oxide nanoparticles synthesized via novel calcination method." *Journal of Thermoplastic Composite Materials,* 28(9), 1343–1358.

Basha, Syed Peer, Koteswara P. Rao, and Chakravarthi, Vedantham (2013). "A brief introduction to methods of preparation, applications and characterization of nanoemulsion drug delivery systems." *Indian Journal of Research in Pharmacy and Biotechnology*, 1(1), 25.

Bernardi, Daniela S., Tatiana A. Pereira, Naira R. Maciel, Josiane, Bortoloto, Gisely S. Viera, Gustavo C. Oliveira, and Pedro A. Rocha-Filho (2011). "Formation and stability of oil-in-water nanoemulsions containing rice bran oil: In vitro and in vivo assessments." *Journal of Nanobiotechnology,* 9(1), 44.

Bhargava, Kanika, Denise S. Conti, Sandro R. P. da Rocha, and Yifan Zhang (2015). "Application of an oregano oil nanoemulsion to the control of foodborne bacteria on fresh lettuce." *Food Microbiology,* 47, 69–73.

Borthakur, Priyakshree, Purna K. Boruah, Bhagyasmeeta, Sharma, and Manash R. Das (2016). Nanoemulsion: Preparation and its application in food industry. In *Emulsions*, pp. 153–191. Elsevier, London, Academic Press.

Brayner, Roberta, Roselyne, Ferrari-Iliou, Nicolas, Brivois, Shakib, Djediat, Marc F. Benedetti, and Fernand, Fiévet (2006). "Toxicological impact studies based on Escherichia coli bacteria in ultrafine ZnO nanoparticles colloidal medium." *Nano Letters,* 6(4), 866–870.

Buranasuksombat, Umaporn, Yun Joong, Kwon, Mark, Turner, and Bhesh, Bhandari (2011). "Influence of emulsion droplet size on antimicrobial properties." *Food Science and Biotechnology,* 20(3), 793–800.

Butnariu, Monica, and Ioan, Sarac (2018). "Essential oils from plants." *Journal of Biotechnology and Biomedical Science,* 1(4), 35.

Chang, Yuhua, Lynne, McLandsborough, and David Julian, McClements (2013). "Physicochemical properties and antimicrobial efficacy of carvacrol nanoemulsions formed by spontaneous emulsification." *Journal of Agricultural and Food Chemistry,* 61(37), 8906–8913.

Chawla, P., Kumar, N., Bains, A., Dhull, S. B., Kumar, M., Kaushik, R., and Punia, S. (2020). "Gum arabic capped copper nanoparticles: Synthesis, characterization, and applications." *International Journal of Biological Macromolecules*, 146, 232–242.

Chawla, P., Kumar, N., Kaushik, R., and Dhull, S. B. (2019). "Synthesis, characterization and cellular mineral absorption of nanoemulsions of Rhododendron arboreum flower extracts stabilized with gum arabic." *Journal of Food Science and Technology*, 56(12), 5194–5203.

Che, Marzuki, Nur, Haziqah, Roswanira Abdul, Wahab, and Mariani, Abdul Hamid (2019). "An overview of nanoemulsion: Concepts of development and cosmeceutical applications." *Biotechnology & Biotechnological Equipment,* 33(1), 779–797.

CheMarzuki, N. H., R. A. Wahab, and M. Abdul Hamid (2019). "An overview of nanoemulsion: Concepts of development and cosmeceutical applications." *Biotechnology & Biotechnological Equipment*, 33(1), 779–797.

Chen, Huanle, Xiaorong, Hu, Enmin, Chen, Shan, Wu, David Julian, McClements, Shilin, Liu, Bin, Li, and Yan, Li (2016). "Preparation, characterization, and properties of chitosan films with cinnamaldehyde nanoemulsions." *Food Hydrocolloids,* 61 , 662–671.

Chime, S. A., F. C. Kenechukwu, and A. A. Attama (2014). *Nanoemulsions—Advances in Formulation, Characterization and Applications in Drug Delivery*, Intechopen, Vol. 3.

Chircov, C., and Grumezescu, A. M. (2019). Nanoemulsion preparation, characterization, and application in the field of biomedicine. In *Nanoarchitectonics in Biomedicine* (pp. 169–188). William Andrew Publishing.

Dasgupta, Nandita, Shivendu, Ranjan, and Mansi, Gandhi (2019). "Nanoemulsion ingredients and components." *Environmental Chemistry Letters,* 17(2), 917–928.

Dasgupta, Nandita, Shivendu, Ranjan, Shraddha, Mundra, Chidambaram, Ramalingam, and Ashutosh, Kumar (2016). "Fabrication of food grade vitamin E nanoemulsion by low energy approach, characterization and its application." *International Journal of Food Properties,* 19(3), 700–708.

deOca-Ávalos, Juan Manuel Montes, Roberto Jorge, Candal, and María Lidia, Herrera (2017). "Nanoemulsions: Stability and physical properties." *Current Opinion in Food Science,* 16, 1–6.

Dhull, S. B., M. Anju, S. Punia, R. Kaushik, and P. Chawla (2019). Application of gum Arabic in nanoemulsion for safe conveyance of bioactive components. In *Nanobiotechnology in Bioformulations* (pp. 85–98). Cham: Springer.

Dima, Cristian, and Stefan, Dima (2015). "Essential oils in foods: Extraction, stabilization, and toxicity." *Current Opinion in Food Science,* 5, 29–35.

Donsì, Francesco, Marianna, Annunziata, Mariarenata, Sessa, and Giovanna, Ferrari (2011). "Nanoencapsulation of essential oils to enhance their antimicrobial activity in foods." *LWT-Food Science and Technology,* 44(9), 1908–1914.

Donsì, Francesco, and Giovanna, Ferrari, (2016). "Essential oil nanoemulsions as antimicrobial agents in food." *Journal of Biotechnology,* 233, 106–120.

El Asbahani, A., K. Miladi, W. Badri, M. Sala, E.H.A. Addi, H. Casabianca, A. ElMousadik, D. Hartmann, A. Jilale, F.N.R. Renaud, and A. Elaissari (2015). "Essential oils: From extraction to encapsulation." *International Journal of Pharmacy*, 483, 220–243, http://dx.doi.org/10.1016/j.ijpharm.2014.12.069.

Gharibzahedi, Seyed Mohammad Taghi, César, Hernández-Ortega, Jorge, Welti-Chanes, Predrag, Putnik, Francisco J. Barba, Kumar, Mallikarjunan, Zamantha, Escobedo-Avellaneda, and Shahin, Roohinejad (2019). "High pressure processing of food-grade emulsion systems: Antimicrobial activity, and effect on the physicochemical properties." *Food Hydrocolloids,* 87, 307–320.

Ghosh, Vijayalakshmi, Amitava, Mukherjee, and Natarajan, Chandrasekaran (2014). "Eugenol-loaded antimicrobial nanoemulsion preserves fruit juice against, microbial spoilage." *Colloids and Surfaces B: Biointerfaces,* 114, 392–397.

Gonçalves, Antónia, Shahin Roohinejad, Nooshin Nikmaram, Berta N. Estevinho, Fernando, Rocha, Ralf, Greiner, and David Julian, McClements (2018). "Production, properties, and applications of solid self-emulsifying delivery systems (S-SEDS) in the food and pharmaceutical industries." *Colloids and Surfaces A: Physicochemical and Engineering Aspects,* 538, 108–126.

Guo, Zhiqian, Sookil, Park, Juyoung, Yoon, and Injae, Shin (2014). "Recent progress in the development of near-infrared fluorescent probes for bioimaging applications." *Chemical Society Reviews,* 43(1), 16–29.

Gupta, Ankur, H. BurakEral, T. Alan Hatton, and Patrick S. Doyle (2016). "Nanoemulsions: Formation, properties and applications." *Soft Matter,* 12(11), 2826–2841.

Gurpreet, K., and S. K. Singh (2018). "Review of nanoemulsion formulation and characterization techniques." *Indian Journal of Pharmaceutical Sciences,* 80(5), 781–789.

Gutiérrez, J. M., C. González, A. Maestro, I. M. P. C. Solè, C. M. Pey, and J. Nolla (2008). "Nano-emulsions: New applications and optimization of their preparation." *Current Opinion in Colloid & Interface Science,* 13(4), 245–251.

Gyawali, Rabin, and Salam A. Ibrahim (2014). "Natural products as antimicrobial agents." *Food Control,* 46, 412–429.

Hamid, A. A., O. O. Aiyelaagbe, and L. A. Usman (2011). "Essential oils: Its medicinal and pharmacological uses." *International Journal of Current Research,* 33(2), 86–98.

Hariri, Ahmed, Nawel, Ouis, Djilali, Bouhadi, and KarimaOuld, Yerou (2017). "Evaluation of the quality of the date syrups enriched by cheese whey during the period of storage." *Banat's Journal of Biotechnology*, 8(16, 75–82.

Hassan, SonaAyadi, and Tayebeh, Soleimani (2016). "Improvement of artemisinin production by different biotic elicitors in artemisiaannua by elicitation-infiltration method." *Banat's Journal of Biotechnology*, 7(13), 82–94.

Hwang, Yoon Y., Karthikeyan, Ramalingam, Diane R. Bienek, Valerie, Lee, Tao, You, and Rene, Alvarez (2013). "Antimicrobial activity of nanoemulsion in combination with cetylpyridinium chloride in multidrug-resistant Acinetobacter baumannii." *Antimicrobial Agents and Chemotherapy,* 57(8), 3568–3575.

Jo, Yeon-Ji, Ji-Yeon, Chun, Yun-Joong, Kwon, Sang-Gi, Min, Geun-Pyo, Hong, and Mi-Jung, Choi (2015). "Physical and antimicrobial properties of trans-cinnamaldehyde nanoemulsions in water melon juice." *LWT-Food Science and Technology,* 60(1), 444–451.

Joe, Manoharan Melvin, K. Bradeeba, Rengasamy, Parthasarathi, Palanivel Karpagavinaya, Sivakumaar, Puneet Singh, Chauhan, Sherlyn, Tipayno, Abitha, Benson, and Tongmin, Sa (2012). "Development of surfactin based nanoemulsion formulation from selected cooking oils: Evaluation for antimicrobial activity against selected food associated microorganisms." *Journal of the Taiwan Institute of Chemical Engineers,* 43(2), 172–180.

Karthik, P., P. N. Ezhilarasi, and C. Anandharamakrishnan (2017). "Challenges associated in stability of food grade nanoemulsions." *Critical Reviews in Food Science and Nutrition,* 57(7), 1435–1450.

Karton-Lifshin, Naama, Ehud, Segal, Liora, Omer, Moshe, Portnoy, Ronit, Satchi-Fainaro, and Doron, Shabat (2011). "A unique paradigm for a Turn-ON near-infrared cyanine-based probe: Noninvasive intravital optical imaging of hydrogen peroxide." *Journal of the American Chemical Society,* 133(28), 10960–10965.

Komaiko, Jennifer S., and David Julian, McClements (2016). "Formation of food-grade nanoemulsions using low-energy preparation methods: A review of available methods." *Comprehensive Reviews in Food Science and Food Safety,* 15(2), 331–352.

Krishnamoorthy, Rajapandiyan, Jegan, Athinarayanan, Vaiyapuri Subbarayan, Periasamy, Abdulraheem R. Adisa, Mohammed A. Al-Shuniaber, Mustafa A. Gassem, and Ali A. Alshatwi (2018). "Antimicrobial activity of nanoemulsion on drug-resistant bacterial pathogens." *Microbial Pathogenesis*, 120, 85–96.

Landry, Kyle S., Yuhua, Chang, David Julian, McClements, and Lynne, McLandsborough (2014). "Effectiveness of a novel spontaneous carvacrol nanoemulsion against Salmonella enterica Enteritidis and Escherichia coli O157: H7 on contaminated mung bean and alfalfa seeds." *International Journal of Food Microbiology,* 187, 15–21.

Li, Xiuxiu, Xi, Yang, Hong, Deng, Yurong, Guo, and Jia, Xue (2020). "Gelatin films incorporated with thymol nanoemulsions: Physical properties and antimicrobial activities." *International Journal of Biological Macromolecules,* 150, 161–168.

Liang, Rong, Shiqi, Xu, Charles F. Shoemaker, Yue, Li, Fang, Zhong, and Qingrong, Huang (2012). "Physical and antimicrobial properties of peppermint oil nanoemulsions." *Journal of Agricultural and Food Chemistry,* 60(30), 7548–7555.

Lu, Wen-Chien, Da-Wei, Huang, Chiun-CR, Wang, Ching-Hua, Yeh, Jen-Chieh, Tsai, Yu-Ting, Huang, and Po-Hsien, Li (2018). "Preparation, characterization, and antimicrobial activity of nanoemulsions incorporating citral essential oil." *Journal of Food and Drug Analysis,* 26(1), 82–89.

Luo, X, Y. Zhou, L. Bai, F. Liu, Y. Deng, D. J. McClements (2017). "Fabrication of β-carotene nanoemulsion-based delivery systems using dual-channel microfluidization: Physical and chemical stability." *Journal of Colloid and Interface Science.*, 490, Mar 15), 328–335.

Ma, Qiumin, P. Michael, Davidson, and Qixin, Zhong (2016). "Nanoemulsions of thymol and eugenol co-emulsified by lauric arginate and lecithin." *Food Chemistry,* 206,167–173.

Malik, Parth, R. K. Ameta, and Man, Singh (2014). "Preparation and characterization of bionanoemulsions for improving and modulating the antioxidant efficacy of natural phenolic antioxidant curcumin." *Chemico-Biological Interactions,* 222, 77–86.

McClements, David Julian (2011). "Edible nanoemulsions: Fabrication, properties, and functional performance." *Soft Matter,* 7(6), 2297–2316.

McClements, David Julian (2015). *Food Emulsions: Principles, Practices, and Techniques.* London, New York, CRC Press.

McClements, David Julian, and Jiajia, Rao (2011). "Food-grade nanoemulsions: Formulation, fabrication, properties, performance, biological fate, and potential toxicity." *Critical Reviews in Food Science and Nutrition,* 51(4), 285–330.

Moghimi, Roya, Lida, Ghaderi, Hasan, Rafati, Atousa, Aliahmadi, and David Julian, McClements (2016). "Superior antibacterial activity of nanoemulsion of *Thymus daenensis* essential oil against *E. coli*." *Food Chemistry*, 194, 410–415.

Naeem, A., T. Abbas, T. M. Ali, and A. Hasnain (2018). "Essential oils: Brief background and uses." *Annals of Short Reports,* 1(1), 1006.

Nguyen, Van Thang, Viet Tien, Vu, Tuan Anh, Nguyen, Van Khanh, Tran, and Phuong, Nguyen-Tri (2019). "Antibacterial activity of TiO_2-and ZnO-decorated with silver nanoparticles." *Journal of Composites Science,* 3(2), 61.

Panghal, A., N. Chhikara, V. Anshid, M. V. S. Charan, V. Surendran, A. Malik, and S. B. Dhull (2019). Nanoemulsions: A promising tool for dairy sector. In *Nanobiotechnology in Bioformulations* (pp. 99–117). Cham: Springer.

Pannu, J., A. McCarthy, A. Martin, T. Hamouda, S. Ciotti, A. Fothergill, and J. Sutcliffe (2009). "NB-002, a novel nanoemulsion with broad antifungal activity against dermatophytes, other filamentous fungi, and Candida albicans." *Antimicrobial Agents and Chemotherapy,* 53(8), 3273–3279.

Punia, S., K. S. Sandhu, A. K. Siroha, and S. B. Dhull (2019). "Omega 3-Metabolism, Absorption, Bioavailability and health benefits-A review." *PharmaNutrition,* 1(10), 100162.

Ranjan, Shivendu, Nandita, Dasgupta, and Eric, Lichtfouse (2016), eds. *Nanoscience in Food and Agriculture*. Cham: Springer International Publishing.

Rehman, Rafia, Muhammad Asif, Hanif, Zahid, Mushtaq, and Abdullah Mohammed, Al-Sadi (2016). "Biosynthesis of essential oils in aromatic plants: A review." *Food Reviews International*, 32(2), 117–160.

Ryu, Victor, David J. McClements, Maria G. Corradini, and Lynne, McLandsborough (2018). "Effect of ripening inhibitor type on formation, stability, and antimicrobial activity of thyme oil nanoemulsion." *Food Chemistry,* 245, 104–111.

Ryu, Victor, David J. McClements, Maria G. Corradini, Jason Szuhao, Yang, and Lynne, McLandsborough (2018). "Natural antimicrobial delivery systems: Formulation, antimicrobial activity, and mechanism of action of quillajasaponin-stabilized carvacrol nanoemulsions." *Food Hydrocolloids,* 82, 442–450.

Saidi, Abbas, Yazdan, Eghbalnegad, and Zahra, Hajibarat. (2017) "Study of genetic diversity in local rose varieties (Rosa spp.) using molecular markers." *Banat's Journal of Biotechnology,* 8(16), 148–157.

Salvia-Trujillo, Laura, M. Alejandra, Rojas-Graü, Robert, Soliva-Fortuny, and Olga, Martín-Belloso (2013). "Effect of processing parameters on physicochemical characteristics of microfluidized lemongrass essential oil-alginate nanoemulsions." *Food Hydrocolloids,* 30(1), 401–407.

Sangwan, N. S., Farooqi, A. H. A., Shabih, F., and Sangwan, R. S. (2001). "Regulation of essential oil production in plants." *Plant Growth Regulation*, 34(1), 3–21.

Saranya, S., N. Chandrasekaran, and A. M. I. T. A. V. A. Mukherjee (2012). "Antibacterial activity of eucalyptus oil nanoemulsion against Proteus mirabilis." *International Journal of Pharm Pharm Sci,* 4(3), 668–671.

Setya, Sonal, Sushama, Talegaonkar, and B. K. Razdan (2014). "Nanoemulsions: Formulation methods and stability aspects." *World Journal of Pharm Pharm Science,* 3(2), 2214–2228.

Shams, Najmeh, and Mohammad Ali, Sahari (2016). "Nanoemulsions: Preparation, Structure, Functional Properties and Their Antimicrobial Effects." *Applied Food Biotechnology*, 3 (pp.138–149).

Shim, Jaehong, Young-Seok, Seo, Byung-Taek, Oh, and Min, Cho (2016). "Microbial inactivation kinetics and mechanisms of carbon-doped TiO_2 (C-TiO_2) under visible light." *Journal of Hazardous Materials,* 306, 133–139.

Silva, Hélder Daniel, Miguel, ÂngeloCerqueira, and António A. Vicente (2012). "Nanoemulsions for food applications: Development and characterization." *Food and Bioprocess Technology,* 5(3), 854–867.

Singh, Y., J. G. Meher, K. Raval, F. A. Khan, M. Chaurasia, N. K. Jain, and M. K. Chourasia (2017a). "Nanoemulsion: Concepts, development and applications in drug delivery." *Journal of Controlled Release*, 252, 28–49.

Singh, Yuvraj, Jaya Gopal, Meher, Kavit, Raval, Farooq Ali, Khan, Mohini, Chaurasia, Nitin K. Jain, and Manish K. Chourasia (2017b). "Nanoemulsion: Concepts, development and applications in drug delivery." *Journal of Controlled Release,* 252, 28–49.

Sirelkhatim, Amna, Shahrom, Mahmud, Azman, Seeni, Noor Haida Mohamad, Kaus, Ling Chuo, Ann, Siti Khadijah Mohd, Bakhori, Habsah, Hasan, and Dasmawati, Mohamad (2015). "Review on zinc oxide nanoparticles: Antibacterial activity and toxicity mechanism." *Nano-Micro Letters,* 7(3), 219–242.

Solans, Conxita, and Isabel, Solé (2012). "Nano-emulsions: Formation by low-energy methods." *Current Opinion in Colloid & Interface Science,* 17(5), 246–254.

Subhan, MdAbdus (2020). Antibacterial property of metal oxide-based nanomaterials. In *Nanotoxicity* (pp. 283–300). Elsevier, Netherlands.

Sugumar, Saranya, Amitava, Mukherjee, and Natarajan, Chandrasekaran (2015a). "Nanoemulsion formation and characterization by spontaneous emulsification: Investigation of its antibacterial effects on Listeria monocytogenes." *Asian Journal of Pharmaceutics,* 9(1), 23–28.

Sugumar, Saranya, Amitava, Mukherjee, and Natarajan, Chandrasekaran (2015b). "Eucalyptus oil nanoemulsion-impregnated chitosan film: Antibacterial effects against a clinical pathogen, Staphylococcus aureus, in vitro." *International Journal of Nanomedicine,* 10(Suppl 1), 67.

Sugumar, Saranya, Sanjay, Singh, Amitava, Mukherjee, and N. Chandrasekaran (2016). "Nanoemulsion of orange oil with non ionic surfactant produced emulsion using ultrasonication technique: Evaluating against food spoilage yeast." *Applied Nanoscience,* 6(1), 113–120.

Teixeira, Paula C., Gonçalo M. Leite, Ricardo J. Domingues, Joana, Silva, Paul A. Gibbs, and João Paulo, Ferreira (2007). "Antimicrobial effects of a microemulsion and a nanoemulsion on enteric and other pathogens and biofilms." *International Journal of Food Microbiology*, 118(1), 15–19.

Topuz, O. K., E. B. Özvural, Q. Zhao, Q. Huang, M. Chikindas, and Me. Gölükçü (2016). "Physical and antimicrobial properties of anise oil loaded nanoemulsions on the survival of foodborne pathogens." *Food Chemistry*, 203, 117–123.

Vega-Jiménez, Alejandro L., América R. Vázquez-Olmos, Enrique, Acosta-Gío, and Marco Antonio, Álvarez-Pérez (2019). "In vitro antimicrobial activity evaluation of metal oxide nanoparticles." In *Nanoemulsions-Properties, Fabrications and Applications*. IntechOpen, London, UK.

Weiss, Jochen, Sylvia, Gaysinsky, Michael, Davidson, and Julian, McClements (2009). Nanostructured encapsulation systems: Food antimicrobials. In *Global Issues in Food Science and Technology*, (pp. 425-479). Elsevier Inc., Academic Press, USA.

Wik, Johanna, Kuldeep K. Bansal, Tatu, Assmuth, Ari, Rosling, and Jessica M. Rosenholm (2019). "Facile methodology of nanoemulsion preparation using oily polymer for the delivery of poorly soluble drugs." *Drug Delivery and Translational Research,* 19, 1–13.

Xue, Jia, P. Michael Davidson, and Qixin, Zhong (2015). "Antimicrobial activity of thyme oil co-nanoemulsified with sodium caseinate and lecithin." *International Journal of Food Microbiology,* 210, 1–8.

Zerkaoui, Laidia, Mohamed, Benslimane, and Abderrahmane, Hamimed (2018). "The purification performances of the lagooning process, case of the BeniChougrane region in Mascara (Algerian NW)." *Banat's Journal of Biotechnology*, 9(18, 20–28.

19 Nano-Starch Films as Effective Antimicrobial Packaging Materials

Samriti Guleria
Lovely Professional University, India
Mukul Kumar and Shailja Kumari
Shoolini University, India
Ashwani Kumar
Lovely Professional University, India

CONTENTS

19.1 INTRODUCTION

Packaging materials are those that provide protection, communication, containment, and convenience to food products while maintaining their safety and quality during the period of their storage and transportation. A decent food package has the ability to extend the shelf life of food products by preventing them from spoilage, chemical contamination, moisture, oxygen, and external force. In past 20 years, the food-packaging industry, plastic has been used as a packaging material, which increases the problem of the waste disposal. So, recently, the food-packaging industry has grown its interest in the biodegradable and antimicrobial packaging for food products. Nowadays, nanotechnology is advanced field to fight against several problems like packaging, medicine, formulations, and cosmetics (Nakazato et al., 2017; dos Santos Caetano et al., 2018). Several food ingredients are used for the preparation of nanoparticles (NPs), such as protein, lipids, surfactants, minerals, and polysaccharides (Chang and McClements, 2014; Agarwal et al., 2020; Dhull et al., 2020). The physico-chemical, releasing, protective, and encapsulation properties affect the composition of nanoparticles (McClements and Yan, 2010). To overcome this problem, new techniques such as nanoemulsion, nanospheres, nanocapsules, nanoliposomes, and nanofibers are being used for the production of packaging material in food industries, as shown in Figure 19.1.

Nano-sized starch particles have gained much attention due to their unique properties from bulk material. Starch is a naturally existing biopolymer that is extracted from plants in the form of small-sized granules. Starch is composed of amylose and amylopectin units linked by α (1–4) and α (1–6) (Punia et al., 2019). The amount of starch depends on the type of source (Punia et al., 2019). Starch is a source of energy that is obtained from the carbohydrates (Zhao et al., 2018; Punia et al., 2020). Starch NPs are also known as nanocrystals, but they differ in their structure. Nanocrystals include a crystalline structure whereas NPs include an amorphous region (Yang et al., 2018). They both show effective properties in different areas such as food

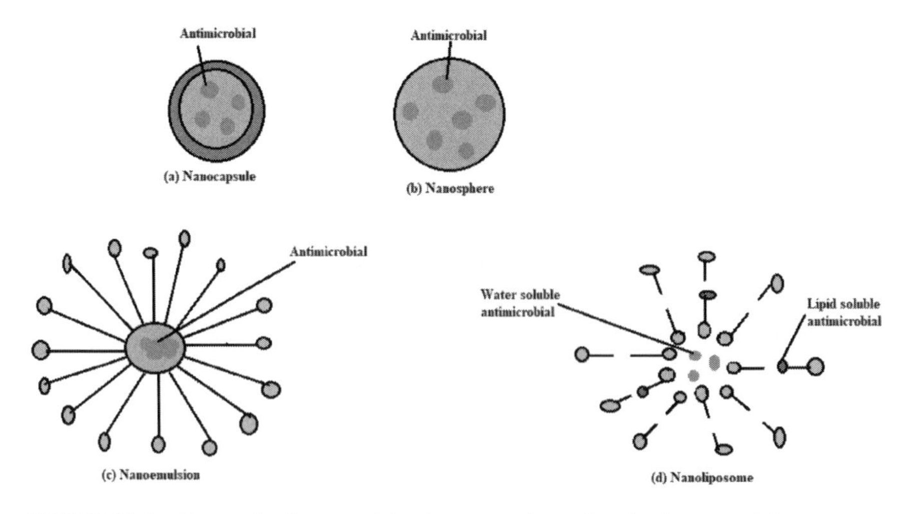

FIGURE 19.1 New technology used for the preparations of packaging material.

additives, coating binder, flavor binders, pigment carrier, vitamins, and emulsion stability (Ahmad et al., 2019). Nano-starch-based packaging film is used more nowadays due to the its biocompatible, biodegradable material, edibility, sustainability, and low cost (BeMiller, 2018). It is prepared with the blend of non-starch biopolymers such as chitosan (Bonilla et al., 2013), gelatin, and carboxymethyl cellulose (Ghanbarzadeh et al., 2011). The starch matrix is also filled with starch nanocrystals, cellulose nanocrystals (Ali et al., 2018), montmorillonites (Li et al., 2018), metal NPs (zinc oxide [ZnO] and silver [Ag] NPs; Nafchi et al., 2012), lysozyme (Fabra et al., 2014), and antimicrobial compounds such as potassium sorbate (Shen et al., 2010). Several commercial starches are available for the preparation of films, such as Mater BiÕ, Bioplast (Kaur et al., 2018), Biopar, Cereplast, Biostarch, Ever Corn TM, and Terraloy TM (Hafsa et al., 2016).

19.2 SYNTHESIS PROCESS OF NANO-STARCH

Nano-starch can be synthesized by variety of methods such as acid hydrolysis, regeneration (LeCorre and Bra, 2012), and enzymatic treatment (Le Corre and Hélène, 2014); physical, mechanical, and thermal treatments include nanoprecipitation, high-pressure homogenization, ultrasonication, gamma irradiation, and reactive extrusion, among others (Kim et al., 2015; Liu et al., 2016).

19.2.1 High-Pressure Homogenization

In this process, liquid is passed through microfabricated channels by external pressure, such as electro-kinetic mechanisms or micropumps (Hale and Mitchell, 2001). Due to the high shearing in the product responsible to speed up the velocity using an electronic hydraulic pump, (Banderas et al., 2005), the size of the particles is reduced through the shear force due to breakage of hydrogen bonds (Liu et al., 2016). High-speed techniques (ultrasonication) also help in the degradation of the crystalline structure (Ding and Kan, 2017).

19.2.2 Extrusion Technique

Higher temperatures, pressure, and shear forces are responsible for the structural change with change in melting point, fragmentation, and gelatinization (Song et al., 2011). In this process, the amount of water is minimal, which affects the gelatinization process. The higher shear force affects the structure of the starch granules due an interruption in the molecular bonds (Ding and Kan, 2017).

19.2.3 Acid Hydrolysis

This process used for the preparation of starch nanocrystals which are isolated from the sulfuric and hydrochloric acid of amorphous regions. The size of the granules is dependent on the lower temperature as compared to the gelatinization temperature, which responsible for the crystals that are not soluble in the water (Wang et al.,

2014). After the process, the residues left are highly crystalline and platelet-like in structure with a nano-size which can be stable at the higher temperatures during food processing (Aldao et al., 2018).

19.2.4 Enzymatic Hydrolysis

It is the pretreatment whereby the rate of hydrolysis is increased in native starch granules before the acid hydrolysis. For this process, various enzymes are used, such as glucoamylase, β-amylase, pullulanase, and α-amylase, to reduce the time of acid hydrolysis (LeCorre et al., 2012). These enzymes affect the structure of starch granules, which make the amorphous region more hydrolysable and selectively. Enzymes like glucoamylase treated with starch are helpful in decreasing the acid hydrolysis time from 5 days to 45 hours (Kim and Lim, 2008). However, this process is more effective compared to acid hydrolysis because it reduces the time, increases the yield to about 55%, and uses no chemical reagents (Sun et al., 2014).

19.2.5 Ultrasonication

This is the advance technology for the processing of food due to the low-energy consumption and short processing time (Kim et al., 2015). In this process, ultrasonic cavitation was used to generate microbubbles when the high energy is released. It helps degrade the polymer due to the higher temperature and high pressure. It also disturbs the surface of the starch, which reduces the crystals of the NPs with an increase in the amorphous state. Ultrasonication also affected the production of NPs and hydrolysis efficiency, and the NPs' size varied between 30 and 200 nm (Amini and Seyed, 2016).

19.2.6 Gamma Irradiation

Polymers are modified through cross-linking, degradation, and grafting with the use gamma irradiation, which help in the reduction of the starch size (Akhavan and Ataeevarjovi, 2012; Punia et al., 2020). In study by Singh et al. (2011), larger molecules of the starch reportedly were converted into the smaller molecules in the amorphous state, and the physical properties and structure of the starch were affected. This method is similar with the acid hydrolysis process, whereby the active free radicals are generated which reduce the hydrolysis of starch (Lin et al., 2011).

19.3 CHARACTERIZATION OF THE NANO-STARCH

Several techniques are used for the characterization of nano-starch-based films such as transmission electron microscopy (TEM), energy-dispersive X-ray spectroscopy (EDX), X-ray diffraction (XRD), scanning electron microscopy (SEM), FESEM morphology, Fourier-transform infrared (FTIR) spectroscopy; (Meraat et al., 2016). TEM and SEM show the image of nano-starch on the basis of scattered and transmitted electrons in the presence of light (Wang et al., 2014). However, EDX was used to known about the chemical characterization of sample whereas XRD informed about the interference and the structure of the crystal (Jiang et al., 2016). FESEM morphology tell about the pore size, shape, topography, and porous pattern of films, and FTIR spectroscopy informed about the quantity of the compound present in sample (Yang et al., 2018).

Kaur et al. (2018) reported in its study that cereal starch nano-film is prepared by from micro-granules of starch which are isolated from plants and are the most abundant biopolymers in nature. Due to some disadvantages raw starch cannot be used, and main disadvantage is it is not soluble in cold water. To increase its functionality, some alteration is required. Studying in detail starch NPs' physical, chemical, and molecular structures helps us synthesize starch NPs of proper size and shape. They mention several techniques of characterization of starch NPs in which the morphological characterization of NPs was done through as TEM, SEM, and FTIR. On the other hand, they also discussed the chemical method for the characterization of starch NPs, such as acid hydrolysis, emulsion cross-linking, and nanoprecipitation. Physical methods, such as high-pressure homogenization, ultrasonication, reactive extrusion (REX), and gamma irradiation, were discussed. The result of this study is that cereal starch nano-film showed highly mechanical property (tensile strength and rupture), good moisture barrier, and higher antimicrobial property against different microorganism such as *Staphylococcus aureus*, *Escherichia coli*, and *Salmonella typhimurium.*

Song et al. (2011) reported nanoprecipitation method being used to synthesize the size of NPs. The size of synthesis sago starch NPs range occurs between 300 to 400 nm. The starch NPs were formed through drop-wise accumulation of a liquefied native starch solution to surplus absolute ethanol. The researchers performed the characterization of the starch NPs through SEM. Their results showed the effect of adding surfactants, hexadecyl (cetyl) trimethyl ammonium bromide (CTAB) and Tween 80, onto the mean particle sizes of starch NPs formed. They reported that of the two surfactants, Tween-80 is more hydrophilic and would interact stronger with starch molecules as compare to CTAB; hence, starch NPs of smaller sizes were created. This technique of characterization of the nano-starch is convenient and easy. Hebeish et al. (2013) did the characterization of starch NPs through an ultra-fine technique. They synthesized NPs via native maize starch. Their work addressed the productive method for the synthesis of starch NPs. They used the solvent displacement method; as a solvent, they use an alkaline aqueous solution. Through their experiment, they trying to prove that, in the form of NPs, (nano-form) starch will have probable applications in medical areas, particularly as a drug carrier. Liu et al. (2018) reported that there are two process for the preparation of starch NPs, such as the top-down and bottom-up methods. In the top-down process, the starch NPs are synthesize by means of size through a breakdown of starch granules, and in bottom-up process, the starch NPs are synthesized by means of controlled shape by thermodynamic process. They use different techniques, such as acid hydrolysis, and physical techniques, such as high-pressure homogenization, ultrasonication, reactive extrusion, gamma irradiation, steam-jet cooking, Some other methods include nanoprecipitation, enzymatic debranching and recrystallization, emulsion, polyelectrolyte complex formation, electrospinning, and electrospray. The researchers concluded that to prepare starch NP in pilot-scale, short-chain amylose, retrogradation and antisolvent precipitation of starch are the most auspicious approaches. Besides that, in foods, to form novel structures or textures, starch NPs were utilized as building blocks which are useful in the development of reduced fat, sodium, or carbohydrate foods. In study by Abioye et al. (2019), starch NPs synthesized through matrix reinforcements for low-density polyethylene (LDPE) through acid hydrolysis process was reported. They utilized the starches of different crops such as *Zea mays* (corn), *Ipomoea batatas* (potato), and *Manihot esculenta* (cassava). The characterization of the synthesized NPs was done through TEM, EDX, FTIR, XRD and SEM. For the purpose of modifying the matrix of LDPE to enhance its biodegradability they convert locally sourced feedstock starches into NPs. According to their research, they concluded that the acid hydrolysis is an appropriate method for synthesizing NPs; at the same time, NPs still maintaining the characteristic properties of the original substance. Haaj et al. (2013) synthesized starch NPs via physical method names as ultrasonification they were use this technique in high intensity. To ensure the crystallinity of the starch NPs and to study the morphology of the synthesized starch NPs, they used a technique named Raman spectrophotometry. They compared the acid hydrolysis technique with ultrasonification technique, and they concluded that high-intensity ultrasonification is the best technique to yield high-intensity NPs. They reported this technique for first time without using any chemical essences. Starch NPs synthesized by using enzymolys is and recrystallization (Dufresne et al., 2010). Their research leads to good yield of starch NPs that are environmentally-friendly. They concluded that they yield starch NPs through advanced methods. Therefore, their technique is suitable for the synthesis of NPs at an industrial scale.

The characterization of starch NPs through different techniques was done by Szymońska et al. (2009). They synthesize NPs through potato and cassava starch. They were done the microscopic investigation through SEM. They also check some physiochemical properties of starch NPs, such as aqueous solubility and iodine-binding capacity. Due to the defencelessness of starch NPs towards chemical reagents, as well as properties qualitatively dissimilar from those of natural starch granules, they could be used in novel applications, such as being appropriate carriers in the drug and/or chemical delivery systems.

The characterization and synthesis of starch NPs through gamma radiation technique were done by Lamanna et al. (2013). Gamma radiation technique is very efficient and synthesize NPs at low cost, and besides that, it is a very simple technique. They did the morphological characterization through FESEM, TEM, and XRD. They also discussed the mechanical and thermal characterizations of starch NPs. Finally, their research concluded that for the mass production of starch NPs, the gamma radiation technique is very efficient technique which is simple and helpful in the large-scale production of starch NPs.

19.4 ANTIMICROBIAL EFFICACY OF THE NANO-STARCH-BASED PACKAGING FILMS

Nano-starch-based packaging film showed high antimicrobial property which help increase the shelf life, nutritive value, and freshness of food products (Ortega et al., 2019). Several antimicrobial compounds, such as potassium sorbate, Benzoic acid,

Nisin, lauric acid, organic acids, essential oils, enzymes, bacteriocins, oxygen absorbers, fungicides, natural extracts, polymers and gas, are attached to the starch matrix (Malhotra et al., 2015; Table 19.1).

In market, the demand of nano-based antimicrobial starch film has increased due to the maintaining of the product quality by the protection from microbial contamination through natural sources and metal oxides (Sung et al., 2013).

19.5 NATURAL SOURCES

Several natural sources such as pomegranate peel, mung bean, curcumin, sago and garlic nanoparticles are embedded with the starch film which shown higher antimicrobial properties as shown in Table 19.2.

19.5.1 Cinnamon Essential Oil

The antibacterial activity of nano-based starch film was increased due to the phytochemical compound present in cinnamon essential oil (CEO; Zhang et al., 2016). It has been shown to have effective antibacterial properties against *Staphylococcus aureus* and *Escherichia coli* which responsible for the foodborne spoilage (Shen et al., 2015). It also effects the cell wall permeability and properties of different microorganisms (Ma et al., 2016). CEO-base nano-starch films increase the shelf life of meat, fish, and bread products (Van et al., 2016). They also have the ability to reduce the gas permeability (Hafsa et al., 2016).

19.5.2 Pomegranate Peel (*Punicag ranatum L*)

Nano-starch films are embedded with the pomegranate peel which show higher bacterial activity and inhibits the microbial growth due to the presence of higher bioactive compounds (Ali et al., 2018). It inhibits the growth of gram positive and gram-negative bacteria (*Staphylococcus aureus* and *Salmonella*). Pomegranate peel nano-starch film was used more to increase the shelf life of meat and chicken products (Abid et al., 2017).

19.5.3 Nutmeg Oil

Nutmeg oil is used to coated nano-starch-based films due to its higher antimicrobial properties. It is a rich source of sabinene (25.4%) as well as myristicine and pinene (14.8 and 15.8%, respectively; Chang et al., 2010). The non-polar compound in nutmeg oil shows a low solubility rate which affects the stability of the nanoemulsion. As an edible film, it also improves the storage modulus, tensile strength, glass transition temperature, and water vapour with the combination of nutmeg oil. It also inhibits the growth of *S. aureus* and *E. coli* (10.40 and 10.76 mm, respectively; Aisyah et al., 2018b).

TABLE 19.1
Several Antimicrobial Compounds were Embedded with the Starch Matrix Which Shown Antimicrobial Properties of Nano-Starch-Based Packaging Film

Antimicrobial Compound	Type of Antimicrobial	Microorganism	References
Benzoic acid	Organic acid	Total bacteria assay	Nakazato et al. (2017)
Sorbates	Organic acid	Yeast, filamentous fungi	Ghanbarzadeh et al. (2011)
Nisin	Bacteriocin	*Staphylococcus aureus* and *Brochothrix thermosphacta*	Aldao et al. (2018)
Immobilized lysozyme	Enzyme	Lysozyme activity assay	Liu et al. (2016)
Chitosan	Polymer	*Escherichia coli*	Zhang et al. (2016)
Chitosan and herb extract	Polymer	*E. coli*, *Lactobacillus plantarum, Fusarium oxysporum*, and *Saccharomyces cerevisiae*	Kim et al. (2015)
Sorbic anhydride	Organic acid	*Saccharomyces cerevisiae* and filamentous fungi	Zhang et al. (2018)
Imazadil	Fungicide	Filamentous fungi	Amini and Seyed (2016)

TABLE 19.2
Different Studies Showing Higher Antimicrobial Properties of Nano-Based Starch Film when it is Bound With A Natural Source

Natural Source	Microorganism	Zone of Inhibition (mm or %)	Result	Reference
Sago starch	*Escherichia coli*	23	Higher antimicrobial property due presence of higher phenolic compounds	Nafchi et al. (2012)
Corn starch	*Staphylococcus aureus* *E. coli*	38 32	Gram-positive bacteria were more sensitive to edible film than were gram-negative bacteria due to an impermeable outer membrane that surround the gram-negative bacteria.	Song et al. (2018)
Curcumin	*Salmonella*, *S. aureus*	804 201	Antibacterial resistance inside the structure	Khodaeimehr et al. (2018)
Cinnamon essential oil	*Mucor* spp.		Good antimicrobial activity	Zhang et al. (2018)
Nutmeg oil	*Salmonella typhimurium*	30	Inhibits growth of foodborne pathogen	Xie et al. (2011)
Pomegranate peel	*S. aureus* *Salmonella*	12% 14%	Pomegranate peel particles are released from starch and have shown antimicrobial properties due to the interaction between phenolic toxicity and sulfhydryl group of proteins in microorganisms	Ali et al. (2019)

19.5.4 Pea Starch

The pea starch nano-based film shows effective results against several microorganisms such as *S. aureus*, *E. coli, Aspergillus niger*, *P. fluorescence*, and *Penicillium italicum* due to the presence of lysozyme, isoflavones, luteolin, apigenin, and phenolic compounds. The pea starch nano-film shows higher hydrophobicity (Aisyah et al., 2018a).

19.5.5 Ginger Starch

Nano-starch-based film was prepared with ginger starch, which shows higher antimicrobial, anti-inflammatory and anti-analgesic activity due to the presence of terpenes, oleoresin, gingerol, and gingerol-related compound (Aisyah et al., 2018a). It shows effective antimicrobial activity against *E. coli*, *Salmonella typhi*, *Candida albicans*, and *Bacillus subtilis* (Chen et al., 2008).

19.5.6 Oregano Essential Oil

Oregano essential oil is embedded with starch to produce a nano-based film (Soylu et al., 2007). It shows effective antimicrobial properties by inhibiting foodborne pathogens and fungi (*Zygosaccharo mycesbailii*, *P. digitatum*, *P. italicum*, and *C. albicans*) due to the presence of volatile compounds such as carvacrol and thymol (Avila et al., 2010).

19.5.7 Nisin

Grower et al. (2004) reported that the cellulose, nisin was released, and it affected packaging films by inhibiting *Listeria monocytogenes* strains. The pH was affecting the growth of bacteria through neutralization, and it overcome the stress of antibacterial compound (Grower et al., 2004).

19.6 METAL SOURCES

Several metal NPs such silver, zinc oxide, titanium dioxide, and magnesium oxide are used with starch to prepared a nano-starch film and shown high antimicrobial properties as shown in Table 19.3. Aging, growth, and nucleation mechanisms are also responsible for the determination of NPs (Oskam, 2006).

19.6.1 Silver Nanoparticles

Silver NPs (AgNPs) embedded in the starch matrix and show an effective antimicrobial property with the gram-positive and gram-negative microorganisms due to the large surface area and volume ratio (Azeredo, 2013). The different mechanisms of AgNPs are the inhibition of DNA replication and adenosine triphosphate (ATP) production from the releasing of ions, AgNPs responsible for the damage of cell membranes, and AgNPs are helpful in the generation of reactive oxygen species (ROS;

TABLE 19.3
Several Antimicrobial Studies Occur on Nano-Based Starch Film which Coated With the Metal Source

Metal Source	Microorganism	Zone of Inhibition (mm)	Result	Reference
Zinc oxide (ZnO) nanoparticle (NP)	*Salmonella typhimurium*	28	Nontoxic antimicrobial which are effective against gram-positive and -negative bacteria due to the increased surface area of ZnONPs Disturbing of bacterial cell membrane or by antimicrobial properties	Meraat et al. (2016)
Silver NP	*Escherichia coli* *Staphylococcus aureus* *Candida albicans*	2.2 1.8 1.0	Excellent antimicrobial activity Highly feasible for packaging (Biodegradable) Good entrapment of the active compound in nanocomposite starch film matrix.	Ana et al. (2015)
Silicon oxide NP	*E. coli*, *S. aureus*	0.8	Higher antibacterial activity	Zhang et al. (2018)

Dallas et al., 2011; Jung et al., 2018). Sulfur, oxygen, phosphates, amines, and thiols are attached to electron donor group with silver ion which has shown an antimicrobial activity (Kumar and Munstedt, 2005).

19.6.2 Zinc Oxide Nanoparticles

In starch-based films, zinc oxide NPs (ZnONPs) help enhance antibacterial activity with the contraction in microbial cells and enhance ROS production (Premanathan et al., 2011). The concentration of the ZnO responsible for the bactericidal and bacteriostatic property (Jin et al., 2009). ZnONPs shown effective result against several microorganism such as *Salmonella enteritidis*, *E. coli* and *Listeria* (Premanathan et al., 2011).

19.6.3 Titanium Dioxide Nanoparticles

Titanium dioxide (TiO_2) has been used in the preparation of nano-starch films due to its photocatalytic disinfecting properties. It helps in enhancing the reduction of the phospholipids which are present in microbial cell membranes and is responsible for the loss of membrane due to the photocatalysis process (Jing et al., 2011).

19.6.4 Magnesium Oxide Nanoparticles

Magnesium oxide (MgO) showed higher antibacterial activity with *E. coli* and *Bacillus* species due to the bactericidal mechanism. It helps in producing of superoxide ions on the surface of MgONP-coated starch film with carbonyl groups of peptide linkage which destroyed the cell wall of bacteria (Al-Hazmi et al., 2012).

19.7 CONCLUSION

Nanotechnology is an advanced technique used in food packaging due to its unique properties. Starch-based nano-films are prepared with natural source (essential oil, fruit peel, compounds) and metal sources (silver, zinc, magnesium) and show higher antimicrobial properties against gram-positive and gram-negative microorganism such as *E. coli*, *S. aureus*, *S. typhimurium*, and *Mucor* spp. These films not only enhance the quality and shelf life of the food products but also fight against several environmental factors. In future, the trends of nano-based starch films are increased due its edible nature.

REFERENCES

Abid, Mouna, S. Cheikhrouhou, Catherine M. G. C. Renard, Sylvie, Bureau, Gérard, Cuvelier, Hamadi, Attia, and M. A. Ayadi. (2017). "Characterization of pectins extracted from pomegranate peel and their gelling properties." *Food Chemistry,* 215, 318–325.

Abioye, Abiodun Ayodeji, Oreofe Praise, Oluwadare, and Oluwabunmi Pamilerin, Abioye. (2019). "Environmental impact on biodegradation speed and biodegradability of polyethylene and Ipomoea Batatas starch blend. "*International Journal of Engineering Research in Africa*, 41, 145–154. Trans Tech Publications.

Abreu, Ana S., Manuel, Oliveira, Arsénio de, Sá, Rui M. Rodrigues, Miguel A. Cerqueira, António A. Vicente, and A. V. Machado. (2015). "Antimicrobial nanostructured starch-based films for packaging." *Carbohydrate Polymers,* 129, 127–134.

Agarwal, A., A. K. Pathera, R. Kaushik, N. Kumar, S. B. Dhull, S. Arora, and P. Chawla. (2020). "Succinylation of milk proteins: Influence on micronutrient binding and functional indices." *Trends in Food Science & Technology*, 97, 54–64.

Ahmad, Mudasir, Priti, Mudgil, Adil, Gani, Fathalla, Hamed, Farooq A. Masoodi, and Sajid, Maqsood. (2019). "Nano-encapsulation of catechin in starch nanoparticles: Characterization, release behavior and bioactivity retention during simulated in-vitro digestion." *Food Chemistry,* 270, 95–104.

Aisyah, Y., L. P. Irwanda, S. Haryani, and N. Safriani. (2018a). "Characterization of corn starch-based edible film incorporated with nutmeg oil nanoemulsion." *IOP Conference Series: Materials Science and Engineering*, 352 (1), 012050. IOP Publishing.

Aisyah, Y., L. P. Irwanda, S. Haryani, and N. Safriani. (2018b). "Characterization of corn starch-based edible film incorporated with nutmeg oil nanoemulsion." *IOP Conference Series: Materials Science and Engineering*, 352 (1), 012050. IOP Publishing.

Akhavan, A., and E. Ataeevarjovi. (2012). "The effect of gamma irradiation and surfactants on the size distribution of nanoparticles based on soluble starch." *Radiation Physics and Chemistry,* 81 (7), 913–914.

Aldao, David Chena, Šárka, Evžen, Ulbrich, Pavel, and Menšíková, Eva. (2018). "Starch nanoparticles–two ways of their preparation." *Czech Journal of Food Sciences,* 36 (2), 133–138.

Al-Hazmi, F., F. Alnowaiser, A. A. Al-Ghamdi, A. A. Al-Ghamdi, M. M. Aly, R. M. Al-Tuwirqi, and F. El-Tantawy. (2012). "A new large–scale synthesis of magnesium oxide nanowires: structural and antibacterial properties." *Superlattices and Microstructures*, 52 (2), 200–209.

Ali, Amjad, Ying, Chen, Hongsheng, Liu, Long, Yu, Zulqarnain, Baloch, Saud, Khalid, Jian, Zhu, and Ling, Chen. (2019). "Starch-based antimicrobial films functionalized by pomegranate peel." *International Journal of Biological Macromolecules,* 129, 1120–1126.

Ali, Amjad, Fengwei, Xie, Long, Yu, Hongsheng, Liu, Linghan, Meng, Saud, Khalid, and Ling, Chen. (2018). "Preparation and characterization of starch-based composite films reinforced by polysaccharide-based crystals." *Composites Part B: Engineering,* 133, 122–128.

Amini, Asad Mohammad, and Seyed Mohammad Ali, Razavi. (2016). "A fast and efficient approach to prepare starch nanocrystals from normal corn starch." *Food Hydrocolloids,* 57, 132–138.

Avila-Sosa, Raúl, Erika, Hernández-Zamoran, Ingrid, López-Mendoza, Enrique, Palou, María Teresa Jiménez, Munguía, Guadalupe Virginia, Nevárez-Moorillón, and Aurelio, López-Malo. (2010). "Fungal inactivation by Mexican oregano (*Lippia berlandieri* Schauer) essential oil added to amaranth, chitosan, or starch edible films." *Journal of Food Science,* 75 (3) , M127–M133.

BeMiller, James N. (2018) *Carbohydrate Chemistry for Food Scientists*. United Kingdom, Elsevier.

Bonilla, J., E. L. E. N. A. Fortunati, Maria, Vargas, A. Chiralt, and José, María Kenny. (2013). "Effects of chitosan on the physicochemical and antimicrobial properties of PLA films." *Journal of Food Engineering,* 119 (2), 236–243.

Chang, Peter R., Ruijuan, Jian, Jiugao, Yu, and Xiaofei, Ma. (2010). "Starch-based composites reinforced with novel chitin nanoparticles." *Carbohydrate Polymers,* 80 (2), 420–425.

Chang, Y., and D. J. McClements. (2014). "Optimization of orange oil nanoemulsion formation by isothermal low-energy methods: influence of the oil phase, surfactant, and temperature." *Journal of Agricultural and Food Chemistry*, 62 (10), 2306–2312.

Chen, I-Nan, Chen-Chin, Chang, Chang-Chai, Ng, Chung-Yi, Wang, Yuan-Tay, Shyu, and Tsu-Liang, Chang. (2008). "Antioxidant and antimicrobial activity of Zingiberaceae plants in Taiwan." *Plant Foods for Human Nutrition,* 63 (1), 15–20.

Dallas, Panagiotis, Virender K. Sharma, and Radek, Zboril. (2011). "Silver polymeric nanocomposites as advanced antimicrobial agents: classification, synthetic paths, applications, and perspectives." *Advances in Colloid and Interface Science,* 166 (1–2), 119-135.

de Azeredo, Henriette M. C.. (2013). "Antimicrobial nanostructures in food packaging." *Trends in Food Science & Technology,* 30 (1), 56–69.

Dhull, S. B., K. S. Sandhu, S. Punia, M. Kaur, P. Chawla, and A. Malik. (2020). "Functional, thermal and rheological behavior of fenugreek (*Trigonella foenum–graecum* L.) gums from different cultivars: A comparative study." *International Journal of Biological Macromolecules*, 159, 406–414.

Ding, Yongbo, and Jianquan, Kan. (2017). "Optimization and characterization of high pressure homogenization produced chemically modified starch nanoparticles." *Journal of Food Science and Technology,* 54 (13), 4501–4509.

dos Santos Caetano, Karine, Nathalie, Almeida Lopes, Tania Maria, Haas Costa, Adriano, Brandelli, Eliseu, Rodrigues, Simone, Hickmann Flôres, and Florencia, Cladera-Olivera. (2018). "Characterization of active biodegradable films based on cassava starch and natural compounds." *Food Packaging and Shelf Life,* 16, 138–147.

Dufresne, A., D. Corre, and J. Bras. (2010). "Starch nanoparticles: A review." *Biomacromolecules,* 11, 1139–1153.

Fabra, M.J., Sánchez-González, L., Chiralt, A.. (2014). "Lysozyme release from isolate pea protein and starch based films and their antimicrobial properties." *LWT-Food Science and Technology*, 55 (1), 22–26.

Ghanbarzadeh, Babak, Hadi, Almasi, and Ali A. Entezami. (2011). "Improving the barrier and mechanical properties of corn starch-based edible films: Effect of citric acid and carboxymethyl cellulose." *Industrial Crops and Products,* 33 (1), 229–235.

Grower, J. L., K. Cooksey, and K. Getty. (2004). "Release of nisin from methylcellulose-hydroxypropyl methylcellulose film formed on low-density polyethylene film." *Journal of Food Science,* 69 (4), FMS107–FMS111.

Haaj, Sihem Bel, Albert, Magnin, Christian, Pétrier, and Sami, Boufi. (2013). "Starch nanoparticles formation via high power ultrasonication." *Carbohydrate Polymers,* 92 (2), 1625–1632.

Hafsa, Jawhar, Med ali, Smach, Med Raâfet, Ben Khedher, Bassem, Charfeddine, Khalifa, Limem, Hatem, Majdoub, and Sonia, Rouatbi. (2016). "Physical, antioxidant and antimicrobial properties of chitosan films containing *Eucalyptus globulus* essential oil." *LWT-Food Science and Technology,* 68, 356–364.

Hale, Michelle S., and James G. Mitchell. (2001). "Motion of submicrometer particles dominated by Brownian motion near cell and microfabricated surfaces." *Nano Letters,* 1 (11), 617–623.

Hebeish, A., M. H. El-Rafie, M. A. El-Sheikh, and M. E. El-Naggar. (2013). "Nanostructural features of silver nanoparticles powder synthesized through concurrent formation of the nanosized particles of both starch and silver." *Journal of Nanotechnology*, (2013), 1–10.

Jiang, Suisui, Chenzhen, Liu, Zhongjie, Han, Liu, Xiong, and Qingjie, Sun. (2016). "Evaluation of rheological behavior of starch nanocrystals by acid hydrolysis and starch nanoparticles by self-assembly: A comparative study." *Food Hydrocolloids,* 52, 914–922.

Jin, T., D. Sun, J. Y. Su, H. Zhang, and H-J. Sue. (2009). "Antimicrobial efficacy of zinc oxide quantum dots against Listeria monocytogenes, Salmonella enteritidis, and Escherichia coli O157: H7." *Journal of Food Science,* 74 (1), M46–M52.

Jing, Zhihong, Daojun, Guo, Weihua, Wang, Shufang, Zhang, Wei, Qi, and Baoping, Ling. (2011). "Comparative study of titania nanoparticles and nanotubes as antibacterial agents." *Solid State Sciences,* 13 (9), 1797–1803.

Jung, Jeyoung, Gownolla Malegowd, Raghavendra, Dowan, Kim, and Jongchul, Seo. (2018). "One-step synthesis of starch-silver nanoparticle solution and its application to antibacterial paper coating." *International Journal of Biological Macromolecules,* 107, 2285–2290.

Kaur, Jashandeep, Gurkirat, Kaur, Savita, Sharma, and Kiran, Jeet. (2018). "Cereal starch nanoparticles—A prospective food additive: A review." *Critical Reviews in Food Science and Nutrition,* 58 (7), 1097–1107.

Khodaeimehr, Rouhollah, Seyed Jamaleddin, Peighambardoust, and Seyed Hadi, Peighambardoust. (2018). "Preparation and characterization of corn starch/clay nanocomposite films: effect of clay content and surface modification." *Starch-Stärke,* 70 (3–4), 1700251.

Kim, Jong-Yea, Dong-June, Park, and Seung-Taik, Lim. (2008). "Fragmentation of waxy rice starch granules by enzymatic hydrolysis." *Cereal Chemistry,* 85 (2), 182–187.

Kim, Hee-Young, Sung Soo, Park, and Seung-Taik, Lim. (2015). "Preparation, characterization and utilization of starch nanoparticles." *Colloids and Surfaces B: Biointerfaces,* 126, 607–620.

Kumar, Radhesh, and Helmut, Münstedt. (2005). "Silver ion release from antimicrobial polyamide/silver composites." *Biomaterials,* 26 (14), 2081–2088.

Lamanna, Melisa, Noé J. Morales, Nancy Lis, García, and Silvia, Goyanes. (2013). "Development and characterization of starch nanoparticles by gamma radiation: Potential application as starch matrix filler." *Carbohydrate Polymers,* 97 (1), 90–97.

Le Corre, D., and A. Coussy, Hélène. (2014). "Preparation and application of starch nanoparticles for nanocomposites: A review." *Reactive and Functional Polymers*, 85, 97–120.

LeCorre, Déborah, Julien, Bras, and Alain, Dufresne. (2012). "Influence of native starch's properties on starch nanocrystals thermal properties." *Carbohydrate Polymers,* 87 (1), 658–666.

Lin, Ning, Jiahui, Yu, Peter R. Chang, Junli, Li, and Jin, Huang. (2011). "Poly (butylene succinate)-based biocomposites filled with polysaccharide nanocrystals: Structure and properties." *Polymer Composites,* 32 (3), 472–482.

Liu, Chengzhen, Suisui, Jiang, Zhongjie, Han, Liu, Xiong, and Qingjie, Sun. (2016). "In vitro digestion of nanoscale starch particles and evolution of thermal, morphological, and structural characteristics." *Food Hydrocolloids,* 61, 344–350.

Liu, Qing, Man, Li, Liu, Xiong, Lizhong, Qiu, Xiliang, Bian, Chunrui, Sun, and Qingjie, Sun. (2018). "Oxidation modification of debranched starch for the preparation of starch nanoparticles with calcium ions." *Food Hydrocolloids,* 85, 86–92.

Ma, Qiumin, Yue, Zhang, Faith, Critzer, P. Michael, Davidson, Svetlana, Zivanovic, and Qixin, Zhong. (2016). "Physical, mechanical, and antimicrobial properties of chitosan films with microemulsions of cinnamon bark oil and soybean oil." *Food Hydrocolloids,* 52, 533–542.

Malhotra, Bhanu, Anu, Keshwani, and Harsha, Kharkwal. (2015). "Antimicrobial food packaging: potential and pitfalls." *Frontiers in Microbiology,* 6, 611.

Martín-Banderas, Lucía, María, Flores-Mosquera, Pascual, Riesco-Chueca, Alfonso, Rodríguez-Gil, Ángel, Cebolla, Sebastián, Chávez, and Alfonso M. Gañán-Calvo. (2005). "Flow focusing: a versatile technology to produce size-controlled and specific-morphology microparticles." *Small,* 1 (7), 688–692.

McClements, D. J., and Yan, Li. (2010). "Structured emulsion-based delivery systems: controlling the digestion and release of lipophilic food components." *Advances in Colloid and Interface Science*, 159, 213–228.

Meraat, Rafieh, Ali Abdolahzadeh, Ziabari, Khosro, Issazadeh, Nima, Shadan, and Kamyar Mazloum, Jalali. (2016). "Synthesis and characterization of the antibacterial activity of zinc oxide nanoparticles against Salmonella typhi." *Acta Metallurgica Sinica (English Letters),* 29 (7), 601–608.

Nafchi, Abdorreza Mohammadi, Abd Karim, Alias, Shahrom, Mahmud, and Marju, Robal. (2012). "Antimicrobial, rheological, and physicochemical properties of sago starch films filled with nanorod-rich zinc oxide." *Journal of Food Engineering,* 113 (4), 511–519.

Nakazato, Gerson, Renata, K. T., Kobayashi, Amedea, B. Seabra, and Nelson, Duran. (2017) "Use of nanoparticles as a potential antimicrobial for food packaging." In *Food Preservation*, (pp. 413–447). London, Taylor and Francis Group, Academic Press.

Ortega, F., García, M. A., and Arce, V. B. (2019). "Nanocomposite films with silver nanoparticles synthesized in situ: Effect of corn starch content." *Food Hydrocolloids*, 97, 105–200.

Oskam, Gerko. (2006). "Metal oxide nanoparticles: synthesis, characterization and application." *Journal of Sol-Gel Science and Technology*, 37, 161e164.

Premanathan, Mariappan, Krishnamoorthy, Karthikeyan, Kadarkaraithangam, Jeyasubramanian, and Govindasamy, Manivannan. (2011). "Selective toxicity of ZnO nanoparticles toward Gram-positive bacteria and cancer cells by apoptosis through lipid peroxidation." *Nanomedicine: Nanotechnology, Biology and Medicine,* 7 (2), 184–192.

Punia, S., S. B. Dhull, P. Kunner and S. Rohilla. (2020). "Effect of γ-radiation on physicochemical, morphological and thermal characteristics of lotus seed *(Nelumbo nucifera)* starch." *International Journal of Biological Macromolecules,* 157, 584–590.

Punia, S., S. B. Dhull, K. S. Sandhu, and M. Kaur. (2019). "Faba bean *(Vicia faba)* starch: Structure, properties, and in vitro digestibility—A review." *Legume Science,* 1 (1), e18.

Punia, S., K. S. Sandhu, S. B. Dhull, and M. Kaur. (2019). "Dynamic, shear and pasting behaviour of native and octenyl succinic anhydride (OSA) modified wheat starch and their utilization in preparation of edible films." *International Journal of Biological Macromolecules,* 133, 110–116.

Punia, S., K. S. Sandhu, S. G. Dhull, A. K. Siroha, S. S. Purewal, M. Kaur, and M. K. Kidwai. (2020). "Oat starch: Physico-chemical, morphological, rheological characteristics and its application-A review." *International Journal of Biological Macromolecules,* 154, 493–498

Shen, Suxia, Tiehua, Zhang, Yuan, Yuan, Songyi, Lin, Jingyue, Xu, and Haiqing, Ye. (2015). "Effects of cinnamaldehyde on Escherichia coli and Staphylococcus aureus membrane." *Food Control,* 47, 196-202.

Singh, Sandeep, Narpinder, Singh, Rajarathnam, Ezekiel, and Amritpal, Kaur. (2011). "Effects of gamma-irradiation on the morphological, structural, thermal and rheological properties of potato starches." *Carbohydrate Polymers,* 83 (4), 1521–1528.

Song, Delong, Yonathan S. Thio, and Yulin, Deng. (2011). "Starch nanoparticle formation via reactive extrusion and related mechanism study." *Carbohydrate Polymers,* 85 (1), 208–214.

Song, X., G. Zuo, and Chen, F. (2018). "Effect of essential oil and surfactant on the physical and antimicrobial properties of corn and wheat starch films." *International Journal of Biology Macromolecule*, 107, 1302–1309.

Soylu, Soner, H. Yigitbas, E. M. Soylu, and Ş. Kurt. (2007). "Antifungal effects of essential oils from oregano and fennel on *Sclerotinia sclerotiorum*." *Journal of Applied Microbiology,* 103 (4), 1021–1030.

Sun, Qingjie, Guanghua, Li, Lei, Dai, Na, Ji, and Liu, Xiong. (2014). "Green preparation and characterisation of waxy maize starch nanoparticles through enzymolysis and recrystallisation." *Food chemistry,* 162, 223–228.

Sung, Suet-Yen, Lee Tin, Sin, Tiam-Ting, Tee, Soo-Tueen, Bee, A. R. Rahmat, W. A. W. A. Rahman, Ann-Chen, Tan, and M. Vikhraman. (2013). "Antimicrobial agents for food packaging applications." *Trends in Food Science & Technology,* 33 (2), 110–123.

Szymońska, J., Marta, Targosz-Korecka, and Franciszek, Krok. (2009). "Characterization of starch nanoparticles." *Journal of Physics: Conference Series*, 146 (1), 012027. IOP Publishing.

Van Haute, Sam, Katleen Raes, Paul Van Der Meeren, and Imca Sampers. (2016). "The effect of cinnamon, oregano and thyme essential oils in marinade on the microbial shelf life of fish and meat products." *Food Control,* 68, 30–39.

Wang, C., Z. L. Pan, and J. L. Zeng. (2014). "Structure, morphology, and properties of benzyl starch nanocrystals". *Arabian Journal of Science and Engineering,* 39,6703–6710.

Xie, Yanping, Yiping, He, Peter L. Irwin, Tony, Jin, and Xianming, Shi. (2011). "Antibacterial activity and mechanism of action of zinc oxide nanoparticles against *Campylobacter jejuni*." *Applied Environment Microbiology,* 77 (7), 2325–2331.

Yang, Tao, Jie, Zheng, Bi-Sheng, Zheng, Fu, Liu, Shujun, Wang, and Chuan-He, Tang. (2018). "High internal phase emulsions stabilized by starch nanocrystals." *Food Hydrocolloids,* 82, 230–238.

Zhang, Yunbin, Xiaoyu, Liu, Yifei, Wang, Pingping, Jiang, and Siew Young, Quek. (2016). "Antibacterial activity and mechanism of cinnamon essential oil against Escherichia coli and *Staphylococcus aureus*." *Food Control,* 59, 282–289.

Zhang, Rongfei, Xiangyou, Wang, and Meng, Cheng. (2018). "Preparation and characterization of potato starch film with various size of nano-SiO_2." *Polymers,* 10 (10), 1172.

Zhao, Xiang-fei, Lan-qin, Peng, Hong-ling, Wang, Yan-bin, Wang, and Hong, Zhang. (2018). "Environment-friendly urea-oxidized starch adhesive with zero formaldehyde-emission." *Carbohydrate Polymers,* 181, 1112–1118.

Index

C

D

E

F

G

H

I

M

O

T

Y

Z